2019年四川省重点出版专项资金资助项目
"一带一路"输电线路建设管理专业丛书
OBOR Transmission Line Construction and Management - Book Series

特高压直流输电线路运检专业培训及考核标准

Standard for Professional Training and Assessment
for Operation Maintenance of UHV DC Transmission Line

丛书主编／汤晓青
本册主编／陈　立　祝　捷
副 主 编／全昌前　赵　强　郑和平
编　 者／范　宇　刘　意　李　陈　覃　浩
　　　　　饶建彬　邱中华　杜印官　卢国栋
　　　　　郭定海　周炜刚　陈　楠　隆　茂
　　　　　胡永银　刘正军　安　锋　张　杰
　　　　　张　柯　王　磊　王如鹏　周刘育
　　　　　廖家俊　钟　鑫
译　 者／成都优译信息技术股份有限公司

·成都·

图书在版编目(CIP)数据

特高压直流输电线路运检专业培训及考核标准 / 陈立，祝捷主编. -- 成都：电子科技大学出版社，2019.12
ISBN 978-7-5647-7549-0

Ⅰ.①特… Ⅱ.①陈… ②祝… Ⅲ.①特高压输电－直流输电线路－电力系统运行－检修－标准 Ⅳ.①TM726.1-65

中国版本图书馆CIP数据核字(2019)第287196号

特高压直流输电线路运检专业培训及考核标准
TEGAOYA ZHILIU SHUDIANXIANLU YUNJIANZHUANYE PEIXUN JI KAOHE BIAOZHUN

陈　立　祝　捷　主编

策划编辑	陈松明　陈　亮　罗国良
责任编辑	岳　慧　罗国良

出版发行	电子科技大学出版社
	成都市一环路东一段159号电子信息产业大厦　邮编 610051
主　页	www.uestcp.com.cn
服务电话	028-83203399
邮购电话	028-83201495
印　刷	成都市火炬印务有限公司
成品尺寸	185mm×260mm
印　张	43.75
字　数	1165千字
版　次	2019年12月第一版
印　次	2019年12月第一次印刷
书　号	ISBN 978-7-5647-7549-0
定　价	175.00元

版权所有，侵权必究

前　言

"'一带一路'输电线路建设管理专业丛书"共分十册，该丛书全面系统地介绍了国家电网公司在输电线路上的建设、运行及维护知识。每册图书都由中文版和英译文版组成，可作为"一带一路"上的国家和地区的电力职业教育规划教材，职业教育电力技术类专业培训用书。该丛书的出版对于推动"一带一路"的输电线路管理具有重要的意义。

特高压输电是目前世界上最先进的输电技术。中国特高压技术的出现，使得"煤从空中走，电送全国"成为现实。使"以电代煤，以电代油、电从远方来、来的是清洁电"成为中国能源和电力发展的新常态，为构建"全球能源互联网"、落实国家"一带一路"发展战略提供了强大的基础支撑。

直流输电工程是实施能源资源大范围优化配置的重要方式。国家电网公司建设了一大批直流输电工程，对促进西部、北部地区的能源资源优势转化为经济优势，满足中东部地区经济社会发展对电力的需求，推动清洁能源的集约开发与利用，保障国民经济持续健康发展发挥了十分重要的作用。从上世纪90年代开始，国内先后开展了葛沪直流、三广直流和二沪直流、德宝直流等±500kV的直流工程建设。葛洲坝　上海南桥±500kV直流输电工程是我国发展高压直流输电的起步工程，工程的成功建设解决了葛洲坝水电站向上海运距离送电问题，实现了华中与华东电网的非同期联网；三峡—上海±500kV直流输电工程解决了三峡大型电站向华东地区送电问题，加强了华中与华东电网的非同期联网。±800kV的直流工程建设，包括向上、锦苏线等西南水电外送工程,这些特高压直流工程的建成投运促进了金沙江流域向家坝、溪洛渡以及雅砻江流域锦屏、官地大型梯级水电基地的集约开发与规模外送，打通了西南水电送往华东等负荷中心的电力大通道。宁东工程、呼辽工程等直流输电工程的建成投运，推动了宁夏东部、呼伦贝尔等煤电基地的规模开发与外送，将当地资源优势转化经济优势，满足了山东、辽宁等地区经济发展对电力的需求。准东-华东（皖南）±1100千伏特高压直流输电工程是这是目前世界上电压等级最高、输送容量最大、输送距离最远、技术水

·1·

平最先进的特高压输电工程。随着多条特高压直流线路投运，为保障线路正常运行，线路检修、运行和维护工作显得非常重要。

目前从事特高压线路检修和运行人员大部分来自超高压线路检修、运维人员，大部分均很少接受过特高压线路检修、运维的专项培训，同时，特高压线路带电检修作业也逐渐开展起来，从业人员很需要进行规范性培训，从而提升作业水平并储备技能人才。国网四川省电力公司技能培训中心依托输配电线路带电作业基地和特高压交直流实训基地，长期从事输电线路运检、输电线路带电作业培训和资质认证工作，在教学过程中积累了丰富的培训资源。目前，已经制定了国网四川省电力公司多个输配电线路专业培训和考核标准，培训及考核标准编制经验丰富；国网四川省电力公司检修公司长期从事四川过境特高压线路的运行维护，实践经验丰富，加强本标准实用性。基于以上现状，国网四川省电力公司技能培训中心联合国网四川省电力公司检修公司及其他运检单位共同编写了这本《特高压直流输电线路运检及考核标准》。

本标准分两部分，第一部分为特高压直流输电线路运检带电作业培训及考核标准，第二部分为特高压直流输电线路运检停电检修培训及考核标准，共17个模块，其中带电检修作业11个、停电检修作业6个。各模块均由培训标准和考核标准两部分组成，涵盖了培训和考核的全部元素。

本标准的制定得到了国网四川省电力公司技能培训中心、国网四川省电力公司检修公司的大力支持和帮助，在此一并致谢。

由于编者水平有限，书中难免有不足之处，请广大读者批评、指正。

编 者

2019年9月

Foreword

OBOR Transmission Line Construction and Management - Book Series totally include ten volumes. The books comprehensively and systematically introduce the knowledge of the State Grid in the construction, operation and maintenance of power transmission lines. Each volume of the books consists of Chinese version and English translation version, which can be used as the electric power vocational education planning textbooks for the countries and regions on the "Belt and Road", and the professional training books for vocational education in electric power technologies. The publication of the book series is of great significance for promoting the management of the power transmission lines on the "Belt and Road".

UHV power transmission is most advanced power transmission technique in the world at present. The emergence of UHV technique in China makes "air transportation of coal, and power transmission to the whole country" become a reality. It makes "replacing coal by electric power, replacing oil by electric power, transmitting power from distant places, and delivering clean electric power" as the new normal of China's energy and power development, which provides strong foundation support for structuring the "Global Energy Internet", and implementing the national "Belt and Road" development strategy.

DC transmission project is the important approach to implement the wide-range optimal configuration of energy resources. State Grid Corporation of China has constructed a large batch of DC transmission projects, which play a very important role in facilitating the conversion of energy resources in Western and Northern China into economic advantages, meeting the demands in electric power for the economic and social development in Central and Eastern China, driving the intensive development and utilization of clean energy, and ensure the sustainable and healthy development of the national economy. Since 1990s, several ±500kV DC projects including Gezhouba-Shanghai DC Project, Sanxia-Guangdong DC Project, Sanxia-Shanghai DC Project, and Deyang-Baoji DC Project have been constructed in China. Gezhouba-Shanghai Nanqiao ±500kV DC Transmission Project is the starting project of China's development of HV DC transmission, the successful construction of which has

solved the long-distance power transmission from Gezhouba Hydropower Station to Shanghai, and realized the asynchronous networking between the power grids in Central China and Eastern China; Sanxia-Shanghai ±500kV DC Transmission Project has solved the power transmission from Three Gorges Hydropower Station to Eastern China, and strengthen the asynchronous networking between the power grids in Central China and Eastern China. The completion and operation of ±800kV DC projects including Xiangjiaba-Shanghai, Jinping-Tongli Line and other southwest hydropower outward transmission projects have facilitated the intensive development scale outward transmission of Xiangjiaba in Jinsha River Area, Xiluodu, Jinping in Yalong River Area, and Guandi large-scale cascade hydropower base, and opened up the great channel of power for transmitting hydropower from Southwest China to Eastern China and other load centers. The completion and operation of Ningdong Project, Hulun Buir-Liaoning Project and other DC transmission projects have promoted the scale development and outward transmission of the coal power base in Eastern Ningxia, Hulun Buir, etc., converted the local resource advantages into economic advantages, and met the demands in electric power for the regional economic development in Shandong, Liaoning and so on. Zhundong-Huadong(Wannan) ±1100 KV UHV DC Transmission Project is the UHV power transmission project currently having the highest voltage class, largest transmission capacity, longest transmission distance, and most advanced technical level in the world. As multiple UHV DC lines have been put into operation, the line repair, operation and maintenance become very important in order to guarantee the normal operation of the lines.

Currently, most of the personnel engaged in the repair and operation maintenance of UHV lines are from the personnel for maintenance and operation of EHV lines. Most of them have rarely received the special training for repair and operation maintenance of UHV lines. Meanwhile, as the live maintenance of UHV lines has been carried out gradually, the engaged personnel are in great need of normative training, so as to improve the operation level and reserve skill-based talents. The Skill Training Center of State Grid Sichuan Electric Power Company, relying on the power transmission and distribution line live working base and UHV AC & DC practical training base, has been engaged in the training and qualification for operation maintenance of power transmission lines, live working on power transmission lines for a long time, with rich training resources accumulated in the process of teaching. Currently, it has formulated the standards for professional training and assessment of multi-

ple power transmission lines under State Grid Sichuan Electric Power Company, and has rich experiences in preparation of training and assessment standards; the Maintenance Company of State Grid Sichuan Electric Power Company has been engaged in the operation and maintenance of Sichuan cross-border UHV lines for a long time, and is rich in practical experiences, which have strengthened the practicability of this standard.Based on above situation, the Skill Training Center of State Grid Sichuan Electric Power Company has compiled this Standard for Professional Training and Assessment for Operation Maintenance of UHV DC Power Transmission Line jointly with the Maintenance Company of State Grid Sichuan Electric Power Company and other operation maintenance units.

This standard is divided into two parts which are the standard for training and assessment on live working for operation maintenance of UHV power transmission lines as the first part and the standard for training and assessment on interruption maintenance for operation maintenance of UHV power transmission lines as the second parts, with totally 17 modules, including 11 modules for live maintenance and 6 modules for interruption maintenance. Each module consists of two parts: training standard and assessment standard, covering all elements of training and assessment.

The strong support and help for the preparation of this standard from the Skill Training Center of State Grid Sichuan Electric Power Company and the Maintenance Company of State Grid Sichuan Electric Power Company is greatly acknowledged.

Due to the limited level of the editor, there are inevitably some places that can be improved in the book, and the criticism and correction from readers are appreciated.

Editor

September 2019

目　录

第一部分　±800kV 特高压输电线路运检带电作业培训及考核标准 ……1

　模块一　带电更换±800kV 特高压输电线路直线塔单 V 型复合绝缘子培训及考核标准 …………………………………………………………………2

　模块二　带电更换±800kV 特高压输电线路耐张塔横担侧 1-3 片玻璃绝缘子培训及考核标准 …………………………………………………………18

　模块三　带电更换±800kV 特高压输电线路耐张塔导线侧 1-3 片玻璃绝缘子培训及考核标准 …………………………………………………………33

　模块四　带电更换±800kV 特高压输电线路耐张玻璃绝缘子串任意单片绝缘子培训及考核标准 ………………………………………………………50

　模块五　带电更换±800kV 特高压输电线路导线间隔棒培训及考核标准 ……66

　模块六　带电更换±800kV 特高压输电线路导线防振锤培训及考核标准 ……84

　模块七　带电修补±800kV 特高压输电线路分裂导线培训及考核标准 ……103

　模块八　带电处理±800kV 特高压输电线路导线引流板发热缺陷培训及考核标准 ………………………………………………………………………117

　模块十　带电修补±800kV 特高压输电线路接地极线路导线培训及考核标准 ……143

　模块十一　带电更换±800kV 特高压输电线路接地极线路间隔棒培训及考核标准 ………………………………………………………………………157

第二部分　±800kV 特高压输电线路运检停电检修培训及考核标准 ……171

　模块一　停电更换±800kV 特高压输电线路直线塔双 V 型瓷质绝缘子培训及考核标准 …………………………………………………………………172

　模块二　停电更换±800kV 特高压输电线路耐张整串绝缘子培训及考核标准 …184

　模块三　停电修补±800kV 特高压输电线路架空地线培训及考核标准 ……196

　模块四　停电更换±800kV 特高压输电线路架空地线培训及考核标准 ……209

　模块五　停电更换±800kV 特高压输电线路接地极耐张整串绝缘子培训及考核标准 ………………………………………………………………………224

　模块六　停电更换±800kV 特高压输电线路接地极导线培训及考核标准 ……236

CONTENTS

Part I Standard for Training and Assessment on Live Working for Operation Maintenance of ±800kV UHV Power Transmission Line ·················249

    Module 1 Standard for Training and Assessment on Live Replacement of Single-V Composite Insulator for ±800kV UHV Power Transmission Line Tangent Tower ··250

    Module 2 Standards for Training and Assessment on Live Replacement of 1-3 Glass Insulators on the Cross Arm Side of ±800kV UHV Transmission Line Resisting-tensile Tower ···280

    Module 3 Standards for Training and Assessment on Live Replacement of 1-3 Glass Insulators on the Conductor Side of ±800kV UHV Transmission Line Resisting-tensile Tower ···304

    Module 4 Standards for Training and Assessment on Live Replacement of Arbitrary Single Insulator of Tensile Glass Insulator String of ±800kV UHV Transmission Line ··333

    Module 5 Standards for Training and Assessment on Live Replacement of ± 800kV UHV Transmission Line Conductor Spacer ···361

    Module 6 Standards for Training and Assessment on Live Replacement of ± 800kV UHV Transmission Line Vibration Damper ···394

    Module 7 Standards for Training and Assessment on Live Repair of ± 800kV UHV Transmission Line Bundle Conductor ···426

    Module 8 Standards for Training and Assessment on Live Treatment of Heating Defects of Drainage Plate for Conductor of ±800kV UHV Transmission Line ··451

  Module 9 Standards for Training and Assessment on Live Replacement of Whole String of Insulators in a Straight Line of Grounding Electrode of ±800kV UHV Transmission Line ·················468

  Module 10 Standards for Training and Assessment on Live Repair of ±800kV UHV Transmission Line Grounding Electrode Line Conductor ···············498

  Module 11 Standards for Training and Assessment on Live Replacement of ±800kV UHV Transmission Line Grounding Electrode Line Spacer ············523

Part II Standard for Training and Assessment on Interruption Maintenance for Operation Maintenance of ±800kV UHV Power Transmission Line·········549

  Module 1 Standard for Training and Assessment on Power-cut Replacement of Double-V Porcelain Insulator for ±800kV UHV Power Transmission Line Tangent Tower ·················550

  Module 2 Standards for Training and Assessment on Power-cut Replacement of Whole Strain Insulators String on ±800kV UHV Transmission Line ······573

  Module 3 Training and Assessment Standard for Power-cut Repair of ±800kV UHV Transmission Line Overhead Ground Wire ················596

  Module 4 Training and Assessment Standard for Power-cut Replacement of ±800kV UHV Transmission Line Overhead Ground Wire ··················620

  Module 5 Standards for Training and Assessment on Power-cut Replacement of Whole Strain Insulators String on ±800kV UHV Transmission Line Earth Electrode ·················646

  Module 6 Standards for Training and Assessment on Power-cut Replacement of ±800kV UHV Transmission Line Grounding Electrode Conductor ···············667

第一部分

±800kV 特高压输电线路运检带电作业培训及考核标准

模块一 带电更换±800kV特高压输电线路直线塔单V型复合绝缘子培训及考核标准

一、培训标准

（一）培训要求

模块名称	带电更换±800kV特高压输电线路直线塔单V型复合绝缘子	培训类别	操作类
培训方式	实操培训	培训学时	21学时
培训目标	1.掌握直线塔进、出±800kV强电场时采用"吊篮法"作业方式的电学意义。 2.能完成采用"吊篮法"进入±800kV等电位作业点。 3.能独立完成带电更换±800kV特高压输电线路直线塔单V型复合绝缘子的操作（等电位作业法）		
培训场地	特高压±800kV直流实训线路		
培训内容	等电位和地电位配合，通过"吊篮法"进入等电位作业，采用等电位作业法带电更换±800kV特高压输电线路直线塔单V型复合绝缘子		
适用范围	特高压±800kV直流输电线路检修人员		

（二）引用规程规范

（1）《±800kV直流线路带电作业技术规范》（DL/T 1242-2013）

（2）《±800kV直流架空输电线路运行规程》（GB/T 28813-2012）

（3）《±800kV直流架空输电线路检修规程》（DL/T 251-2012）

（4）《±800kV直流输电线路带电作业技术导则》（Q/GDW 302-2009）

（5）《交流线路带电作业安全距离计算方法》（GB/T 19185-2008）

（6）《带电作业用绝缘配合导则》（DL/T 867-2004）

（7）《带电作业用绝缘工具试验导则》（DL/T 878-2004）

（8）《国家电网公司带电作业工作管理规定（试行）》（国家电网生〔2007〕751号）

（9）《国家电网公司电力安全工作规程（线路部分）》（Q/GDW 1799.2-2013）

（10）《电工术语架空线路》（GB/T 2900.51-1998）

（11）《电工术语带电作业》（GB/T 2900.55-2016）

（12）《带电作业工具设备术语》（GB/T 14286-2008）

（13）《带电作业用工具、装置和设备使用的一般要求》（DL/T 877-2004）

（14）《带电作业工具、装置和设备预防性试验规程》（DL/T 976-2005）

（15）《带电作业用绝缘滑车》（GB/T 13034-2008）

（16）《带电作业用绝缘绳索》（GB 13035-2008）

（17）《带电作业用屏蔽服装》（GB/T 6568-2008）

（18）《1000kV交流带电作业用屏蔽服装》（GB/T 25726-2010）

（三）培训教学设计

本设计以完成"带电更换±800kV特高压输电线路直线塔单V型复合绝缘子"为工作任务，按工作任务完成的标准化作业流程来设计各个培训阶段，每个阶段包括了具体的培训目标、培训内容、培训学时、培训方法（培训资源）、培训环境和考核评价等内容，如表1-1-1所示。

表1-1-1 带电更换±800kV特高压输电线路直线塔单V型复合绝缘子

培训流程	培训目标	培训内容	培训学时	培训方法与资源	培训环境	考核评价
1.理论教学	1.初步掌握"吊篮法"进出±800kV电场基本方法。2.熟悉±800kV输电线路直线杆塔单V型复合绝缘子更换方法。	1."吊篮法"进出强电场作业方式的电学意义。2.±800kV输电线路直线塔单V型复合绝缘子更换方法和质量标准	2	培训方法：讲授法。培训资源：PPT、相关规程规范	多媒体教室	考勤、课堂提问和作业
2.准备工作	能完成作业前准备工作	1.作业现场查勘。2.编制培训标准化作业卡。3.填写培训操作工作票。4.完成本操作的工器具及材料准备	2	培训方法：1.现场查勘和工器具及材料清理采用现场实操方法。2.编写作业卡和填写工作票采用讲授方法。培训资源：1.特高压实训线路（±800kV实训线路）。2.特高压工器具库房。3.空白工作票	1.特高压输电实训线路。2.多媒体教室	

续表

培训流程	培训目标	培训内容	培训学时	培训方法与资源	培训环境	考核评价
3.作业现场准备	能完成作业现场准备工作	1.作业现场复勘。2.工作申请。3.作业现场布置。4.班前会。5.工器具检查	1	培训方法：演示与角色扮演法。资源：特高压实训线路（±800kV实训线路）	±800kV实训线路	
4.培训师演示	通过现场观摩，使学员初步领会本任务操作流程	1.塔上工器具布置安装。2.等电位电工采用"吊篮法"进、出强电场。3.地电位电工与等电位电工相互配合利用荷载转移装置完成单V型复合绝缘子更换	2	培训方法：演示法。资源：特高压实训线路（±800kV实训线路）	±800kV实训线路	
5.学员分组训练	通过培训：1.能完成进、出±800kV强电场操作。2.能完成±800kV输电线路直线塔单V型复合绝缘子更换	1.学员分组（12人一组）训练进出±800kV强电场和直线塔单V型复合绝缘子更换。2.培训师对学员操作进行指导和安全监护	12	培训方法：角色扮演法。资源：特高压实训线路（±800kV实训线路）	±800kV实训线路	采用技能考核评分细则对学员操作评分
6.工作终结	通过培训：1.使学员进一步认识操作过程不足处，便于后期提升。2.培训学员安全文明生产的工作作风	1.作业现场清理。2.向调度汇报工作。3.班后会，对今天工作任务进行点评总结	1	培训方法：讲授和归纳法	±800kV实训线路	

（四）作业流程

1. 工作任务

等电位和地电位配合，采用"吊篮法"进入等电位作业，带电更换±800kV特高压输电线路直线塔单V型复合绝缘子。

2. 天气及作业现场要求

（1）带电更换±800kV特高压输电线路直线塔单V型复合绝缘子应在良好的天

气进行。

如遇雷电（听见雷声、看见闪电）、雪、雹、雨、雾等，禁止进行带电作业。风力大于5级，或空气相对湿度大于80%时，不宜进行带电作业；恶劣天气下必须开展带电抢修时，应组织有关人员充分讨论并编制必要的安全措施，经本单位批准后方可进行。

（2）作业人员精神状态良好，熟悉工作中保证安全的组织措施和技术措施；应持有在有效期内的带电作业资质证书。

（3）工作负责人应事先组织相关人员完成现场勘察，根据勘察结果确定本次作业方法和所需工器具，以及应采取的必要措施，并办理带电作业工作票。

（4）作业现场应合理设置围栏，并妥当布置警示标示牌，禁止非工作人员入内。

（5）本项目需停用线路直流再启动装置。

（6）工作中安全距离及有效绝缘长度如表1-1-2所示。

表1-1-2　带电更换±800kV特高压输电线路直线塔单V型复合绝缘子的安全距离（m）

海拔高度	等电位电工与接地构架之间的最小安全距离	绝缘工器具的最小有效绝缘长度	最小组合间隙
$H \leqslant 1000$	6.8	6.8	6.7
$1000 < H \leqslant 2000$	7.3	7.3	7.3
$2000 < H \leqslant 2500$	7.9	7.8	7.8

注：表中最小安全距离、最小组合间隙包括人体占位间隙0.5m。

3. 准备工作

3.1　危险点及其预控措施

（1）危险点——触电伤害

预控措施：

①工作前，工作负责人应与值班调控人员联系，停用线路直流再启动装置，并履行许可手续。

②塔上地电位作业人员登塔前，必须仔细核对线路名称、杆塔编号、相别，确认无误后方可上塔。

③工作中，如遇线路突然停电，作业人员应视其仍然带电。工作负责人应尽快与调控人员联系，值班调控人员未与工作负责人取得联系前不准强送电。

④绝缘工具及绝缘绳索不得损坏、受潮、变形、失灵，不准使用非绝缘绳索（如棉纱绳、白棕绳、钢丝绳）。

⑤等电位作业人员应穿着阻燃内衣，衣服外面应穿戴全套屏蔽服（包括帽、面罩、衣裤、手套、袜和鞋），且各部分应连接良好。

⑥等电位作业人员在电位转移前，应得到工作负责人的许可，人体裸露部分与带电体的最小距离不小于0.5m；电位转移时，动作应迅速，严禁用头部充放电；与地电位作业人员传递工具和材料时，使用绝缘工具或绝缘绳索的有效长度不准小于表1-1-2的规定。

⑦用绝缘绳索传递大件金属物品时，地电位电工及地面作业人员应将金属物品接地后再接触。

⑧专责监护人应对作业人员进行不间断监护，随时纠正其不规范或违章动作。重点关注高处作业人员，使其保持足够的安全距离（符合表1-1-2的规定），禁止同时接触两个非连通的带电体或带电体与接地体。

（2）危险点——高处坠落

预控措施：

①高处作业人员登高前，必须具备符合本项作业要求的身体状况、精神状态和技能素质。

②地电位电工登塔至作业点位后应打好安全绳并检查确认牢固，吊篮轨迹绳、等电位电工人体后备保护绳应布置合理且可靠锚固。吊篮四周应使用四根绝缘绳索稳固悬吊。吊拉绝缘绳索长度应准确计算或实地测量，使等电位作业人员头部高度不超过导线侧均压环。等电位电工系好绝缘保护绳、主保护绳，对吊篮做冲击试验合格后，方可进入吊篮。

③监护人员应随时纠正其不规范或违章动作，重点关注作业人员在转位的过程中不得失去安全带或绝缘后备保护绳的保护，严禁低挂高用。

④人员登塔时检查脚钉和塔材的紧固情况，登塔时手抓主材，严禁手抓脚钉。

（3）危险点——高处坠物伤人

预控措施：

①高处作业人员的个人工具及零星材料应装入工具袋，严禁在高处浮置物件、口中含物。

②地面作业人员必须正确佩戴安全帽，正确使用绳结，与作业点垂直下方距离不得小于坠落半径。

③作业现场设置围栏并挂好警示标示牌。监护人员应随时注意，禁止非工作人员及车辆进入作业区域。

3.2 工器具及材料选择

带电更换±800kV特高压输电线路直线塔单V型复合绝缘子所需工器具及材料见表1-1-3。工器具出库前，应认真核对工器具的使用电压等级和试验周期，并检查确认外观良好、连接牢固、转动灵活,且符合本次工作任务的要求;工器具出库后,应存放在工具袋或工具箱内进行运输,防止脏污、受潮;金属工具和绝缘工器具应分开装运,防止因混装运输导致工器具变形、损伤等现象发生。

表1-1-3　带电更换±800kV输电线路直线杆塔单V型复合绝缘子所需工器具及材料表

序号	名称	规格型号	单位	数量	备注
1	绝缘传递绳	TJS-10	根	1	绝缘工具
2	绝缘传递绳	TJS-16	根	3	绝缘工具
3	绝缘保护绳	TJS-16	根	2	绝缘工具
4	吊篮轨迹绳	TJS-16	根	1	绝缘工具
5	绝缘滑车	JH10-1	个	5	绝缘工具
6	绝缘吊杆		套	2	绝缘工具
7	绝缘紧线器	0.5T	把	1	绝缘工具
8	绝缘绳套		根	若干	绝缘工具
9	吊篮		付	1	绝缘工具
10	液压丝杠		根	2	金属工具
11	提线卡具		副	2	金属工具
12	机动绞磨		台	1	金属工具
13	屏蔽服	屏蔽效率≥60dB(屏蔽面罩屏蔽效率≥20dB)	套	4	个人防护用具
14	安全带		副	4	个人防护用具
15	安全帽		顶	12	个人防护用具
16	绝缘电阻测试仪	5000V,电极宽2cm、极间宽2cm	块	1	其他工具
17	风速风向仪		块	1	其他工具
18	温湿度仪		块	1	其他工具
19	万用表		块	1	其他工具
20	防潮帆布	2m×4m	块	4	其他工具
21	红马甲	"工作负责人"	件	1	其他工具
22	对讲机		台	4	其他工具
23	防坠器	与杆塔防坠器装置型号对应	只	4	其他工具
24	安全围栏		套	若干	其他工具
25	个人工具		套	2	其他工具
26	警示标示牌	"在此工作""从此进出""从此上下"	套	1	其他工具
27	复合绝缘子		套	1	材料

3.3 作业人员分工

本任务作业人员分工如表1-1-4所示。

表1-1-4 带电更换±800kV特高压输电线路直线塔单V型复合绝缘子人员分工表

序号	工作岗位	数量(人)	工作职责
1	工作负责人	1	负责作业现场的各项工作
2	专责监护人	1	负责作业现场的安全把控
3	等电位电工	2	负责进入等电位完成带电端绝缘子更换工作
4	塔上地电位电工	2	负责完成接地端绝缘子更换工作、协助等电位进出电场
5	地面电工	6	负责传递工具、材料配合等电位电工进出等电位

4. 工作程序

本任务工作流程如表1-1-5所示。

表1-1-5 带电更换±800kV特高压输电线路直线塔单V型复合绝缘子工作流程表

序号	作业内容	作业步骤及标准	安全措施及注意事项	责任人
1	现场复勘	工作负责人负责完成以下工作： (1)现场核对线路名称、杆塔编号，相别无误；基础及杆塔完好无异常；交叉跨越距离符合安全要求；确认缺陷情况等。 (2)检测风速、湿度等现场气象条件符合作业要求。 (3)检查地形环境符合作业要求。 (4)检查工作票所列安全措施与现场实际情况相符，必要时予以补充	(1)正确穿戴安全帽、工作服、工作鞋、劳保手套。 (2)不得在危及作业人员安全的气象条件下作业。 (3)严禁非工作人员、车辆进入作业现场	
2	工作许可	(1)工作负责人负责联系值班调控人员，按工作票内容申请停用线路直流再启动装置。 (2)经值班调控人员许可后，方可开始带电作业工作	不得未经值班调控人员许可即开始工作	
3	现场布置	正确装设安全围栏并悬挂标示牌： (1)安全围栏范围应充分考虑高处坠物，以及对道路交通的影响。 (2)安全围栏出入口设置合理。 (3)妥当布置"从此进出""在此工作""从此上下"等标示	对道路交通安全影响不可控时，应及时联系交通管理部门强化现场交通安全管控	

续表

序号	作业内容	作业步骤及标准	安全措施及注意事项	责任人
4	召开班前会	(1)全体工作成员列队。 (2)工作负责人宣读工作票,明确工作任务及人员分工;讲解工作中的安全措施和技术措施;查(问)全体工作成员精神状态;告知工作中存在的危险点及采取的预控措施。 (3)全体工作成员在工作票上签名确认	(1)工作票填写、签发和许可手续规范,签名完整。 (2)全体工作成员精神状态良好。 (3)全体工作成员明确任务分工、安全措施和技术措施	
5	检查工具	(1)塔上地电位电工和等电位电工正确地穿戴好屏蔽服并检测合格,由负责人监督检查。 (2)正确佩戴个人安全用具(大小合适,锁扣自如),由负责人监督检查。 (3)测量风速风向、湿度,检查绝缘工具的绝缘性能,并做好记录	(1)金属、绝缘工具使用前,应仔细检查其是否损坏、变形、失灵。绝缘工具应使用2500V及以上绝缘电阻表进行分段绝缘检测,阻值应不低于700MΩ,并用清洁干燥的毛巾将其擦拭干净。 (2)用万用表测量屏蔽服衣裤最远端点之间的电阻值不得大于20Ω。工作负责人认真检查作业电工屏蔽服的连接情况。 (3)检查工具组装情况并确认连接可靠。 (4)现场所使用的带电作业工具应放置在防潮帆布上	
6	登塔	(1)核对线路名称、杆塔编号无误后,塔上地电位电工和等电位冲击检查安全带、防坠器受力情况。 (2)塔上地电位电工携带绝缘传递绳登塔,等电位电工随后登塔,至横担作业点,选择合适位置系好安全带,塔上地电位电工将绝缘滑车和绝缘传递绳安装在横担合适位置。然后配合地面电工将绝缘传递绳分开作起吊准备	(1)核对线路名称和杆塔编号无误后,方可登塔作业。 (2)登塔过程中应使用杆塔上安装的防坠装置;杆塔上移动及转位时,不准失去安全保护,作业人员必须攀抓牢固构件。 (3)作业电工必须穿全套合格的屏蔽服,且全套屏蔽服必须连接可靠。在横担进入等电位前,等电位电工要检查确认屏蔽服各个部位连接可靠后方能进行下一步操作	
7	安装滑车组及吊篮	(1)地面电工利用绝缘传递绳将吊篮、吊篮轨迹绳、2-2绝缘滑车组传至横担。 (2)塔上电工将2-2绝缘滑车组及吊篮安装在横担上平面合适位置,并将吊篮与组二滑车组可靠连接;将吊篮轨迹绳一端安装在横担合适位置,一端与吊篮可靠连接	(1)传递时绝缘吊绳要起吊平稳、无磕碰、无缠绕。 (2)吊篮安装好后由塔上电工对吊篮情况进行认真检查核对。 (3)2-2滑车组及吊篮应在横担上合适位置可靠安装	

续表

序号	作业内容	作业步骤及标准	安全措施及注意事项	责任人
8	进入强电场	(1)一名等电位电工系好绝缘保护绳后进入吊篮,地面电工缓慢松出2-2绝缘滑车组控制绳,待吊篮距带电导线约2m处放慢速度。 (2)在吊篮向导线继续移动过程中、等电位电工面向带电导线,同时向工作负责人申请电位转移,得到同意后,等电位电工待吊篮距导线0.5m时迅速伸手抓住最近的子导线进行电位转移。 (3)等电位电工进入强电场后检查绝缘后备保护绳,同时等电位电工要控制头部不超过导线侧均压环。 (4)地面电工收紧2-2绝缘滑车组控制绳,将吊篮向上传至横担部位。2号等电位电工系好绝缘保护绳进入吊篮,用同样的方法进入强电场	(1)进入等电位前,等电位电工要再次检查确认屏蔽服各部位、电位转移棒与绝缘屏蔽服连接可靠后方能进行下一步操作。 (2)等电位电工进入电位前必须得到工作负责人的许可。 (3)等电位电工进入吊篮前必须系好绝缘保护绳。 (4)地面电工配合等电位电工进入等电位过程中收放滑车组时控制绳应平稳。 (5)等电位电工在进入电位过程中与接地体和带电体两部分间隙所组成的组合间隙不得小于表1-1-2要求。 (6)专责监护人负责监护等电位电工进入强电场的安全注意事项。对塔上作业人员的危险、不规范动作应及时提醒,必要时应制止。 (7)等电位电工进入电场后不得解除绝缘保护绳,安全带不得系在子导线上。	
9	安装工具并转移导线荷载	(1)地面电工将绝缘吊杆、提线卡具、液压丝杠传递至工作位置,由等电位电工和地电位电工配合将复合绝缘子更换工具进行正确安装。 (2)检查各构件连接可靠,得到工作负责任人同意后,先收紧机械丝杠,待机械丝杠适当受力后等电位电工收紧液压丝杠,使之稍受力,检查受力点情况。 (3)地面电工将复合绝缘子串控制绳传递给等电位电工,等电位电工将其安装在复合绝缘子串尾部。地面电工收紧复合绝缘子串控制绳。 (4)检查承力工具受力正常,得到工作负责人同意后,等电位电工拆开导线侧碗头挂板螺栓,地面电工缓缓放松复合绝缘子串控制绳,使之自然垂直。 (5)地电位电工将绝缘传递绳系在复合绝缘子串上端,然后取出复合绝缘子串与球头挂环连接的锁紧销。地面电工与地电位电工配合脱开复合绝缘子串与球头挂环的连接	(1)上、下作业电工要密切配合,所有作业电工要听从等工作负责人的统一指挥。 (2)地电位电工对带电体、等电位电工对接地体的最小安全距离不得小于表1-1-2。 (3)杆塔上、下传递工具绑扎绳扣应正确可靠,塔上电工不得高空落物。 (4)工具受力后应试冲击检查无误后,报告工作负责人,在得到工作负责人许可后,方可继续作业。 (5)专责监护人对塔上作业人员的危险、不规范动作应及时提醒,必要时应制止。 (6)在转移荷载过程中,作业人员应时刻注意工器具受力情况,出现异常立即报告工作负责人	

续表

序号	作业内容	作业步骤及标准	安全措施及注意事项	责任人
10	更换绝缘子串	(1)地面电工控制好复合绝缘子串控制绳,利用机动绞磨缓慢将复合绝缘子串放至地面。注意控制好复合绝缘子串的控制绳,不得碰撞承力工具、导线及杆塔。 (2)地面电工将绝缘传递绳和复合绝缘子控制绳分别转移到新复合绝缘子串上,然后利用机动绞磨将新复合绝缘子串传递至塔上工作位置,地电位电工恢复新复合绝缘子串与球头挂环的连接,并恢复锁紧销。 (3)地面电工缓慢松出机动绞磨使复合绝缘子串自然垂直,然后收紧复合绝缘子串控制绳将复合绝缘子串尾部拉至导线侧等电位电工工作位置。等电位电工利用绝缘紧线器恢复绝缘子串碗头挂板与联板的连接,装好开口销	(1)绝缘子串在退出运行时,详细检查受力部件是否正常良好,检查确认无问题后征得工作负责人同意方可拆除。 (2)绝缘子串起吊过程中不得与杆塔发生碰撞,控制好绝缘子串尾绳。 (3)利用机动绞磨起吊绝缘子串时应安置平稳,尾绳应有有作业经验人员控制,不可疏忽放松。 (4)必须检查绞磨及转向滑车的受力情况,无误后方可进行作业。 (5)杆塔上、下传递工具绑扎绳扣应正确可靠	
11	拆除工具	(1)经检查复合绝缘子串连接可靠,得到工作负责人同意后,地电位电工松出液压丝杆。 (2)经检查复合绝缘子串受力正常,得到工作负责人的同意后,地电位电工与等电位电工配合拆除绝缘吊杆、液压丝杆等,并传至地面	(1)复合绝缘子安装复位后,应详细检查各部位连接正常无误,并得到工作负责人的同意后方可拆除提线工具。 (2)工具在传递过程中不得碰撞杆塔,绑扎绳扣应正确可靠	
12	退出电场	(1)一名等电位电工系好绝缘保护绳,进入吊篮,然后保持手臂伸直状态使吊篮距子导线0.5m。 (2)等电位电工向工作负责人申请退出电位,得到同意后,等电位电工迅速脱开子导线。 (3)地面电工同时迅速收紧2-2绝缘滑车组控制绳,将吊篮向上拉至横担部位停住,然后等电位电工登上横担,并系好安全带。 (4)地面电工利用绝缘传递绳将吊篮传至另一名等电位电工处,等电位电工检查导线上无遗留物后进入吊篮,用同样的方法退出电位	(1)上、下作业电工要密切配合,听从工作负责人的指挥。 (2)等电位电工退出电位前必须得到工作负责人的许可。 (3)等电位电工进入吊篮前必须系好保护绳。 (4)地面电工配合等电位电工进入等电位时收放滑车组,控制绳应平稳。 (5)等电位电工在退出电位过程中与接地体和带电体两部分间隙所组成的组合间隙不得小于表1-1-2规定。 (6)专责监护人负责监护等电位电工退出强电场的安全注意事项。对塔上作业人员的危险、不规范动作应及时提醒,必要时应制止	

续表

序号	作业内容	作业步骤及标准	安全措施及注意事项	责任人
13	拆除吊篮返回地面	(1)塔上电工配合拆除吊篮轨迹绳、绝缘保护绳、2-2绝缘滑车组及吊篮并传至地面。 (2)塔上电工检查塔上无遗留物后,向工作负责人汇报,得到工作负责人同意后携带绝缘传递绳下塔	(1)工具在传递过程中不得碰撞,绑扎绳扣应正确可靠。 (2)登塔过程中应使用塔上安装的防坠装置;杆塔上移动及转位时,不准失去安全保护,作业人员必须抓牢固构件	
14	工作结束	(1)清理现场及工具,认真检查杆(塔)上有无遗留物,工作负责人全面检查工作完成情况,清点人数,无误后、宣布工作结束,撤离施工现场。 (2)通知调度工作完毕,履行工作票完工手续	不得约时恢复线路再启动装置	

二、考核标准

表1-1-6 国网四川省电力公司特高压直流输电线路运检技能考核评分细则

考生填写栏	编号: 姓名: 所在岗位: 单位: 日期: 年 月 日						
考评员填写栏	成绩: 考评员: 考评组长: 开始时间: 结束时间: 操作时长:						
考核模块	带电更换±800kV特高压输电线路直线塔单V型复合绝缘子		考核对象	特高压±800kV直流输电线路检修人员	考核方式	操作	考核时限	120min
任务描述	带电更换±800kV特高压输电线路直线塔单V型复合绝缘子						
工作规范及要求	1. 带电作业工作应在良好天气下进行。如遇雷、雨、雪、雾天气不得进行带电作业。风力大于5级、湿度大于80%时,一般不宜进行带电作业。 2. 本项作业需工作负责人1名,专责监护人1人,塔上地电工2人,等电位电工2人,地面辅助电工6人,采用吊篮摆渡法进入强电场进行绝缘子更换工作。 3. 工作负责人职责:负责本次工作任务的人员分工、工作票的宣读、办理线路停用直流再启动、办理工作许可手续、召开工作班前会、工作中突发情况的处理、工作质量的监督、工作后的总结。 4. 专责监护人:负责作业现场的安全把控。 5. 等电位电工职责:配合地电位电工安装提线系统,操作液压丝杠转移导线荷载,拆装复合绝缘子串。 6. 地电位电工职责:负责安装吊篮、提线系统、绝缘磨绳及配合等电位电工进出电位,拆装复合绝缘子串。 7. 地面电工职责:负责传递工具、材料配合等电位电工进出等电位。 8. 在带电作业中,如遇雷、雨、大风或其他任何情况威胁到工作人员的安全时,工作负责人或监护人可根据情况,临时停止工作。						

续表

工作规范及要求	给定条件： 1. 培训基地：特高压交流±800kV线路塔。 2. 工作票已办理，安全措施已经完备（直流再启动已停用），工作开始、工作终结时应口头提出申请（调度或考评员）。 3. 安全、正确地使用仪器对绝缘工具进行检测。 4. 必须按工作程序进行操作，工序错误扣除应做项目分值，出现重大人身、器材和操作安全隐患，考评员可下令终止操作（考核）
考核情景准备	1. 线路：特高压交流±800kV线路003#塔，工作内容：±800kV输电线路直线杆塔单V型复合绝缘子。 2. 所需作业工器具：绝缘传递绳1根（TJS-12），绝缘传递绳3根（TJS-16），绝缘保护绳2根（TJS-16），吊篮轨迹绳1根（TJS-16），吊篮1个，液压丝杠2只，绝缘吊杆2组，提线卡具2副，机动绞磨1台，绝缘滑车6个（JH10-1），2-2绝缘滑车2个（JH20-2），屏蔽服（屏蔽效率≥60dB）4套，绝缘检测仪1台，万用表1块，温湿度仪1台，速仪1台，防潮帆布2块，个人工具2套。 3. 作业现场做好监护工作，作业现场安全措施（围栏等）已全部落实；禁止非作业人员进入现场，工作人员进入作业现场必须戴安全帽。 4. 考生自备工作服，阻燃纯棉内衣，安全帽，线手套，安全带（含二保绳）
备注	1. 各项目得分均扣完为止，出现重大人身、器材和操作安全隐患，考评员可下令终止操作。 2. 设备、作业环境、安全带、安全帽、工器具、屏蔽服等不符合作业条件考评员可下令终止操作

表1-1-7　国网四川省电力公司特高压直流输电线路运检技能考核评分标准

序号	项目名称	质量要求	分值	扣分标准	扣分原因	扣分	得分
1	现场复勘	（1）工作负责人到作业现场核对线路名称和杆塔编号、现场工作条件、缺陷部位等。 （2）检测风速、湿度等现场气象条件符合作业要求。 （3）检查工作票填写完整，无涂改，检查是否所列安全措施与现场实际情况相符，必要时予以补充	5	（1）未进行核对线路称号，扣1分。 （2）未核实现场工作条件（气象）、缺陷部位，扣1分。 （3）工作票填写出现涂改，每项扣0.5分；工作票编号有误，扣1分，工作票填写不完整，扣1.5分。			
2	工作许可	（1）工作负责人联系值班调控人员，按工作票内容申请停用线路直流再启动装置。 （2）汇报内容规范、完整	2	（1）未联系调度部门（裁判）停用直流再启动装置，扣2分。 （2）汇报专业用语不规范或不完整的，各扣0.5分			
3	现场布置	正确装设安全围栏并悬挂标示牌： （1）安全围栏范围应充分考虑高处坠物，以及对道路交通的影响。 （2）安全围栏出入口设置合理。 （3）妥当布置"从此进出""在此工作""从此上下"等标示	3	（1）作业现场未装设围栏，扣0.5分。 （2）未设立警示牌，扣0.5分。 （3）未悬挂登塔作业标志，扣0.5分			

续表

序号	项目名称	质量要求	分值	扣分标准	扣分原因	扣分	得分
4	召开班前会	(1)全体工作成员全体人员正确佩戴安全帽、工作服。(2)工作负责人佩戴红色背心,宣读工作票,明确工作任务及人员分工;讲解工作中的安全措施和技术措施;查(问)全体工作成员精神状态;告知工作中存在的危险点及采取的预控措施。(3)全体工作成员在工作票上签名确认	3	(1)工作人员着装不整齐,扣0.5分;工作人员着装不整齐,每人次扣0.5分。(2)未进行分工,本项不得分;分工不明,扣1分。(3)现场工作负责人未穿佩安全监护背心,扣0.5分。(4)工作票上工作班成员未签字或签字不全的,扣1分。			
5	工器具检查	(1)工作人员按要求将工器具放在防潮帆布上;防潮帆布应清洁、干燥。(2)工器具应按定置管理要求分类摆放;绝缘工器具不能与金属工具、材料混放;对工器具进行外观检查。(3)绝缘工具表面不应磨损、变形损坏,操作应灵活。绝缘工具应使用2500V及以上绝缘电阻表进行分段绝缘检测,阻值应不低于700MΩ,并用清洁干燥的毛巾将其擦拭干净。(4)塔上地电位和等电位人员按要求正确穿戴全套合格的屏蔽服、导电鞋,各部分连接应良好,屏蔽服内不得贴身穿着化纤类衣服,并系好安全带;工作负责人应认真检查是否穿戴正确。(5)登塔人员再次核对线路名称、杆号、相别并报告	7	(1)未使用防潮布并定置摆放工器具,扣1分。(2)未检查工器具试验合格标签及外观检查,每项扣0.5分。(3)未正确使用检测仪器对工器具进行检测,每项扣1分。(4)作业人员未正确穿戴屏蔽服且各部位连接良好,每人次扣2分。(5)现场工作负责人未对登塔作业人员进安全防护装备进行检查,扣1分。(6)登塔人员未核对线路名称、杆号、相别,每人扣2分。(7)登塔人员未报告核对结果,每人扣2分			
6	登塔	(1)塔上地电位电工、等电位电工穿好全套合格的屏蔽服,将安全带做冲击试验后,系好安全带后携带绝缘传递绳相继登塔。		(1)未系安全带或安全带及后备保护绳未进行冲击试验,各扣2分。(2)手抓脚钉,扣2分。(3)滑车传递绳悬挂位置不便工具取用,扣1分。			

第一部分
±800kV 特高压输电线路运检带电作业培训及考核标准

序号	项目名称	质量要求	分值	扣分标准	扣分原因	扣分	得分
6	登塔	(2)登塔过程中系好防坠落保护装置，登塔至合适位置，系好安全带，布置好绝缘传递绳，然后配合地面电工将绝缘传递绳分开作起吊准备。 (3)登塔过程中应系好防坠落保护装置，匀速登塔，手抓主材，将安全带挂在肩上并与带电体保持6.8m以上安全距离，工作负责人加强作业监护。	5	(4)传递时金属工具难以保证安全距离，扣2分；工具绑扎不牢，扣2分。 (5)传递时高空落物，扣2分。 (6)传递过程工具与塔身磕碰，扣2分。 (7)传递工具绳索打结混乱，扣1分。 (8)工作负责人监护不到位，扣2分。 (9)塔上电工操作不正确，扣2分			
7	安装滑车组及吊篮	(1)传递时绝缘吊绳起吊要平稳、无磕碰、无缠绕。 (2)吊篮安装好后由塔上电工对吊篮情况进行认真检查核对。 (3)2-2滑车组及吊篮应在横担上合适位置可靠	5	(1)2-2滑车组绳子缠绕，扣0.5分。 (2)轨迹绳安装位置不合理，扣1分。 (3)绝缘保护绳、长度不合适，扣1分。 (4)传递工器具不平稳、磕碰，扣1分			
8	进入强电场	(1)等电位再次检查确认屏蔽服各部位连接可靠后对吊篮进行冲击实验，汇报工作负责人后系好保护绳登上吊篮行。 (2)地面电工缓慢松出2-2绝缘滑车组控制绳，当距离导线约2m处放慢速度，等电位电工在距离导线0.5m处向工作负责人申请电位转移，得到同意后，迅速伸手抓住最近的子导线进行电位转移。 (3)等电位电工进入强电场后做好人体后保护，要控制头部不超过导线侧均压环。 (4)电位电工在进入电位过程中与接地体和带电体两部分间隙所组成的组合间隙不得小于6.8m	13	(1)等电位电工未对吊篮进行冲击，扣1分。 (2)未系好绝缘保护绳，扣1分。 (3)地电位电工未检查等电位电工安全措施，扣2分。 (4)地面电工控制滑车尾绳不平稳，扣1分。 (5)等电位电工进入强电场前未向工作负责人申请，扣2分；申请了但未得同意即开始，扣1分。 (6)转移电位动作不熟练，扣1分。 (7)等电位电工进入强电场后解开人体后备保护绳，扣2分。 (8)等电位电工进入强电场后头部超过导线侧均压环，扣2分			

续表

序号	项目名称	质量要求	分值	扣分标准	扣分原因	扣分	得分
9	安装工具并转移导线荷载	(1)地面电工将绝缘吊杆、提线卡具、液压丝杠传递至工作位置,由等电位电工和地电位电工配合将复合绝缘子更换工具进行正确安装。(2)检查各部构件可靠得到工作负责任人同意后,先收紧机械丝杠,待机械丝杠适当受力后等电位电工收紧液压丝杠,使之稍受力,检查受力点情况。(3)地面电工将复合绝缘子串控制绳传递给等电位电工,等电位电工将其安装在复合绝缘子串尾部。地面电工收紧复合绝缘子串控制绳。(4)检查承力工具受力正常得到工作负责人同意后,等电位电工拆开导线侧连接,地面电工缓缓放松复合绝缘子串控制绳,使之自然垂直。(5)地电位电工将绝缘传递绳系在复合绝缘子串上端,地面电工与地电位电工配合脱开复合绝缘子串接地端连接	15	(1)工器具传递过程不平稳,扣1分;有磕碰每次,扣1分。(2)未检查承力工具安装可靠、受力良好,扣2分;未汇报并取得工作负责人同意,扣2分。(3)地电位电工与等电位电工沟通无效,扣1分。(4)拆除绝缘子串前未对承力工具进行检查,扣5分;检查结果未汇报,扣2分。(5)绝缘绳系复合绝缘子位置不合适,扣2分。(6)绝缘子串传递时有磕碰每次,扣2分。(7)地面电工机动绞磨控制不到位,扣1分			
10	更换绝缘子串	(1)地面电工控制好复合绝缘子串控制绳,利用机动绞磨缓慢将复合绝缘子串放至地面。注意控制好复合绝缘子串的控制绳,不得碰撞承力工具、导线及杆塔。(2)地面电工将绝缘传递绳和复合绝缘子串控制绳分别转移到新复合绝缘子上。(3)地面电工启动机动绞磨,将新复合绝缘子串传递至塔上工作位置。地电位电工恢复新复合绝缘子串与球头挂环的连接,并复位锁紧销。(4)地面电工缓慢松出机动绞磨使复合绝缘子串自然垂直,等电位电工恢复碗头挂板与联板的连接,并装好开口销	17	(1)地面电工未控制好绝缘子尾绳,扣1分。(2)绑扎绳扣不合理,扣1分。(3)未检查绞磨转向及滑车的受力情况,扣2分。(4)绝缘子串传递时有磕碰每次,扣2分。(5)绝缘子串安装不到位,扣5分。(6)作业人员未检查绝缘子串连接,扣5分。(7)作业人员未检查销子安装到位,扣2分。(8)专责监护人未尽监护职责,扣2分			

续表

序号	项目名称	质量要求	分值	扣分标准	扣分原因	扣分	得分
11	退出电位	(1)一名等电位电工系好绝缘保护绳,进入吊篮,然后保持手臂伸直状态使吊篮距子导线0.5m,得到工作负责人同意后,等电位电工迅速脱开子导线。(2)地面电工同时迅速收紧2-2绝缘滑车组控制绳,将吊篮向上拉至横担部位停住,然后等电位电工登上横担,并系好安全带。(3)地面电工利用绝缘传递绳将吊篮传至另一名等电位电工处,等电位电工检查导线上无遗留物后进入吊篮,用同样的方法退出电位	10	(1)未报告工作结束,扣2分,强电场内有遗留物每件,扣1分。(2)等电位未系好绝缘保护绳,扣1分。(3)地面电工控制滑车尾绳不平稳,扣1分。(4)等电位电工退出强电场前未向工作负责人申请,扣2分;申请了但未得同意即开始,扣1分。(5)转移电位动作不熟练,扣1分			
12	拆除吊篮返回地面	(1)塔上电工配合拆除吊篮轨迹绳、绝缘保护绳、2-2绝缘滑车组及吊篮并传至地面。(2)塔上电工检查塔上无遗留物后,向工作负责人汇报,得到工作负责人同意后携带绝缘传递绳下塔	5	(1)下塔过程未使用防坠装置,扣2分。(2)塔上移位失去安全带保护的,扣2分。(3)下塔抓塔钉,每处,扣1分。(4)塔上有遗留物的,扣2分			
13	工作结束	(1)工作负责人组织全体工作成员整理工器具和材料,将工器具清洁后放入专用的箱(袋)中;清理现场,做到"工完料尽场地清"。(2)召开班后会,工作负责人进行工作总结和点评工作。点评本次工作的施工质量;点评全体工作成员的安全措施落实情况。(3)工作负责人向值班调控人员汇报工作结束,申请恢复线路重合闸,终结工作票	10	(1)工器具未清理,扣2分。(2)工器具有遗漏,扣2分。(3)未开班后会,扣2分。(4)未拆除围栏,扣2分。(5)未向调度汇报,扣2分			
	合计		100				

模块二 带电更换±800kV特高压输电线路耐张塔横担侧1-3片玻璃绝缘子培训及考核标准

一、培训标准

(一) 培训要求

模块名称	带电更换±800kV特高压输电线路耐张塔横担侧1-3片玻璃绝缘子	培训类别	操作类
培训方式	实操培训	培训学时	21学时
培训目标	1.掌握地电位作业法中电磁防护的电学意义。 2.能独立完成带电更换±800kV特高压输电线路耐张塔横担侧1-3片玻璃绝缘子的操作(地电位作业法)		
培训场地	特高压±800kV直流实训线路		
培训内容	采用地电位作业法带电更换±800kV特高压输电线路耐张塔横担侧1-3片玻璃绝缘子的操作		
适用范围	特高压±800kV直流输电线路检修人员		

(二) 引用规程规范

(1)《±800kV直流线路带电作业技术规范》(DL/T 1242-2013)

(2)《±800kV直流架空输电线路运行规程》(GB/T 28813-2012)

(3)《±800kV直流架空输电线路检修规程》(DL/T 251-2012)

(4)《±800kV直流输电线路带电作业技术导则》(Q/GDW 302-2009)

(5)《交流线路带电作业安全距离计算方法》(GB/T 19185-2008)

(6)《带电作业用绝缘配合导则》(DL/T 867-2004)

(7)《带电作业用绝缘工具试验导则》(DL/T 878-2004)

(8)《国家电网公司带电作业工作管理规定(试行)》(国家电网生〔2007〕751号)

(9)《国家电网公司电力安全工作规程(线路部分)》(Q/GDW 1799.2-2013)

(10)《电工术语架空线路》(GB/T 2900.51-1998)

（11）《电工术语带电作业》（GB/T 2900.55-2016）

（12）《带电作业工具设备术语》（GB/T 14286-2008）

（13）《带电作业用工具、装置和设备使用的一般要求》（DL/T 877-2004）

（14）《带电作业工具、装置和设备预防性试验规程》（DL/T 976-2005）

（15）《带电作业用绝缘滑车》（GB/T 13034-2008）

（16）《带电作业用绝缘绳索》（GB 13035-2008）

（17）《带电作业用屏蔽服装》（GB/T 6568-2008）

（18）《1000kV交流带电作业用屏蔽服装》（GB/T 25726-2010）

（三）培训教学设计

本设计以完成"带电更换±800kV特高压输电线路耐张塔横担侧1-3片玻璃绝缘子"为工作任务，按工作任务完成的标准化作业流程来设计各个培训阶段，每个阶段包括了具体的培训目标、培训内容、培训学时、培训方法（培训资源）、培训环境和考核评价等内容，如表1-2-1所示。

表1-2-1 带电更换±800kV特高压输电线路耐张塔横担侧1-3片玻璃绝缘子培训内容设计

培训流程	培训目标	培训内容	培训学时	培训方法与资源	培训环境	考核评价
1.理论教学	1.初步掌握地电位作业法中电磁防护的基本方法。2.熟悉±800kV输电线路耐张杆塔横担侧1-3片玻璃绝缘子更换方法	1.地电位作业法中电磁防护的电学意义。2.±800kV输电线路耐张杆塔横担侧1-3片玻璃绝缘子更换方法和质量标准	2	培训方法：讲授法。培训资源：PPT、相关规程规范	多媒体教室	考勤、课堂提问和作业
2.准备工作	能完成作业前准备工作	1.作业现场查勘。2.编制培训标准化作业卡。3.填写培训操作工作票。4.完成本操作的工器具及材料准备	2	培训方法：1.现场查勘和工器具及材料清理采用现场实操方法。2.编写作业卡和填写工作票采用讲授方法。培训资源：1.特高压实训线路（±800kV实训线路）。2.特高压工器具库房。3.空白工作票	1.特高压输电实训线路；2.多媒体教室	

续表

培训流程	培训目标	培训内容	培训学时	培训方法与资源	培训环境	考核评价
3.作业现场准备	能完成作业现场准备工作	1.作业现场复勘。 2.工作申请。 3.作业现场布置。 4.班前会。 5.工器具检查	1	培训方法:演示与角色扮演法。 资源:特高压实训线路(±800kV实训线路)	±800kV实训线路	
4.培训师演示	通过现场观摩,使学员初步领会本任务操作流程	1.地电位电工组装工器具。 2.地电位电工完成单片玻璃绝缘子的更换工作	2	培训方法:演示法。 资源:特高压实训线路(±800kV实训线路)	±800kV实训线路	
5.学员分组训练	能完成±800kV输电线路耐张杆塔横担侧1-3片玻璃绝缘子的更换工作	1.学员分组(7人一组)训练更换绝缘子技能操作。 2.培训师对学员操作进行指导和安全监护	12	培训方法:角色扮演法。 资源:特高压实训线路(±800kV实训线路)	±800kV实训线路	采用技能考核评分细则对学员操作评分
6.工作终结	通过培训: 1.使学员进一步认识操作过程不足处,便于后期提升。 2.培训学员安全文明生产的工作作风	1.作业现场清理。 2.向调度汇报工作。 3.班后会,对今天工作任务进行点评总结	1	培训方法:讲授和归纳法	±800kV实训线路	

(四)作业流程

1.工作任务

带电更换±800kV特高压输电线路耐张塔横担侧1-3片玻璃绝缘子。

2.天气及作业现场要求

(1)带电更换±800kV特高压输电线路耐张塔横担侧1-3片玻璃绝缘子应在良好的天气进行。如遇雷电(听见雷声、看见闪电)、雪、雹、雨、雾等,禁止进行带电作业。风力大于5级,或空气相对湿度大于80%时,不宜进行带电作业;恶劣天气下必须开展带电抢修时,应组织有关人员充分讨论并编制必要的安全措施,经本单位批准后方可进行。

(2)作业人员精神状态良好,熟悉工作中保证安全的组织措施和技术措施;应持有在有效期内的带电作业资质证书。

（3）工作负责人应事先组织相关人员完成现场勘察，根据勘察结果确定本次作业方法和所需工器具，以及应采取的必要措施，并办理带电作业工作票。

（4）作业现场应合理设置围栏，并妥当布置警示标示牌，禁止非工作人员入内。

（5）本项目需停用线路直流再启动装置。

（6）工作中安全距离及有效绝缘长度如表1-2-2所示。

表1-2-2 带电更换±800kV特高压输电线路耐张塔横担侧1-3片玻璃绝缘子的安全距离（m）

海拔高度	等电位电工与接地构架之间的最小安全距离	绝缘工器具的最小有效绝缘长度	最小组合间隙
$H \leqslant 1000$	6.8	6.8	6.7
$1000 < H \leqslant 2000$	7.3	7.3	7.3
$2000 < H \leqslant 2500$	7.9	7.8	7.8

注：表中最小安全距离、最小组合间隙包括人体占位间隙0.5m。

（7）地电位电工进入耐张绝缘子串横担侧时，人体短接绝缘子片数不得多于4片。耐张绝缘子串中扣除人体短接和不良绝缘子片数后，良好绝缘子最少片数应满足表1-2-3要求。

表1-2-3 耐张绝缘子串良好绝缘子最少片数

海拔高度（m）	单片玻璃绝缘子结构高度（mm）	良好绝缘子串的总长度最小值（m）	良好绝缘子的最少片数
$H \leqslant 1000$	170	6.2	37
	195		32
	205		31
	240		26
$1000 < H \leqslant 2000$	170	7.1	42
	195		37
	205		35
	240		30
$2000 < H \leqslant 2500$	170	7.55	45
	195		39
	205		37
	240		32

3. 准备工作

3.1 危险点及其预控措施

（1）危险点——触电伤害

预控措施：

①工作前，工作负责人应与值班调控人员联系，停用线路直流再启动装置，并履行许可手续。

②塔上地电位作业人员登塔前，必须仔细核对线路名称、杆塔编号、相别，确认无误后方可上塔。

③工作中，如遇线路突然停电，作业人员应视其仍然带电。工作负责人应尽快与调控人员联系，值班调控人员未与工作负责人取得联系前不准强送电。

④绝缘工具及绝缘绳索不得损坏、受潮、变形、失灵，不准使用非绝缘绳索（如棉纱绳、白棕绳、钢丝绳）。

⑤地面电工操作绝缘工具时应戴清洁、干燥的手套，进入作业现场应将使用的带电作业工具放置在防潮的帆布或绝缘垫上，防止绝缘工具在使用中脏污和受潮。

⑥地电位电工应穿着阻燃内衣，衣服外面应穿戴合格全套屏蔽服（包括帽、面罩、衣裤、手套、袜和鞋），且各部分应连接良好。

⑦地电位电工进入耐张绝缘子串横担侧时，手与脚的移动必须保持对应一致，且人体和工具短接的绝缘子片数不得超过4片。

⑧用绝缘绳索传递大件金属物品时，地电位电工及地面电工应将金属物品接地后再接触。

⑨带电作业过程中，工作负责人（监护人）应对作业人员进行不间断监护，随时纠正其不规范或违章动作。重点关注高处作业人员，使其保持足够的安全距离（符合表1-2-2的规定），禁止同时接触两个非连通的带电体或带电体与接地体。

（2）危险点——高处坠落

预控措施：

①高处作业人员登高前，必须具备符合本项作业要求的身体状况、精神状态和技能素质。

②监护人员应随时纠正其不规范或违章动作，重点关注作业人员在转位的过程中不得失去安全带或绝缘后备保护绳的保护，严禁低挂高用。

（3）危险点——高处坠物伤人

预控措施：

①高处作业人员的个人工具及零星材料应装入工具袋，严禁在高处浮置物件、口

中含物。

②地面作业人员必须正确佩戴安全帽，正确使用绳结，与作业点垂直下方距离不得小于坠落半径。

③作业现场设置围栏并挂好警示标示牌。监护人员应随时注意，禁止非工作人员及车辆进入作业区域。

3.2 工器具及材料选择

带电更换±800kV特高压输电线路耐张塔横担侧1-3片玻璃绝缘子工器具及材料表见表1-2-4。工器具出库前，应认真核对工器具的使用电压等级和试验周期，并检查确认外观良好、连接牢固、转动灵活，且符合本次工作任务的要求；工器具出库后，应存放在工具袋或工具箱内进行运输，防止脏污、受潮；金属工具和绝缘工器具应分开装运，防止因混装运输导致工器具变形、损伤等现象发生。

表1-2-4 带电更换±800kV特高压输电线路耐张塔横担侧1-3片玻璃绝缘子所需工器具及材料表

序号	名称	规格型号	单位	数量	备注
1	绝缘传递绳	TJS-12	根	2	绝缘工具
2	绝缘保护绳	TJS-16	根	2	绝缘工具
3	绝缘绳套	Φ14mm	根	2	绝缘工具
4	绝缘滑车	JH10-1	个	2	绝缘工具
5	耐张端部卡		个	1	金属工具
6	液压丝杠		根	2	金属工具
7	闭式卡（后卡）		个	1	金属工具
8	屏蔽服	屏蔽效率≥60dB（屏蔽面罩屏蔽效率≥20dB）	套	2	个人防护用具
9	导电鞋	尺码视穿着人员而定	双	2	个人防护用具
10	阻燃内衣	纯桑蚕丝	套	2	个人防护用具
11	双保险安全带	背带式	根	2	个人防护用具
12	安全帽		顶	7	个人防护用具
13	护目镜		副	2	个人防护用具
14	绝缘电阻测试仪	5000V，电极宽2cm、极间宽2cm	套	1	其他工具
15	风速风向仪		块	1	其他工具
16	温湿度仪		块	1	其他工具
17	万用表		块	1	其他工具
18	防潮帆布	2m×4m	块	2	其他工具

续表

序号	名称	规格型号	单位	数量	备注
19	防坠器	与杆塔防坠器装置型号对应	只	2	其他工具
20	安全围栏		套	若干	其他工具
21	警示标示牌	"在此工作""从此进出""从此上下"	套	1	其他工具
22	红马甲	"工作负责人"	件	1	其他工具
23	个人工具	扳手、老虎钳	套	1	其他工具
24	拔销器		把	1	其他工具
25	防坠器	与杆塔防坠落装置型号对应	只	2	其他工具
26	清洁毛巾	棉质	条	1	其他工具
27	对讲机		台	3	其他工具
28	绝缘子		片	1	材料

3.3 作业人员分工

本任务作业人员分工如表1-2-5所示。

表1-2-5 带电更换±800kV特高压输电线路耐张塔横担侧1-3片玻璃绝缘子人员分工表

序号	工作岗位	数量(人)	工作职责
1	工作负责人	1	负责作业现场的各项工作
2	专责监护人	1	负责作业现场的安全把控
3	地电位电工	2	负责工器具安装及绝缘子更换工作
4	地面电工	3	负责传递工具、材料配合地电位电工更换绝缘子

4.工作程序

本任务工作流程如表1-2-6所示。

表1-2-6 带电更换±800kV特高压输电线路耐张塔横担侧1-3片玻璃绝缘子工作流程表

序号	作业内容	作业步骤及标准	安全措施及注意事项	责任人
1	现场复勘	工作负责人负责完成以下工作： (1)现场核对线路名称、杆塔编号，相别无误；基础及杆塔完好无异常；交叉跨越距离符合安全要求；确认缺陷情况等。 (2)检测风速、湿度等现场气象条件符合作业要求。 (3)检查地形环境符合作业要求。 (4)检查工作票所列安全措施与现场实际情况相符，必要时予以补充	(1)正确穿戴安全帽、工作服、工作鞋、劳保手套。 (2)不得在危及作业人员安全的气象条件下作业。 (3)严禁非工作人员、车辆进入作业现场	

续表

序号	作业内容	作业步骤及标准	安全措施及注意事项	责任人
2	工作许可	(1)工作负责人负责联系值班调控人员,按工作票内容申请停用线路直流再启动装置。 (2)经值班调控人员许可后,方可开始带电作业工作。	不得未经值班调控人员许可即开始工作	
3	现场布置	正确装设安全围栏并悬挂标示牌: (1)安全围栏范围应充分考虑高处坠物,以及对道路交通的影响。 (2)安全围栏出入口设置合理。 (3)妥当布置"从此进出""在此工作""从此上下"等标示。	对道路交通安全影响不可控时,应及时联系交通管理部门强化现场交通安全管控	
4	召开班前会	(1)全体工作成员列队。 (2)工作负责人宣读工作票,明确工作任务及人员分工;讲解工作中的安全措施和技术措施;查(问)全体工作成员精神状态;告知工作中存在的危险点及采取的预控措施。 (3)全体工作成员在工作票上签名确认。	(1)工作票填写、签发和许可手续规范,签名完整。 (2)全体工作成员精神状态良好。 (3)全体工作成员明确任务分工、安全措施和技术措施	
5	检查工具	(1)地电位电工正确地穿戴好屏蔽服并检测合格,由负责人监督检查。 (2)正确佩戴个人安全用具(大小合适,锁扣自如),由负责人监督检查。 (3)测量风速风向、湿度,检查绝缘工具的绝缘性能,并做好记录。	(1)金属、绝缘工具使用前,应仔细检查其是否损坏、变形、失灵。绝缘工具应使用2500V及以上绝缘电阻表进行分段绝缘检测,阻值应不低于700MΩ,并用清洁干燥的毛巾将其擦拭干净。 (2)用万用表测量屏蔽服衣裤最远端点之间的电阻值不得大于20Ω。工作负责人认真检查作业电工屏蔽服的连接情况。 (3)检查工具组装情况并确认连接可靠。 (4)现场所使用的带电作业工具应放置在防潮帆布上	

续表

序号	作业内容	作业步骤及标准	安全措施及注意事项	责任人
6	登塔	(1)核对线路名称、杆塔编号无误后,塔上地电位电工检查安全带、防坠器受力情况。 (2)地电位电工携带绝缘传递绳登塔,两人至横担作业点,选择合适位置系好安全带,塔上地电位电工将绝缘滑车和绝缘传递绳安装在横担合适位置。然后配合地面电工将绝缘传递绳分开作起吊准备	(1)核对线路名称和杆塔编号无误后,方可登塔作业。 (2)登塔过程中应使用塔上安装的防坠装置;杆塔上移动及转位时,不准失去安全保护,作业人员必须攀抓牢固构件。 (3)地电位电工必须穿全套合格的屏蔽服,且全套屏蔽服必须连接可靠。在横担进入绝缘子串前,地电位电工要检查确认屏蔽服各个部位连接可靠后方能进行下一步操作	
7	安装工具并转移导线张力	(1)地面电工使用绝缘传递绳将闭式卡(后卡)、液压丝杠、耐张端部卡等分别传至地电位电工。 (2)地电位电工先在牵引板上安装耐张端部卡,后将闭式卡(后卡)安装在横担侧第4片绝缘子上,并连接好液压丝杠。 (3)检查并确认承力工具各部分安装情况良好,经工作负责人许可后,操作液压丝杠使其逐渐受力,使需更换的绝缘子松弛	(1)起吊过程平稳、无磕碰、无缠绕,正确使用绳结,防止高空落物。 (2)工器具安装完成后应再次确认各部分安装牢固可靠;操作过程中,人体和工器具短接绝缘子不得超过4片,人体与带电体的最小安全距离不得小于表1-2-2规定。 (3)在收紧液压丝杠的过程中应保持两边均匀同步受力,摇动丝杠应平稳有力。	
8	更换绝缘子	(1)地电位电工做冲击试验,检查并确认承力工具受力正常,经工作负责人许可后,用绝缘传递绳系好旧绝缘子,取出旧绝缘子两端锁紧销,继续操作并收紧液压丝杠,直至拆除旧绝缘子。两根液压丝杠的受力应均匀,操作手柄不得敲击绝缘子。 (2)地面电工用绝缘传递绳的另一端系好新绝缘子,采用旧下、新上的方法,将新绝缘子起吊给地电位电工。 (3)地电位电工安装新绝缘子,并复位其两端锁紧销,确认安装到位	(1)在调节液压丝杠过程中应保持两边均匀同步受力,手柄不得敲击玻璃绝缘子。 (2)起吊过程平稳、无磕碰、无缠绕,正确使用绳结.地面电工相互配合,防止绝缘子发生碰撞。 (3)作业过程中时刻注意各部件受力情况。	

续表

序号	作业内容	作业步骤及标准	安全措施及注意事项	责任人
9	拆除工具	(1)地电位电工检查并确认新绝缘子连接可靠,经工作负责人许可后,操作并松出液压丝杠,使更换的绝缘子逐渐受力。 (2)荷载转移完毕后,地电位电工做冲击试验,检查并确认新绝缘子受力情况良好,经工作负责人许可后,拆除系在绝缘子上的绝缘传递绳,并将其系牢于承力工具适当位置,拆除闭式卡(后卡)、液压丝杠、耐张端部卡等承力工具,在地面电工配合下传递至地面。	(1)在拆除工器具前应检查绝缘子锁紧销是否安装到位。 (2)传递过程平稳、无磕碰、无缠绕,正确使用绳结,防止高空落物伤人	
10	返回地面	塔上电工检查塔上无遗留物后,向工作负责人汇报,得到工作负责人同意后携带绝缘传递绳下塔	下塔过程中应使用塔上安装的防坠装置,杆塔上移动及转位时,不准失去安全保护,作业人员必须攀抓牢固构件	
11	工作结束	(1)工作负责人组织全体工作成员整理工器具和材料,将工器具清洁后放入专用的箱(袋)中;清理现场,做到"工完料尽场地清"。 (2)召开班后会,工作负责人进行工作总结和点评工作;点评本次工作的施工质量;点评全体工作成员的安全措施落实情况。 (3)工作负责人向值班调控人员汇报工作结束,申请恢复线路直流再启动装置,终结工作票	不得约时恢复线路直流再启动装置	

二、考核标准

表1-2-7 国网四川省电力公司特高压直流输电线路运检技能考核评分细则

考生填写栏	编号： 姓 名： 所在岗位： 单位： 日 期： 年 月 日					
考评员填写栏	成绩： 考评员： 考评组长： 开始时间： 结束时间： 操作时长：					
考核模块	带电更换±800kV特高压输电线路耐张塔横担侧1-3片玻璃绝缘子	考核对象	特高压±800kV直流输电线路检修人员	考核方式	操作	考核时限 90min
任务描述	带电更换±800kV特高压输电线路耐张塔横担侧1-3片玻璃绝缘子。					
工作规范及要求	1. 带电作业工作应在良好天气下进行。如遇雷、雨、雪、雾天气不得进行带电作业。风力大于5级、湿度大于80%时，一般不宜进行带电作业。 2. 本项作业需工作负责人1名，专责监护人1人，塔上地电工2人，地面辅助电工3人。 3. 工作负责人职责：负责本次工作任务的人员分工、工作票的宣读、办理停用直流再启动装置、办理工作许可手续、召开工作班前会、工作中突发情况的处理、工作质量的监督.工作后的总结。 4. 专责监护人：负责作业现场的安全把控。 5. 地电位电工职责：负责工器具安装及绝缘子更换工作。 6. 地面电工职责：负责传递工具、材料配合地电位电工更换绝缘子。 7. 在带电作业中，如遇雷、雨、大风或其他任何情况威胁到工作人员的安全时，工作负责人或监护人可根据情况，临时停止工作。 给定条件： 1. 培训基地：特高压直流±800kV线路铁塔，绝缘子型号：U500BP。 2. 工作票已办理，安全措施已经完备（再启动装置已停用），工作开始、工作终结时应口头提出申请（调度或考评员）。 3. 安全、正确地使用仪器对绝缘工具进行检测。 4. 必须按工作程序进行操作，工序错误扣除应做项目分值，出现重大人身、器材和操作安全隐患，考评员可下令终止操作（考核）					
考核情景准备	1. 线路：特高压直流±800kV线路铁塔，工作内容：更换±800kV输电线路耐张杆塔横担侧1-3片玻璃绝缘子，绝缘子型号：U500BP。 2. 所需作业工器具：绝缘传递绳2根（TJS-12），绝缘保护绳2根（TJS-16），绝缘滑车2个（JH10-1），闭式卡具1套，液压丝杠2根，屏蔽服2套，绝缘电阻测试仪1台，万用表1台，温湿度仪1台，风速仪1台，防潮帆布2块，个人工具1套。 3. 作业现场做好监护工作，作业现场安全措施（围栏等）已全部落实；禁止非作业人员进入现场，工作人员进入作业现场必须戴安全帽。 4. 考生自备工作服，阻燃纯棉内衣，安全帽，线手套，安全带（含二保绳）					
备注	1. 各项目得分均扣完为止，出现重大人身、器材和操作安全隐患，考评员可下令终止操作。 2. 设备、作业环境、安全带、安全帽、工器具、屏蔽服等不符合作业条件考评员可下令终止操作					

表1-2-8 国网四川省电力公司特高压直流输电线路运检技能考核评分标准

序号	项目名称	质量要求	分值	扣分标准	扣分原因	扣分	得分
1	现场复勘	(1)工作负责人到作业现场核对线路名称和杆塔编号、现场工作条件、缺陷部位等。 (2)检测风速、湿度等现场气象条件符合作业要求。 (3)检查工作票填写完整,无涂改,检查是否所列安全措施与现场实际情况相符,必要时予以补充	5	(1)未进行核对线路称号,扣1分。 (2)未核实现场工作条件(气象)、缺陷部位,扣1分。 (3)工作票填写出现涂改,每项,扣0.5分;工作票编号有误,扣1分;工作票填写不完整,扣1.5分			
2	工作许可	(1)工作负责人联系值班调控人员,按工作票内容申请停用线路直流再启动装置。 (2)汇报内容规范、完整	2	(1)未联系调度部门(裁判)停用再启动装置,扣2分。 (2)汇报专业用语不规范或不完整的,各扣0.5分			
3	现场布置	正确装设安全围栏并悬挂标示牌: (1)安全围栏范围应充分考虑高处坠物,以及对道路交通的影响。 (2)安全围栏出入口设置合理。 (3)妥当布置"从此进出""在此工作""从此上下"等标示	3	(1)作业现场未装设围栏,扣0.5分。 (2)未设立警示牌,扣0.5分。 (3)未悬挂登塔作业标志,扣0.5分			
4	召开班前会	(1)全体工作成员全体人员正确佩戴安全帽、工作服。 (2)工作负责人佩戴红色背心,宣读工作票,明确工作任务及人员分工;讲解工作中的安全措施和技术措施;查(问)全体工作成员精神状态;告知工作中存在的危险点及采取的预控措施。 (3)全体工作成员在工作票上签名确认	3	(1)工作人员着装不整齐,扣0.5分;工作人员着装不整齐每人次,扣0.5分。 (2)未进行分工,本项不得分;分工不明,扣1分。 (3)现场工作负责人未穿佩安全监护背心,扣0.5分。 (4)工作票上工作班成员未签字或签字不全的,扣1分			

续表

序号	项目名称	质量要求	分值	扣分标准	扣分原因	扣分	得分
5	工器具检查	(1)工作人员按要求将工器具放在防潮帆布上;防潮帆布应清洁、干燥。 (2)工器具应按定置管理要求分类摆放;绝缘工器具不能与金属工具、材料混放;对工器具进行外观检查。 (3)绝缘工具表面不应磨损、变形损坏,操作应灵活。绝缘工具应使用2500V及以上绝缘电阻表进行分段绝缘检测,阻值应不低于700MΩ,并用清洁干燥的毛巾将其擦拭干净。 (4)塔上地电位人员按要求正确穿戴全套合格的屏蔽服、导电鞋,且各部分连接应良好,屏蔽服内不得贴身穿着化纤类衣服,并系好安全带;工作负责人应认真检查是否穿戴正确。 (5)登塔人员再次核对线路名称、杆号、相别并报告	7	(1)未使用防潮布并定置摆放工器具,扣1分。 (2)未检查工器具试验合格标签及外观检查,每项扣0.5分。 (3)未正确使用检测仪器对工器具进行检测,每项,扣1分。 (4)作业人员未正确穿戴屏蔽服且各部位连接良好,每人次扣2分。 (5)现场工作负责人未对登塔作业人员进安全防护装备进行检查,扣1分。 (6)登塔人员未核对线路名称、杆号、相别,每人扣2分。 (7)登塔人员未报告核对结果,每人扣2分			
6	登塔	(1)塔上地电位电工穿好全套合格的屏蔽服,将安全带做冲击试验后,系好安全带后携带绝缘传递绳相继登塔。 (2)登塔过程中系好防坠落保护装置,登塔至合适位置,系安全带,布置好绝缘传递绳,然后配合地面电工将绝缘传递绳分开作起吊准备。 (3)登塔过程中应系好防坠落保护装置,匀速登塔,手抓主材,将安全带挂在肩上并与带电体保持6.8m以上安全距离,工作负责人加强作业监护。	5	(1)未系安全带或安全带及后备保护绳未进行冲击试验,各扣2分。 (2)手抓脚钉,扣2分。 (3)滑车传递绳悬挂位置不便工具取用,扣1分。 (4)传递时金属工具难以保证安全距离,扣2分;工具绑扎不牢,扣2分。 (5)传递时高空落物,扣2分。 (6)传递过程工具与塔身磕碰,扣2分。 (7)传递工具绳索打结混乱,扣1分。 (8)工作负责人监护不到位,扣2分。 (9)塔上电工操作不正确,扣2分			

续表

序号	项目名称	质量要求	分值	扣分标准	扣分原因	扣分	得分
7	安装工具	(1)地面电工使用绝缘传递绳将闭式卡(后卡)、液压丝杠、耐张端部卡等分别起吊给地电位电工。起吊过程平稳、无磕碰、无缠绕,正确使用绳结。 (2)地电位电工先在牵引板上安装耐张端部卡,后将闭式卡(后卡)安装在横担侧第4片绝缘子上,并连接好液压丝杠。承力工具各部分安装牢固可靠。 (3)检查并确认承力工具各部分安装情况良好,经工作负责人许可后,操作液压丝杠使其逐渐受力,使需更换的绝缘子松弛。两根液压丝杠的受力应均匀	20	(1)起吊过程不平稳,出现磕碰、缠绕,扣1分/次。 (2)高处坠物,扣2分/次。 (3)卡具安装不正确、固定不到位,扣2分。 (4)未检查承力工具安装情况,扣3分;检查了未报告,扣1分;报告了但工作负责人未同意即开始收紧丝杠,扣1分。 (5)作业过程短接绝缘子片数超过4片,扣3分/次。 (6)安装卡具出现绝缘子碰撞破损,扣2分。 (7)未均衡收紧丝杠,扣2分			
8	更换绝缘子	(1)地电位电工做冲击试验,检查并确认承力工具受力正常,经工作负责人许可后,用绝缘传递绳系好旧绝缘子,取出旧绝缘子两端锁紧销,继续操作并收紧液压丝杠,直至拆除旧绝缘子。两根液压丝杠的受力应均匀,操作手柄不得敲击绝缘子。 (2)地面电工用绝缘传递绳的另一端系好新绝缘子,采用旧下、新上的方法,将新绝缘子起吊给地电位电工。起吊过程平稳、无磕碰、无缠绕,正确使用绳结。 (3)地电位电工安装新绝缘子,并复位其两端锁紧销	20	(1)未检查承力工具受力情况,扣3分;检查了未报告,扣2分;报告了但工作负责人未同意即取出旧绝缘子两端锁紧销,扣1分。 (2)未均衡收紧丝杆,扣2分。 (3)操作手柄敲击绝缘子,扣1分/次。 (4)绳结错误,扣1分。 (5)高处坠物,扣2分/次。 (6)新旧绝缘子相互碰撞,扣1分。 (7)传递绝缘子与塔身相互碰撞,扣1分/次。 (8)绝缘绳缠绕,扣2分			

续表

序号	项目名称	质量要求	分值	扣分标准	扣分原因	扣分	得分
9	拆除工具	(1)地电位电工检查新绝缘子连接可靠,经工作负责人许可后,操作并松出液压丝杠,使更换的绝缘子逐渐受力。 (2)荷载转移完毕后,等电位电工做冲击试验,检查并确认新绝缘子受力情况良好,经工作负责人许可后,拆除系在绝缘子上的绝缘传递绳,并将其系牢于承力工具适当位置,拆除闭式卡(后卡)、液压丝杠、耐张端部卡等承力工具,在地面电工配合下传递至地面。传递过程平稳、无磕碰、无缠绕,正确使用绳结	20	(1)未检查新绝缘子连接情况,扣3分;检查了未报告,扣2分;报告了但工作负责人未同意即松液压丝杠,扣1分。 (2)未检查新绝缘子受力情况,扣3分;检查了未报告,扣2分;报告了但工作负责人未同意即拆除液压丝杠,扣1分。 (3)捆扎工具时,未正确使用绳结,扣1分。 (4)高处坠物,扣2分/次。 (5)工器具相互碰撞扣,扣1分/次;工器具与带电体或塔身相互碰撞,扣1分/次;绝缘绳缠绕,扣2分			
10	返回地面	地电位电工检查塔上无遗留物,拆除绝缘传递绳。经工作负责人许可后,挂好防坠器,解开并整理好安全带,正确携带绝缘传递绳,脚踩脚钉、手抓主材,匀步下塔至地面	5	(1)下塔过程未使用防坠器,扣5分。 (2)塔上移位失去安全带保护,扣5分。 (3)下塔手抓脚钉,扣1分/次。 (4)塔上有遗留物,扣2分			
11	工作结束	(1)工作负责人组织全体工作成员整理工器具和材料,将工器具清洁后放入专用的箱(袋)中;清理现场,做到"工完料尽场地清"。 (2)召开班后会,工作负责人进行工作总结和点评工作。点评本次工作的施工质量;点评全体工作成员的安全措施落实情况。 (3)工作负责人向值班调控人员汇报工作结束,申请恢复线路再启动,终结工作票	10	(1)工器具未清理,扣2分。 (2)工器具有遗漏,扣2分。 (3)未开班后会,扣10分。 (4)未拆除围栏,扣2分。 (5)未向调度汇报,扣2分			
	合计		100				

模块三　带电更换±800kV特高压输电线路耐张塔导线侧1-3片玻璃绝缘子培训及考核标准

一、培训标准

（一）培训要求

模块名称	带电更换±800kV特高压输电线路耐张塔导线侧1-3片玻璃绝缘子	培训类别	操作类
培训方式	实操培训	培训学时	21学时
培训目标	1.掌握沿耐张绝缘子串进、出±800kV强电场时采用"跨二短三"作业方式，以及电位转移的电学意义。 2.能完成沿耐张绝缘子串进入±800kV等电位作业点。 3.能独立完成带电更换±800kV特高压输电线路耐张塔导线侧1-3片玻璃绝缘子（等电位作业法）。		
培训场地	特高压直流实训线路		
培训内容	采用"跨二短三"作业方式沿耐张绝缘子串进入等电位，采用等电位作业法带电更换±800kV特高压输电线路耐张塔导线侧1-3片玻璃绝缘子的操作		
适用范围	特高压直流输电线路检修人员		

（二）引用规程规范

（1）《±800kV直流线路带电作业技术规范》（DL/T 1242-2013）

（2）《±800kV直流架空输电线路运行规程》（GB/T 28813-2012）

（3）《±800kV直流架空输电线路检修规程》（DL/T 251-2012）

（4）《±800kV直流输电线路带电作业技术导则》（Q/GDW 302-2009）

（5）《交流线路带电作业安全距离计算方法》（GB/T 19185-2008）

（6）《带电作业用绝缘配合导则》（DL/T 867-2004）

（7）《带电作业用绝缘工具试验导则》（DL/T 878-2004）

（8）《国家电网公司带电作业工作管理规定（试行）》（国家电网生〔2007〕751号）

（9）《国家电网公司电力安全工作规程（线路部分）》（Q/GDW 1799.2-2013）

（10）《电工术语架空线路》（GB/T 2900.51-1998）

(11)《电工术语带电作业》(GB/T 2900.55-2016)

(12)《带电作业工具设备术语》(GB/T 14286-2008)

(13)《带电作业用工具、装置和设备使用的一般要求》(DL/T 877-2004)

(14)《带电作业工具、装置和设备预防性试验规程》(DL/T 976-2005)

(15)《带电作业用绝缘滑车》(GB/T 13034-2008)

(16)《带电作业用绝缘绳索》(GB 13035-2008)

(17)《带电作业用屏蔽服装》(GB/T 6568-2008)

(18)《1000kV交流带电作业用屏蔽服装》(GB/T 25726-2010)

(三)培训教学设计

本设计以完成"带电更换±800kV特高压输电线路耐张塔导线侧1-3片玻璃绝缘子"为工作任务,按工作任务完成的标准化作业流程来设计各个培训阶段,每个阶段包括了具体的培训目标、培训内容、培训学时、培训方法(培训资源)、培训环境和考核评价等内容,如表1-3-1所示。

表1-3-1 带电更换±800kV特高压输电线路耐张塔导线侧1-3片玻璃绝缘子

培训流程	培训目标	培训内容	培训学时	培训方法与资源	培训环境	考核评价
1.理论教学	1.初步掌握沿绝缘子串进出±800kV强电场基本方法。2.熟悉转移电位的方法。3.熟悉输电线路耐张单片绝缘子更换方法	1.沿绝缘子进出强电场"跨二短三"作业方式的电学意义。2.电位转移棒的使用方法。3.输电线路耐张单片绝缘子更换方法和质量标准	2	培训方法:讲授法。培训资源:PPT、相关规程规范	多媒体教室	考勤、课堂提问和作业
2.准备工作	能完成作业前准备工作	1.作业现场查勘。2.编制培训标准化作业卡。3.填写培训操作工作票。4.完成本操作的工器具及材料准备	1	培训方法:1.现场查勘和工器具及材料清理采用现场实操方法。2.编写作业卡和填写工作票采用讲授方法。培训资源:1.±800kV实训线路。2.特高压工器具库房。3.空白工作票	1.特高压输电实训线路。2.多媒体教室	

续表

培训流程	培训目标	培训内容	培训学时	培训方法与资源	培训环境	考核评价
3.作业现场准备	能完成作业现场准备工作	1.作业现场复勘。 2.工作申请。 3.作业现场布置。 4.班前会。 5.工器具及材料检查	1	培训方法：演示与角色扮演法。 资源：±800kV实训线路	±800kV实训线路	
4.培训师演示	通过现场观摩，使学员初步领会本任务操作流程	1.等电位电工沿耐张绝缘子串进、出强电场。 2.等电位电工组装工器具。 3.等电位电工完成单片玻璃绝缘子的更换工作	2	培训方法：演示法。 资源：±800kV实训线路	±800kV实训线路	
5.学员分组训练	1.能完成进、出±800kV强电场操作。 2.能完成单片玻璃绝缘子的更换工作	1.学员分组（6人一组）训练进、出±800kV强电场和更换绝缘子技能操作。 2.培训师对学员操作进行指导和安全监护	14	培训方法：角色扮演法。 资源：±800kV实训线路	±800kV实训线路	采用技能考核评分细则对学员操作评分
6.工作终结	1.使学员进一步辨析操作过程不足之处，便于后期提升。 2.培训学员安全文明生产的工作作风	1.作业现场清理。 2.向调度汇报工作。 3.班后会，对本次工作任务进行点评总结	1	培训方法：讲授和归纳法	±800kV实训线路	

（四）作业流程

1. 工作任务

带电更换±800kV特高压输电线路耐张塔导线侧1-3片玻璃绝缘子。

2. 天气及作业现场要求

（1）带电更换±800kV特高压输电线路耐张塔导线侧1-3片玻璃绝缘子应在良好的

天气进行。如遇雷电（听见雷声、看见闪电）、雪、雹、雨、雾等，禁止进行带电作业。风力大于5级，不宜进行带电作业；相对湿度大于80%的天气，若需进行带电作业，应采用具有防潮性能的绝缘工具。恶劣天气下必须开展带电抢修时，应组织有关人员充分讨论并编制必要的安全措施，经本单位批准后方可进行。带电作业过程中如遇天气突然变化，有可能危及人身或设备安全时，应立即停止工作；在保证人身安全的情况下，尽快恢复设备正常状态，或采取其他措施。

（2）作业人员精神状态良好，熟悉工作中保证安全的组织措施和技术措施，掌握高处应急救援及触电急救的方法；应持有在有效期内的带电作业资质证书。

（3）工作负责人应事先组织相关人员完成现场勘察，根据勘察结果确定本次作业方法和所需工器具，以及应采取的必要措施，并办理带电作业工作票。

（4）作业现场应合理设置围栏，并妥当布置警示标示牌，禁止非工作人员入内。

（5）本项目须停用直流再启动保护装置。

（6）作业方式：等电位作业

（7）工作中安全距离及有效绝缘长度如表1-3-2所示。

表1-3-2　带电更换±800kV特高压输电线路耐张塔导线侧1-3片玻璃绝缘子的安全距离（m）

海拔高度	等电位电工与接地构架之间的最小安全距离	绝缘工器具的最小有效绝缘长度	最小组合间隙
$H \leq 1000$	6.8	6.8	6.7
$1000 < H \leq 2000$	7.3	7.3	7.3
$2000 < H \leq 2500$	7.9	7.8	7.8

注：表中最小安全距离、最小组合间隙包括人体占位间隙0.5m。

（8）等电位电工沿耐张绝缘子串进入等电位时，人体短接绝缘子片数不得多于4片。耐张绝缘子串中扣除人体短接和不良绝缘子片数后，良好绝缘子最少片数应满足表1-4-3的规定。

表1-3-3　最小组合间隙和良好绝缘子的最小片数

海拔高度（m）	单片玻璃绝缘子结构高度（mm）	良好绝缘子串的总长度最小值(m)	良好绝缘子的最少片数
$H \leq 1000$	170	6.2	37
	195		32
	205		31
	240		26

续表

海拔高度（m）	单片玻璃绝缘子结构高度（mm）	良好绝缘子串的总长度最小值(m)	良好绝缘子的最少片数
1000＜H≤2000	170	7.1	42
	195		37
	205		35
	240		30
2000＜H≤2500	170	7.55	45
	195		39
	205		37
	240		32

注：表中数值不包括人体占位间隙，作业中需考虑人体占位间隙不得小于0.5 m。

3. 准备工作

3.1 危险点及其预控措施

（1）危险点——触电伤害

预控措施：

①工作前，工作负责人应与值班调控人员联系，停用直流再启动保护装置，并履行许可手续，严禁约时停用或恢复直流再启动保护装置。工作结束后应及时向调度汇报。

②塔上作业人员登塔前，必须仔细核对线路双重命名、杆塔编号，确认无误后方可上塔。

③工作中，如遇线路突然停电，作业人员应视其仍然带电。工作负责人应尽快与调控人员联系，值班调控人员未与工作负责人取得联系前不准强送电。

④绝缘工具及绝缘绳索不得损坏、受潮、变形、失灵，其有效长度不准小于表1-3-2规定。不准使用非绝缘绳索（如棉纱绳、白棕绳、钢丝绳）。

⑤地面电工操作绝缘工具时应戴清洁、干燥的手套，进入作业现场应将使用的带电作业工具放置在防潮的帆布或绝缘垫上，防止绝缘工具在使用中脏污和受潮。

⑥等电位电工应穿着阻燃内衣，衣服外面应穿戴合格全套屏蔽服（包括帽、衣裤、手套、袜和鞋），且各部分应连接良好。

⑦等电位电工沿绝缘子串移动时，手与脚的位置必须保持对应一致，且人体和工具短接后的完好绝缘子片数应符合表1-3-3规定。

⑧等电位电工采用"跨二短三"作业方式（也称自由作业法）进入强电场。当作业人员平行移动至距导线侧均压环三片绝缘子处，应停止移动，利用电位转移棒进行

电位转移,电位转移棒长度为0.4m,电位转移时,人体面部与带电体距离不得小于0.5m。

⑨用绝缘绳索传递大件金属物品时,地面作业人员应将金属物品接地后再接触。

⑩带电作业过程中,工作负责人(监护人)应对作业人员进行不间断监护,随时纠正其不规范或违章动作。重点关注高处作业人员,使其保持足够的安全距离及组合间隙(符合表1-3-2的规定),禁止同时接触两个非连通的带电体或带电体与接地体。

(2)危险点——高处坠落

预控措施:

①高处作业人员登高前,必须具备符合本项作业要求的身体状况、精神状态和技能素质。

②高处作业人员应使用双保险安全带。上、下塔时,应手抓主材、脚踩脚钉,匀速行进。

③等电位电工作业前应认真检查液压丝杠、闭式卡、导线端部卡等,确保承力工具合格;沿绝缘子串移动时,手与脚的位置必须保持对应一致,安全带应系挂在手扶的绝缘子串上,并同步移动;更换绝缘子时,承力工具安装应可靠,荷载转移前、后应做冲击试验判定其可靠性,并及时向工作负责人汇报,得到工作负责人许可后方可实施。

④监护人员应随时纠正其不规范或违章动作,重点关注高处作业人员在转位的过程中不得失去安全带或绝缘后备保护绳的保护,严禁低挂高用。

(3)危险点——高处落物伤人

预控措施:

①高处作业人员的个人工具及零星材料应装入工具袋,严禁在高处浮置物件、口中含物。

②以上下循环交换方式传递较重的工器具时,均应系好控制绳,防止被传递物品相互碰撞及误碰处于工作状态的承力工器具。

③地面作业人员必须正确佩戴安全帽,正确使用绳结,与作业点垂直下方距离不得小于坠落半径。

④作业现场设置围栏并挂好警示标示牌。监护人员应随时注意,禁止非工作人员及车辆进入作业区域。

3.2 工器具及材料选择

带电更换±800kV特高压输电线路耐张塔导线侧1-3片玻璃绝缘子所需工器具及材料见表1-3-4。工器具出库前,应认真核对工器具的使用电压等级和试验周期,并检查确认外观良好、连接牢固、转动灵活,且符合本次工作任务的要求;工器具出库后,应存放在工具袋或工具箱内进行运输,防止脏污、受潮;金属工具和绝缘工器具应分开装运,防止因混装运输导致工器具变形、损伤等现象发生。

表1-3-4 带电更换±800kV特高压输电线路耐张塔导线侧1-3片玻璃绝缘子所需工器具及材料表

序号	名称	规格型号	单位	数量	备注
1	屏蔽服	屏蔽效率≥60dB（屏蔽面罩屏蔽效率≥20dB）	套	2	个人防护用具
2	导电鞋	尺码视穿着人员而定	双	2	个人防护用具
3	阻燃内衣	纯桑蚕丝	套	2	个人防护用具
4	双保险安全带	背带式	根	2	个人防护用具
5	安全帽		顶	6	个人防护用具
6	护目镜		副	2	个人防护用具
7	绝缘传递绳	TJS-12	根	1	绝缘工具
8	绝缘后备保护绳	φ16mm	根	2	绝缘工具
9	绝缘绳套	φ14 mm	根	1	绝缘工具
10	绝缘滑车	JH10-1B	个	1	绝缘工具
11	导线端部卡		个	1	金属工具
12	液压丝杠	8T	根	2	金属工具
13	闭式卡（前卡）	Tc4	个	1	金属工具
14	电位转移棒		根	2	其他工具
15	拔销器		把	1	其他工具
16	绝缘电阻测试仪	5000V,电极宽2cm、极间宽2cm	套	1	其他工具
17	万用表		套	1	其他工具
18	风速、温湿度测试仪		只	1	其他工具
19	安全围网		套	若干	其他工具
20	警示标示牌	"在此工作""从此进出""车辆慢行""车辆绕行"	套	1	其他工具
21	红马甲	"工作负责人""专责监护人"	件	1	其他工具
22	防潮苫布	3m×3m	块	2	其他工具
23	个人工具	扳手、老虎钳	套	1	其他工具
24	防坠器	与杆塔防坠落装置型号对应	只	2	其他工具
25	毛巾	棉质	条	1	其他工具
26	绝缘子		片	1	材料

3.3 作业人员分工

本任务作业人员分工如表1-3-5所示。

表1-3-5 带电更换±800kV特高压输电线路耐张塔导线侧1-3片玻璃绝缘子人员分工表

序号	工作岗位	数量(人)	工作职责
1	工作负责人	1	负责本次工作任务的人员分工、工作票的宣读、办理线路停用重合闸、办理工作许可手续、召开工作班前会、工作中突发情况的处理、工作质量的监督、工作后的总结
2	专责监护人	1	负责作业现场的安全把控
3	等电位电工	2	负责工器具安装及绝缘子更换工作
4	地面电工	2	负责本次作业过程的地面辅助工作

4. 工作程序

本任务工作流程如表1-3-6所示。

表1-3-6 带电更换±800kV特高压输电线路耐张塔导线侧1-3片玻璃绝缘子工作流程表

序号	作业内容	作业步骤及标准	安全措施及注意事项	责任人
1	现场复勘	工作负责人负责完成以下工作： (1)现场核对线路名称、杆塔编号、双重编号无误；基础及杆塔完好无异常；交叉跨越距离符合安全要求；确认缺陷情况及导地线规格型号等。 (2)检测风速、湿度等现场气象条件符合作业要求。 (3)检查地形环境符合作业要求。 (4)检查工作票所列安全措施与现场实际情况相符，必要时予以补充。	(1)正确穿戴安全帽、工作服、工作鞋、劳保手套。 (2)不得在危及作业人员安全的气象条件下作业。 (3)严禁非工作人员、车辆进入作业现场。	
2	工作许可	(1)工作负责人负责联系值班调控人员，按工作票内容申请停用直流再启动保护装置。 (2)经值班调控人员许可后，方可开始带电作业工作。	不得未经值班调控人员许可即开始工作。	
3	现场布置	正确装设安全围栏并悬挂标示牌： (1)安全围栏范围应充分考虑高处坠物，以及对道路交通的影响。 (2)安全围栏出入口设置合理。 (3)妥当布置"从此进出""在此工作""车辆慢行"或"车辆绕行"等标示。	对道路交通安全影响不可控时，应及时联系交通管理部门强化现场交通安全管控。	

第一部分 ±800kV 特高压输电线路运检带电作业培训及考核标准

续表

序号	作业内容	作业步骤及标准	安全措施及注意事项	责任人
4	召开班前会	(1)全体工作成员列队。 (2)工作负责人宣读工作票,明确工作任务及人员分工;讲解工作中的安全措施和技术措施;查(问)全体工作成员精神状态;告知工作中存在的危险点及采取的预控措施。 (3)全体工作成员在工作票上签名确认。	(1)工作票填写、签发和许可手续规范,签名完整。 (2)全体工作成员精神状态良好。 (3)全体工作成员明确任务分工、安全措施和技术措施。	
5	检查工器具	(1)在防潮苫布上,将工器具按作业要求准备齐备,并分类定置摆放整齐。检查工器具外观和试验合格证,无遗漏。 (2)使用绝缘电阻测试仪检测绝缘工具及绝缘绳索的表面绝缘电阻值,方法正确,不得低于700MΩ。 (3)将新绝缘子擦拭干净,外观检查完好,不得有锈蚀、裂纹及破损。使用绝缘电阻测试仪测试其绝缘电阻值,方法正确,不得低于 700 MΩ。 (4)使用万用表检测全套屏蔽服内阻,方法正确,不得大于20Ω。 (5)检查人员向工作负责人汇报各项检查结果符合作业要求。	(1)防潮苫布数量足够,设置位置合理,保持清洁、干燥。 (2)金属、绝缘工器具在使用前,应仔细检查其是否无损伤、受潮、变形、失灵现象,合格证在有效期内。 (3)绝缘工具及绝缘绳索检测合格。 (4)屏蔽服外观完好且检测合格。	
6	登塔	(1)1号、2号等电位电工再次核对线路双重名称和相别,检查并确认脚钉齐全、牢固;系好安全带、加挂防坠器;对安全带、后备保护绳、防坠器做冲击检查,方法正确;工作负责人检查并确认等电位电工穿戴的双保险安全带合屏蔽服各连接点各部件的连接情况良好,大小合适,锁扣自如。 (2)背上工具包,携带绝缘传递绳(含绝缘滑车),方法正确。 (3)将安全带主带和后备保护绳斜跨肩上。 (4)清洁鞋底,经工作负责人许可后登塔。 (5)脚踩脚钉、手抓主材,匀步登塔至横担适当位置,系好安全带,脱离防坠器。地面电工控制好尾绳,避免传递绳与脚钉、塔材打绞。	(1)安全带、后备保护绳、防坠器冲击检查合格。 (2)防止安全带、绝缘传递绳钩挂塔材。 (3)人体与导线保持的最小安全距离应符合表1-3-2的规定。 (4)禁止手抓脚钉。 (5)正确使用防坠器。 (6)转位时,不得失去安全带的保护。	

续表

序号	作业内容	作业步骤及标准	安全措施及注意事项	责任人
7	进入强电场	(1)1号等电位电工携带绝缘传递绳,转位至耐张绝缘子串挂点处,将安全带主带系挂在绝缘子串连接金具上,将后备保护绳系留在横担适当位置。 (2)再次检查并确认屏蔽服各部分连接良好、绝缘子串连接良好及故障绝缘子位置,经工作负责人许可后,双手抓扶一串,双脚踩另一串,采用"跨二短三"作业方式,沿绝缘子串进入强电场,短接绝缘子不得超过4片。当作业人员到达导线侧均压环外三片绝缘子处时,应停止移动,使用电位转移棒勾住导线端连接金具进行电位转移,实现作业人员与导线等电位。 (3)2号等电位电工按相同方式进入电场	(1)防止安全带、绝缘传递绳钩挂塔材。 (2)转位时,不得失去安全带的保护。 (3)人体与接地体之间的安全距离、人体与接地体和带电体间的组合间隙不得小于表1-3-2的规定。 (4)进行电位转移前,应得到工作负责人许可,电位转移棒应与屏蔽服电气连接,动作应平稳、准确、快速。 (5)电位转移时,人体面部与带电体距离不得小于0.5m。	
8	安装承力工具	(1)1号、2号等电位电工进入电场后系好安全带,在适当位置布置绝缘传递绳,牢固可靠,方便作业。地面电工使用绝缘传递绳将闭式卡(前卡)、液压丝杠、导线端部卡等分别起吊给等电位电工。起吊过程平稳、无磕碰、无缠绕,正确使用绳结。 (2)1号、2号等电位电工相互配合,在导线侧合适位置上安装导线端部卡,将闭式卡(前卡)安装在导线侧第3片绝缘子上,连接好液压丝杠。承力工具各部分安装牢固可靠。 (3)检查并确认承力工具各部分安装情况良好,经工作负责人许可后,操作液压丝杠使其逐渐受力,使需更换的绝缘子松弛。两根液压丝杠的受力应均匀。	(1)人体与接地体之间的安全距离、绝缘传递绳的有效绝缘长度不得小于表1-3-2的规定。 (2)上下传递物件应用绳索栓牢传递,防止高处坠物。 (3)扣除劣质绝缘子、人体操作和工具短接的绝缘子后,良好绝缘子片数应符合表1-3-3的规定。	
9	更换绝缘子	(1)1号、2号等电位电工做冲击试验,检查并确认承力工具受力正常,经工作负责人许可后,用绝缘传递绳系牢旧绝缘子,取出绝缘子两端锁紧销,继续操作并收紧液压丝杠,直至拆除旧绝缘子。两根液压丝杠的受力应均匀,操作手柄不得敲击绝缘子。 (2)地面电工用绝缘传递绳的另一端系牢新绝缘子,采用旧下、新上的方法,将新绝缘子传给中间电位电工。起吊过程平稳、无磕碰、无缠绕,正确使用绳结。 (3)安装新绝缘子,并复位其两端锁紧销。	(1)人体与接地体之间的安全距离不得小于表1-3-2的规定。 (2)上下传递工具过程中不得碰撞,绑扎绳应正确可靠,防止高处坠物。	

续表

序号	作业内容	作业步骤及标准	安全措施及注意事项	责任人
10	拆除工具	(1)等电位电工检查新绝缘子连接可靠,经工作负责人许可后,操作并松出液压丝杠,使更换的绝缘子逐渐受力。 (2)荷载转移完毕后,等电位电工做冲击试验,检查并确认新绝缘子受力情况良好,经工作负责人许可后,拆除系在绝缘子上的绝缘传递绳,并将其系牢于承力工具适当位置,拆除液压丝杠、闭式卡、导线端部卡等承力工具,在地面电工配合下传递至地面。传递过程平稳、无磕碰、无缠绕,正确使用绳结。	(1)人体与接地体之间的安全距离不得小于表1-3-2的规定。 (2)上下传递工具过程中不得碰撞,绑扎绳应正确可靠,防止高处坠物。	
11	退出强电场	(1)等电位电工检查作业部位无遗留物后,拆除并整理绝缘传递绳。 (2)转移安全带主带,系挂在绝缘子串适当位置,越过均压环回到绝缘子串上,将电位转移棒钩紧均压环适当位置,沿绝缘子串向横担侧移动到均压环外三片绝缘子时,停止移动;一只手抓紧绝缘子,另一只手握紧电位转移棒,利用电位转移棒快速脱离等电位。 (3)按照"跨二短三"作业方式沿绝缘子串到达横担。	(1)人体与接地体和带电体间的组合间隙不得小于表1-3-2的规定。 (2)转位时,不得失去安全带的保护	
12	撤离杆塔	等电位电工检查塔上无遗留物,经工作负责人许可后,挂好防坠器,解开并整理好安全带,正确携带绝缘传递绳,脚踩脚钉、手抓主材,匀步下塔至地面	(1)转位时不得失去安全带保护。 (2)防止手滑脱、脚踏空,禁止手抓脚钉。 (3)正确使用防坠器。 (4)防止绝缘传递绳、安全带钩挂塔材或脚钉	
13	工作结束	(1)工作负责人组织全体工作成员整理工器具和材料,将工器具清洁后放入专用的箱(袋)中;清理现场,做到"工完料尽场地清"。 (2)召开班后会,工作负责人进行工作总结和点评工作。点评本次工作的施工质量;点评全体工作成员的安全措施落实情况。 (3)工作负责人向值班调控人员汇报工作结束,终结工作票		

二、考核标准

表1-3-7 国网四川省电力公司特高压直流输电线路运检技能考核评分细则

考生填写栏	编号：	姓名：	所在岗位：	单位：	日期：	年 月 日
考评员填写栏	成绩：	考评员：	考评组长：	开始时间：	结束时间：	操作时长：
考核模块	带电更换±800kV特高压输电线路耐张塔导线侧1-3片玻璃绝缘子	考核对象	特高压直流输电线路检修人员	考核方式	操作	考核时限 90min
任务描述	带电更换±800kV输电线路耐张塔导线侧1-3片玻璃绝缘子					
工作规范及要求	1. 带电作业工作应在良好天气下进行。如遇雷、雨、雪、雾天气不得进行带电作业。风力大于5级时，不宜进行带电作业。湿度大于80%时，若需进行带电作业，应采用具有防潮性能的绝缘工具。 2. 本项作业需6人，其中工作负责人1名，专责监护人1名，等电位电工2人，地面电工2名。 3. 工作负责(监护)人职责：负责本次工作任务的人员分工、工作票的宣读、办理线路停用重合闸、办理工作许可手续、召开工作班前会、负责作业过程中的安全监督、工作中突发情况的处理、工作质量的监督、工作后的总结。 4. 在带电作业中，遇雷、雨、大风或其他任何情况威胁到工作人员的安全时，工作负责人或监护人可根据情况，临时停止工作 给定条件： 1. 工作票已办理，安全措施已经完备(重合闸已停用)，工作开始、工作终结时应口头提出申请(调度或考评员)。 2. 安全、正确地使用仪器对绝缘工具进行检测。 3. 必须按工作程序进行操作，工序错误扣应做项目分值，出现重大人身、器材和操作安全隐患，考评员可下令终止操作(考核)					
考核情景准备	1. 塔形：±800kV耐张塔。 2. 所需作业工器具：安全带(含二保绳)2根，屏蔽服2套，防潮布2张，万用表1，绝缘电阻检测仪1个，风速仪、温湿度二合一1台，液压丝杠2根，绝缘绳1根，绝缘套1根，导线端部卡1个，闭式卡(前卡)1个，滑车1个，电位转移棒2把，护目镜2个，拔销器1把，手动工具1套。 3. 作业现场做好监护工作，作业现场安全措施(围栏等)已全部落实；禁止非作业人员进入现场，工作人员进入作业现场必须戴安全帽。 4. 考生自备工作服，安全帽，线手套					
备注	1. 各项目得分均扣完为止，出现重大人身、器材和操作安全隐患，考评员可下令终止操作。 2. 设备、作业环境、安全带、安全帽、工器具、屏蔽服等不符合作业条件考评员可下令终止操作					

表1-3-8　国网四川省电力公司特高压直流输电线路运检技能考核评分标准

序号	项目名称	质量要求	分值	扣分标准	扣分原因	扣分	得分
1	现场复勘	(1)工作负责人到作业现场核对线路名称、杆塔编号、现场工作条件、缺陷部位等无误。 (2)检测风速、湿度等现场气象条件符合作业要求。 (3)检查工作票填写完整，无涂改，检查是否所列安全措施与现场实际情况相符，必要时予以补充	5	(1)未核对线路名称、杆塔编号、现场工作条件、缺陷部位等，扣1分/项。 (2)未检测风速、湿度等现场气象条件，扣1分/项。 (3)工作票填写出现涂改，扣0.5分/处；工作票编号有误，扣1分；工作票填写不完整，扣1.5分			
2	工作许可	(1)工作负责人联系值班调控人员(裁判)，按工作票内容申请停用线路重合闸。 (2)汇报内容规范、完整	2	(1)未联系调度部门(裁判)停用重合闸，扣2分。 (2)汇报专业用语不规范或不完整，扣1分			
3	现场布置	正确装设安全围栏并悬挂标示牌： (1)安全围栏范围应充分考虑高处坠物，以及对道路交通的影响。 (2)安全围栏出入口设置合理。 (3)妥当布置"从此进出""在此工作""从此上下"等标示	3	(1)作业现场未装设围栏，扣1分。 (2)未设立警示牌，扣1分。 (3)未悬挂登塔作业标志，扣1分			
4	召开班前会	(1)全体工作成员全体人员正确佩戴安全帽、工作服。 (2)工作负责人佩戴红色背心，宣读工作票，明确工作任务及人员分工；讲解工作中的安全措施和技术措施；查(问)全体工作成员精神状态；告知工作中存在的危险点及采取的预控措施。 (3)全体工作成员在工作票上签名确认	3	(1)工作人员着装不整齐，扣0.5分/人。 (2)未进行分工，扣3分；分工不明确，扣1分。 (3)现场工作负责人未穿佩安全监护背心，扣1分。 (4)工作票上工作班成员未签字或签字不全，扣1分			

续表

序号	项目名称	质量要求	分值	扣分标准	扣分原因	扣分	得分
5	工器具检查	(1)在防潮苫布上,将工器具按作业要求准备齐备,并分类定置摆放整齐。检查工器具外观和试验合格证,无遗漏。 (2)使用绝缘电阻测试仪检测绝缘工具及绝缘绳索的表面绝缘电阻值,方法正确,不得低于700MΩ。 (3)将新绝缘子擦拭干净,外观检查完好,不得有锈蚀、裂纹及破损。使用绝缘电阻测试仪测试其绝缘电阻值,方法正确,不得低于700 MΩ。 (4)使用万用表检测全套屏蔽服内阻,方法正确,不得大于20Ω。 (5)检查人员向工作负责人汇报各项检查结果符合作业要求。	7	(1)未使用防潮苫布并定置摆放工器具,扣1分。 (2)未检查工器具外观及试验合格证,扣0.5分/项。 (3)未正确使用检测仪器对工器具进行检测,扣1分/项。 (4)汇报检测结果不规范,扣1分;不完整,扣0.5分/项。			
6	登塔	(1)1号、2号等电位电工再次核对线路双重名称及相别,检查并确认脚钉齐全、牢固;系好安全带、加挂防坠器;对安全带、后备保护绳、防坠器做冲击检查,方法正确;工作负责人检查并确认等电位电工穿戴的双保险安全带合屏蔽服各连接点各部件的连接情况良好,大小合适,锁扣自如。 (2)背上工具包,携带绝缘传递绳(含绝缘滑车),方法正确。 (3)将安全带主带和后备保护绳斜跨肩上。 (4)清洁鞋底,经工作负责人许可后登塔。 (5)脚踩脚钉、手抓主材,匀步登塔至横担适当位置,系好安全带,脱离防坠器。地面电工控制好尾绳,避免传递绳与脚钉、塔材打绞。	5	(1)等电位电工未核对线路双重名称、杆号、相别、塔材情况,扣1分/项;核对完未汇报,扣1分。 (2)双保险安全带及防坠器未进行冲击试验,扣2分/项。 (3)现场工作负责人未对等电位电工进行安全防护装备进行检查,扣1分。 (4)手抓脚钉,扣0.5分/次。 (5)滑车传递绳悬挂位置不合理,扣1分。 (6)转位时失去安全带保护,扣5分			

续表

序号	项目名称	质量要求	分值	扣分标准	扣分原因	扣分	得分
7	进入强电场	(1)1号等电位电工携带绝缘传递绳,转位至耐张绝缘子串挂点处,将安全带主带系挂在绝缘子串连接金具上,将后备保护绳系留在横担适当位置。 (2)再次检查并确认屏蔽服各部分连接良好、绝缘子串连接良好及故障绝缘子位置,经工作负责人许可后,双手抓扶一串,双脚踩另一串,采用"跨二短三"作业方式,沿绝缘子串进入强电场,短接绝缘子不得超过4片。当作业人员到达导线侧均压环外三片绝缘子处时,应停止移动,使用电位转移棒勾住导线端连接金具进行电位转移,实现作业人员与导线等电位。 (3)2号等电位电工按相同方式进入电场	5	(1)安全带后背保护绳系留位置不合理、使用不规范,扣2分。 (2)等电位电工未检查屏蔽服连接情况、绝缘子串连接情况及故障绝缘子位置,扣1分/项。 (3)未得到工作负责人许可就进入强电场,扣5分。 (4)等电位电工进入等电位动作不正确,反复放电,扣2分/次。 (5)电位转移动作不正确,扣3分;未汇报,扣2分,汇报了工作负责人未同意即进行电位转移,扣1分。 (6)绝缘传递绳安装位置不合理,扣1分。 (7)高处坠物,扣2分/次。 (8)转位时失去安全带保护,扣5分			
8	安装工具	(1)1号、2号等电位电工进入电场后系好安全带,在适当位置布置绝缘传递绳,牢固可靠,方便作业。地面电工使用绝缘传递绳将闭式卡(前卡)、液压丝杠、导线端部卡等分别起吊给等电位电工。起吊过程平稳、无磕碰、无缠绕,正确使用绳结。 (2)1号、2号等电位电工相互配合,在导线侧合适位置上安装导线端部卡,将闭式卡(前卡)安装在导线侧第3片绝缘子上,连接好液压丝杠。承力工具各部分安装牢固可靠。 (3)检查并确认承力工具各部分安装情况良好,经工作负责人许可后,操作液压丝杠使其逐渐受力,使需更换的绝缘子松弛。两根液压丝杠的受力应均匀。	15	(1)起吊过程不平稳,出现磕碰、缠绕,扣1分/次。 (2)高处坠物,扣2分/次。 (3)卡具安装不正确、固定不到位,扣2分。 (4)未检查承力工具安装情况,扣3分;检查了未报告,扣1分;报告了但工作负责人未同意即开始收紧丝杠,扣1分。 (5)作业过程短接绝缘子片数超过4片,扣3分/次。 (6)安装卡具出现绝缘子碰撞破损,扣2分。 (7)未均衡收紧丝杠,扣2分			

续表

序号	项目名称	质量要求	分值	扣分标准	扣分原因	扣分	得分
9	更换绝缘子	(1)1号、2号等电位电工做冲击试验,检查并确认承力工具受力正常,经工作负责人许可后,用绝缘传递绳系牢旧绝缘子,取出旧绝缘子两端锁紧销,继续操作并收紧液压丝杠,直至拆除旧绝缘子。两根液压丝杠的受力应均匀,操作手柄不得敲击绝缘子。(2)地面电工用绝缘传递绳的另一端系牢新绝缘子,采用旧下、新上的方法,将新绝缘子传给中间电位电工。起吊过程平稳、无磕碰、无缠绕,正确使用绳结。(3)安装新绝缘子,并复位其两端锁紧销。	20	(1)未检查承力工具受力情况,扣3分;检查了未报告,扣2分;报告了但工作负责人未同意即取出旧绝缘子两端锁紧销,扣1分。(2)未均衡收紧丝杆,扣2分。(3)操作手柄敲击绝缘子,扣1分/次。(4)绳结错误,扣1分。(5)高处坠物,扣2分/次。(6)新旧绝缘子相互碰撞,扣1分。(7)传递绝缘子与塔身相互碰撞,扣1分/次。(8)绝缘绳缠绕,扣2分			
10	拆除工具	(1)等电位电工检查新绝缘子连接可靠,经工作负责人许可后,操作并松出液压丝杠,使更换的绝缘子逐渐受力。(2)荷载转移完毕后,等电位电工做冲击试验,检查并确认新绝缘子受力情况良好,经工作负责人许可后,拆除系在绝缘子上的绝缘传递绳,并将其系牢于承力工具适当位置,拆除液压丝杠、闭式卡、导线端部卡等承力工具,在地面电工配合下传递至地面。传递过程平稳、无磕碰、无缠绕,正确使用绳结。	15	(1)未检查新绝缘子连接情况,扣3分;检查了未报告,扣2分;报告了但工作负责人未同意即松液压丝杠,扣1分。(2)未检查新绝缘子受力情况,扣3分;检查了未报告,扣2分;报告了但工作负责人未同意即拆除液压丝杠,扣1分。(3)捆扎工具时,未正确使用绳结,扣1分。(4)高处坠物,扣2分/次。(5)工器具相互碰撞扣,扣1分/次;工器具与带电体或塔身相互碰撞,扣1分/次;绝缘绳缠绕,扣2分			

续表

序号	项目名称	质量要求	分值	扣分标准	扣分原因	扣分	得分
11	退出强电场	(1)等电位电工检查作业部位无遗留物后,拆除并整理绝缘传递绳。 (2)转移安全带主带,系挂在绝缘子串适当位置,越过均压环回到绝缘子串上,将电位转移棒钩紧均压环适当位置,沿绝缘子串向横担侧移动到均压环外三片绝缘子时,停止移动;一只手抓紧绝缘子,另一只手握紧电位转移棒,利用电位转移棒快速脱离等电位。 (3)按照"跨二短三"作业方式沿绝缘子串到达横担。	5	(1)未向工作负责人申请即进行电位转移,扣2分;申请了但未得同意即开始,扣1分。 (2)申请电位转移位置不合适,扣1分。 (3)等电位电工退出强电场动作不正确,反复放电,扣2分。 (4)未有效控制后备保护绳,扣1分。			
12	返回地面	塔上电工检查塔上无遗留物后,向工作负责人汇报,得到工作负责人同意后携带绝缘传递绳下塔	5	(1)下塔过程未使用防坠装置,扣2分。 (2)塔上移位失去安全带保护的,扣2分。 (3)下塔抓塔钉,每处扣1分。 (4)塔上有遗留物的,扣2分			
13	工作结束	(1)工作负责人组织全体工作成员整理工器具和材料,将工器具清洁后放入专用的箱(袋)中;清理现场,做到"工完料尽场地清"。 (2)召开班后会,工作负责人进行工作总结和点评工作。点评本次工作的施工质量;点评全体工作成员的安全措施落实情况。 (3)工作负责人向值班调控人员汇报工作结束,申请恢复线路重合闸,终结工作票	10	(1)工器具未清理,扣2分。 (2)工器具有遗漏,扣2分。 (3)未开班后会,扣2分。 (4)未拆除围栏,扣2分。 (5)未向调度汇报,扣2分			
	合计		100				

模块四 带电更换±800kV特高压输电线路耐张玻璃绝缘子串任意单片绝缘子培训及考核标准

一、培训标准

(一) 培训要求

模块名称	带电更换±800kV特高压输电线路耐张玻璃绝缘子串任意单片绝缘子	培训类别	操作类
培训方式	实操培训	培训学时	21学时
培训目标	1.掌握沿耐张绝缘子串进、出±800kV强电场时采用"跨二短三"作业方式的电学意义。 2.能完成沿耐张绝缘子串进入±800kV中间电位作业点。 3.能独立完成带电更换±800kV特高压输电线路耐张玻璃绝缘子串任意单片绝缘子的操作(中间电位作业法)		
培训场地	特高压实训线路		
培训内容	采用"跨二短三"作业方式沿耐张绝缘子串进入强电场,采用中间电位作业法带电更换±800kV特高压输电线路耐张玻璃绝缘子串任意单片绝缘子		
适用范围	特高压输电线路检修人员		

(二) 引用的规程规范

(1)《±800kV直流线路带电作业技术规范》(DL/T 1242-2013)

(2)《±800kV直流架空输电线路运行规程》(GB/T 28813-2012)

(3)《±800kV直流架空输电线路检修规程》(DL/T 251-2012)

(4)《±800kV直流输电线路带电作业技术导则》(Q/GDW 302-2009)

(5)《交流线路带电作业安全距离计算方法》(GB/T 19185-2008)

(6)《带电作业用绝缘配合导则》(DL/T 867-2004)

(7)《带电作业用绝缘工具试验导则》(DL/T 878-2004)

(8)《国家电网公司带电作业工作管理规定(试行)》(国家电网生〔2007〕751号)

(9)《国家电网公司电力安全工作规程(线路部分)》(Q/GDW1799.2-2013)

(10)《电工术语架空线路》(GB/T 2900.51-1998)

(11)《电工术语带电作业》(GB/T 2900.55-2016)
(12)《带电作业工具设备术语》(GB/T 14286-2008)
(13)《带电作业用工具、装置和设备使用的一般要求》(DL/T 877-2004)
(14)《带电作业工具、装置和设备预防性试验规程》(DL/T 976-2005)
(15)《带电作业用绝缘滑车》(GB/T 13034-2008)
(16)《带电作业用绝缘绳索》(GB 13035-2008)
(17)《带电作业用屏蔽服装》(GB/T 6568-2008)
(18)《1000kV交流带电作业用屏蔽服装》(GB/T 25726-2010)

(三)培训教学设计

本设计以完成"带电更换±800kV特高压输电线路耐张玻璃绝缘子串任意单片绝缘子"为工作任务,按工作任务完成的标准化作业流程来设计各个培训阶段,每个阶段包括了具体的培训目标、培训内容、培训学时、培训方法(培训资源)、培训环境和考核评价等内容,如表1-4-1所示。

表1-4-1 带电更换±800kV特高压输电线路耐张玻璃绝缘子串任意单片绝缘子

培训流程	培训目标	培训内容	培训学时	培训方法与资源	培训环境	考核评价
1.理论教学	1.初步掌握沿绝缘子串进出±800kV强电场基本方法。2.熟悉输电线路耐张单片玻璃绝缘子更换方法	1.沿绝缘子进出强电场"跨二短三"作业方式的电学意义。2.输电线路耐张单片玻璃绝缘子更换方法和质量标准	2	培训方法:讲授法。培训资源:PPT、相关规程规范	多媒体教室	考勤、课堂提问和作业
2.准备工作	能完成作业前准备工作	1.作业现场查勘。2.编制培训标准化作业卡。3.填写培训操作工作票。4.完成本操作的工器具及材料准备	1	培训方法:1.现场查勘和工器具及材料清理采用现场实操方法。2.编写作业卡和填写工作票采用讲授方法。培训资源:1.±800kV实训线路。2.特高压工器具库房。3.空白工作票	1.特高压输电实训线路。2.多媒体教室	

续表

培训流程	培训目标	培训内容	培训学时	培训方法与资源	培训环境	考核评价
3.作业现场准备	能完成作业现场准备工作	1.作业现场复勘。 2.工作申请。 3.作业现场布置。 4.班前会。 5.工器具及材料检查	1	培训方法：演示与角色扮演法。 资源：±800kV 实训线路	±800kV 实训线路	
4.培训师演示	通过现场观摩，使学员初步领会本任务操作流程	1.中间电位电工沿耐张绝缘子串进、出强电场。 2.中间电位电工组装工器具。 3.中间电位电工完成单片玻璃绝缘子的更换工作	2	培训方法：演示法。 资源：±800kV 实训线路	±800kV 实训线路	
5.学员分组训练	1.能完成进、出±800kV 强电场操作。 2.能完成单片玻璃绝缘子的更换工作	1.学员分组(6人一组)训练进、出±800kV 强电场和更换绝缘子技能操作。 2.培训师对学员操作进行指导和安全监护	14	培训方法：角色扮演法。 资源：±800kV 实训线路	±800kV 实训线路	采用技能考核评分细则对学员操作评分
6.工作终结	1.使学员进一步辨析操作过程不足之处，便于后期提升。 2.培训学员安全文明生产的工作作风	1.作业现场清理。 2.向调度汇报工作终结。 3.班后会，对本次工作任务进行点评总结	1	培训方法：讲授和归纳法	±800kV 实训线路	

（四）作业流程

1. **工作任务**

带电更换±800kV 特高压输电线路耐张玻璃绝缘子串任意单片绝缘子。

2. **天气及作业现场要求**

（1）带电更换±800kV 特高压输电线路耐张玻璃绝缘子串任意单片绝缘子应在良好的天气进行。如遇雷电（听见雷声、看见闪电）、雪、雹、雨、雾等，禁止进行带电作业。风力大于5级，不宜进行带电作业；相对湿度大于80%的天气，若需进行带电作业，应采用具有防潮性能的绝缘工具。恶劣天气下必须开展带电抢修时，应组织有关人员充分讨论并编制必要的安全措施，经本单位批准后方可进行。带电作业过程中如遇天气突然变化，有可能危及人身或设备安全时，应立即停止工作；在保证人身安全的情况下，尽快恢复设备正常状态，或采取其他措施。

（2）作业人员精神状态良好，熟悉工作中保证安全的组织措施和技术措施，掌握高处应急救援及触电急救的方法；应持有在有效期内的带电作业资质证书。

（3）工作负责人应事先组织相关人员完成现场勘察，根据勘察结果确定本次作业方法和所需工器具，以及应采取的必要措施，并办理带电作业工作票。

（4）作业现场应合理设置围栏，并妥当布置警示标示牌，禁止非工作人员入内。

（5）本项目须停用线路再启动装置。

（6）作业方式：中间电位作业。

（7）工作中安全距离及有效绝缘长度如表1-4-2所示。

表1-4-2　带电更换±800kV特高压输电线路耐张玻璃绝缘子串任意单片绝缘子的安全距离（m）

海拔高度	等电位电工与接地构架之间的最小安全距离	绝缘工器具的最小有效绝缘长度	最小组合间隙
$H \leqslant 1000$	6.8	6.8	6.7
$1000 < H \leqslant 2000$	7.3	7.3	7.3
$2000 < H \leqslant 2500$	7.9	7.8	7.8

注：表中最小安全距离包括人体占位间隙0.5m。

（8）中间电位作业人员沿耐张绝缘子串进入±800kV强电场时，人体短接绝缘子片数不得多于4片。耐张绝缘子串中扣除人体短接和不良绝缘子片数后，良好绝缘子最少片数应满足表1-4-3的规定。

表1-4-3最小组合间隙和良好绝缘子的最小片数

海拔高度（m）	单片玻璃绝缘子结构高度（mm）	良好绝缘子串的总长度最小值(m)	良好绝缘子的最少片数
$H \leqslant 1000$	170	6.2	37
	195		32
	205		31
	240		26
$1000 < H \leqslant 2000$	170	7.1	42
	195		37
	205		35
	240		30
$2000 < H \leqslant 2500$	170	7.55	45
	195		39
	205		37
	240		32

注：表中数值不包括人体占位间隙，作业中需考虑人体占位间隙不得小于0.5m。

3. 准备工作

3.1 危险点及其预控措施

（1）危险点——触电伤害

预控措施：

①工作前，工作负责人应与值班调控人员联系，停用直流再启动保护装置，并履行许可手续，严禁约时停用或恢复直流再启动保护装置。工作结束后应及时向调度汇报。

②塔上作业人员登塔前，必须仔细核对线路双重命名、杆塔编号，确认无误后方可上塔。

③工作中，如遇线路突然停电，作业人员应视其仍然带电。工作负责人应尽快与调控人员联系，值班调控人员未与工作负责人取得联系前不准强送电。

④绝缘工具及绝缘绳索不得损坏、受潮、变形、失灵，其有效长度不准小于表1-4-2规定。不准使用非绝缘绳索（如棉纱绳、白棕绳、钢丝绳）。

⑤地面电工操作绝缘工具时应戴清洁、干燥的手套，进入作业现场应将使用的带电作业工具放置在防潮的帆布或绝缘垫上，防止绝缘工具在使用中脏污和受潮。

⑥中间电位电工应穿着阻燃内衣，衣服外面应穿戴合格全套屏蔽服（包括帽、衣裤、手套、袜和鞋），且各部分应连接良好。

⑦中间电位作业人员沿绝缘子串移动时，手与脚的位置必须保持对应一致，且人体和工具短接的绝缘子片数应符合表1-4-3规定。

⑧采用"跨二短三"作业方式（也称自由作业法）进入强电场。

⑨用绝缘绳索传递大件金属物品时，地面电工应将金属物品接地后再接触。

⑩带电作业过程中，工作负责人（监护人）应对作业人员进行不间断监护，随时纠正其不规范或违章动作。重点关注高处作业人员，使其保持足够的安全距离及组合间隙（符合表1-4-2的规定），禁止同时接触两个非连通的带电体或带电体与接地体。

（2）危险点——高处坠落

预控措施：

①高处作业人员登高前，必须具备符合本项作业要求的身体状况、精神状态和技能素质。

②高处作业人员应使用双保险安全带。上、下塔时，应手抓主材、脚踩脚钉，匀速行进。

③中间电位作业人员作业前应认真检查液压丝杠、闭式卡等，确保承力工具合格；沿绝缘子串移动时，手与脚的位置必须保持对应一致，安全带应系挂在手扶的绝

缘子串上，并同步移动；更换绝缘子时，承力工具安装应可靠，荷载转移前、后应做冲击试验判定其可靠性，并及时向工作负责人汇报，得到工作负责人许可后方可实施。

④监护人员应随时纠正其不规范或违章动作，重点关注高处作业人员在转位的过程中不得失去安全带或绝缘后备保护绳的保护，严禁低挂高用。

（3）危险点——高处坠物伤人

预控措施：

①高处作业人员的个人工具及零星材料应装入工具袋，严禁在高处浮置物件、口中含物。

②以上下循环交换方式传递较重的工器具时，均应系好控制绳，防止被传递物品相互碰撞及误碰处于工作状态的承力工器具。

③地面作业人员必须正确佩戴安全帽，正确使用绳结，与作业点垂直下方距离不得小于坠落半径。

④作业现场设置围栏并挂好警示标示牌。监护人员应随时注意，禁止非工作人员及车辆进入作业区域。

3.2 工器具及材料选择

带电更换±800kV特高压输电线路耐张玻璃绝缘子串任意单片绝缘子所需工器具及材料见表1-4-4。工器具出库前，应认真核对工器具的使用电压等级和试验周期，并检查确认外观良好、连接牢固、转动灵活，且符合本次工作任务的要求；工器具出库后，应存放在工具袋或工具箱内进行运输，防止脏污、受潮；金属工具和绝缘工器具应分开装运，防止因混装运输导致工器具变形、损伤等现象发生。

表1-4-4 带电更换±800kV特高压输电线路耐张玻璃绝缘子串任意单片绝缘子所需工器具及材料表

序号	名称	规格型号	单位	数量	备注
1	屏蔽服	屏蔽效率≥40dB（屏蔽面罩屏蔽效率≥20dB）	套	2	个人防护用具
2	导电鞋	尺码视穿着人员而定	双	2	个人防护用具
3	阻燃内衣	纯桑蚕丝	套	2	个人防护用具
4	双保险安全带	背带式	根	2	个人防护用具
5	安全帽		顶	6	个人防护用具
6	护目镜		副	2	个人防护用具
7	绝缘传递绳	Φ14mm，长度与起吊高度匹配	根	1	绝缘工具
8	绝缘后备保护绳	Φ16mm	根	2	绝缘工具
9	绝缘绳套	Φ14mm	根	2	绝缘工具

续表

序号	名称	规格型号	单位	数量	备注
10	绝缘滑车	1T	个	1	绝缘工具
11	液压丝杠	8T	根	2	金属工具
12	闭式卡	Tc4	套	1	金属工具
13	拔销器		把	1	其他工具
14	绝缘电阻测试仪	5000V，电极宽2cm、极间宽2cm	套	1	其他工具
15	万用表		套	1	其他工具
16	风速、温湿度测试仪		只	1	其他工具
17	安全围网		套	若干	其他工具
18	警示标示牌	"在此工作""从此进出""车辆慢行""车辆绕行"	套	1	其他工具
19	红马甲	"工作负责人""专责监护人"	件	1	其他工具
20	防潮苫布	3m×3m	块	2	其他工具
21	个人工具	扳手、老虎钳	套	1	其他工具
22	防坠器	与杆塔防坠落装置型号对应	只	2	其他工具
23	毛巾	棉质	条	1	其他工具
24	绝缘子	与被更换绝缘子同型号	片	1	材料

3.3 作业人员分工

本任务作业人员分工如表1-4-5所示。

表1-4-5 带电更换±800kV特高压输电线路耐张玻璃绝缘子串任意单片绝缘子人员分工表

序号	工作岗位	数量(人)	工作职责
1	工作负责人	1	负责本次工作任务的人员分工、工作票的宣读、办理线路停用再启动、办理工作许可手续、召开工作班前会、工作中突发情况的处理、工作质量的监督、工作后的总结
2	专责监护人	1	负责作业现场的安全把控
3	中间电位电工	2	负责工器具安装及绝缘子更换工作
4	地面电工	2	负责本次作业过程的地面辅助工作

4. 工作程序

本任务工作流程如表1-4-6所示。

表1-4-6 带电更换±800kV特高压输电线路耐张玻璃绝缘子串任意单片绝缘子工作流程表

序号	作业内容	作业步骤及标准	安全措施及注意事项	责任人
1	现场复勘	工作负责人负责完成以下工作： (1)现场核对线路名称、杆塔编号，双重编号无误；基础及杆塔完好无异常；交叉跨越距离符合安全要求；确认缺陷情况及导地线规格型号等。 (2)检测风速、湿度等现场气象条件符合作业要求。 (3)检查地形环境符合作业要求。 (4)检查工作票所列安全措施与现场实际情况相符，必要时予以补充	(1)正确穿戴安全帽、工作服、工作鞋、劳保手套。 (2)不得在危及作业人员安全的气象条件下作业。 (3)严禁非工作人员、车辆进入作业现场	
2	工作许可	(1)工作负责人负责联系值班调控人员，按工作票内容申请直流再启动保护装置。 (2)经值班调控人员许可后，方可开始带电作业工作	不得未经值班调控人员许可即开始工作	
3	现场布置	正确装设安全围栏并悬挂标示牌： (1)安全围栏范围应充分考虑高处坠物，以及对道路交通的影响。 (2)安全围栏出入口设置合理。 (3)妥当布置"从此进出""在此工作""车辆慢行"或"车辆绕行"等标示	对道路交通安全影响不可控时，应及时联系交通管理部门强化现场交通安全管控	
4	召开班前会	(1)全体工作成员列队。 (2)工作负责人宣读工作票，明确工作任务及人员分工；讲解工作中的安全措施和技术措施；查(问)全体工作成员精神状态；告知工作中存在的危险点及采取的预控措施。 (3)全体工作成员在工作票上签名确认	(1)工作票填写、签发和许可手续规范，签名完整。 (2)全体工作成员精神状态良好。 (3)全体工作成员明确任务分工、安全措施和技术措施	
5	检查工器具	(1)在防潮苫布上，将工器具按作业要求准备齐备，并分类定置摆放整齐。检查工器具外观和试验合格证，无遗漏。 (2)使用绝缘电阻测试仪检测绝缘工具及绝缘绳索的表面绝缘电阻值，方法正确，不得低于700MΩ。 (3)将新绝缘子擦拭干净，外观检查完好，不得有锈蚀、裂纹及破损。使用绝缘电阻测试仪测试其绝缘电阻值，方法正确，不得低于700 MΩ。 (4)使用万用表检测全套屏蔽服内阻，方法正确，不得大于20Ω。 (5)检查人员向工作负责人汇报各项检查结果符合作业要求	(1)防潮苫布数量足够，设置位置合理，保持清洁、干燥。 (2)金属、绝缘工器具在使用前，应仔细检查其是否无损伤、受潮、变形、失灵现象，合格证在有效期内。 (3)绝缘工具及绝缘绳索检测合格	

续表

序号	作业内容	作业步骤及标准	安全措施及注意事项	责任人
6	登塔	(1)中间电位电工再次核对线路双重名称及相别，检查并确认脚钉齐全、牢固；系好安全带、加挂防坠器；对安全带、防坠器做冲击试验，方法正确；工作负责人检查并确认中间电位电工穿戴的双保险安全带各部件的连接情况良好，包括肩带、胸带、腰带、腿带、后背保护绳、扣和环。 (2)背上工具包，携带绝缘传递绳(含绝缘滑车)，方法正确。 (3)将安全带主带和后备保护绳斜跨肩上。 (4)清洁鞋底，经工作负责人许可后登塔。 (5)脚踩脚钉，手抓主材，匀步登塔至横担适当位置，系好安全带，脱离防坠器。	(1)安全带、防坠器冲击试验合格。 (2)防止安全带、绝缘传递绳钩挂塔材。 (3)人体与导线保持的最小安全距离应符合表1-4-2的规定。 (4)禁止手抓脚钉。 (5)正确使用防坠器。 (6)转位时，不得失去安全带的保护	
7	进入强电场	(1)中间电位电工携带绝缘传递绳，转位至作业相耐张绝缘子串挂点处，将安全带主带系挂在绝缘子串连接金具上，将安全带后备保护绳系留在横担适当位置。 (2)中间电位电工再次检查并确认屏蔽服各部分连接良好、绝缘子串连接良好及故障绝缘子位置，经工作负责人许可后，双手抓扶一串，双脚踩另一串，采用"跨二短三"作业方式，沿绝缘子串平稳移动到作业点；手与脚的位置必须保持对应一致，安全带主带系挂在手扶的绝缘子串上，并同步移动。 (3)中间电位电工到达作业点后，在绝缘子串适当位置用绝缘绳套固定绝缘滑车，穿入绝缘传递绳；安装牢固可靠，便于工作。	(1)防止安全带、绝缘传递绳钩挂塔材。 (2)转位时，不得失去安全带的保护。 (3)人体与带电体之间的安全距离，人体与接地体和带电体间的组合间隙不得小于表1-4-2的规定	
8	安装工具并转移导线张力	(1)地面电工使用绝缘传递绳将闭式卡、液压丝杠等分别传给中间电位电工。起吊过程平稳、无磕碰、无缠绕，正确使用绳结。 (2)中间电位电工将闭式卡前卡安装在需要更换绝缘子后两片绝缘子的卡槽内，后卡安装在需要更换绝缘子前一片绝缘子的钢帽上，并连接好液压丝杠。承力工具各部分安装牢固可靠。 (3)检查并确认承力工具各部分安装情况良好，经工作负责人许可后，操作液压丝杠使其逐渐受力，使更换的绝缘子松弛。两根液压丝杠的受力应均匀	(1)人体与接地体和带电体间的组合间隙不得小于表1-4-2的规定。 (2)防止高处坠物。 (3)扣除劣质绝缘子、人体操作和工具短接的绝缘子后，良好绝缘子片数应符合表1-4-3的规定	

续表

序号	作业内容	作业步骤及标准	安全措施及注意事项	责任人
9	更换绝缘子	(1)中间电位电工做冲击试验,检查并确认承力工具受力正常;经工作负责人许可后,用绝缘传递绳系好旧绝缘子,取出旧绝缘子两端锁紧销,继续操作并收紧液压丝杠,直至拆除旧绝缘子。两根液压丝杠的受力应均匀,操作手柄不得敲击绝缘子。 (2)地面电工用绝缘传递绳的另一端系好新绝缘子,采用旧下、新上的方法,将新绝缘子传给中间电位电工。起吊过程平稳、无磕碰、无缠绕,正确使用绳结。 (3)中间电位电工安装新绝缘子,复位其两端锁紧销,并确认其安装到位	(1)人体与接地体和带电体间的组合间隙不得小于表1-4-2的规定。 (2)防止高处坠物	
10	拆除工具	(1)中间电位电工检查新绝缘子连接可靠,经工作负责人许可后,操作并松出液压丝杠,使更换的绝缘子逐渐受力。 (2)荷载转移完毕后,中间电位电工做冲击试验;检查并确认新绝缘子受力情况良好,经工作负责人许可后,拆除系在绝缘子上的绝缘传递绳,并将其牢于承力工具适当位置,拆除液压丝杠、闭式卡等承力工具,在地面电工配合下传递至地面。传递过程平稳、无磕碰、无缠绕,正确使用绳结	(1)人体与接地体和带电体间的组合间隙不得小于表1-4-2的规定。 (2)防止高处坠物	
11	退出强电场	(1)中间电位电工检查作业部位无遗留物后,拆除绝缘传递绳。 (2)携带绝缘传递绳,按照"跨二短三"的作业方式沿绝缘子串回到横担上	(1)人体与接地体和带电体间的组合间隙不得小于表1-4-2的规定。 (2)转位时,不得失去安全带的保护	
12	撤离杆塔	中间电位电工检查塔上无遗留物,经工作负责人许可后,挂好防坠器,解开并整理好安全带,正确携带绝缘传递绳,脚踩脚钉、手抓主材,匀步下塔至地面	(1)转位时不得失去安全带保护。 (2)防止手滑脱、脚踏空,禁止手抓脚钉。 (3)正确使用防坠器。 (4)防止绝缘传递绳、安全带钩挂塔材或脚钉	
13	工作结束	(1)工作负责人组织全体工作成员整理工器具和材料,将工器具清洁后放入专用的箱(袋)中;清理现场,做到"工完料尽场地清"。 (2)召开班后会,工作负责人进行工作总结和点评工作。点评本次工作的施工质量;点评全体工作成员的安全措施落实情况。 (3)工作负责人向值班调控人员汇报工作结束,并申请恢复线路再启动,终结工作票		

二、考核标准

表1-4-7　国网四川省电力公司特高压直流输电线路运检技能考核评分细则

考生填写栏	编号：	姓名：	所在岗位：	单位：	日期：	年	月	日
考评员填写栏	成绩：	考评员：	考评组长：	开始时间：	结束时间：	操作时长：		
考核模块	带电更换±800kV特高压输电线路耐张玻璃绝缘子串任意单片绝缘子		考核对象	特高压直流输电线路检修人员	考核方式	操作	考核时限	90min
任务描述	带电更换±800kV输电线路耐张塔玻璃绝缘子串任意单片玻璃绝缘子							
工作规范及要求	1. 带电作业工作应在良好天气下进行。如遇雷、雨、雪、雾天气不得进行带电作业。风力大于5级时，不宜进行带电作业。湿度大于80%时，若需进行带电作业，应采用具有防潮性能的绝缘工具。 2. 本项作业需6人，其中工作负责人1名，专责监护人1名，中间电位电工2人，地面电工2名。 3. 工作负责(监护)人职责：负责本次工作任务的人员分工、工作票的宣读、办理线路停用再启动、办理工作许可手续、召开工作班前会、负责作业过程中的安全监督、工作中突发情况的处理、工作质量的监督、工作后的总结。 4. 在带电作业中，遇雷、雨、大风或其他任何情况威胁到工作人员的安全时，工作负责人或监护人可根据情况临时停止工作。 给定条件： 1. 工作票已办理，安全措施已经完备(再启动已停用)，工作开始、工作终结时应口头提出申请(调度或考评员)。 2. 安全、正确地使用仪器对绝缘工具进行检测。 3. 必须按工作程序进行操作，工序错误扣除应做项目分值，出现重大人身、器材和操作安全隐患，考评员可下令终止操作(考核)							
考核情景准备	1. 塔形：±800kV耐张塔。 2. 所需作业工器具：双保险安全带2根，屏蔽服2套，防潮布2张，万用表1块，绝缘电阻检测仪1个，风速仪、温湿度二合一1台，液压丝杠2根，绝缘绳2根，闭式卡1套，滑车1个，护目镜2个，拔销器1把，手动工具1套。 3. 作业现场做好监护工作，作业现场安全措施(围栏等)全部落实；禁止非作业人员进入现场，工人员进入作业现场必须戴安全帽。 4. 考生自备工作服，安全帽，线手套							
备注	1. 各项目得分均扣完为止，出现重大人身、器材和操作安全隐患，考评员可下令终止操作 2. 设备、作业环境、安全带、安全帽、工器具、屏蔽服等不符合作业条件考评员可下令终止操作							

表1-4-8　国网四川省电力公司特高压直流输电线路运检技能考核评分标准

序号	项目名称	质量要求	分值	扣分标准	扣分原因	扣分	得分
1	现场复勘	(1)工作负责人到作业现场核对线路名称、杆塔编号、现场工作条件、缺陷部位等无误。 (2)检测风速、湿度等现场气象条件符合作业要求。 (3)检查工作票填写完整，无涂改，检查是否所列安全措施与现场实际情况相符，必要时予以补充	5	(1)未核对线路名称、杆塔编号、现场工作条件、缺陷部位等，扣1分/项。 (2)未检测风速、湿度等现场气象条件，扣1分/项。 (3)工作票填写出现涂改，扣0.5分/处；工作票编号有误，扣1分；工作票填写不完整，扣1.5分			
2	工作许可	(1)工作负责人联系值班调控人员（裁判），按工作票内容申请停用线路再启动。 (2)汇报内容规范、完整	2	(1)未联系调度部门（裁判）停用再启动，扣2分。 (2)汇报专业用语不规范或不完整，扣1分			
3	现场布置	正确装设安全围栏并悬挂标示牌： (1)安全围栏范围应充分考虑高处坠物，以及对道路交通的影响。 (2)安全围栏出入口设置合理。 (3)妥当布置"从此进出""在此工作""从此上下"等标示	3	(1)作业现场未装设围栏，扣1分。 (2)未设立警示牌，扣1分。 (3)未悬挂登塔作业标志，扣1分			
4	召开班前会	(1)全体工作成员全体人员正确佩戴安全帽、工作服。 (2)工作负责人佩戴红色背心，宣读工作票，明确工作任务及人员分工；讲解工作中的安全措施和技术措施；查(问)全体工作成员精神状态；告知工作中存在的危险点及采取的预控措施。 (3)全体工作成员在工作票上签名确认	3	(1)工作人员着装不整齐，扣0.5分/人。 (2)未进行分工，扣3分；分工不明确，扣1分。 (3)现场工作负责人未穿佩安全监护背心，扣1分。 (4)工作票上工作班成员未签字或签字不全，扣1分			

续表

序号	项目名称	质量要求	分值	扣分标准	扣分原因	扣分	得分
5	工器具检查	(1)在防潮苫布上,将工器具按作业要求准备齐备,并分类定置摆放整齐。检查工器具外观和试验合格证,无遗漏。 (2)使用绝缘电阻测试仪检测绝缘工具及绝缘绳索的表面绝缘电阻值,方法正确,不得低于700MΩ。 (3)将新绝缘子擦拭干净,外观检查完好,不得有锈蚀、裂纹及破损。使用绝缘电阻测试仪测试其绝缘电阻值,方法正确,不得低于700 MΩ。 (4)使用万用表检测全套屏蔽服内阻,方法正确,不得大于20Ω。 (5)检查人员向工作负责人汇报各项检查结果符合作业要求	7	(1)未使用防潮苫布并定置摆放工器具,扣1分。 (2)未检查工器具外观及试验合格证,扣0.5分/项。 (3)未正确使用检测仪器对工器具进行检测,扣1分/项。 (4)汇报检测结果不规范,扣1分;不完整,扣0.5分/项			
6	登塔	(1)中间电位电工再次核对线路双重名称及相别,检查并确认脚钉齐全、牢固;系好安全带、加挂防坠器;对双保险安全带、防坠器做冲击试验,方法正确;并进行汇报。工作负责人检查并确认中间电位电工穿戴的双保险安全带各部件的连接情况良好,包括肩带、胸带、腰带、腿带、后背保护绳、扣和环。 (2)背上工具包,携带绝缘传递绳(含绝缘滑车),方法正确。 (3)登塔过程中系好防坠落保护装置,脚踩脚钉、手抓主材、匀步登塔至合适位置,系好安全带,脱离防坠器	5	(1)中间电位电工未核对线路双重名称、杆号、相别、塔材情况,扣1分/项;核完未汇报,扣1分。 (2)双保险安全带及防坠器未进行冲击试验,扣2分/项。 (3)现场工作负责人未对中间电位电工进行安全防护装备进行检查,扣1分。 (4)手抓脚钉,扣0.5分/次。 (5)滑车传递绳悬挂位置不合理,扣1分。 (6)转位时失去安全带保护,扣5分			

序号	项目名称	质量要求	分值	扣分标准	扣分原因	扣分	得分
7	进入强电场	(1)中间电位电工携带绝缘传递绳,转位至作业相耐张绝缘子串挂点处,将安全带主带系挂在绝缘子串连接金具上,将安全带后备保护绳系留在横担适当位置。(2)中间电位电工再次检查并确认屏蔽服各部分连接良好、绝缘子串连接良好及故障绝缘子位置,经工作负责人许可后,双手抓扶一串,双脚踩另一串,采用"跨二短三"作业方式,沿绝缘子串平稳移动到作业点;手与脚的位置必须保持对应一致,安全带主带挂在手扶的绝缘子串上,并同步移动。(3)中间电位电工到达作业点后,在绝缘子串适当位置用绝缘绳套固定绝缘滑车,穿入绝缘传递绳;安装牢固可靠,便于工作	5	(1)安全带后背保护绳系留位置不合理、使用不规范,扣2分。(2)中间电位电工未检查屏蔽服连接情况、绝缘子串连接情况及故障绝缘子位置,扣1分/项。(3)未得到工作负责人许可就进入强电场,扣5分。(4)中间电位电工进入强电场动作不正确、反复放电,扣2分/次。(5)绝缘传递绳安装位置不合理,扣1分。(6)高处坠物,扣2分/次。(7)转位时失去安全带保护,扣5分			
8	安装工具	(1)地面电工使用绝缘传递绳将闭式卡、液压丝杠等工具分别传至中间电位电工,起吊过程平稳、无磕碰、无缠绕,正确使用绳结。(2)中间电位电工将闭式卡前卡安装在需要更换绝缘子后两片绝缘子的卡槽内,后卡安装在需要更换绝缘子前一片绝缘子的钢帽上,并连接好液压丝杠。承力工具各部分安装牢固可靠。(3)检查并确认承力工具各部分安装情况良好,经工作负责人许可后,操作液压丝杠使其逐渐受力,使需更换的绝缘子松弛。两根液压丝杠的受力应均匀	15	(1)起吊过程不平稳、出现磕碰、缠绕,扣1分/次。(2)高处坠物,扣2分/次。(3)卡具安装不正确、固定不到位,扣2分。(4)未检查承力工具安装情况,扣3分;检查了未报告,扣1分;报告了但工作负责人未同意即开始收紧丝杠,扣1分。(5)作业过程短接绝缘子片数超过4片,扣3分/次。(6)安装卡具出现绝缘子碰撞破损,扣15分。(7)未均衡收紧丝杠,扣2分			

续表

序号	项目名称	质量要求	分值	扣分标准	扣分原因	扣分	得分
9	更换绝缘子	(1)中间电位电工做冲击试验,检查并确认承力工具受力正常,经工作负责人许可后,用绝缘传递绳系好旧绝缘子,取出旧绝缘子两端锁紧销,继续操作并收紧液压丝杠,直至拆除旧绝缘子。两根液压丝杠的受力应均匀,操作手柄不得敲击绝缘子。 (2)地面电工用绝缘传递绳的另一端系好新绝缘子,采用旧下、新上的方法,将新绝缘子传给中间电位电工。起吊过程平稳、无磕碰、无缠绕,正确使用绳结。 (3)安装新绝缘子,并复位其两端锁紧销	20	(1)未检查承力工具受力情况,扣3分;检查了未报告,扣2分;报告了但工作负责人未同意即取出旧绝缘子两端锁紧销,扣1分。 (2)未均衡收紧丝杆,扣2分。 (3)操作手柄敲击绝缘子,扣1分/次。 (4)绳结错误,扣1分。 (5)高处坠物,扣2分/次。 (6)新旧绝缘子相互碰撞,扣1分。 (7)传递绝缘子与塔身相互碰撞,扣1分/次。 (8)绝缘绳缠绕,扣2分			
10	拆除工具	(1)中间电位电工检查新绝缘子连接可靠,经工作负责人许可后,操作并松出液压丝杠,使更换的绝缘子逐渐受力。 (2)荷载转移完毕后,中间电位电工做冲击试验,检查并确认新绝缘子受力情况良好,经工作负责人许可后,拆除系在绝缘子上的绝缘传递绳,并将其系牢于承力工具适当位置,拆除液压丝杠、闭式卡等承力工具,在地面电工配合下传递至地面。传递过程平稳、无磕碰、无缠绕,正确使用绳结	15	(1)未检查新绝缘子连接情况,扣3分;检查了未报告,扣2分;报告了但工作负责人未同意即松液压丝杠,扣1分。 (2)未检查新绝缘子受力情况,扣3分;检查了未报告,扣2分;报告了但工作负责人未同意即拆除液压丝杠,扣1分。 (3)捆扎工具时,未正确使用绳结,扣1分。 (4)高处坠物,扣2分/次。 (5)工器具相互碰撞,扣1分/次;工器具与带电体或塔身相互碰撞,扣1分/次;绝缘绳缠绕,扣2分			

续表

序号	项目名称	质量要求	分值	扣分标准	扣分原因	扣分	得分
11	退出强电场	(1)中间电位电工检查作业部位无遗留物后,带好绝缘传递绳,作退出电位准备。 (2)中间电位电工按照"跨二短三"作业方式退出等电位	5	(1)未向工作负责人申请即退出强电场,扣2分;申请了但未得同意即开始退出强电场,扣1分。 (2)中间电位电工退出强电场动作不正确,反复放电,扣2分/次。 (3)未有效控制后备保护绳,扣1分			
12	返回地面	塔上电工检查塔上无遗留物后,经工作负责人许可后携带绝缘传递绳下塔	5	(1)下塔过程未使用防坠器,扣5分。 (2)塔上移位失去安全带保护,扣5分。 (3)下塔手抓脚钉,扣1分/次。 (4)塔上有遗留物,扣2分			
13	工作结束	(1)工作负责人组织全体工作成员整理工器具和材料,将工器具清洁后放入专用的箱(袋)中;清理现场,做到"工完料尽场地清"。 (2)召开班后会,工作负责人进行工作总结和点评工作。点评本次工作的施工质量;点评全体工作成员的安全措施落实情况。 (3)工作负责人向值班调控人员汇报工作结束,申请恢复线路再启动,终结工作票	10	(1)工器具未清理,扣2分。 (2)工器具有遗漏,扣2分。 (3)未开班后会,扣10分。 (4)未拆除围栏,扣2分。 (5)未向调度汇报,扣2分			
	合计		100				

模块五 带电更换±800kV特高压输电线路导线间隔棒培训及考核标准

一、培训标准

（一）培训要求

模块名称	带电更换±800kV特高压输电线路导线间隔棒	培训类别	操作类
培训方式	实操培训	培训学时	21学时
培训目标	1.掌握沿耐张绝缘子串进、出±800kV强电场时采用"跨二短三"作业方式的电学意义。 2.能完成沿耐张绝缘子串进入±800kV等电位作业点。 3.能独立完成更换导线间隔棒的操作（等电位作业法）		
培训场地	特高压直流实训线路		
培训内容	采用"跨二短三"作业方式沿耐张绝缘子串进入强电场，采用等电位作业法带电更换±800kV特高压直流输电线路八分裂导线间隔棒		
适用范围	特高压直流输电线路检修人员		

（二）引用规程规范

（1）《±800kV直流架空输电线路设计规范》（GB/T 50790-2013）

（2）《±800kV直流架空输电线路检修规程》（DL/T 251-2012）

（3）《±800kV直流架空输电线路运行规程》（GB/T 28813-2012）

（4）《±800kV直流线路带电作业技术规范》（DL/T 1242-2013）

（5）《±800kV特高压输电线路金具技术规范》（GB/T 31235-2014）

（6）《国家电网公司带电作业工作管理规定（试行）》（国家电网生〔2007〕751号）

（7）《国家电网公司电力安全工作规程（线路部分）》（Q/GDW1799.2-2013）

（8）《电工术语 架空线路》（GB/T 2900.51-1998）

（9）《电工术语 带电作业》（GB/T 2900.55-2016）

（10）《带电作业工具设备术语》（GB/T 14286-2002）

（11）《带电作业用绝缘滑车》（GB/T 13034-2008）

（12）《带电作业用绝缘绳索》（GB 13035-2008）

（13）《带电作业用工具、装置和设备使用的一般要求》（DL/T 877-2004）

（14）《带电作业工具、装置和设备预防性试验规程》（DL/T 976-2005）

（15）《±800kV 特高压输电线路带电作业技术导则》（Q/GDW 302-2009）

（16）《带电作业用屏蔽服装》（GB/T 6568-2008）

（17）《带电作业工具基本技术要求与设计导则》（GB 18037-2008）

（18）《带电设备红外线诊断应用规范》（DL/T 664-2016）

（三）培训教学设计

本设计以完成"带电更换±800kV 特高压输电线路导线间隔棒"为工作任务，按工作任务完成的标准化作业流程来设计各个培训阶段，每个阶段包括了具体的培训目标、培训内容、培训学时、培训方法（培训资源）、培训环境和考核评价等内容，如表1-5-1 所示。

表1-5-1　带电更换±800kV 特高压输电线路导线间隔棒培训内容设计

培训流程	培训目标	培训内容	培训学时	培训方法与资源	培训环境	考核评价
1. 理论教学	1.初步掌握沿绝缘子串进出±800kV 强电场基本方法。2.熟悉电位转移的方法。3.熟悉输电线路受损导线间隔棒更换方法。4.熟悉特高压直流线路带电作业的安全距离、危险点辨识及预控	1.沿绝缘子进出强电场"跨二短三"作业方式的电学意义。2.进、出特高压强电场时电位转移棒的使用方法。3.输电线路导线间隔棒更换方法和质量标准。4.特高压直流线路带电作业安全距离、危险点分析及预控措施	2	培训方法：讲授法。培训资源：PPT、相关规程、规范及技术导则	多媒体教室	考勤、课堂提问和作业
2. 准备工作	能完成作业前准备工作	1.作业现场查勘。2.编制培训标准化作业卡。3.填写培训操作工作票。4.完成本操作的工器具及材料准备	1	培训方法：1.现场查勘和工器具及材料准备采用现场实操方法。2.编写作业卡和填写工作票采用讲授方法。培训资源：1. ±800kV 实训线路。2.特高压工器具库房。3.空白工作票	1. ±800kV 实训线路2.多媒体教室	

续表

培训流程	培训目标	培训内容	培训学时	培训方法与资源	培训环境	考核评价
3.作业现场准备	能完成作业现场准备工作	1.作业现场复勘。2.工作申请。3.作业现场布置。4.班前会。5.工器具及材料检查。6.间隔棒专用扳手使用方法	1	培训方法：演示与角色扮演法。资源：±800kV实训线路	±800kV实训线路	
4.培训师演示	通过现场观摩，使学员初步领会本任务操作流程	1.等电位电工沿耐张绝缘子串进、出强电场及电位转移。2.等电位电工采用走线方式到达间隔棒更换位置。3.等电位电工用专用工具完成导线间隔棒更换	2	培训方法：演示法。资源：±800kV实训线路	±800kV实训线路	
5.学员分组训练	1.能完成沿绝缘子串进、出±800kV强电场及电位转移操作。2.能完成±800kV输电线路导线间隔棒更换作业	1.学员分组（6人一组）训练进、出±800kV强电场、电位转移和更换导线间隔棒技能操作。2.培训师对学员操作进行指导和安全监护	14	培训方法：角色扮演法。资源：±800kV实训线路	±800kV实训线路	采用技能考核评分细则对学员操作评分
6.工作终结	1.使学员进一步辨析操作过程不足之处，便于后期提升。2.培养学员安全文明生产的工作作风	1.作业现场清理。2.向调度汇报工作。3.班后会，对本次工作任务进行点评总结	1	培训方法：讲授和归纳法	±800kV实训线路	

（四）作业流程

1. 工作任务

采用"跨二短三"的作业方式沿耐张绝缘子串进入强电场、到达作业点，采用等电位作业法带电更换±800kV特高压输电线路八分裂导线间隔棒。

（本作业任务适用于海拔1000m及以下地区±800kV直流单回输电线路耐张塔边相导线第一个间隔棒作业点位）

2. 天气及作业现场要求

（1）带电更换±800kV特高压输电线路导线八分裂间隔棒应在良好的天气进行。

如遇雷电（听见雷声、看见闪电）、雪、雹、雨、雾等，不应进行带电作业。风力大于5级时，不宜进行带电作业；相对湿度大于80%的天气，如需进行带电作业应采用具有防潮性能的绝缘工具；恶劣天气下必须开展带电抢修时，应组织有关人员充分讨论并编制必要的安全措施，经本单位批准后方可进行。

（2）作业人员精神状态良好，熟悉工作中保证安全的组织措施和技术措施；应持有在有效期内的带电作业资质证书。

（3）工作负责人应事先组织相关人员完成现场勘察，根据勘察结果确定本次作业方法和所需工器具，以及应采取的必要措施，并办理带电作业工作票。

（4）作业现场应合理设置围栏，并妥当布置警示标示牌，禁止非工作人员入内。

（5）本项目需停用直流再启动装置。

（6）工作中安全距离及有效绝缘长度如表1-5-2所示。

表1-5-2　带电更换±800kV特高压输电线路导线间隔棒的安全距离（m）

电压等级	人身与带电体安全距离	与邻相导线的最小距离	最小有效绝缘长度		最小组合间隙	转移电位时人体裸露部分与带电体的最小距离
			绝缘操作杆	绝缘承力工具、绝缘绳索		
±800kV	6.8（7.3）	6.8（7.3）	6.8	6.8	6.6	0.5

注：①海拔高度1000m以上时，±800kV直流单回输电线路带电作业中的安全距离采用括号内7.3m的数据、绝缘工具最小有效绝缘长度采用括号内7.3m的数据、组合间隙采用括号内7.2m的数据。

②因为±800kV特高压线路相间距离足够大（一般控制相地距离即可），不做重要安全因素考虑，所以《安规》没有给出"与邻相导线的最小距离"的数据，实际工作中可以参考750kV的数据。

（7）在±800kV输电线路上作业，应保证作业相良好绝缘子片数不少于37片（单片绝缘子高度170mm）、32片（单片绝缘子高度195mm）、31片（单片绝缘子高度205mm）、26片（单片绝缘子高度240mm）。

3. 准备工作

3.1　危险点及其预控措施

（1）危险点——触电伤害

预控措施：

①工作前，工作负责人应与值班调控人员联系，停用线路直流再启动装置，并履行许可手续。

②塔上等电位作业人员登塔前，必须仔细核对线路名称、杆塔编号、相别，确认

无误后方可上塔。

③工作中，如遇线路突然停电，作业人员应视其仍然带电。工作负责人应尽快与调控人员联系，值班调控人员未与工作负责人取得联系前不准强送电。

④地面电工操作绝缘工具时应戴清洁、干燥的防汗手套，绝缘工具及绝缘绳索不得损坏、受潮、变形、失灵，不准使用非绝缘绳索（如棉纱绳、白棕绳、钢丝绳），现场所使用的带电作业工具应放置在防潮苫布上，防止绝缘工具在使用中脏污和受潮。

⑤等电位作业人员应穿着阻燃内衣，衣服外面应穿戴全套±800kV带电作业用屏蔽服（包括连衣裤帽、面罩、手套、导电袜和导电鞋），且各部分应连接良好，全套屏蔽服衣裤最远端点之间的电阻值不得大于20Ω。

⑥等电位作业人员在电位转移前，应得到工作负责人的许可，人体裸露部分与带电体的最小距离不小于0.5m；电位转移时，应使用电位转移棒，动作应迅速，严禁用头部充放电；与地电位作业人员传递工具和材料时，使用绝缘工具或绝缘绳索的有效长度不准小于表1-5-2的规定。

⑦用绝缘绳索传递大件金属物品时，地电位作业人员应将金属物品接地后再与绝缘绳接触。

⑧专责监护人应对作业人员进行不间断监护，随时纠正其不规范或违章动作。重点关注高处作业人员，使其保持足够的安全距离（符合表1-5-2的规定），禁止同时接触两个非连通的带电体或带电体与接地体。

（2）危险点——高处坠落

预控措施：

①高处作业人员登高前，必须具备符合本项作业要求的身体状况、精神状态和技能素质。

②高处作业人员登塔前对安全带和防坠器进行外观检查和冲击试验检查，确保其机械强度符合要求。

③高处作业人员应先检查脚钉是否齐全牢固、鞋底是否清洁，防坠装置是否牢固可靠并加挂防坠器；上下塔时，手抓主材、脚踩脚钉、匀步登（下）塔。

④监护人员应随时纠正其不规范或违章动作，重点关注高处作业人员在转位的过程中不得失去安全带或绝缘后备保护绳的保护，安全带系在牢固部件上，严禁低挂高用。

⑤等电位作业人员在绝缘子串上平行移动通常采取双手抓扶一串，双脚踩另一串的姿势匀速进入强电场，移动过程中，后备保护绳兜住二串绝缘子，避免大挥手、大迈步等动作发生。

⑥等电位作业人员沿导线走线，必须系好安全带，并使后备保护绳需将子导线全

部兜住；走线过程中应控制重心，防止导线翻转。

（3）危险点——高处坠物伤人

预控措施：

①高处作业人员的个人工具及零星材料应装入工具袋，严禁在高处浮置物件、口中含物。

②地面作业人员必须正确佩戴安全帽，正确使用绳结传递工器具及材料，与作业点垂直下方距离不得小于坠落半径。

③作业现场设置围栏并挂好警示标示牌。监护人员应时刻注意，禁止非工作人员及车辆进入作业区域。

3.2 工器具及材料选择

带电更换±800kV特高压输电线路八分裂导线间隔棒所需工器具及材料见表1-5-3。工器具出库前，应认真核对工器具的使用电压等级和试验周期，并检查确认外观良好、连接牢固、转动灵活，且符合本次工作任务的要求；工器具出库后，应存放在工具袋或工具箱内进行运输，防止脏污、受潮；金属工具和绝缘工器具应分开装运，防止因混装运输导致工器具变形、损伤等现象发生。

表1-5-3 带电更换±800kV特高压输电线路导线间隔棒所需工器具及材料表

序号	名称	规格型号	单位	数量	备注
1	绝缘传递绳	TJS-12,长度与起吊高度匹配	根	2	绝缘工具
2	绝缘后备保护绳	TJS-16,加缓冲器	根	2	绝缘工具
3	绝缘滑车	JH10-0.5	只	2	绝缘工具
4	绝缘绳套	TJS-14	根	2	绝缘工具
5	电位转移棒	0.4m	根	1	绝缘工具
6	绝缘千斤		根	4	绝缘工具
7	I型屏蔽服（连衣裤帽、面罩、手套和导电袜）	屏蔽效率≥60dB（屏蔽面罩屏蔽效率≥20dB）	套	2	个人防护用具
8	导电鞋	尺码视穿着人员而定	双	2	个人防护用具
9	阻燃内衣	纯桑蚕丝	套	2	个人防护用具
10	双保险安全带	全身背带式	副	2	个人防护用具
11	防坠器	与杆塔防坠器装置型号对应	只	2	个人防护用具
12	安全帽		顶	6	个人防护用具
13	间隔棒专用扳手	八分裂间隔棒用	个	1	专用工具
14	绝缘电阻测试仪	5000V、电极宽2cm、极间宽2cm	套	1	其他工具

续表

序号	名称	规格型号	单位	数量	备注
15	风速、温湿度测试仪	HT-8321	套	1	其他工具
16	万用表		只		其他工具
17	对讲机	视工作需要	套	2	其他工具
18	防潮苫布	2m×4m	块	2	其他工具
19	安全围栏		套	若干	其他工具
20	警示标示牌	"在此工作""从此进出""从此上下"	套	1	其他工具
21	红马甲	"工作负责人"	件	1	其他工具
22	清洁毛巾	棉质	条	1	其他工具
23	鞋套		双	若干	其他工具
24	工作手套		双	若干	其他工具
25	个人工具	工具袋、平口钳、记号笔	套	2	其他工具
26	八分裂间隔棒	与被更换间隔棒同型号	只	1	材料

注：绝缘工器具的电气及机械强度应满足《电力安全工作规程（线路部分）》要求，试验合格并在有效期内。

3.3 作业人员分工

本任务作业人员分工如表1-5-4所示。

表1-5-4 带电更换±800kV特高压输电线路导线间隔棒人员分工表

序号	工作岗位	数量(人)	工作职责
1	工作负责人	1	负责本次工作任务的人员分工、工作票的宣读、办理线路停用重合闸、办理工作许可手续、召开工作班前会、工作中突发情况的处理、工作质量的监督、工作后的总结
2	专责监护人	1	负责作业过程中的安全监督及把控
3	等电位电工	1	负责进入等电位更换八分裂导线间隔棒工作
4	塔上地电位电工	1	协助等电位电工进出强电场
5	地面电工	2	负责执行现场安全措施、布置作业现场、检查工器具、传递工具及材料，配合等电位电工进出等电位

4. 工作程序

本任务工作流程如表1-5-5所示。

表1-5-5 带电更换±800kV特高压输电线路导线间隔棒工作流程表

序号	作业内容	作业步骤及标准	安全措施及注意事项	责任人
1	现场复勘	工作负责人负责完成以下工作： (1)现场核对线路名称、杆塔编号、相别无误；基础及杆塔完好无异常；交叉跨越距离符合安全要求；确认缺陷情况及导地线规格型号等。 (2)检测风速、湿度等现场气象条件符合作业要求。 (3)检查地形环境符合作业要求。 (4)检查工作票所列安全措施与现场实际情况相符，必要时予以补充	(1)正确穿戴安全帽、工作服、工作鞋、劳保手套。 (2)不得在危及作业人员安全的气象条件下作业。 (3)严禁非工作人员、车辆进入作业现场	
2	工作许可	(1)工作负责人负责联系值班调控人员，按工作票内容申请停用线路重合闸。 (2)经值班调控人员许可后，方可开始带电作业工作	不得未经值班调控人员许可即开始工作	
3	现场布置	正确装设安全围栏并悬挂标示牌： (1)安全围栏范围应充分考虑高处坠物，以及对道路交通的影响。 (2)安全围栏出入口设置合理。 (3)妥当布置齐备"从此进出""在此工作""从此上下"等标示	对道路交通安全影响不可控时，应及时联系交通管理部门强化现场交通安全管控	
4	召开班前会	(1)全体工作成员列队。 (2)工作负责人宣读工作票，明确工作任务及人员分工；讲解工作中的安全措施和技术措施；查(问)全体工作成员精神状态；告知工作中存在的危险点及采取的预控措施。 (3)全体工作成员在工作票上签名确认	(1)工作票填写、签发和许可手续规范，签名完整。 (2)全体工作成员精神状态良好。 (3)全体工作成员明确任务分工、安全措施和技术措施	
5	检查工器具	(1)塔上地电位电工和等电位电工正确地穿戴好屏蔽服并检测合格，由负责人监督检查。 (2)正确佩戴个人安全用具(大小合适，锁扣自如)，由负责人监督检查。 (3)测量风速风向、湿度，检查绝缘工具的绝缘性能，并做好记录	(1)金属、绝缘工具使用前，应仔细检查其是否损坏、变形、失灵。绝缘工具应使用5000V及以上绝缘电阻测试仪进行分段绝缘检测，阻值应不低于700MΩ，并用清洁干燥的毛巾将其擦拭干净。 (2)用万用表测量屏蔽服衣裤最远端点之间的电阻值不得大于20Ω。工作负责人认真检查作业电工屏蔽服的连接情况。 (3)检查工具组装情况并确认连接可靠。 (4)现场所使用的带电作业工具应置在防潮苫布上	

续表

序号	作业内容	作业步骤及标准	安全措施及注意事项	责任人
6	登塔	(1)核对线路名称、杆塔编号无误后,塔上地电位电工和等电位电工冲击检查安全带、防坠器受力情况。 (2)塔上地电位电工携带绝缘传递绳登塔、等电位电工随后登塔,两人至横担作业点,选择合适位置系好安全带,塔上地电位电工将绝缘滑车和绝缘传递绳安装在横担合适位置。然后配合地面电工将绝缘传递绳分开作起吊准备	(1)核对线路名称和杆塔编号无误后,方可登塔作业。 (2)登塔过程中应使用塔上安装的防坠装置;杆塔上移动及转位时,不准失去安全保护,作业人员必须攀抓牢固构件。 (3)作业电工必须穿全套合格的屏蔽服,且全套屏蔽服必须连接可靠。在横担进入等电位前,等电位电工再次检查确认屏蔽服各部位连接可靠后方能进行下一步操作	
7	进入强电场	(1)等电位电工将安全带转移到绝缘子连接金具上,并携带电位转移棒、绝缘滑车和绝缘传递绳。 (2)等电位电工检查屏蔽服各部分连接良好后报经工作负责人同意,双手抓扶一串,双脚踩另一串,采用"跨二短三"作业方式沿绝缘子串进入强电场。 (3)当作业人员平行移动至距导线侧均压环3片绝缘子时,应停止移动,利用电位转移棒进行电位转移	(1)等电位电工进入电位前必须得到工作负责人的许可。 (2)等电位电工进入绝缘子串时应交替使用安全带和后备保护绳(用后备保护绳兜住两串绝缘子、手抓扶其中一串,脚踩另一串),不得失去安全带的保护;并调整好绝缘传递绳和电位转移棒。 (3)等电位电工在进入电位过程中手和脚应协调配合,速度均匀,避免大挥手、大迈步等动作发生;与接地体和带电体两部分间隙所组成的组合间隙边相大于6.6m。 (4)与相邻导线的最小距离大于6.8m。 (5)等电位电工进行电位转移前应检查电位转移棒与屏蔽服的电气连接是否可靠,人体裸露部分与带电体的最小距离大于0.5m,并得到工作负责人的许可;电位转移时不得失去安全带的保护,进入强电场瞬间动作准确、平稳、迅速	
8	更换导线八分裂间隔棒	(1)等电位电工进入等电位后,将安全带系在上子导线上,并装好走线绝缘保护绳(需将子导线全部兜住)。 (2)等电位电工携带绝缘传递绳沿导线走线至更换间隔棒作业点,先将绝缘绳套安装在子导线上合适	(1)等电位电工不得失去安全带的保护。 (2)等电位电工与地面电工要密切配合,听从工作负责人的指挥	

序号	作业内容	作业步骤及标准	安全措施及注意事项	责任人
		位置,其次连接绝缘滑车和绝缘传递绳,再将绝缘滑车钩挂在绝缘绳套内。 (3)等电位电工对导线上旧间隔棒安装点使用记号笔4点对称画印进行标记。 (4)等电位电工在旧间隔棒旁合适位置采用二二对应的方式安装4根绝缘千斤,将子导线可靠固定。 (5)等电位电工先利用绝缘传递绳采用活结的方式绑牢旧间隔棒,然后利用间隔棒专用扳手将旧间隔棒拆除,与地面电工配合利用绝缘传递绳将其放至地面。 (6)地面电工起吊新间隔棒至等电位电工处,等电位电工对应画印标记,正确安装新间隔棒,安装完毕后,应保持间隔棒的平面与子导线垂直。 (7)等电位电工依次拆除绝缘千斤、绝缘滑车、绝缘传递绳及绳套。 (8)等电位作业过程中不得掉落工器具和材料	(3)与相邻导线的最小距离大于6.8m。 (4)导线间隔棒在上下传递过程中,不得磕碰,绝缘传递绳索不得相互缠绕。 (5)上下传递工具时,绑扎绳结应正确可靠,防止高处坠物。 (6)传递工器具及材料过程中,地面电工禁止站立在等电位电工工作点位正下方	
9	退出强电场	(1)经检查间隔棒安装牢固,作业点无遗留物后经工作负责人许可,等电位电工携带绝缘传递绳沿导线返回均压环处,作退出电位准备。 (2)等电位电工利用电位转移钩紧均压环,并进入距均压环的第3片绝缘子,一只手抓紧绝缘子,另一只手握电位转移棒,利用电位转移棒快速脱离等电位。 (3)等电位电工按照"跨二短三"作业方式退出强电场	(1)等电位电工退出电位前必须得到工作负责人的许可。 (2)等电位电工返回绝缘子串时应交替使用安全带和后备保护绳,电位转移时不得失去安全带的保护,退出强电场瞬间动作准确、平稳、迅速。 (3)等电位电工在脱离电位过程中手和脚应协调配合,速度均匀,避免大挥手、大迈步等动作发生;与接地体和带电体两部分间隙所组成的组合间隙边相大于6.6m;人体裸露部分与带电体的最小距离大于0.5m。 (4)等电位电工沿绝缘子串移动时,用后备保护绳兜住两串绝缘子,手要抓牢,脚要踏实。 (5)与相邻导线的最小距离大于6.8m。 (6)等电位电工返回横担时不得同时失去安全带或后备保护绳的保护	

续表

序号	作业内容	作业步骤及标准	安全措施及注意事项	责任人
10	返回地面	塔上电工检查塔上无遗留物后,向工作负责人汇报,得到工作负责人同意后携带绝缘传递绳下塔	下塔过程中应使用塔上安装的防坠装置,杆塔上移动及转位时,不得失去安全保护,作业人员必须攀抓牢固构件	
11	工作结束	(1)工作负责人组织全体工作成员整理工器具和材料,将工器具清洁后放入专用的箱(袋)中;清理现场,做到"工完料尽场地清"。 (2)召开班后会,工作负责人进行工作总结和点评工作。点评本次工作的施工质量;点评全体工作成员的安全措施落实情况。 (3)工作负责人向值班调控人员汇报工作结束,申请恢复线路重合闸,终结工作票	严禁约时恢复线路再启动装置	

二、考核标准

表1-5-6 特高压直流输电线路运检技能考核评分细则

考生填写栏	编号: 姓名: 所在岗位: 单位: 日期: 年 月 日				
考评员填写栏	成绩: 考评员: 考评组长: 开始时间: 结束时间: 操作时长:				
考核模块	带电更换±800kV特高压输电线路导线间隔棒	考核对象	特高压直流输电线路检修人员	考核方式	操作	考核时限	90min
任务描述	沿耐张绝缘子串进入强电场对±800kV特高压输电线路受损导线八分裂间隔棒进行带电更换(等电位作业法)				
工作规范及要求	1. 带电作业工作应在良好天气下进行。如遇雷、雨、雪、雾天气不得进行带电作业。风力大于5级、湿度大于80%时,一般不宜进行带电作业。 2. 本项作业需工作负责人1名,专责监护人1人,塔上地电工1人,等电位电工1人,地面辅助电工2人,采用沿绝缘子串进入强电场对±800kV特高压输电线路受损导线八分裂间隔棒进行带电更换。 3. 工作负责人职责:负责本次工作任务的人员分工、工作票的宣读、办理线路停用重合闸、办理工作许可手续、召开工作班前会、工作中突发情况的处理、工作质量的监督、工作后的总结。 4. 专责监护人:负责作业过程中的安全监督及把控。 5. 等电位电工职责:负责沿绝缘子串进入强电场对导线八分裂间隔棒进行更换。 6. 塔上地电工职责:协助等电位电工进、出强电场。 7. 地面电工职责:负责执行现场安全措施、布置作业现场、检查工器具、传递工具及材料,配合等电位电工进出等电位。 8. 在带电作业中,如遇雷、雨、大风或其他任何情况威胁到工作人员的安全时,工作负责人或监护人可根据情况,临时停止工作。				

续表

考核情景准备	给定条件： 1. 培训基地：特高压直流±800kV实训线路耐张塔大号侧A相导线第1个八分裂间隔棒，导线型号：8×JL/G1A-630/45。 2. 工作票已办理，安全措施已经完备（重合闸已停用），工作开始、工作终结时应口头提出申请（调度或考评员）。 3. 作业现场装设安全围栏，悬挂"在此工作""从此进出"等标示牌，安全措施已完备。 4. 安全、正确地使用仪器仪表对绝缘工具进行检测。 5. 上下塔过程中应使用防坠落装置，防止高处坠落。 6. 必须按标准化作业程序进行操作，工序错误扣除应做项目分值，出现重大人身、器材和操作安全隐患，考评员可下令终止操作（考核），本模块考核成绩记为"不合格"	
	1. 线路：特高压直流±800kV实训线路耐张塔大号侧A相导线，工作内容：带电更换±800kV受损导线间隔棒，导线型号：8×JL/G1A-630/45。 2. 所需作业工器具：绝缘传递绳2根（TJS-12），绝缘后备保护绳2根（TJS-16、加缓冲器），绝缘滑车2只（JH10-0.5），绝缘绳套2根（TJS-14），绝缘千斤4根，电位转移棒1根（0.4m），I型屏蔽服2套（连衣裤帽、面罩、手套和导电袜），导电鞋2双，防坠器2只，间隔棒专用扳手1个，绝缘电阻测试仪1套（5000V型），风速、温湿度测试仪1套（HT-8321），万用表1只，防潮苫布2块（2m×4m），红马甲1件（工作负责人），清洁毛巾2条，个人工具2套（工具袋、平口钳、记号笔），同型号八分裂间隔棒1只。 3. 作业现场做好监护工作，作业现场安全措施（围栏等）已全部落实；禁止非作业人员进入现场，工作人员进入作业现场必须戴安全帽。 4. 考生自备工作服，阻燃纯棉内衣，安全帽，线手套，安全带（含二保绳）	
备注	1. 本模块总分为100分，各项得分均以对应分值扣完即止，在规定时间内不能完成任务应立即终止考试，本模块成绩按已完成项实际得分统计，未完成项不得分。 2. 考核过程中因设备、作业环境、安全措施、安全防护、安全距离等不符作业要求，或人为误操作，出现可能危及作业安全的任意情况，考评员应下令终止操作。 3. 考试前统一组织参考人员进行现场查勘，并提前办理工作票	

表1-5-7 特高压直流输电线路运检技能考核评分标准

序号	项目名称	质量要求	分值	扣分标准	扣分原因	扣分	得分
1	现场复勘	（1）工作负责人到作业现场核对线路名称和杆塔编号、现场工作条件、缺陷部位等。 （2）检测风速、湿度等现场气象条件符合作业要求。 （3）检查工作票填写完整，无涂改，检查是否所列安全措施与现场实际情况相符，必要时予以补充	5	（1）无工作票，本项不得分。 （2）未核对双重称号，扣1分。 （3）未核实现场工作条件（气象）、缺陷部位，每项扣1分。 （4）工作票填写出现涂改、不整洁，每处扣0.5分；工作票编号有误，扣1分；工作票填写漏项，每项扣1分			

续表

序号	项目名称	质量要求	分值	扣分标准	扣分原因	扣分	得分
2	工作许可	(1)工作负责人联系值班调控人员,按工作票内容申请停用线路重合闸。 (2)汇报内容规范、完整,声音清楚洪亮。 (3)及时履行相关许可手续	2	(1)未取得调度部门(考评员)工作许可擅自开工,本项不得分。 (2)汇报专业用语不规范、不完整或声音清楚洪亮,各扣0.5分。 (3)未申请停用重合闸,扣1分。 (4)未复诵许可内容,扣1分。 (5)复诵内容漏项,每项扣0.5分(许可人姓名、许可时间、工作任务、重合闸状态)。 (6)未及时完善工作票,扣1分			
3	现场布置	正确装设安全围栏并悬挂标示牌: (1)安全围栏范围应充分考虑高处坠物,以及对道路交通的影响。 (2)安全围栏出入口设置合理。 (3)妥当布置"从此进出""在此工作""从此上下"等标示	3	(1)未装设作业现场安全围栏,扣2分。 (2)作业现场安全围栏设置不合理,每处扣1分。 (3)未设悬挂标示牌,扣1.5分。 (4)悬挂标示牌不齐,每块0.5分。 (5)非作业人员进入围栏区,每人扣0.5分			
4	召开班前会	(1)全体工作成员正确佩戴安全帽、工作服。 (2)工作负责人穿(戴)安全红马甲,宣读工作票,明确工作任务及人员分工;讲解工作中的安全措施和技术措施;查(问)全体工作成员精神状态;告知工作中存在的危险点及采取的预控措施。 (3)全体工作成员在工作票上签名确认	3	(1)工作成员安全帽佩戴不正确,每人扣0.5分;着装不整齐,每人次扣0.5分。 (2)工作负责人和专责监护人未穿(戴)安全红马甲,每人扣0.5分。 (3)未明确工作任务及分工,本项不得分。 (4)人员分工不明确,扣1分。 (5)安全措施、预控措施交代不全,扣1分。 (6)未告知工作中存在的危险点,扣1分。 (7)未确认工作班成员精神状态,扣1分。 (8)工作班成员未签字或签字不全,扣1分			

续表

序号	项目名称	质量要求	分值	扣分标准	扣分原因	扣分	得分
5	工器具检查	(1)工作班成员在合适位置正确设置防潮苫布,防潮苫布应清洁、干燥,严禁踩踏苫布。 (2)工器具应按定置管理要求分类、整齐摆放于防潮苫布上;绝缘工器具不能与金属工具、材料混放;对工器具及仪器仪表进行外观检查。 (3)各种工具均试验合格,并在试验有效时间内。绝缘工具表面不应磨损、变形损坏,操作应灵活。绝缘工具应使用5000V及以上绝缘电阻表进行分段绝缘检测,阻值应不低于700MΩ,并用清洁干燥的毛巾将其擦拭干净。 (4)塔上地电位和等电位人员按要求正确穿戴全套合格的屏蔽服、导电鞋,且各部分连接应良好,屏蔽服内不得贴身穿着化纤类衣服,并系好安全带;工作负责人应认真检查确认是否穿戴正确、各部连接良好。 (5)全套屏蔽服应使用万用表进行测试,其最远两点之间阻值不大于20Ω,单件不大于15Ω。 (6)登塔电工对安全带及后备保护绳、防坠器进行外观检查,并经冲击试验合格。	7	(1)防潮苫布设置位置不合适,扣1分。 (2)踩踏防潮苫布,每次扣0.5分。 (3)工器具未分类定置摆放,扣1分。 (4)未检查工器具及仪器仪表试验合格标签和外观检查,每件扣1分。 (5)工器具及仪器仪表检查漏项,每件扣0.5分。 (6)仪器仪表、工器具检查方法不正确,每件扣0.5分。 (7)未对硬质绝缘工具进行清洁、擦拭,每件扣0.5分。 (8)未戴清洁干燥的棉线手套持、拿绝缘工具,每次扣0.5分。 (9)未正确使用检测仪器对工器具及全套屏蔽服进行检测,每项扣1分。 (10)登塔电工未正确穿戴屏蔽服或连接部分未检查,每人扣2分。 (11)安全带及后备保护绳、防坠器未外观检查和冲击试验(或方式不正确),每项扣1分 (12)工作负责人未检查或漏查登塔电工安全防护装备,每项扣1分			
6	登塔	(1)登塔人员再次核对双重名称、杆号、相别并向工作负责人报告及申请登塔。 (2)塔上地电位电工、等电位电工携带绝缘传递绳相继登塔。	5	(1)登塔电工未核对线路双重名称、塔号、相别,每项扣1分。 (2)登塔电工未报告核对结果,未申请登塔,每项扣1分。 (3)未使用防坠落装置登塔,考评员应下令终止操作(考核)。			

续表

序号	项目名称	质量要求	分值	扣分标准	扣分原因	扣分	得分
6	登塔	(3)登塔过程中必须使用防坠落装置,手抓主材、匀步登塔,安全带及后备保护绳挂在肩上并与带电体保持6.8m以上安全距离,工作负责人加强作业监护。 (4)登塔至合适位置,正确使用安全带,布置好绝缘传递绳,然后塔上地电位电工配合地面电工将绝缘传递绳分开作起吊准备。 (5)工作负责人认真监护并提醒整个登塔过程	5	(4)手抓脚钉登塔,每次扣0.5分。 (5)登塔踩滑、踏空,每次扣1分。 (6)安全带主带和后备保护绳未挂肩上,每人扣1分。 (7)安全带及后备保护绳发生缠绕、勾住,每次扣1分。 (8)安全带及后备保护绳低挂高用,每次扣1分。 (9)安全带及后备保护绳系在同一构件上,每次扣1分。 (10)高处作业失去安全带的保护,本项不得分。 (11)安装滑车未使用绝缘绳套,扣1分。 (12)滑车传递绳悬挂位置不便于工具取用,扣1分。 (13)工作负责人监护及提醒不到位,扣1分。 (14)安全距离不够,考评员应下令终止操作(考核)			
7	进入强电场	(1)进入电位前检查屏蔽服各部分连接良好后,报经工作负责人同意。 (2)等电位电工将安全带转移到绝缘子连接金具上,并携带电位转移棒、绝缘滑车和绝缘传递绳。 (3)等电位电工进入绝缘子串前必须系好保护绳(用后备保护绳兜住两串绝缘子、双手抓扶其中一串,脚踩另一串)。 (4)采用"跨二短三"作业方式沿绝缘子串平行移动至距导线侧均压环3片绝缘子时,应停止移动,得到工作负责人许可后,利用电位转移棒进行电位转移。	10	(1)等电位电工未检查屏蔽服各连接部分,扣2分。 (2)等电位电工转移到绝缘子连接金具上失去安全带的保护,扣2分。 (3)等电位电工与接地体和带电体两部分间隙所组成的组合间隙不够,扣2分。 (4)等电位电工进入强电场动作不正确、不熟练,每次扣2分。 (5)等电位电工电位转移过程中裸露部分距离不够,导致反复放电,扣2分。 (6)转移电位动作不熟练,扣1分。			

续表

序号	项目名称	质量要求	分值	扣分标准	扣分原因	扣分	得分
7	进入强电场	(5)等电位电工在进入电位过程中与接地体和带电体两部分间隙所组成的组合间隙边相不得小于6.6m,进入强电场必须用电位转移棒进行电位转移,人体裸露部分与带电体的最小距离不得小于0.5m。 (6)进入强电场过程不得失去安全带的保护	10	(7)电位转移过程未使用电位转移棒,扣5分。 (8)未得到工作负责人许可就进行电位转移,扣3分。 (9)等电位电工进入强电场过程失去安全带的保护,扣2分。 (10)工作负责人监护及提醒不到位,扣2分。			
8	更换导线八分裂间隔棒	(1)等电位电工进入等电位后,将围杆带系在上子导线上,并装好走线绝缘保护绳(需将子导线全部兜住)。 (2)等电位电工携带绝缘传递绳走线至更换间隔棒作业点,将绝缘绳套正确装在子导线合适位置,并挂好绝缘滑车和绝缘传递绳。 (3)等电位电工使用记号笔对旧间隔棒固定点对称画印,做好标记。 (4)等电位电工在旧间隔棒旁合适位置采用二二对应的方式安装4根绝缘千斤,将子导线可靠固定。 (5)等电位电工首先采用活结的方式将绝缘传递绳绑牢于旧间隔棒上,然后利用间隔棒专用扳手拆除旧间隔棒,与地面电工配合将其放至地面防潮苫布上。 (6)地面电工起吊新间隔棒传递给等电位电工。 (7)等电位电工对应画印及标记,正确使用间隔棒专用扳手安装新间隔棒,安装质量应保持间隔棒的平面与子导线垂直。 (8)操作中不得掉落工器具和材料	40	(1)绝缘保护绳未将子导线全部兜住,扣3分。 (2)走线过程围杆带未系在上子,扣2分。 (3)走线动作不熟练,扣2分。 (4)绝缘滑车直接钩挂在导线上,扣5分。 (5)绝缘绳套安装方式错误或位置不合适,每项扣1分。 (6)绝缘滑车钩挂绝缘绳套内后未闭锁,扣1分。 (7)未画印及标记,扣3分。 (8)未安装4根绝缘千斤进行子导线固定,每根扣2分。 (9)未系绳结就开始拆除间隔棒,扣5分。 (10)拆除工器具时动作慌乱,扣2分。 (11)发生高处坠物,每件扣3分。 (12)发生高处抛掷旧间隔棒,本模块不得分。 (13)地面电工站立在等电位电工工作点位正下方,每人次扣2分。 (14)旧间隔棒未落放在防潮苫布上,扣1分。 (15)上下传递工具时绑扎绳结方式错误,扣1分。			

续表

序号	项目名称	质量要求	分值	扣分标准	扣分原因	扣分	得分
8	更换导线八分裂间隔棒	(9)等电位电工依次拆除绝缘千斤、绝缘滑车及绝缘传递绳、绳套。 (10)等电位作业过程	40	(16)未安装完新间隔棒就解开绳结,扣5分。 (17)安装新间隔棒偏离原间隔棒位置,扣3分。 (18)安装完后,新间隔棒的平面与子导线不垂直,扣3分。 (19)未拆除绝缘千斤、绝缘滑车及绝缘传递绳、绳套,每件扣1分。 (20)未按顺序拆除绝缘千斤、绝缘滑车及绝缘传递绳、绳套,扣2分。 (21)未完成新旧间隔棒的更换工作,本模块不得分。 (22)与相邻导线安全距离不够,考评员应下令终止操作			
9	退出强电场	(1)经检查间隔棒安装牢固,作业点无遗留物后经工作负责人许可,等电位电工携带绝缘传递绳沿导线返回均压环处,作退出电位准备。 (2)等电位电工利用电位转移棒钩紧均压环,并进入距均压环的第3片绝缘子,一只手抓紧绝缘子,另一只手握电位转移棒,利用电位转移棒快速脱离等电位。 (3)退出强电场过程不得失去安全带的保护。 (4)等电位电工按照"跨二短三"作业方式沿绝缘子串退出强电场,转移到横担上。 (5)等电位电工在退出电位过程中与接地体和带电体两部分间隙所组成的组合间隙边相不得小于6.6m,退出强电场必须用电位转移棒进行电位转移,人体裸露部分与带电体的最小距离不得小于0.5m	10	(1)作业点有遗留物,每件扣2分。 (2)未向工作负责人申请即进行电位转移,扣3分。 (3)未得到工作负责人许可就进行电位转移,扣2分。 (4)电位转移过程未使用电位转移棒,扣5分。 (5)转移电位动作不熟练,扣1分。 (6)等电位电工退出强电场过程失去安全带的保护,扣2分。 (7)等电位电工电位转移过程中裸露部分距离不够,导致反复放电,扣2分。 (8)等电位电工退出强电场动作不正确、不熟练,每次扣2分。 (9)等电位电工与接地体和带电体两部分间隙所组成的组合间隙不够,扣2分。 (10)等电位电工转移到横担上失去安全带的保护,扣2分。 (11)工作负责人监护及提醒不到位,扣2分			

续表

序号	项目名称	质量要求	分值	扣分标准	扣分原因	扣分	得分
10	返回地面	(1)塔上电工检查塔上无遗留物后,向工作负责人汇报,得到工作负责人同意后,携带绝缘传递绳相继下塔。 (2)下塔过程中必须使用防坠落装置,手抓主材、匀步下塔,安全带及后备保护绳挂在肩上并与带电体保持6.8m以上安全距离,工作负责人加强作业监护	5	(1)塔上有遗留物,每件扣2分。 (2)登塔电工未报告遗留物检查结果,未申请下塔,每项扣1分。 (3)未使用防坠落装置下塔,考评员应下令终止操作(考核)。 (4)手抓脚钉下塔,每次扣0.5分。 (5)下塔踩滑、踏空,每次扣1分。 (6)安全带主带和后备保护绳未挂肩上,每人扣1分			
11	工作结束	(1)工作负责人组织全体工作成员整理工器具和材料,将工器具清洁后放入专用的箱(袋)中;清理现场,做到"工完料尽场地清"。 (2)召开班后会,工作负责人进行工作总结和点评。点评本次工作的施工质量;点评全体工作成员的安全措施落实情况。 (3)工作负责人向值班调控人员汇报工作结束,申请恢复线路重合闸,终结工作票	10	(1)对绝缘工器具未进行清洁、擦拭,每件扣0.5分。 (2)工器具未归类整理摆放,每件扣1分。 (3)工器具乱丢乱扔或踩踏防潮苫布,每次扣0.5分。 (4)未拆除围栏及标示牌或有遗留物,每件扣1分。 (5)未开班后会,扣2分。 (6)集合站队不整齐、注意力不集中,扣1分。 (7)工作班成员参加班后会人员不齐,缺一人扣1分。 (8)点评不到位,扣1分。 (9)未向调度部门(考评员)汇报工作结束,申请恢复线路重合闸,扣1分。 (10)汇报专业用语不规范、不完整或声音不洪亮,各扣0.5分。 (11)未复诵许可内容,扣1分。 (12)复诵内容漏项,每项扣0.5分(单位名称、负责人姓名、时间、线路名称、工作完成情况、设备已恢复正常、人员已撤离、可恢复重合闸)。 (13)未及时完善工作票终结手续或填写错误,每项扣1分			
	合计		100				

模块六 带电更换±800kV特高压输电线路导线防振锤培训及考核标准

一、培训标准

（一）培训要求

模块名称	带电更换±800kV特高压输电线路导线防振锤	培训类别	操作类
培训方式	实操培训	培训学时	21学时
培训目标	1.掌握沿耐张绝缘子串进、出±800kV强电场时采用"跨二短三"作业方式的电学意义。 2.能完成沿耐张绝缘子串进入±800kV等电位作业点。 3.能独立完成更换导线防振锤的操作（等电位作业法）		
培训场地	特高压直流实训线路		
培训内容	采用"跨二短三"作业方式沿耐张绝缘子串进入强电场，采用等电位作业法带电更换±800kV特高压直流输电线路导线防振锤		
适用范围	特高压直流输电线路检修人员		

（二）引用规程规范

（1）《±800kV直流架空输电线路设计规范》（GB/T 50790-2013）

（2）《±800kV直流架空输电线路检修规程》（DL/T 251-2012）

（3）《±800kV直流架空输电线路运行规程》（GB/T 28813-2012）

（4）《±800kV直流线路带电作业技术规范》（DL/T 1242-2013）

（5）《±800kV特高压输电线路金具技术规范》（GB/T 31235-2014）

（6）《国家电网公司带电作业工作管理规定（试行）》（国家电网生〔2007〕751号）

（7）《国家电网公司电力安全工作规程（线路部分）》（Q/GDW1799.2-2013）

（8）《电工术语 架空线路》（GB/T 2900.51-1998）

（9）《电工术语 带电作业》（GB/T 2900.55-2016）

（10）《带电作业工具设备术语》（GB/T 14286-2002）

（11）《带电作业用绝缘滑车》（GB/T 13034-2008）

(12)《带电作业用绝缘绳索》(GB 13035-2008)

(13)《带电作业用工具、装置和设备使用的一般要求》(DL/T 877-2004)

(14)《带电作业工具、装置和设备预防性试验规程》(DL/T 976-2005)

(15)《±800kV特高压输电线路带电作业技术导则》(Q/GDW 302-2009)

(16)《带电作业用屏蔽服装》(GB/T 6568-2008)

(17)《带电作业工具基本技术要求与设计导则》(GB 18037-2008)

(18)《带电设备红外线诊断应用规范》(DL/T 664-2016)

(三)培训教学设计

本设计以完成"带电更换±800kV特高压输电线路导线防振锤"为工作任务,按工作任务完成的标准化作业流程来设计各个培训阶段,每个阶段包括了具体的培训目标、培训内容、培训学时、培训方法(培训资源)、培训环境和考核评价等内容,如表1-6-1所示。

表1-6-1 带电更换±800kV特高压输电线路导线防振锤培训内容设计

培训流程	培训目标	培训内容	培训学时	培训方法与资源	培训环境	考核评价
1.理论教学	1.初步掌握沿绝缘子串进出±800kV强电场基本方法。2.熟悉电位转移的方法。3.熟悉输电线路受损导线防振锤更换方法。4.熟悉特高压直流线路带电作业的安全距离、危险点辨识及预控	1.沿绝缘子进出强电场"跨二短三"作业方式的电学意义。2.进、出特高压强电场时电位转移棒的使用方法。3.输电线路导线防振锤更换方法和质量标准。4特高压直流线路带电作业安全距离、危险点分析及预控措施	2	培训方法:讲授法。培训资源:PPT、相关规程、规范及技术导则	多媒体教室	考勤、课堂提问和作业
2.准备工作	能完成作业前准备工作	1.作业现场查勘。2.编制培训标准化作业卡。3.填写培训操作工作票。4.完成本操作的工器具及材料准备	1	培训方法:1.现场查勘和工器具及材料准备采用现场实操方法。2.编写作业卡和填写工作票采用讲授方法。培训资源:1.±800kV实训线路。2.特高压工器具库房。3.空白工作票	1.±800kV实训线路。2.多媒体教室	

续表

培训流程	培训目标	培训内容	培训学时	培训方法与资源	培训环境	考核评价
3.作业现场准备	能完成作业现场准备工作	1.作业现场复勘。2.工作申请。3.作业现场布置。4.班前会。5.工器具及材料检查。6.防振锤专用扳手使用方法	1	培训方法：演示与角色扮演法。资源：±800kV实训线路	±800kV实训线路	
4.培训师演示	通过现场观摩，使学员初步领会本任务操作流程	1.个人工器具穿戴、登塔。2.等电位电工沿耐张绝缘子串进、出强电场及电位转移。3.等电位电工采用走线方式到达防振锤更换位置。4.等电位电工用专用工具完成导线防振锤更换	2	培训方法：演示法。资源：±800kV实训线路	±800kV实训线路	
5.学员分组训练	1.能完成沿绝缘子串进、出±800kV强电场及电位转移操作。2.能完成±800kV输电线路导线防振锤更换作业	1.学员分组（6人一组）训练进、出±800kV强电场、电位转移和更换导线防振锤技能操作。2.培训师对学员操作进行指导和安全监护	14	培训方法：角色扮演法。资源：±800kV实训线路	±800kV实训线路	采用技能考核评分细则对学员操作评分
6.工作终结	1.使学员进一步辨析操作过程不足之处，便于后期提升。2.培养学员安全文明生产的工作作风	1.作业现场清理。2.向调度汇报工作。3.班后会，对本次工作任务进行点评总结	1	培训方法：讲授和归纳法	±800kV实训线路	

（四）作业流程

1. 工作任务

采用"跨二短三"的作业方式沿耐张绝缘子串进入强电场到达作业点，采用等电位作业法带电更换±800kV特高压输电线路导线防振锤。

2. 天气及作业现场要求

(1) 带电更换±800kV特高压输电线路导线防振锤应在良好的天气进行。

如遇雷电（听见雷声、看见闪电）、雪、雹、雨、雾等，不应进行带电作业。风力大于5级时，不宜进行带电作业；相对湿度大于80%的天气，如需进行带电作业应采用具有防潮性能的绝缘工具；恶劣天气下必须开展带电抢修时，应组织有关人员充分讨论并编制必要的安全措施，经本单位批准后方可进行。

(2) 作业人员精神状态良好，熟悉工作中保证安全的组织措施和技术措施；应持有在有效期内的带电作业资质证书。

(3) 工作负责人应事先组织相关人员完成现场勘察，根据勘察结果确定本次作业方法和所需工器具，以及应采取的必要措施，并办理带电作业工作票。

(4) 作业现场应合理设置围栏，并妥当布置警示标示牌，禁止非工作人员入内。

(5) 本项目需停用直流再启动装置。

(6) 工作中安全距离及有效绝缘长度如表1-6-2所示。

表1-6-2　带电更换±800kV特高压输电线路导线防振锤的安全距离（m）

电压等级	人身与带电体安全距离	与邻相导线的最小距离	绝缘工具最小有效绝缘长度	最小组合间隙	转移电位时人体裸露部分与带电体的最小距离
±800kV	6.8（7.3）	6.9	6.8（7.3）	6.6（7.2）	0.5

注：①海拔高度1000m以上时，±800kV直流单回输电线路带电作业中的安全距离采用括号内7.3m的数据、绝缘工具最小有效绝缘长度采用括号内7.3m的数据、组合间隙采用括号内7.2m的数据。
②因为±800kV特高压线路相间距离足够大（一般控制相地距离即可），不做重要安全因素考虑，所以《安规》没有给出"与邻相导线的最小距离"的数据，实际工作中可以参考750kV的数据。

(7) 在±800kV输电线路上作业，应保证作业相良好绝缘子片数不少于37片（单片绝缘子高度170mm）、32片（单片绝缘子高度195mm）、31片（单片绝缘子高度205mm）、26片（单片绝缘子高度240mm）。

3. 准备工作

3.1　危险点及其预控措施

(1) 危险点——触电伤害

预控措施：

①工作前，工作负责人应与值班调控人员联系，停用直流再启动装置，并履行许可手续。

②塔上等电位作业人员登塔前，必须仔细核对线路名称、杆塔编号、相别，确认

无误后方可上塔。

③工作中,如遇线路突然停电,作业人员应视其仍然带电。工作负责人应尽快与调控人员联系,值班调控人员未与工作负责人取得联系前不准强送电。

④地面电工操作绝缘工具时应戴清洁、干燥的防汗手套,绝缘工具及绝缘绳索不得损坏、受潮、变形、失灵,不准使用非绝缘绳索(如棉纱绳、白棕绳、钢丝绳),现场所使用的带电作业工具应放置在防潮苫布上,防止绝缘工具在使用中脏污和受潮。

⑤等电位作业人员应穿着阻燃内衣,衣服外面应穿戴全套±800kV带电作业用屏蔽服(包括连衣裤帽、面罩、手套、导电袜和导电鞋),且各部分应连接良好,全套屏蔽服衣裤最远端点之间的电阻值不得大于20Ω。

⑥等电位作业人员在电位转移前,应得到工作负责人的许可,人体裸露部分与带电体的最小距离不小于0.5m;电位转移时,应使用电位转移棒,动作应迅速,严禁用头部充放电;与地电位作业人员传递工具和材料时,使用绝缘工具或绝缘绳索的有效长度不得小于表1-6-2的规定。

⑦用绝缘绳索传递大件金属物品时,地电位作业人员应将金属物品接地后再与绝缘绳接触。

⑧专责监护人应对作业人员进行不间断监护,随时纠正其不规范或违章动作。重点关注高处作业人员,使其保持足够的安全距离(符合表1-6-2的规定),禁止同时接触两个非连通的带电体或带电体与接地体。

(2)危险点——高处坠落

预控措施:

①高处作业人员登高前,必须具备符合本项作业要求的身体状况、精神状态和技能素质。

②高处作业人员登塔前对安全带和防坠器进行外观检查和冲击试验检查,确保其机械强度符合要求。

③高处作业人员应先检查脚钉是否齐全牢固、鞋底是否清洁,防坠装置是否牢固可靠并加挂防坠器;上下塔时,手抓主材、脚踩脚钉、匀步登(下)塔。

④监护人员应随时纠正其不规范或违章动作,重点关注高处作业人员在转位的过程中不得失去安全带或绝缘后备保护绳的保护,安全带系在牢固部件上,严禁低挂高用。

⑤等电位作业人员在绝缘子串上平行移动通常采取双手抓扶一串,双脚踩另一串的姿势匀速进入强电场,移动过程中,后备保护绳兜住二串绝缘子,避免大挥手、大迈步等动作发生。

⑥等电位作业人员沿导线走线,必须系好安全带,并使后备保护绳需将子导线全

部兜住；走线过程中应控制重心，防止导线翻转。

（3）危险点——高处坠物伤人

预控措施：

①高处作业人员的个人工具及零星材料应装入工具袋，严禁在高处浮置物件、口中含物。

②地面作业人员必须正确佩戴安全帽，正确使用绳结传递工器具及材料，与作业点垂直下方距离不得小于坠落半径。

③作业现场设置围栏并挂好警示标示牌。监护人员应时刻注意，禁止非工作人员及车辆进入作业区域。

3.2 工器具及材料选择

带电更换±800kV特高压输电线路导线防振锤所需工器具及材料见表1-6-3。工器具出库前，应认真核对工器具的使用电压等级和试验周期，并检查确认外观良好、连接牢固、转动灵活，且符合本次工作任务的要求；工器具出库后，应存放在工具袋或工具箱内进行运输，防止脏污、受潮；金属工具和绝缘工器具应分开装运，防止因混装运输导致工器具变形、损伤等现象发生。

表1-6-3 带电更换±800kV特高压输电线路导线防振锤所需工器具及材料表

序号	名称	规格型号	单位	数量	备注
1	绝缘传递绳	TJS-12，长度与起吊高度匹配	根	2	绝缘工具
2	绝缘后备保护绳	TJS-16，加缓冲器	根	2	绝缘工具
3	绝缘滑车	JH10-0.5	只	2	绝缘工具
4	绝缘绳套	TJS-14	根	2	绝缘工具
5	电位转移棒	0.4m	根	1	绝缘工具
6	I型屏蔽服（连衣裤帽、面罩、手套和导电袜）	屏蔽效率≥60dB（屏蔽面罩屏蔽效率≥20dB）	套	2	个人防护用具
7	导电鞋	尺码视穿着人员而定	双	2	个人防护用具
8	阻燃内衣	纯桑蚕丝	套	2	个人防护用具
9	双保险安全带	全身背带式	副	2	个人防护用具
10	防坠器	与杆塔防坠器装置型号对应	只	2	个人防护用具
11	安全帽		顶	6	个人防护用具
12	防振锤专用扳手		个	1	专用工具
13	绝缘电阻测试仪	5000V，电极宽2cm、极间宽2cm	套	1	其他工具
14	风速、温湿度测试仪	HT-8321	套	1	其他工具

续表

序号	名称	规格型号	单位	数量	备注
15	万用表		只		其他工具
16	对讲机	视工作需要	套	2	其他工具
17	防潮苫布	2m×4m	块	2	其他工具
18	安全围栏		套	若干	其他工具
19	警示标示牌	"在此工作""从此进出""从此上下"	套	1	其他工具
20	红马甲	"工作负责人"	件	1	其他工具
21	清洁毛巾	棉质	条	1	其他工具
22	鞋套		双	若干	其他工具
23	工作手套		双	若干	其他工具
24	个人工具	工具袋、平口钳、记号笔	套	2	其他工具
25	防振锤	与被更换防振锤同型号	只	1	材料
26	铝包带		米	适量	材料

注：绝缘工器具的电气及机械强度应满足《电力安全工作规程（线路部分）》要求，试验合格并在有效期内。

3.3 作业人员分工

本任务作业人员分工如表1-6-4所示。

表1-6-4 带电更换±800kV特高压输电线路导线防振锤人员分工表

序号	工作岗位	数量（人）	工作职责
1	工作负责人	1	负责本次工作任务的人员分工、工作票的宣读、办理停用直流再启动装置手续、办理工作许可手续、召开工作班前会、工作中突发情况的处理、工作质量的监督、工作后的总结
2	专责监护人	1	负责作业过程中的安全监督及把控
3	等电位电工	1	负责进入等电位更换导线防振锤工作
4	塔上地电位电工	1	负责协助等电位进出强电场
5	地面电工	2	负责执行现场安全措施、布置作业现场、检查工器具、传递工具及材料，配合等电位电工进出等电位

4. 工作程序

本任务工作流程如表1-6-5所示。

表1-6-5 带电更换±800kV特高压输电线路导线防振锤工作流程表

序号	作业内容	作业步骤及标准	安全措施及注意事项	责任人
1	现场复勘	工作负责人负责完成以下工作： (1)现场核对线路名称、杆塔编号，相别无误；基础及杆塔完好无异常；交叉跨越距离符合安全要求；确认缺陷情况及导地线规格型号等。 (2)检测风速、湿度等现场气象条件符合作业要求。 (3)检查地形环境符合作业要求。 (4)检查工作票所列安全措施与现场实际情况相符，必要时予以补充	(1)正确穿戴安全帽、工作服、工作鞋、劳保手套。 (2)不得在危及作业人员安全的气象条件下作业。 (3)严禁非工作人员、车辆进入作业现场	
2	工作许可	(1)工作负责人负责联系值班调控人员，按工作票内容申请停用直流再启动装置。 (2)经值班调控人员许可后，方可开始带电作业工作	不得未经值班调控人员许可即开始工作	
3	现场布置	正确装设安全围栏并悬挂标示牌： (1)安全围栏范围应充分考虑高处坠物，以及对道路交通的影响。 (2)安全围栏出入口设置合理。 (3)妥当布置齐备"从此进出""在此工作""从此上下"等标示	对道路交通安全影响不可控时，应及时联系交通管理部门强化现场交通安全管控	
4	召开班前会	(1)全体工作成员列队。 (2)工作负责人宣读工作票，明确工作任务及人员分工；讲解工作中的安全措施和技术措施；查(问)全体工作成员精神状态；告知工作中存在的危险点及采取的预控措施。 (3)全体工作成员在工作票上签字确认	(1)工作票填写、签发和许可手续规范，签字完整。 (2)全体工作成员精神状态良好。 (3)全体工作成员明确任务分工、安全措施和技术措施	
5	检查工器具	(1)塔上地电位电工和等电位电工正确地穿戴好屏蔽服并检测合格，由负责人监督检查。 (2)正确佩戴个人安全用具(大小合适，锁扣自如)，由负责人监督检查。 (3)测量风速风向、湿度，检查绝缘工具的绝缘性能，并做好记录	(1)金属、绝缘工具使用前，应仔细检查其是否损坏、变形、失灵。绝缘工具应使用2500V及以上绝缘电阻测试仪进行分段绝缘检测，阻值应不低于700MΩ，并用清洁干燥的毛巾将其擦拭干净。 (2)用万用表测量屏蔽服衣裤最远端点之间的电阻值不得大于20Ω。工作负责人认真检查作业电工屏蔽服的连接情况。 (3)检查工具组装情况并确认连接可靠。 (4)现场所使用的带电作业工器具应放置在防潮苫布上	

续表

序号	作业内容	作业步骤及标准	安全措施及注意事项	责任人
6	登塔	(1)核对线路名称、杆塔编号无误后,塔上地电位电工和等电位电工冲击检查安全带、防坠器受力情况。 (2)塔上地电位电工携带绝缘传递绳登塔、等电位电工随后登塔,两人至横担作业点,选择合适位置系好安全带,塔上地电位电工将绝缘滑车和绝缘传递绳安装在横担合适位置。然后配合地面电工将绝缘传递绳分开作起吊准备	(1)核对线路名称和杆塔编号无误后,方可登塔作业。 (2)登塔过程中应使用塔上安装的防坠装置;杆塔上移动及转位时,不准失去安全保护,作业人员必须攀抓牢固构件。 (3)作业电工必须穿全套合格的屏蔽服,且全套屏蔽服必须连接可靠。在横担进入等电位前,等电位电工再次检查确认屏蔽服各部位连接可靠后方能进行下一步操作	
7	进入强电场	(1)等电位电工将安全带转移到绝缘子连接金具上,并携带电位转移棒、绝缘滑车和绝缘传递绳。 (2)等电位电工检查屏蔽服各部分连接良好后报经工作负责人同意,双手抓扶一串,双脚踩另一串,采用"跨二短三"作业方式沿绝缘子串进入等电位。 (3)当作业人员平行移动至距导线侧均压环3片绝缘子时,应停止移动,利用电位转移棒进行电位转移	(1)等电位电工进入电位前必须得到工作负责人的许可。 (2)等电位电工进入绝缘子串时应交替使用安全带和后备保护绳(用后备保护绳兜住两串绝缘子,手抓扶其中一串,脚踩另一串),不得失去安全带的保护;并调整好绝缘传递绳和电位转移棒。 (3)等电位电工在进入强电场过程中手和脚应协调配合,速度均匀,避免大挥手、大迈步等危险动作;与接地体和带电体两部分间隙所组成的组合间隙应大于6.6m。 (4)与相邻导线的最小距离大于6.9m。 (5)等电位电工进行电位转移前应检查电位转移棒与屏蔽服的电气连接是否可靠,人体裸露部分与带电体的最小距离应大于0.5m,并得到工作负责人的许可;电位转移时不得失去安全带的保护,进入强电场瞬间动作准确、平稳、迅速	

续表

序号	作业内容	作业步骤及标准	安全措施及注意事项	责任人
8	更换子导线防振锤	(1)等电位电工进入等电位后,将安全带系在上子导线上,并装好走线绝缘保护绳(需将子导线全部兜住)。 (2)等电位电工携带绝缘传递绳沿导线走线至更换防振锤作业点,先将绝缘绳套安装在子导线上合适位置,其次连接绝缘滑车和绝缘传递绳,再将绝缘滑车钩挂在绝缘绳套内。 (3)等电位电工对导线上旧防振锤安装点使用记号笔两端画印进行标记。 (4)等电位电工先利用绝缘传递绳采用活结的方式绑牢旧防振锤,然后利用防振锤专用扳手将旧防振锤拆除,与地面电工配合利用绝缘传递绳将其放至地面。 (5)等电位电工拆除铝包带,放于工具包中;对应画印标记缠绕新铝包带,缠绕应平整、紧密,其绕向应与外层导线绞制方向一致。 (6)地面电工起吊新防震锤至等电位电工处,等电位电工正确安装新防振锤。 (7)对安装质量进行检查,防震锤方向竖直向下,锤球与主线平行;安装位移不应超过±30mm;铝包带两端断头应回压到防震锤夹内,两端应露出10mm,螺栓处弹簧垫圈应紧平,螺栓穿向与其他子导线防震锤一致。 (8)等电位电工依次拆除绝缘滑车、绝缘传递绳及绳套。 (9)等电位作业过程中不得掉落工器具和材料	(1)等电位电工不得失去安全带的保护。 (2)等电位电工与地面电工要密切配合,听从工作负责人的指挥。 (3)与相邻导线的最小距离大于6.9m。 (4)导线防振锤在上下传递过程中,不得磕碰,两侧绝缘传递绳不得相互缠绕。 (5)上下传递工具时,绑扎绳结应正确可靠,防止高处坠物。 (6)传递工器具及材料过程中,地面电工禁止站立在等电位电工工作点位正下方	
9	退出强电场	(1)经检查防震锤安装牢固,作业点无遗留物后经工作负责人许可,等电位电工携带绝缘传递绳沿导线返回均压环处,作退出电位准备。	(1)等电位电工退出电位前必须得到工作负责人的许可。 (2)等电位电工返回绝缘子串时应交替使用安全带和后备保护绳,电位转移时不得失去安全带的保护,退出强电场瞬间动作准确、平稳、迅速。	

续表

序号	作业内容	作业步骤及标准	安全措施及注意事项	责任人
9	退出强电场	(2)等电位电工利用电位转移棒钩紧均压环,并进入距均压环的第3片绝缘子,一只手抓紧绝缘子,另一只手握电位转移棒,利用电位转移棒快速脱离等电位。 (3)等电位电工按照"跨二短三"作业方式退出强电场	(3)等电位电工在脱离电位过程中手和脚应协调配合,速度均匀,避免大挥手、大迈步等动作发生;与接地体和带电体两部分间隙所组成的组合间隙应大于6.6m;人体裸露部分与带电体的最小距离大于0.5m。 (4)等电位电工沿绝缘子串移动时,用后备保护绳兜住两串绝缘子、手要抓牢,脚要踏实。 (5)与相邻导线的最小距离大于6.9m。 (6)等电位电工返回横担时不得同时失去安全带或后备保护绳的保护	
10	返回地面	塔上电工检查塔上无遗留物后,向工作负责人汇报,得到工作负责人同意后携带绝缘传递绳下塔	下塔过程中应使用塔上安装的防坠装置,杆塔上移动及转位时,不得失去安全保护,作业人员必须攀抓牢固构件	
11	工作结束	(1)工作负责人组织全体工作成员整理工器具和材料,将工器具清洁后放入专用的箱(袋)中;清理现场,做到"工完料尽场地清"。 (2)召开班后会,工作负责人进行工作总结和点评工作。点评本次工作的施工质量;点评全体工作成员的安全措施落实情况。 (3)工作负责人向值班调控人员汇报工作结束,申请恢复直流再启动装置,终结工作票	严禁约时恢复直流再启动装置	

二、考核标准

表1-6-6 特高压直流输电线路运检技能考核评分细则

考生填写栏	编号:	姓名:	所在岗位:	单位:	日期:	年 月 日			
考评员填写栏	成绩:	考评员:	考评组长:	开始时间:	结束时间:		操作时长:		
考核模块	带电更换±800kV特高压输电线路导线防振锤		考核对象	特高压直流输电线路检修人员		考核方式	操作	考核时限	90min

第一部分
±800kV 特高压输电线路运检带电作业培训及考核标准

续表

任务描述	沿耐张绝缘子串进入强电场对±800kV特高压输电线路受损导线防振锤进行带电更换(等电位作业法)
工作规范及要求	1. 带电作业工作应在良好天气下进行。如遇雷、雨、雪、雾天气不得进行带电作业。风力大于5级、湿度大于80%时，一般不宜进行带电作业。 2. 本项作业需工作负责人1名，专责监护人1人，塔上地电工1人，等电位电工1人，地面辅助电工2人，采用沿绝缘子串进入强电场对±800kV特高压输电线路受损导线防振锤进行带电更换。 3. 工作负责人职责：负责本次工作任务的人员分工、工作票的宣读、办理停用直流再启动装置手续、办理工作许可手续、召开工作班前会、工作中突发情况的处理、工作质量的监督、工作后的总结。 4. 专责监护人：负责作业过程中的安全监督及把控。 5. 等电位电工职责：负责沿绝缘子串进入强电场对导线防振锤进行更换。 6. 塔上地电工职责：负责协助等电位进出强电场。 7. 地面电工职责：负责执行现场安全措施、布置作业现场、检查工器具、传递工具及材料，配合等电位电工进出等电位。 8. 在带电作业中，如遇雷、雨、大风或其他任何情况威胁到工作人员的安全时，工作负责人或监护人可根据情况，临时停止工作。 给定条件： 1. 培训基地：特高压直流±800kV实训线路xx线001#耐张塔大号侧正极子导线防振锤，防振锤型号：FRT-7。 2. 工作票已办理，安全措施已经完备（直流再启动装置已停用），工作开始、工作终结时应口头提出申请（调度或考评员）。 3. 作业现场装设安全围栏，悬挂"在此工作""从此进出"等标示牌，安全措施已完备。 4. 安全、正确地使用仪器仪表对绝缘工器具进行检测。 5. 上下塔过程中应使用防坠落装置，防止高处坠落。 6. 必须按标准化作业程序进行操作，工序错误扣除应做项目分值，出现重大人身、器材和操作安全隐患，考评员可下令终止操作(考核)，本模块考核成绩记为"不合格"
考核情景准备	1. 线路：特高压直流±800kV实训线路xx线001#耐张塔大号侧正极导线，工作内容：带电更换±800kV导线受损防振锤，防振锤型号：FRT-7。 2. 所需作业工器具：绝缘传递绳2根(TJS-12)，绝缘后备保护绳2根(TJS-16、加缓冲器)，绝缘滑车2只(JH10-0.5)，绝缘绳套2根(TJS-14)，电位转移棒1根(0.4m)，I型屏蔽服2套(连衣裤帽、面罩、手套和导电袜)，导电鞋2双，防坠器2只，防振锤专用扳手1个，绝缘电阻测试仪1套(5000V型)，风速、温湿度测试仪1套(HT-8321)，万用表1只，防潮苫布2块(2m×4m)，红马甲1件(工作负责人)，清洁毛巾2条，个人工具2套(工具袋、平口钳、记号笔)，同型号防振锤1只，铝包带若干。 3. 作业现场做好监护工作，作业现场安全措施（围栏等）已全部落实；禁止非作业人员进入现场，工作人员进入作业现场必须戴安全帽。 4. 考生自备工作服，阻燃纯棉内衣，安全帽，线手套，安全带(含后备保护绳)
备注	1. 本模块总分为100分，各项得分均以对应分值扣完即止，在规定时间内不能完成任务应立即终止考试，本模块成绩按已完成项实际得分统计，未完成项不得分。 2. 考核过程中因设备、作业环境、安全措施、安全防护、安全距离等不符合作业要求，或人为误操作，出现可能危及作业安全的任意情况，考评员应下令终止操作。 3. 考试前统一组织参考人员进行现场查勘，并提前办理工作票

表 1-6-7 特高压直流输电线路运检技能考核评分标准

序号	项目名称	质量要求	分值	扣分标准	扣分原因	扣分	得分
1	现场复勘	(1)工作负责人到作业现场核对线路名称和杆塔编号、现场工作条件、缺陷部位等。(2)检测风速、湿度等现场气象条件符合作业要求。(3)检查工作票填写完整,无涂改,检查是否所列安全措施与现场实际情况相符,必要时予以补充	5	(1)无工作票,本项不得分。(2)未核对双重称号,扣1分。(3)未核实现场工作条件(气象)、缺陷部位,每项扣1分。(4)工作票填写出现涂改、不整洁,每处扣0.5分,工作票编号有误,扣1分。工作票填写漏项,每项扣1分			
2	工作许可	(1)工作负责人联系值班调控人员,按工作票内容申请停用直流再启动装置。(2)汇报内容规范、完整,声音清楚洪亮。(3)及时履行相关许可手续	3	(1)未取得调度部门(考评员)工作许可擅自开工,本项不得分。(2)汇报专业用语不规范、不完整或声音清楚洪亮各扣0.5分。(3)未申请停用直流再启动装置,扣1分。(4)未复诵许可内容,扣1分。(5)复诵内容漏项,每项扣0.5分(许可人姓名、许可时间、工作任务、直流再启动装置状态)。(6)未及时完善工作票,扣1分			
3	现场布置	正确装设安全围栏并悬挂标示牌:(1)安全围栏范围应充分考虑高处坠物,以及对道路交通的影响。(2)安全围栏出入口设置合理。(3)妥当布置"从此进出""在此工作""从此上下"等标示	4	(1)未装设作业现场安全围栏,扣2分。(2)作业现场安全围栏设置不合理,每处扣1分。(3)未设悬挂标示牌,扣1.5分。(4)悬挂标示牌不齐,每块0.5分。(5)非作业人员进入围栏区,每人扣0.5分			

第一部分　±800kV特高压输电线路运检带电作业培训及考核标准

续表

序号	项目名称	质量要求	分值	扣分标准	扣分原因	扣分	得分
4	召开班前会	(1)全体工作成员正确佩戴安全帽、工作服。 (2)工作负责人穿(戴)安全红马甲,宣读工作票,明确工作任务及人员分工;讲解工作中的安全措施和技术措施;查(问)全体工作成员精神状态;告知工作中存在的危险点及采取的预控措施。 (3)全体工作成员在工作票上签字确认	4	(1)工作成员安全帽佩戴不正确,每人扣0.5分,着装不整齐,每人次扣0.5分。 (2)工作负责人和专责监护人未穿(戴)安全红马甲,每人扣0.5分。 (3)未明确工作任务及分工,本项不得分。 (4)人员分工不明确,扣1分。 (5)安全措施、预控措施交代不全,扣1分。 (6)未告知工作中存在的危险点,扣1分。 (7)未确认工作班成员精神状态,扣1分。 (8)工作班成员未签字或签字不全,扣1分			
5	工器具检查	(1)工作班成员在合适位置正确设置防潮苫布,防潮苫布应清洁、干燥,严禁踩踏苫布。 (2)工器具应按定置管理要求分类、整齐摆放于防潮苫布上;绝缘工器具不能与金属工具、材料混放;对工器具及仪器仪表进行外观检查。 (3)各种工具均试验合格,并在试验有效时间内。绝缘工具表面不应磨损、变形损坏,操作应灵活。绝缘工具应使用2500V及以上绝缘电阻表进行分段绝缘检测,阻值应不低于700MΩ,并用清洁干燥的毛巾将其擦拭干净。 (4)塔上地电位和等电位人员按要求正确穿戴全套合格的屏蔽服、导电鞋,且各部分连接应良好,屏蔽服内不得贴身穿着化纤类衣服,并系好安全带;工作负责人应认真检查确认是否穿戴正确、各部连接良好。	8	(1)防潮苫布设置位置不合适,扣1分。 (2)踩踏防潮苫布,每次扣0.5分。 (3)工器具未分类定置摆放,扣1分。 (4)未检查工器具及仪器仪表试验合格标签和外观检查,每件扣1分。 (5)工器具及仪器仪表检查漏项,每件扣0.5分。 (6)仪器仪表、工器具检查方法不正确,每件扣0.5分。 (7)未对硬质绝缘工具进行清洁、擦拭,每件扣0.5分。 (8)未戴清洁干燥的棉线手套持、拿绝缘工具,每次扣0.5分。 (9)未正确使用检测仪器对工器具及全套屏蔽服进行检测,每项扣1分。 (10)登塔电工未正确穿戴屏蔽服或连接部分未检查,每人扣2分。			

续表

序号	项目名称	质量要求	分值	扣分标准	扣分原因	扣分	得分
5	工器具检查	(5)全套屏蔽服应使用万用表进行测试,其最远两点之间阻值不大于20Ω,单件不大于15Ω。 (6)登塔电工对安全带及后备保护绳、防坠器进行外观检查,并经冲击试验合格。		(11)安全带及后备保护绳、防坠器未外观检查和冲击试验(或方式不正确),每项扣1分。 (12)工作负责人未检查或漏查登塔电工安全防护装备,每项扣1分。			
6	登塔	(1)登塔人员再次核对双重名称、杆号、相别并向工作负责人报告及申请登塔。 (2)塔上地电位电工、等电位电工携带绝缘传递绳相继登塔。 (3)登塔过程中必须使用防坠落装置,手抓主材、匀步登塔,安全带及后备保护绳挂在肩上并与带电体保持6.8m以上安全距离,工作负责人加强作业监护。 (4)登塔至合适位置,正确使用安全带,布置好绝缘传递绳,然后塔上地电位电工配合地面电工将绝缘传递绳分开作起吊准备。 (5)工作负责人认真监护并提醒整个登塔过程	5	(1)登塔电工未核对线路双重名称、塔号、相别,每项扣1分。 (2)登塔电工未报告核对结果,未申请登塔,每项扣1分。 (3)未使用防坠落装置登塔,考评员应下令终止操作(考核)。 (4)手抓脚钉登塔,每次扣0.5分。 (5)登塔踩滑、踏空,每次扣1分。 (6)安全带主带和后备保护绳未挂肩上,每人扣1分。 (7)安全带及后备保护绳发生缠绕、勾住,每次扣1分。 (8)安全带及后备保护绳低挂高用,每次扣1分。 (9)安全带及后备保护绳系在同一构件上,每次扣1分。 (10)高处作业失去安全带的保护,本项不得分。 (11)安装滑车未使用绝缘绳套,扣1分。 (12)滑车传递绳悬挂位置不便于工具取用,扣1分。 (13)工作负责人监护及提醒不到位,扣1分。 (14)安全距离不够,考评员应下令终止操作(考核)			

第一部分 ±800kV特高压输电线路运检带电作业培训及考核标准

续表

序号	项目名称	质量要求	分值	扣分标准	扣分原因	扣分	得分
7	进入强电场	(1)进入电位前检查屏蔽服各部分连接良好后,报经工作负责人同意。 (2)等电位电工将安全带转移到绝缘子连接金具上,并携带电位转移棒、绝缘滑车和绝缘传递绳。 (3)等电位电工进入绝缘子串前必须系好保护绳(用后备保护绳兜住两串绝缘子、双手抓扶其中一串,脚踩另一串)。 (4)采用"跨二短三"作业方式沿绝缘子串平行移动至距导线侧均压环3片绝缘子时,应停止移动,得到工作负责人许可后,利用电位转移棒进行电位转移。 (5)等电位电工在进入电位过程中与接地体和带电体两部分间隙所组成的组合间隙不得小于6.6m,进入强电场必须用电位转移棒进行电位转移,人体裸露部分与带电体的最小距离不得小于0.5m。 (6)进入强电场过程不得失去安全带的保护	13	(1)等电位电工未检查屏蔽服各连接部分,扣2分。 (2)等电位电工转移到绝缘子连接金具上失去安全带的保护,扣2分。 (3)等电位电工与接地体和带电体两部分间隙所组成的组合间隙不够,扣2分。 (4)等电位电工进入强电场动作不正确、不熟练,每次扣2分。 (5)等电位电工电位转移过程中裸露部分距离不够,导致反复放电,扣2分。 (6)转移电位动作不熟练,扣1分。 (7)电位转移过程未使用电位转移棒,扣5分。 (8)未得到工作负责人许可就进行电位转移,扣3分。 (9)等电位电工进入强电场过程失去安全带的保护,扣2分。 (10)工作负责人监护及提醒不到位,扣2分			
8	更换子导线防振锤	(1)等电位电工进入等电位后,将围杆带系在上子导线上,并装好走线绝缘保护绳(需将子导线全部兜住)。 (2)等电位电工携带绝缘传递绳走线至更换防振锤作业点,将绝缘绳套正确装在子导线合适位置,并挂好绝缘滑车和绝缘传递绳。 (3)等电位电工使用记号笔对旧防振锤安装位置两端画印,做好标记。		(1)绝缘保护绳未将子导线全部兜住,扣3分。 (2)走线过程围杆带未系在上子,扣1分。 (3)走线动作不熟练,扣2分。 (4)绝缘滑车直接钩挂在导线上,扣5分。 (5)绝缘绳套安装方式错误或位置不合适,每项扣1分。 (6)绝缘滑车钩挂绝缘绳套内后未闭锁,扣1分。			

续表

序号	项目名称	质量要求	分值	扣分标准	扣分原因	扣分	得分
8	更换子导线防振锤	(4)等电位电工先利用绝缘传递绳采用活结的方式绑牢旧防振锤,然后利用防振锤专用扳手将旧防振锤拆除,与地面电工配合利用绝缘传递绳将其放到地面防潮苫布上。 (5)等电位电工拆除铝包带,放于工具包中;对应画印标记缠绕新铝包带,缠绕应平整、紧密,其绞制方向应与外层导线绞制方向一致。 (6)地面电工起吊新防震锤至等电位电工处。 (7)等电位电工正确安装新防振锤,并确保防振锤竖直向下,锤球与主线平行,安装位移不应超过±30mm;铝包带两端断头应回压到防震锤夹内,两端应露出10mm,螺栓处弹簧垫圈应紧平,螺栓穿向与其他子导线防震锤一致。 (8)等电位电工依次拆除绝缘滑车、绝缘传递绳及绳套。 (9)等电位作业过程中不得掉落工器具和材料	30	(7)未画印及标记,扣3分。 (8)未系绳结就开始拆除防振锤,扣5分。 (9)拆除工器具时动作慌乱,扣2分。 (10)发生高处坠物,每件扣3分。 (11)发生高处抛掷旧防振锤,本模块不得分。 (12)地面电工站立在等电位电工工作点位正下方,每人次扣2分。 (13)旧防振锤未落放在防潮苫布上,扣1分。 (14)上下传递工具时绑扎绳结方式错误,扣1分。 (15)未安装完新防振锤就解开绳结,扣5分。 (16)安装新防振锤偏离原位置超过允许值,扣3分。 (17)安装完后,防震锤没有竖直向下或锤球与主线不平行,各扣3分。 (18)安装完后,铝包带两端断头应未压到防震锤夹内,扣2分。 (19)安装完后,铝包两端未露出10mm,扣2分。 (20)防振锤螺栓穿向错误,每处扣1分。 (21)未拆除绝缘滑车、绝缘传递绳及绳套,每件扣1分。 (22)未按顺序拆绝缘滑车、绝缘传递绳及绳套,扣2分。 (23)未完成新旧防振锤的更换工作,本模块不得分。 (24)与相邻导线安全距离不够,考评员应下令终止操作			

续表

序号	项目名称	质量要求	分值	扣分标准	扣分原因	扣分	得分
9	退出强电场	(1)经检查防振锤安装牢固,作业点无遗留物后经工作负责人许可,等电位电工携带绝缘传递绳沿导线返回均压环处,作退出电位准备。 (2)等电位电工利用电位转移棒钩紧均压环,并进入距均压环的第3片绝缘子,一只手抓紧绝缘子,另一只手握电位转移棒,利用电位转移棒快速脱离等电位。 (3)退出强电场过程不得失去安全带的保护。 (4)等电位电工按照"跨二短三"作业方式沿绝缘子串退出强电场,转移到横担上。 (5)等电位电工在退出电位过程中与接地体和带电体两部分间隙所组成的组合间隙不得小于6.6m,退出强电场必须用电位转移棒进行电位转移,人体裸露部分与带电体的最小距离不得小于0.5m	13	(1)作业点有遗留物,每件扣2分。 (2)未向工作负责人申请即进行电位转移,扣3分。 (3)未得到工作负责人许可就进行电位转移,扣2分。 (4)电位转移过程未使用电位转移棒,扣5分。 (5)转移电位动作不熟练,扣1分。 (6)等电位电工退出强电场过程失去安全带的保护,扣2分。 (7)等电位电工电位转移过程中裸露部分距离不够,导致反复放电,扣2分。 (8)等电位电工退出强电场动作不正确、不熟练,每次扣2分。 (9)等电位电工与接地体和带电体两部分间隙所组成的组合间隙不够,扣2分。 (10)等电位电工转移到横担上失去安全带的保护,扣2分。 1(1)工作负责人监护及提醒不到位,扣2分			
10	返回地面	(1)塔上电工检查塔上无遗留物后,向工作负责人汇报,得到工作负责人同意后,携带绝缘传递绳相继下塔。 (2)下塔过程中必须使用防坠落装置,手抓主材、匀步下塔,安全带及后备保护绳挂在肩上并与带电体保持6.8m以上安全距离,工作负责人加强作业监护。	5	(1)塔上有遗留物,每件扣2分。 (2)登塔电工未报告遗留物检查结果,未申请下塔,每项扣1分。 (3)未使用防坠落装置下塔,考评员应下令终止操作(考核)。 (4)手抓脚钉下塔,每次扣0.5分。 (5)下塔踩滑、踏空,每次扣1分。 (6)安全带主带和后备保护绳未挂肩上,每人扣1分			

续表

序号	项目名称	质量要求	分值	扣分标准	扣分原因	扣分	得分
11	工作结束	(1)工作负责人组织全体工作成员整理工器具和材料,将工器具清洁后放入专用的箱(袋)中;清理现场,做到"工完料尽场地清"。 (2)召开班后会,工作负责人进行工作总结和点评。点评本次工作的施工质量;点评全体工作成员的安全措施落实情况。 (3)工作负责人向值班调控人员汇报工作结束,申请恢复直流再启动装置,终结工作票	10	(1)对绝缘工器具未进行清洁、擦拭,每件扣0.5分。 (2)工器具未归类整理摆放,每件扣1分。 (3)工器具乱丢乱扔或踩踏防潮苫布,每次扣0.5分。 (4)未拆除围栏及标示牌或有遗留物,每件扣1分。 (5)未开班后会,扣2分。 (6)集合站队不整齐、注意力不集中,扣1分。 (7)工作班成员参加班后会人员不齐,缺一人扣1分。 (8)点评不到位,扣1分。 (9)未向调度部门(考评员)汇报工作结束,申请恢复直流再启动装置,扣1分。 (10)汇报专业用语不规范、不完整或声音不洪亮,各扣0.5分。 (11)未复诵许可内容,扣1分。 (12)复诵内容漏项,每项扣0.5分(单位名称、负责人姓名、时间、线路名称、工作完成情况、设备已恢复正常、人员已撤离、可恢复直流再启动装置)。 (13)未及时完善工作票终结手续或填写错误,每项扣1分			
	合计		100				

模块七 带电修补±800kV特高压输电线路分裂导线培训及考核标准

一、培训标准

(一) 培训要求

模块名称	带电修补±800kV特高压输电线路分裂导线	培训类别	操作类
培训方式	实操培训	培训学时	14学时
培训目标	1.掌握沿耐张绝缘子串进、出±800kV电场时采用"跨二短三"的电学意义； 2.能完成沿耐张绝缘子串进入±800kV等电位作业点； 3.能独立完成用预绞式护线条修补导线的操作（等电位作业法）		
培训场地	特高压直流实训线路		
培训内容	采用"跨二短三"方式沿耐张绝缘子串进入电场，采用等电位作业法带电修补±800kV输电线路分裂导线		
适用范围	特高压±800kV输电线路检修人员		

(二) 引用规程规范

(1)《±800kV直流架空输电线路设计规范》（GB/T 50790-2013）

(2)《±800kV直流架空输电线路检修规程》（DL/T 251-2012）

(3)《±800kV直流架空输电线路运行规程》（GB/T 28813-2012）

(4)《±800kV直流线路带电作业技术规范》（DL/T 1242-2013）

(5)《±800kV特高压输电线路金具技术规范》（GB/T 31235-2014）

(6)《国家电网公司带电作业工作管理规定（试行）》（国家电网生〔2007〕751号）

(7)《国家电网公司电力安全工作规程（线路部分）》（Q/GDW1799.2-2013）

(8)《电工术语 架空线路》（GB/T 2900.51-1998）

(9)《电工术语 带电作业》（GB/T 2900.55-2016）

(10)《带电作业工具设备术语》（GB/T 14286-2002）

(11)《带电作业用工具、装置和设备使用的一般要求》（DL/T 877-2004）

(12)《带电作业工具、装置和设备预防性试验规程》（DL/T 976-2005）

特高压直流输电线路运检专业培训及考核标准

（13）《±800kV特高压输电线路带电作业技术导则》（Q/GDW 302-2009）
（14）《带电作业用屏蔽服装》（GB/T 6568-2008）
（15）《带电作业工具基本技术要求与设计导则》（GB 18037-2008）

（三）培训教学设计

本设计以完成"带电修补±800kV特高压输电线路分裂导线"为工作任务，按工作任务完成的标准化作业流程来设计各个培训阶段，每个阶段包括了具体的培训目标、培训内容、培训学时、培训方法（培训资源）、培训环境和考核评价等内容，如表1-7-1所示。

表1-7-1 带电修补±800kV特高压输电线路分裂导线培训内容设计

培训流程	培训目标	培训内容	培训学时	培训方法与资源	培训环境	考核评价
1.理论教学	1.初步掌握沿绝缘子串进出±800kV电场基本方法；2.熟悉电位转移的方法；3.熟悉输电线路受损害导线修补方法	1.沿绝缘子进出电场"跨二短三"电学意义；2.特高压线路进出电场电位转移棒使用方法；3.输电线路导线修补方法和质量标准	2	培训方法：讲授法 培训资源：PPT、相关规程规范	多媒体教室	考勤、课堂提问和作业
2.准备工作	能完成作业前准备工作	1.作业现场查勘；2.编制培训标准化作业卡；3.填写培训操作工作票；4.完成本操作的工器具及材料准备	1	培训方法：1.现场查勘和工器具及材料清理采用现场实操方法；2.编写作业卡和填写工作票采用讲授方法 培训资源：1.特高压实训线路（±800kV实训线路）；2.特高压工器具库房；3.空白工作票	1.特高压输电实训线路；2.多媒体教室	

续表

培训流程	培训目标	培训内容	培训学时	培训方法与资源	培训环境	考核评价
3.作业现场准备	能完成作业现场准备工作	1.作业现场复勘； 2.工作申请； 3.作业现场布置； 4.班前会； 5.工器具检查	1	培训方法：演示与角色扮演法 资源：特高压实训线路（±800kV 实训线路）	特高压实训线路(±800kV 实训线路)	
4.培训师演示	通过现场观摩，使学员初步领会本任务操作流程	1.等电位电工沿耐张绝缘子串进出电场； 2.等电位电工走线方式进入导线修补位置； 3.等电位电工用预绞丝完成导线修补	1	培训方法：演示法 资源：特高压实训线路	特高压实训线路(±800kV 实训线路)	
5.学员分组训练	通过培训： 1.使学员能完成进、出±800kV电场操作； 2.使学员能完成±800kV输电线路导线修补方法	1.学员分组(6人一组)训练进出±800kV输电训练电场和修补导线技能操作； 2.培训师对学员操作进行指导和安全监护	8	培训方法：角色扮演法 资源：特高压实训线路	特高压实训线路(±800kV 实训线路)	采用技能考核评分细则对学员操作评分
6.工作终结	通过培训： 1.使学员进一步认识操作过程不足处，便于后期提升； 2.培训学员安全文明生产的工作作风	1.作业现场清理； 2.向调度汇报工作； 3.班后会，对今天工作任务进行点评总结	1	培训方法：讲授和归纳法	特高压实训线路(±800kV 实训线路)	

(四) 作业流程

1. 工作任务

采用"跨二短三"方法沿耐张绝缘子串进入电场、到达作业点，采用等电位作业法带电修补±800kV特高压输电线路分裂导线。

2. 天气及作业现场要求

（1）带电修补±800kV特高压输电线路分裂导线应在良好的天气进行。

如遇雷电（听见雷声、看见闪电）、雪、雹、雨、雾等，禁止进行带电作业。风力大于5级，或空气相对湿度大于80%时，不宜进行带电作业；恶劣天气下必须开展带电抢修时，应组织有关人员充分讨论并编制必要的安全措施，经本单位批准后方可进行。

（2）作业人员精神状态良好，熟悉工作中保证安全的组织措施和技术措施；应持有在有效期内的带电作业资质证书。

（3）工作负责人应事先组织相关人员完成现场勘察，根据勘察结果确定本次作业方法和所需工器具，以及应采取的必要措施，并办理带电作业工作票。

（4）作业现场应合理设置围栏，并妥当布置警示标示牌，禁止非工作人员入内。

（5）本项目需停用直流再启动装置。

（6）工作中安全距离及有效绝缘长度如表1-7-2所示。

表1-7-2　带电修补±800kV特高压输电线路分裂导线的安全距离（m）

电压等级	人身与带电体安全距离	最小有效绝缘长度		最小组合间隙	转移电位时人体裸露部分与带电体的最小距离
		绝缘操作杆	绝缘承力工具、绝缘绳索		
±800kV	6.8	6.8	6.8	6.6	0.5

（7）在±800kV输电线路上作业，应保证作业相良好绝缘子片数不少于32片。

3. 准备工作

3.1　危险点及其预控措施

（1）危险点——触电伤害

预控措施：

①工作前，工作负责人应与值班调控人员联系，停用线路直流再启动装置，并履行许可手续。

②塔上地电位作业人员登塔前，必须仔细核对线路名称、杆塔编号、相别，确认无误后方可上塔。

③工作中，如遇线路突然停电，作业人员应视其仍然带电。工作负责人应尽快与调控人员联系，值班调控人员未与工作负责人取得联系前不准强送电。

④绝缘工具及绝缘绳索不得损坏、受潮、变形、失灵，不准使用非绝缘绳索（如棉纱绳、白棕绳、钢丝绳）。

⑤等电位作业人员应穿着阻燃内衣，衣服外面应穿戴全套屏蔽服（包括帽、衣、裤、手套、袜和鞋），且各部分应连接良好。

⑥等电位作业人员在电位转移前，应得到工作负责人的许可，人体裸露部分与带电体的最小距离不小于0.5m；电位转移时，动作应迅速，严禁用头部充放电；与地电位作业人员传递工具和材料时，使用绝缘工具或绝缘绳索的有效长度不准小于表1-7-2的规定。

⑦用绝缘绳索传递大件金属物品时，地电位作业人员应将金属物品接地后再接触。

⑧专责监护人应对作业人员进行不间断监护，随时纠正其不规范或违章动作。重点关注高处作业人员，使其保持足够的安全距离（符合表1-7-2的规定），禁止同时接触两个非连通的带电体或带电体与接地体。

（2）危险点——高处坠落

预控措施：

①高处作业人员登高前，必须具备符合本项作业要求的身体状况、精神状态和技能素质。

②监护人员应随时纠正其不规范或违章动作，重点关注作业人员在转位的过程中不得失去安全带或绝缘后备保护绳的保护，严禁低挂高用。

（3）危险点——高处坠物伤人

预控措施：

①高处作业人员的个人工具及零星材料应装入工具袋，严禁在高处浮置物件、口中含物。

②地面作业人员必须正确佩戴安全帽，正确使用绳结，与作业点垂直下方距离不得小于坠落半径。

③作业现场设置围栏并挂好警示标示牌。监护人员应随时注意，禁止非工作人员及车辆进入作业区域。

3.2 工器具及材料选择

带电修补±800kV输电线路分裂导线所需工器具及材料见表1-7-3。工器具出库前，应认真核对工器具的使用电压等级和试验周期，并检查确认外观良好、连接牢固、转动灵活，且符合本次工作任务的要求；工器具出库后，应存放在工具袋或工具箱内进行运输，防止脏污、受潮；金属工具和绝缘工器具应分开装运，防止因混装运输导致工器具变形、损伤等现象发生。

表1-7-3 带电修补±800kV特高压输电线路分裂导线所需工器具及材料表

序号	名称	规格型号	单位	数量	备注
1	绝缘传递绳	TJS-12	根	2	
2	绝缘保护绳	TJS-16	根	2	
3	绝缘滑车	JH10-1	个	2	
4	安全帽		顶	6	
5	电位转移棒		根	1	
6	绝缘电阻表	5000V	块	1	
7	风速风向仪		块	1	
8	温湿度仪		块	1	
9	万用表		块	1	
10	防潮帆布	2m×4m	块	2	
11	绝缘绳套		根	4	
12	屏蔽服	屏蔽效率≥60dB（屏蔽面罩屏蔽效率≥20dB）	套	2	
13	防坠器	与杆塔防坠器装置型号对应	只	2	
14	安全带		副	2	
15	安全围栏		套	若干	
16	警示标示牌	"在此工作""从此进出""从此上下"	套	1	
17	红马甲	"工作负责人"	件	1	
18	预绞式护线条		套	1	
19	导电膏		盒	1	
20	砂纸		张	1	
21	清洁毛巾		条	1	
22	对讲机		台	4	

3.3 作业人员分工

本任务作业人员分工如表1-7-4所示。

表1-7-4 带电修补±800kV特高压输电线路分裂导线人员分工表

序号	工作岗位	数量(人)	工作职责
1	工作负责人	1	负责本次工作任务的人员分工、工作票的宣读、停用直流线路再启动装置、办理工作许可手续、召开工作班前会、工作中突发情况的处理、工作质量的监督、工作后的总结
2	专责监护人	1	负责作业现场的安全把控
3	等电位电工	1	负责进入等电位补修导线工作
4	塔上地电位电工	1	负责协助等电位进出电场
5	地面电工	2	负责传递工具、材料配合等电位电工进出等电位

4. 工作程序

本任务工作流程如表1-7-5所示。

表1-7-5　带电修补±800kV特高压输电线路分裂导线工作流程表

序号	作业内容	作业步骤及标准	安全措施及注意事项	责任人
1	现场复勘	工作负责人负责完成以下工作： (1)现场核对线路名称、杆塔编号,相别无误；基础及杆塔完好无异常；交叉跨越距离符合安全要求；确认缺陷情况及导地线规格型号等。 (2)检测风速、湿度等现场气象条件符合作业要求。 (3)检查地形环境符合作业要求。 (4)检查工作票所列安全措施与现场实际情况相符,必要时予以补充	(1)正确穿戴安全帽、工作服、工作鞋、劳保手套。 (2)不得在危及作业人员安全的气象条件下作业。 (3)严禁非工作人员、车辆进入作业现场	
2	工作许可	(1)工作负责人负责联系值班调控人员,按工作票内容申请停用直流线路再启动装置。 (2)经值班调控人员许可后,方可开始带电作业工作	不得未经值班调控人员许可即开始工作	
3	现场布置	正确装设安全围栏并悬挂标示牌： (1)安全围栏范围应充分考虑高处坠物,以及道路交通的影响。 (2)安全围栏出入口设置合理。 (3)妥当布置"从此进出""在此工作""从此上下"等标示	对道路交通安全影响不可控时,应及时联系交通管理部门强化现场交通安全管控	
4	召开班前会	(1)全体工作成员列队。 (2)工作负责人宣读工作票,明确工作任务及人员分工；讲解工作中的安全措施和技术措施；查(问)全体工作成员精神状态；告知工作中存在的危险点及采取的预控措施。 (3)全体工作成员在工作票上签名确认	(1)工作票填写、签发和许可手续规范,签名完整。 (2)全体工作成员精神状态良好。 (3)全体工作成员明确任务分工、安全措施和技术措施	

续表

序号	作业内容	作业步骤及标准	安全措施及注意事项	责任人
5	检查工具	(1)塔上地电位电工和等电位电工正确地穿戴好屏蔽服并检测合格,由负责人监督检查。 (2)正确佩戴个人安全用具(大小合适,锁扣自如),由负责人监督检查。 (3)测量风速风向、湿度,检查绝缘工具的绝缘性能,并做好记录	(1)金属、绝缘工具使用前,应仔细检查其是否损坏、变形、失灵。绝缘工具应使用2500V及以上绝缘电阻表进行分段绝缘检测,阻值应不低于700MΩ,并用清洁干燥的毛巾将其擦拭干净。 (2)用万用表测量屏蔽服衣裤最远端点之间的电阻值不得大于20Ω。工作负责人认真检查作业电工屏蔽服的连接情况。 (3)检查工具组装情况并确认连接可靠。 (4)现场所使用的带电作业工具应放置在防潮帆布上	
6	登塔	(1)核对线路名称、杆塔编号无误后,塔上地电位电工和等电位冲击检查安全带、防坠器手里情况。 (2)塔上地电位电工携带绝缘传递绳登塔、等电位电工随后登塔,两人至横担作业点,选择合适位置系好安全带,塔上地电位电工将绝缘滑车和绝缘传递绳安装在横担合适位置。然后配合地面电工将绝缘传递绳分开作起吊准备	(1)核对线路名称和杆塔编号无误后,方可登塔作业。 (2)登塔过程中应使用塔上安装的防坠装置;杆塔上移动及转位时,不准失去安全保护,作业人员必须攀抓牢固构件。 (3)作业电工必须穿全套合格的屏蔽服,且全套屏蔽服必须连接可靠。在横担进入等电位前,等电位电工要检查确认屏蔽服个部位连接可靠后方能进行下一步操作	
7	进入强电场	(1)等电位电工将安全带转移到绝缘子连接金具上,并携带电位转移棒、绝缘滑车和绝缘传递绳。 (2)等电位电工检查屏蔽服各部分连接良好后报经工作负责人同意,双手抓扶一串,双脚踩另一串,采用"跨二短三"方法沿绝缘子串进入等电位。 (3)当作业人员平行移动至距导线侧均压环三片绝缘子时,应停止移动,利用电位转移棒进行电位转移	(1)等电位电工进入电位前必须得到工作负责人的许可。 (2)等电位电工进入绝缘子串前必须系好保护绳(用后备保护绳兜住脚踩绝缘子串),并调整好绝缘传递绳和电位转移棒。 (3)等电位电工在进入电位过程中与接地体和带电体两部分间隙所组成的组合间隙不得小于6.6m	

续表

序号	作业内容	作业步骤及标准	安全措施及注意事项	责任人
8	损伤导线表面处理	(1)等电位电工进入等电位后,将安全带系在上子导线上,并装好走线绝缘保护绳(需将子导线全部兜住)。 (2)等电位电工携带绝缘传递绳走线至作业点,将绝缘滑车和绝缘传递绳安装在子导线上。 (3)等电位电工检查导线损伤情况,并对损伤点进行处理,用0号砂纸将损伤部位毛刺打磨平整。 (4)等电位电工用抹布将打磨后的导线表面处理干净,并将带电膏均匀涂抹在导线受伤处	(1)等电位电工对导线损伤点进行打磨处理时,用力不得过大,不得使损伤程度扩大。 (2)导线打磨后,要将表面充分清洁干净。 (3)导电膏均匀涂抹在导线表面	
9	导线修补	(1)地面电工利用传递绳将预绞丝传给等电位电工。 (2)等电位电工利用预绞丝对导线损伤部位进行补强	(1)预绞式护线条的规格型号应与导线匹配。 (2)预绞式护线条的中心应位于损伤最严重处。 (3)预绞式护线条的长度应将损伤部位全部覆盖,且护线条端部距损伤部位边缘的单位长度不得小于100mm。 (4)预绞式护线条绑扎紧密接触,不得抛股、漏股、散股	
10	退出电场	(1)经检查受损带线已补强良好,作业点无遗留物后经工作负责人许可,等电位电工携带绝缘传递绳走线返回均压环处,作退出电位准备。 (2)等电位电工利用电位转移棒钩紧均压环,并进入距均压环的第3片绝缘子,一只手抓紧绝缘子,另一只手握电位转移棒,利用电位转移棒快速脱离电位。 (3)等电位电工按照"跨二短三"的方法退出等电位	(1)等电位电工退出电位前必须得到工作负责人的许可。 (2)等电位电工在退出电位过程中与接地体和带电体两部分间隙所组成的组合间隙不得小于6.6m (3)沿绝缘子串移动时,手要抓牢,脚要踏实	
11	返回地面	塔上电工检查塔上无遗留物后,向工作负责人汇报,得到工作负责人同意后携带绝缘传递绳下塔	下塔过程中应使用塔上安装的防坠装置,杆塔上移动及转位时,不准失去安全保护,作业人员必须攀抓牢固构件	

续表

序号	作业内容	作业步骤及标准	安全措施及注意事项	责任人
12	工作结束	(1)工作负责人组织全体工作成员整理工器具和材料,将工器具清洁后放入专用的箱(袋)中;清理现场,做到"工完料尽场地清"。 (2)召开班后会,工作负责人进行工作总结和点评工作。点评本次工作的施工质量;点评全体工作成员的安全措施落实情况。 (3)工作负责人向值班调控人员汇报工作结束,申请恢复直流线路再启动装置,终结工作票	不得约时恢复直流线路再启动装置	

二、考核标准

表1-7-6 国网四川省电力公司特高压直流技能培训考核评分细则

考生填写栏	编号: 姓名: 所在岗位: 单位: 日期: 年 月 日					
考评员填写栏	成绩: 考评员: 考评组长: 开始时间: 结束时间: 操作时长:					
考核模块	带电修补±800kV特高压输电线路分裂导线	考核对象	特高压±800kV输电线路检修人员	考核方式	操作	考核时限	60min
任务描述	沿耐张绝缘子串进入电场对±800kV受损导线进行带电修补。					
工作规范及要求	1.带电作业工作应在良好天气下进行。如遇雷、雨、雪、雾天气不得进行带电作业。风力大于5级、湿度大于80%时,一般不宜进行带电作业。 2.本项作业需工作负责人1名,专责监护人1人,塔上地电工1人,等电位电工1人,地面辅助电工2人,采用沿绝缘子串进入电场对±800kV受损导线进行带电修补。 3.工作负责人职责:负责本次工作任务的人员分工、工作票的宣读、办理停用直流线路再启动装置、办理工作许可手续、召开工作班前会、工作中突发情况的处理、工作质量的监督、工作后的总结。 4.专责监护人:负责作业现场的安全把控。 5.等电位电工职责:负责沿绝缘子串进入电场对受损导线进行修补。 5.塔上地电工职责:负责协助等电位电工进出电场。 6.地面电工职责:负责传递工具、材料配合等电位电工进出等电位。 7.在带电作业中,如遇雷、雨、大风或其他任何情况威胁到工作人员的安全时,工作负责人或监护人可根据情况,临时停止工作。 给定条件: 1.培训基地:特高压直流±800kV线路杆塔A相6分裂导线某子导线,导线型号:6×JL/G3A-900/40。 2.工作票已办理,安全措施已经完备(直流线路再启动装置已停用),工作开始、工作终结时应口头提出申请(调度或考评员)。 3.安全、正确地使用仪器对绝缘工具进行检测。 4.必须按工作程序进行操作,工序错误扣除应做项目分值,出现重大人身、器材和操作安全隐患,考评员可下令终止操作(考核)					

续表

考核情景准备	1. 线路:特高压直流±800kV线路001#～002#塔A相6分裂导线某子导线,工作内容:带电修补±800kV输电线路分裂导线,导线型号:6×JL/G3A-900/40 2. 所需作业工器具:绝缘传递绳1根(TJS-12),绝缘保护绳(TJS-16),绝缘滑车1个(JH10-1),绝缘检测仪,电位转移棒(1根),绝缘电阻表(5000V型),屏蔽服(屏蔽效率≥60Db)2套,万用表1块,苫布1块,温湿度表、风速仪各1台,纯棉毛巾2条,预绞丝1组,0号砂纸1张,木榔头1把。 3. 作业现场做好监护工作,作业现场安全措施(围栏等)已全部落实;禁止非作业人员进入现场,工作人员进入作业现场必须戴安全帽。 4. 考生自备工作服,阻燃纯棉内衣,安全帽,线手套,安全带(含二保绳)
备注	1. 各项目得分均扣完为止,出现重大人身、器材和操作安全隐患,考评员可下令终止操作。 2. 设备、作业环境、安全带、安全帽、工器具、屏蔽服等不符合作业条件考评员可下令终止操作

表1-7-7 国网四川省电力公司特高压直流技能培训考核评分标准

序号	项目名称	质量要求	分值	扣分标准	扣分原因	扣分	得分
1	现场复勘	(1)工作负责人到作业现场核对线路名称和杆塔编号、现场工作条件、缺陷部位等。 (2)检测风速、湿度等现场气象条件符合作业要求。 (3)检查工作票填写完整,无涂改,检查是否所列安全措施与现场实际情况相符,必要时予以补充	5	(1)未进行核对双重称号,扣1分。 (2)未核实现场工作条件(气象)、缺陷部位,扣1分。 (3)工作票填写出现涂改,每项扣0.5分;工作票编号有误,扣1分;工作票填写不完整,扣1.5分			
2	工作许可	(1)工作负责人联系值班调控人员,按工作票内容申请停用直流线路再启动装置。 (2)汇报内容规范、完整	2	(1)未联系调度部门(裁判)停用直流线路再启动装置,扣2分。 (2)汇报专业用语不规范或不完整的各,扣0.5分			
3	现场布置	正确装设安全围栏并悬挂标示牌: (1)安全围栏范围应充分考虑高处坠物,以及对道路交通的影响。 (2)安全围栏出入口设置合理。 (3)妥当布置"从此进出""在此工作""从此上下"等标示	3	(1)作业现场未装设围栏,扣0.5分。 (2)未设立警示牌,扣0.5分。 (3)未悬挂登塔作业标志,扣0.5分			

续表

序号	项目名称	质量要求	分值	扣分标准	扣分原因	扣分	得分
4	召开班前会	(1)全体工作成员全体人员正确佩戴安全帽、工作服。 (2)工作负责人佩戴红色背心,宣读工作票,明确工作任务及人员分工;讲解工作中的安全措施和技术措施;查(问)全体工作成员精神状态;告知工作中存在的危险点及采取的预控措施。 (3)全体工作成员在工作票上签名确认	3	(1)工作人员着装不整齐,扣0.5分,工作人员着装不整齐每人次,扣0.5分。 (2)未进行分工本项不得分;分工不明,扣1分。 (3)现场工作负责人未穿佩安全监护背心,扣0.5分。 (4)工作票上工作班成员未签字或签字不全的,扣1分。			
5	工器具检查	(1)工作人员按要求将工器具放在防防潮苫布上;防潮苫布应清洁、干燥。 (2)工器具应按定置管理要求分类摆放;绝缘工器具不能与金属工具、材料混放;对工器具进行外观检查。 (3)绝缘工具表面不应磨损、变形损坏,操作应灵活。绝缘工具应使用2500V及以上绝缘电阻表进行分段绝缘检测,阻值应不低于700MΩ,并用清洁干燥的毛巾将其擦拭干净。 (4)塔上地电位和登电位人员按要求正确穿戴全套合格的屏蔽服、导电鞋,各部分连接应良好;屏蔽服内不得贴身穿着化纤类衣服;并系好安全带;工作负责人应认真检查是否穿戴正确。 (5)登塔人员再次核对双重名称、杆号、相别并报告	7	(1)未使用防潮布并定置摆放工器具,扣1分。 (2)未检查工器具试验合格标签及外观未检查,每项扣0.5分。 (3)未正确使用检测仪器对工器具进行检测,每项扣1分。 (4)作业人员未正确穿戴屏蔽服且各部位连接良好,每人次扣2分。 (5)现场工作负责人未对登塔作业人员进安全防护装备进行检查,扣1分。 (6)登塔人员未核对线路双重名称、杆号、相别,每人扣2分。 (7)登塔人员未报告核对结果,每人扣2分			

续表

序号	项目名称	质量要求	分值	扣分标准	扣分原因	扣分	得分
6	登塔	(1)塔上地电位电工、等电位电工穿好全套合格的屏蔽服,将安全带做冲击试验后,系好安全带后携带绝缘传递绳相继登塔。 (2)登塔过程中系好防坠落保护装置,登塔至合适位置,系好安全带,布置好绝缘传递绳,然后配合地面电工将绝缘传递绳分开作起吊准备。 (3)登塔过程中应系好防坠落保护装置,匀速登塔,手抓主材;将安全带挂在肩上并与带电体保持6.8m以上安全距离,工作负责人加强作业监护。	5	(1)未系安全带或安全带及后备保护绳未进行冲击试验,各扣2分。 (2)手抓脚钉,扣2分。 (3)滑车传递绳悬挂位置不便工具取用,扣1分。 (4)传递时金属工具难以保证安全距离,扣2分;工具绑扎不牢,扣2分。 (5)传递时高空落物,扣2分。 (6)传递过程工具与塔身磕碰,扣2分。 (7)传递工具绳索打结混乱,扣1分。 (8)工作负责人监护不到位,扣2分。 (9)塔上电工操作不正确,扣2分			
7	进入强电场	(1)等电位电工进入绝缘子串前必须系好保护绳(用后备保护绳兜住脚踩绝缘子串),并调整好绝缘传递绳和电位转移棒。 (2)等电位电工检查屏蔽服各部分连接良好后报经工作负责人同意,双手抓扶一串,双脚踩另一串,采用"跨二短三"方法沿绝缘子串进入电场。 (3)等电位电工在进入电位过程中与接地体和带电体两部分间隙所组成的组合间隙不得小于6.6m,进入电场必须用电位转移棒进行电位转移	8	(1)等电位电工电位转移过程中裸露部分距离不够,扣2分。 (2)等电位电工进入电场动作不正确,反复放电,每次扣2分。 (3)转移电位动作不熟练扣1分,电位转移过程未使用电位转移棒的,扣5分。 (4)未得到工作负责人许可就进行电位转移的,扣5分			
8	损伤导线表面处理	(1)等电位电工携带绝缘传递绳走线至作业点,将绝缘滑车和绝缘传递绳安装在子导线上。 (2)等电位电工检查导线损伤情况,并对损伤点进行处理,用0号砂纸将损伤部位毛刺打磨平整。 (3)等电位电工用抹布将打磨后的导线表面处理干净,并将带电膏均匀涂抹在导线受伤处表面	12	(1)未将子导线用绝缘绳索全部箍住的,扣2分。 (2)未向工作负责人汇报损伤情况,扣2分。 (3)对受损导线未处理平整,扣2分。 (4)不清除表面氧化物或未涂抹导电膏的,各扣2分			

续表

序号	项目名称	质量要求	分值	扣分标准	扣分原因	扣分	得分
9	导线修补	(1)地面电工利用传递绳将预绞丝传给等电位电工。等电位电工利用预绞丝对导线损伤部位进行补强。 (2)预绞式护线条的规格型号应与导线匹配，护线条的中心应位于损伤最严重处。 (3)预绞式护线条的长度应将损伤部位全部覆盖，且护线条端部距损伤部位边缘的单位长度不得小于100mm。 (4)预绞式护线条绑扎紧密接触，不得抛股、漏股、散股	30	(1)补修中心误差超过5mm，扣3分。 (2)预绞丝安装一处缝隙，扣2分。 (3)若因操作不当致使预绞丝变形扣8分。 (4)端头一根不平齐，扣1分。 (5)预绞式护线条绑扎不紧密，出现抛股、漏股、散股，每处扣0.5分			
10	退出电场	(1)经检查受损带线已补强良好，作业点无遗留物后经工作负责人许可，等电位电工携带绝缘传递绳走线返回均压环处，作退出电位准备。 (2)等电位电工利用电位转移棒钩紧均压环，并进入距均压环的第3片绝缘子，一只手抓紧绝缘子，另一只手握电位转移棒，利用电位转移棒快速脱离电位。 (3)等电位电工按照"跨二短三"的方法退出等电位	10	(1)未向工作负责人申请即进行电位转移扣2分；申请了但未得同意即开始，扣1分。 (2)等电位电工电位转移过程中裸露部分距离不够，扣2分。 (3)等电位电工退出电场动作不正确，反复放电，扣2分。 (4)转移电位动作不熟练扣2分；电位转移未使用电位转移杆，扣3分			
11	返回地面	塔上电工检查塔上无遗留物后，向工作负责人汇报，得到工作负责人同意后携带绝缘传递绳下塔	5	(1)下塔过程未使用防坠装置，扣2分。 (2)塔上移位失去安全带保护的，扣2分。 (3)下塔抓塔钉，每处扣1分。 (4)塔上有遗留物的，扣2分			
12	工作结束	(1)工作负责人组织全体工作成员整理工器具和材料，将工器具清洁后放入专用的箱(袋)中；清理现场，做到"工完料尽场地清"。 (2)召开班后会，工作负责人进行工作总结和点评工作。点评本次工作的施工质量；点评全体工作成员的安全措施落实情况。 (3)工作负责人向值班调控人员汇报工作结束，申请恢复直流线路再启动装置，终结工作票	10	(1)工器具未清理，扣2分。 (2)工器具有遗漏，扣2分。 (3)未开班后会，不得2分。 (4)未拆除围栏，扣2分。 (5)未向调度汇报，不得2分			
	合计		100				

模块八 带电处理±800kV特高压输电线路导线引流板发热缺陷培训及考核标准

一、培训标准

(一) 培训要求

模块名称	带电处理±800kV特高压输电线路导线引流板发热缺陷	培训类别	操作类
培训方式	实操培训	培训学时	14学时
培训目标	1.掌握采用"跨二短三"的作业方式沿耐张绝缘子串进、出±800kV强电场。 2.能完成沿耐张绝缘子串进入±800kV等电位作业点。 3.能独立完成采用等电位作业法带电处理±800kV特高压输电线路导线引流板发热缺陷(等电位作业法)		
培训场地	特高压直流实训线路		
培训内容	采用"跨二短三"作业方式沿耐张绝缘子串进入强电场,采用等电位作业法带电处理±800kV特高压输电线路导线引流板发热缺陷		
适用范围	特高压直流输电线路带电检修人员		

(二) 引用规程规范

(1)《±800kV直流架空输电线路设计规范》(GB/T 50790-2013)

(2)《±800kV直流架空输电线路检修规程》(DL/T 251-2012)

(3)《±800kV直流架空输电线路运行规程》(GB/T 28813-2012)

(4)《±800kV直流线路带电作业技术规范》(DL/T 1242-2013)

(5)《±800kV特高压输电线路金具技术规范》(GB/T 31235-2014)

(6)《国家电网公司带电作业工作管理规定(试行)》(国家电网生〔2007〕751号)

(7)《国家电网公司电力安全工作规程(线路部分)》(Q/GDW1799.2-2013)

(8)《电工术语 架空线路》(GB/T 2900.51-1998)

(9)《电工术语 带电作业》(GB/T 2900.55-2016)

(10)《带电作业工具设备术语》(GB/T 14286-2002)

(11)《带电作业用绝缘滑车》(GB/T 13034-2008)

(12)《带电作业用绝缘绳索》(GB 13035-2008)

(13)《带电作业用工具、装置和设备使用的一般要求》(DL/T 877-2004)

(14)《带电作业工具、装置和设备预防性试验规程》(DL/T 976-2005)

(15)《±800kV 特高压输电线路带电作业技术导则》(Q/GDW 302-2009)

(16)《带电作业用屏蔽服装》(GB/T 6568-2008)

(17)《带电作业工具基本技术要求与设计导则》(GB 18037-2008)

(18)《带电设备红外线诊断应用规范》(DL/T 664-2016)

(三)培训教学设计

本设计以完成"带电处理±800kV 特高压输电线路导线引流板发热缺陷"为工作任务,按工作任务完成的标准化作业流程来设计各个培训阶段,每个阶段包括了具体的培训目标、培训内容、培训学时、培训方法(培训资源)、培训环境和考核评价等内容,如表1-8-1所示。

表1-8-1 带电处理±800kV 特高压输电线路导线引流板发热缺陷培训内容设计

培训流程	培训目标	培训内容	培训学时	培训方法与资源	培训环境	考核评价
1.理论教学	1.初步掌握沿绝缘子串进出±800kV直流强电场基本方法。2.熟悉电位转移的方法。3.熟悉输电线路导线引流板发热缺陷的处理方法	1.采用"跨二短三"的方式沿绝缘子串进、出强电场。2.进、出特高压强电场时电位转移棒的使用方法。3.输电线路导线引流板发热缺陷的处理方法和质量标准	2	培训方法:讲授法。培训资源:PPT、相关规程规范	多媒体教室	考勤、课堂提问和作业
2.准备工作	能完成作业前准备工作	1.作业现场查勘。2.编制培训标准化作业卡。3.填写培训操作工作票。4.完成本操作的工器具及材料准备。5.值班调控人员联系申请停用工作线路重合闸装置	1	培训方法:1.现场查勘和工器具及材料清理采用现场实操方法。2.编写作业卡和填写工作票采用讲授方法。培训资源:1.±800kV实训线路。2.特高压工器具库房。3.空白工作票	1.特高压输电实训线路;2.多媒体教室	

续表

培训流程	培训目标	培训内容	培训学时	培训方法与资源	培训环境	考核评价
3.作业现场准备	能完成作业现场准备工作	1.重合闸已装置停用,得到调度许可。2.作业现场复勘。3.作业现场布置。4.班前会。5.工器具及材料检查	1	培训方法:演示与角色扮演法。资源:±800kV实训线路	±800kV实训线路	
4.培训师演示	通过现场观摩,使学员初步领会本任务操作流程	1.等电位电工沿耐张绝缘子串进、出强电场,到达引流线连接处。2.等电位电工用套筒扳手对导线引流线螺栓进行紧固	2	培训方法:演示法。资源:±800kV实训线路	±800kV实训线路	
5.学员分组训练	1.能完成进、出±800kV强电场操作。2.能完成±800kV输电线路导线引流线发热带电处理	1.学员分组(6人一组)训练进、出±800kV强电场和导线引流线发热处理技能操作。2.培训师对学员操作进行指导和安全监护	7	培训方法:角色扮演法。资源:±800kV实训线路	±800kV实训线路	采用技能考核评分细则对学员操作评分
6.工作终结	1.使学员进一步辨析操作过程不足之外,便于后期提升。2.培养学员安全文明生产的工作作风	1.作业现场清理。2.向调度汇报工作结束并申请恢复重合闸装置。3.班后会,对本次工作任务进行点评总结	1	培训方法:讲授和归纳法	±800kV实训线路	

(四)作业流程

1. 工作任务

采用"跨二短三"的作业方式沿绝缘子串进入强电场到达作业点,采用等电位作业法带电处理±800kV特高压输电线路导线引流板发热缺陷。

2. 天气及作业现场要求

(1)带电处理±800kV特高压输电线路导线引流板发热缺陷应在良好的天气进行。

如遇雷电（听见雷声、看见闪电）、雪、雹、雨、雾等，不应进行带电作业。风力大于5级时，不宜进行带电作业；相对湿度大于80%的天气，如需进行带电作业应采用具有防潮性能的绝缘工具；恶劣天气下必须开展带电抢修时，应组织有关人员充分讨论并编制必要的安全措施，经本单位批准后方可进行。

（2）作业人员精神状态良好，熟悉工作中保证安全的组织措施和技术措施；应持有在有效期内的带电作业资质证书。

（3）工作负责人应事先组织相关人员完成现场勘察，根据勘察结果确定本次作业方法和所需工器具，以及应采取的必要措施，并办理带电作业工作票。

（4）作业现场应合理设置围栏，并妥当布置警示标示牌，禁止非工作人员入内。

（5）本项目需停用直流再启动装置。

（6）工作中安全距离及有效绝缘长度如表1-8-2所示。

（7）在±800kV输电线路上作业，应保证作业相良好绝缘子片数不少于32片。

表1-8-2 带电补修±800kV特高压输电线路导线的安全距离（m）

海拔高度	等电位电工与接地构架之间的最小安全距离	绝缘工器具的最小有效绝缘长度	最小组合间隙
$H \leqslant 1000$	6.8	6.8	6.7
$1000 < H \leqslant 2000$	7.3	7.3	7.3
$2000 < H \leqslant 2500$	7.9	7.8	7.8

注：表中最小安全距离包括人体占位间隙0.5m。

（8）中间电位作业人员沿耐张绝缘子串进入±800kV强电场时，人体短接绝缘子片数不得多于4片。耐张绝缘子串中扣除人体短接和不良绝缘子片数后，良好绝缘子最少片数应满足表1-8-3的规定。

表1-8-3 最小组合间隙和良好绝缘子的最小片数

海拔高度（m）	单片玻璃绝缘子结构高度（mm）	良好绝缘子串的总长度最小值(m)	良好绝缘子的最少片数
$H \leqslant 1000$	170	6.2	37
	195		32
	205		31
	240		26
$1000 < H \leqslant 2000$	170	7.1	42
	195		37
	205		35
	240		30

续表

海拔高度（m）	单片玻璃绝缘子结构高度（mm）	良好绝缘子串的总长度最小值(m)	良好绝缘子的最少片数
2000＜H≤2500	170	7.55	45
	195		39
	205		37
	240		32

注：表中数值不包括人体占位间隙，作业中需考虑人体占位间隙不得小于0.5m。

3. 准备工作

3.1 危险点及其预控措施

（1）危险点——触电伤害

预控措施：

①工作前，工作负责人应与值班调控人员联系，停用线路直流再启动装置，并履行许可手续。

②塔上地电位作业人员登塔前，必须仔细核对线路名称，确认无误后方可上塔。

③工作中，如遇线路突然停电，作业人员应视其仍然带电。工作负责人应尽快与调控人员联系，值班调控人员未与工作负责人取得联系前不准强送电。

④绝缘工具及绝缘绳索不得损坏、受潮、变形、失灵，不准使用非绝缘绳索（如棉纱绳、白棕绳、钢丝绳）。

⑤等电位作业人员应穿着阻燃内衣，衣服外面应穿戴全套合格的屏蔽服（包括帽、衣裤、手套、袜、面罩和鞋），且各部分应连接良好。

⑥等电位作业人员在电位转移前，应得到工作负责人的许可，人体裸露部分与带电体的最小距离不小于0.5m；电位转移时，动作应迅速，严禁用头部充放电；与地电位作业人员传递工具和材料时，使用绝缘工具或绝缘绳索的有效长度不准小于表1-8-2的规定。

⑦监护人、工作负责人应对作业人员进行不间断监护，随时纠正其不规范动作和行为。重点监护高处作业人员，使其保持足够的安全距离（符合表1-8-2的规定），禁止同时接触两个非连通的带电体或带电体与接地体。

（2）危险点——高处坠落

①高处作业人员登高前，必须具备符合本项作业要求的身体状况、精神状态和技能素质。

②等电位作业人员进入强电场过程中应手脚一致，速度均匀，动作平稳；移动全过程应使用人体后备保护绳。

③监护人员应随时纠正其不规范动作和行为，重点监护作业人员在转位的过程中不得失去安全带或绝缘后备保护绳的保护，严禁低挂高用。

（3）危险点——高处坠物伤人

预控措施：

①高处作业人员的个人工具及零星材料应装入工具袋，严禁在高处浮置物件、口中含物。

②地面作业人员必须正确佩戴安全帽，正确使用绳结，与作业点正下方距离不得小于坠落半径。

③作业现场设置围栏并挂好警示标示牌。监护人员应随时注意，禁止非工作人员及车辆进入作业区域。

3.2 工器具及材料选择

带电处理±800kV特高压输电线路导线引流板发热缺陷所需工器具及材料见表1-8-4。工器具出库前，应认真核对工器具的使用电压等级和试验周期，并检查确认外观良好、连接牢固、转动灵活，且符合本次工作任务的要求；工器具出库后，应存放在工具袋或工具箱内进行运输，防止脏污、受潮；金属工具和绝缘工器具应分开装运，防止因混装运输导致工器具变形、损伤等现象发生。

表1-8-4 带电处理±800kV特高压输电线路导线引流板发热缺陷所需工器具及材料表

序号	名称	规格型号	单位	数量	备注
1	绝缘传递绳	Φ12	根	2	
2	绝缘后备保护绳	Φ16	根	2	
3	绝缘滑车	1T	个	2	
4	套筒扳手	与引流板螺栓型号一致	套	1	
5	电位转移棒		根	1	
6	绝缘电阻测试仪	5000V	台	1	
7	风速风向仪		台	1	
8	温湿度仪		台	1	
9	万用表		台	1	
10	防潮帆布	2m×4m	张	1	
11	屏蔽服	屏蔽效率≥60dB（屏蔽面罩屏蔽效率≥20dB）	套	2	
12	防坠器	与铁塔防坠装置型号对应	只	2	
13	安全围网		套	若干	

续表

序号			套	数量	备注
14	警示标示牌	"在此工作""从此进出""从此上下"	套	1	
15	红马甲	"工作负责人""专责监护人"	件	2	
16	螺栓	与引流板螺栓型号一致	个	若干	备用
17	背负式安全带	带后备保护绳	套	2	
18	清洁毛巾		条	1	
19	绝缘绳套	Φ12	根	2	
20	安全帽		顶	6	
21	工具包		个	2	
22	对讲机		个	4	

3.3 作业人员分工

本任务作业人员分工如表1-8-5所示。

表1-8-5　带电处理±800kV特高压输电线路导线引流板发热缺陷人员分工表

序号	工作岗位	数量(人)	工作职责
1	工作负责人	1	负责组织作业现场的各项工作
2	专责监护人	1	负责作业现场的安全监护
3	等电位电工	1	负责进入等电位处理导线引流板发热缺陷
4	地电位电工	1	负责传递工具、材料配合等电位电工进、出等电位
5	地面电工	2	负责传递工具、材料配合等电位电工进、出等电位

4. 工作程序

本任务工作流程如表1-8-6所示。

表1-8-6　带电处理±800kV特高压输电线路导线引流板发热缺陷工作流程表

序号	作业内容	作业步骤及标准	安全措施及注意事项	责任人
1	现场复勘	工作负责人负责完成以下工作： (1)现场核对线路名称无误；基础及铁塔完好无异常；交叉跨越距离符合安全要求；确认缺陷情况及导地线规格型号等。 (2)检查地形环境符合作业要求。 (3)检查工作票所列安全措施与现场实际情况相符，必要时予以补充	(1)正确穿戴安全帽、工作服、工作鞋、劳保手套。 (2)不得在危及作业人员安全的气象条件下作业。 (3)严禁非工作人员、车辆进入作业现场	

续表

序号	作业内容	作业步骤及标准	安全措施及注意事项	责任人
2	工作许可	(1)工作负责人负责联系值班调控人员,按工作票内容申请停用线路重合闸。 (2)经值班调控人员许可后,方可开始带电作业工作	不得未经值班调控人员许可即开始工作	
3	现场布置	正确装设安全围栏并悬挂标示牌: (1)安全围栏范围应充分考虑高处坠物,以及对道路交通的影响。 (2)安全围栏出入口设置合理。 (3)妥当布置"从此进出""在此工作""从此上下"等标示	对道路交通安全影响不可控时,应及时联系交通管理部门强化现场交通安全管控	
4	召开班前会	(1)全体工作成员列队。 (2)工作负责人宣读工作票,明确工作任务及人员分工;讲解工作中的安全措施和技术措施;查(问)全体工作成员精神状态;告知工作中存在的危险点及采取的预控措施。 (3)全体工作成员在工作票上签名确认	(1)工作票填写、签发和许可手续规范,签名完整。 (2)全体工作成员精神状态良好。 (3)全体工作成员明确任务分工、安全措施和技术措施	
5	检查工具	(1)等电位电工及地电位电工正确地穿戴好屏蔽服并检测合格,由负责人监督检查。 (2)正确佩戴个人安全用具(大小合适,锁扣自如),由负责人监督检查。 (3)测量风速、湿度,检查绝缘工具的绝缘性能,并做好记录	(1)金属、绝缘工具使用前,应仔细检查其是否损坏、变形、失灵。绝缘工具应使用5000V绝缘电阻表进行分段绝缘检测,阻值应不低于700MΩ,并用清洁干燥的毛巾将其擦拭干净。 (2)用万用表测量屏蔽服衣裤最远端点之间的电阻值不得大于20Ω。工作负责人认真检查作业电工屏蔽服的连接情况。 (3)检查工具组装情况并确认连接可靠。 (4)现场所使用的带电作业工器具应放置在防潮帆布上	

续表

序号	作业内容	作业步骤及标准	安全措施及注意事项	责任人
6	登塔	(1)核对线路名称无误后,塔上电工对安全带、防坠器做冲击试验检查。 (2)塔上电工携带绝缘传递绳登塔至横担作业点,选择合适位置系好安全带,将绝缘滑车和绝缘传递绳安装在横担合适位置。然后配合地面电工将绝缘传递绳分开作起吊准备	(1)核对线路名称无误后,方可登塔作业。 (2)登塔过程中应使用塔上安装的防坠装置;塔上移动及转位时,不准失去安全保护,作业人员必须攀抓牢固构件。 (3)作业电工必须穿全套合格的屏蔽服,且全套屏蔽必须连接可靠。在进入等电位前,等电位电工要检查确认屏蔽服个部位连接可靠后方能进行下一步操作	
7	进入强电场	(1)地面电工将电位转移棒传至塔上。 (2)等电位电工携带电位转移棒移动到绝缘子连接金具上,并带好绝缘滑车和绝缘传递绳。 (3)等电位电工检查屏蔽服各部分连接良好后报经工作负责人同意,双手抓扶一串,双脚踩另一串,采用"跨二短三"方法沿绝缘子串进入强电场。 (4)当等电位电工平行移动至距导线侧均压环三片绝缘子时,应停止移动,利用电位转移棒进行电位转移	(1)等电位电工进入电位前必须得到工作负责人的许可。 (2)防止安全带、绝缘传递绳钩挂塔材。 (3)等电位电工进入绝缘子串前必须系好后备保护绳,并调整好绝缘传递绳。 (4)转位时,不得失去安全带的保护。 (5)人体与带电体之间的安全距离,人体与接地体和带电体间的组合间隙不得小于表2-8-2的规定	
8	处理导线引流板发热缺陷	(1)等电位电工进入等电位后,将安全带系在子导线上,地面电工利用传递绳将套筒扳手传给等电位电工。 (2)等电位电工利用套筒扳手对导线引流板连接螺栓进行紧固。 (3)等电位电工利用传递绳将套筒扳手传递至塔下	(1)等电位电工注意避免动作幅度过大,避免将肢体伸向绝缘子。 (2)等电位电工注意避免被烫伤	
9	退出电位	(1)经检查作业点无遗留物后经工作负责人许可,等电位电工带好绝缘传递绳,作退出电位准备。 (2)等电位电工利用电位转移棒钩紧均压环,并进入距均压环的第3片绝缘子,一只手抓紧绝缘子,另一只手握电位转移棒,利用电位转移棒快速脱离电位。 (3)等电位电工按照"跨二短三"的方法退出等电位	(1)等电位电工退出电位前必须得到工作负责人的许可。 (2)等电位电工在退出电位过程中与接地体和带电体两部分间隙所组成的组合间隙不得小于6.7m。 (3)沿绝缘子串移动时,手要抓牢,脚要踏实	

二、考核标准

表1-8-7 特高压直流输电线路运检技能考核评分细则

考生填写栏	编号：		姓名：		所在岗位：		单位：		日期：	年 月 日
考评员填写栏	成绩：		考评员：		考评组长：		开始时间：	结束时间：		操作时长：
考核模块	带电处理±800kV特高压输电线路导线引流板发热缺陷			考核对象		特高压直流输电线路检修人员		考核方式	操作	考核时限	60min
任务描述	采用"跨二短三"沿绝缘子串进强电场带电处理±800kV特高压输电线路导线引流板发热缺陷									
工作规范及要求	1. 带电作业工作应在良好天气下进行。如遇雷、雨、雪、雾天气不得进行带电作业。风力大于5级、湿度大于80%时，一般不宜进行带电作业。 2. 本项作业需6人，其中工作负责人1名，专责监护人1名，等电位电工1名，地电位电工1名，地面电工2名，采用耐张串沿绝缘子串进出强电场法带电处理±800kV特高压输电线路导线引流板发热缺陷。 3. 工作负责(监护)人职责：负责本次工作任务的人员分工、工作票的宣读、办理线路停用重合闸、办理工作许可手续、召开工作班前会、负责作业过程中的安全监督、工作中突发情况的处理、工作质量的监督、工作后的总结。 4. 等电位电工职责：负责本次作业过程中的主要作业，根据作业的位置安装、拆除作业工器具，进入等电位处理导线引流板发热缺陷。 5. 地电位电工职责：负责传递工具、材料配合等电位电工进出等电位。 6. 地面电工职责：负责传递工具、材料。 7. 在带电作业中，如遇雷、雨、大风或其他任何情况威胁到工作人员的安全时，工作负责人或监护人可根据情况，临时停止工作。 给定条件： 1. ±800kV实训线路耐张塔一边相导线。 2. 工作票已办理，安全措施已经完备(重合闸已停用)，工作开始、工作终结时应口头提出申请(调度或考评员)。 3. 安全、正确地使用仪器对绝缘工具进行检测。 4. 必须按工作程序进行操作，工序错误扣除应做项目分值，出现重大人身、器材和操作安全隐患，考评员可下令终止操作(考核)									
考核情景准备	1. 塔形：±800kV实训线路耐张塔一边相导线，工作内容：带电处理±800kV特高压输电线路导线引流板发热缺陷。 2. 所需作业工器具：Φ12绝缘传递绳2根、Φ16绝缘保护绳2根、1T绝缘滑车2个、各不同型号套筒扳手1套、电位转移棒1根、5000V绝缘电阻测试仪1台、风速风向仪1台、温湿度仪1台、万用表1台、2m×4m防潮帆布1张、Ⅱ型屏蔽服2套、防坠器2只、安全围网1套、警示标示牌1套、红马甲2件、各不同型号螺栓若干、各不同型号销子若干、双保险安全带2套、清洁毛巾1条、Φ12绝缘绳套2根、安全帽6顶、操作杆1根、工具包2个、对讲机4个。 3. 作业现场做好监护工作，作业现场安全措施(围栏等)已全部落实；禁止非作业人员进入现场，工作人员进入作业现场必须戴安全帽。 4. 考生自备工作服，阻燃纯棉内衣，安全帽，线手套，安全带(含二保绳)。									
备注	1. 各项目得分均扣完为止，出现重大人身、器材和操作安全隐患，考评员可下令终止操作 2. 设备、作业环境、安全带、安全帽、工器具、屏蔽服等不符合作业条件考评员可下令终止操作									

模块九　带电更换±800kV特高压输电线路接地极直线整串绝缘子培训及考核标准

一、培训标准

(一) 培训要求

模块名称	带电更换±800kV特高压输电线路接地极直线整串绝缘子	培训类别	操作类
培训方式	实操培训	培训学时	14学时
培训目标	1.掌握±800kV直流接地极直线塔进、出电场时采用"软梯法"作业方式的电学意义。 2.能完成采用"软梯法"进入±800kV直流接地极线路等电位作业点。 3.能独立完成±800kV特高压线路直流接地极线路直线整串绝缘子更换的操作（等电位作业法）		
培训场地	特高压直流实训线路		
培训内容	采用从地面攀爬软梯进入电场，采用等电位作业法带电更换±800kV特高压输电线路直流接地极线路直线整串绝缘子		
适用范围	特高压±800kV直流输电线路检修人员		

(二) 引用规程规范

(1)《±800kV直流架空输电线路设计规范》（GB/T 50790 2013）

(2)《±800kV直流架空输电线路检修规程》（DL/T 251-2012）

(3)《±800kV直流架空输电线路运行规程》（GB/T 28813-2012）

(4)《±800kV直流线路带电作业技术规范》（DL/T 1242-2013）

(5)《±800kV特高压输电线路金具技术规范》（GB/T 31235-2014）

(6)《国家电网公司带电作业工作管理规定（试行）》（国家电网生〔2007〕751号）

(7)《国家电网公司电力安全工作规程（线路部分）》（Q/GDW1799.2-2013）

(8)《电工术语　架空线路》（GB/T 2900.51-1998）

(9)《电工术语　带电作业》（GB/T 2900.55-2016）

(10)《带电作业工具设备术语》（GB/T 14286-2002）

(11)《带电作业用绝缘滑车》（GB/T 13034-2008）

(12)《带电作业用绝缘绳索》（GB 13035-2008）

(13)《带电作业用工具、装置和设备使用的一般要求》(DL/T 877-2004)

(14)《带电作业工具、装置和设备预防性试验规程》(DL/T 976-2005)

(15)《±800kV 特高压输电线路带电作业技术导则》(Q/GDW 302-2009)

(16)《带电作业用屏蔽服装》(GB/T 6568-2008)

(17)《带电作业工具基本技术要求与设计导则》(GB 18037-2008)

(18)《带电设备红外线诊断应用规范》(DL/T 664-2016)

(三)培训教学设计

本设计以完成"带电更换±800kV 特高压输电线路接地极直线整串绝缘子"为工作任务,按工作任务完成的标准化作业流程来设计各个培训阶段,每个阶段包括了具体的培训目标、培训内容、培训学时、培训方法(培训资源)、培训环境和考核评价等内容,如表1-9-1所示。

表1-9-1 带电更换±800kV 特高压输电线路接地极直线整串绝缘子培训内容设计

培训流程	培训目标	培训内容	培训学时	培训方法与资源	培训环境	考核评价
1.理论教学	1.初步掌握攀爬绝缘软梯进出±800kV 直流接地极线路电场的方法。 2.熟悉工器具配置及安全注意事项。 3.熟悉±800kV 直流接地极线路直线整串绝缘子更换的方法	1.沿绝缘软梯进出±800kV 直流接地极线路电场的电学意义。 2.工器具配置、危险点分析及安全措施。 3.±800kV 直流接地极线路直线整串绝缘子更换方法和质量标准	2	培训方法:讲授法。 培训资源:PPT、相关规程规范	多媒体教室	考勤、课堂提问和作业
2.准备工作	能完成作业前准备工作	1.作业现场查勘。 2.编制培训标准化作业卡。 3.填写培训操作工作票。 4.完成本操作的工器具及材料准备	1	培训方法: 1.现场查勘和工器具及材料清理采用现场实操方法。 2.编写作业卡和填写工作票采用讲授方法。 培训资源: 1.±800kV 直流接地极实训线路。 2.特高压工器具库房。 3.空白工作票	1.±800kV 直流接地极实训线路; 2.多媒体教室	

续表

培训流程	培训目标	培训内容	培训学时	培训方法与资源	培训环境	考核评价
3.作业现场准备	能完成作业现场准备工作	1.作业现场复勘。 2.工作申请。 3.作业现场布置。 4.班前会。 5.工器具检查	1	培训方法： 演示与角色扮演法。 资源： ±800kV直流接地极实训线路	±800kV直流接地极实训线路	
4.培训师演示	通过现场观摩，使学员初步领会本任务操作流程	1.作业距离测量、导线温度测量。 2.地电位电工攀登铁塔至工作点位安装绝缘软梯。 3.等电位电工攀爬绝缘软梯进入电场。 4.地电位电工安装工器具、布置绝缘子串吊点及导线保护绳。 5.等电位电工安装导线端提线工具，拆除、恢复导线端销子及连接。	3	培训方法： 演示法。 资源： 特高压实训线路	±800kV直流接地极实训线路	
5.学员分组训练	通过培训： 1.使学员能完成进、出±800kV直流接地极线路电场操作。 2.使学员能完成±800kV直流接地极线路直线整串绝缘子更换方法	1.学员分组（6人一组）训练进出±800kV直流接地极线路直线整串绝缘子更换技能操作。 2.培训师对学员操作进行指导和安全监护	8	培训方法： 角色扮演法。 资源： ±800kV直流接地极实训线路	±800kV直流接地极实训线路	采用技能考核评分细则对学员操作评分
6.工作终结	通过培训： 1.使学员进一步认识操作过程不足处，便于后期提升。 2.培训学员安全文明生产的工作作风	1.作业现场清理。 2.向调控汇报工作。 3.班后会，对今天工作任务进行点评总结	1	培训方法： 讲授和归纳法	±800kV直流接地极实训线路	

(四) 作业流程

1. 工作任务

采用攀爬软梯方法沿软梯进入电场、到达作业点，采用等电位作业法带电更换±800kV特高压输电线路接地极直线整串绝缘子。

2. 天气及作业现场要求

（1）带电更换±800kV特高压输电线路接地极直线整串绝缘子应在良好的天气下进行。

工作现场如遇雷电（听见雷声、看见闪电）、雪、雹、雨、雾等情况，应禁止本次作业。风力大于5级，或空气相对湿度大于80%时，不宜进行本次作业；作业过程中，注意天气变化，如遇大风、雨雾等紧急情况，按照规程规定正确采取措施，保证人员和设备安全。

（2）作业人员精神状态良好，熟悉工作中保证安全的组织措施和技术措施；应持有在有效期内的带电作业资质证书。

（3）工作负责人应事先组织相关人员完成现场勘察，根据勘察结果确定本次作业方法和所需工器具，以及应采取的必要措施，并办理带电作业工作票。

（4）作业现场应合理设置围栏，并妥当布置警示标示牌，禁止非工作人员入内。

（5）本项目需停用线路直流再启动装置。

（6）工作中安全距离及有效绝缘长度如表1-9-2所示。

表1-9-2　带电更换±800kV直流接地极线路直线整串绝缘子的安全距离（m）

电压等级	人身与带电体安全距离	最小有效绝缘长度		最小组合间隙	转移电位时人体裸露部分与带电体的最小距离
		绝缘操作杆	绝缘承力工具、绝缘绳索		
±800kV直流接地极	0.75	1.3	1.0	1.2	0.3

3. 准备工作

3.1　危险点及其预控措施

（1）危险点——预防系统过电压

预控措施：本次作业为等电位作业，工作负责人在工作开始前，应与值班调控员联系，申请办理电力线路带电作业工作票，并申请停用直流再启动，由调控值班员履行许可手续。带电作业结束后应及时向调控值班员汇报。严禁约时停用和恢复直流再启动。在带电作业过程中如遇设备突然停电，作业人员应视设备仍然带电。工作负责人应尽快与调控联系，调控值班员未与工作负责人取得联系前不得强送电。

（2）危险点——预防人身触电

预控措施：

①绝缘工具及绝缘绳索不得损坏、受潮、变形、失灵，不准使用非绝缘绳索（如棉纱绳、白棕绳、钢丝绳）。

②等电位作业人员应穿着阻燃内衣，衣服外面应穿戴全套屏蔽服（包括帽、衣裤、手套、袜和鞋），且各部分应连接良好，全套屏蔽服电阻不大于20Ω。

③等电位作业人员在电位转移前，应得到工作负责人的许可，人体裸露部分与带电体的最小距离不小于0.3m；地电位作业人员与带电体（等电位作业人员与接地体）的安全距离小于表1-9-2的规定。

④用绝缘绳索传递大件金属物品时，地电位作业人员应将金属物品接地后再接触。

⑤工作负责人和专责监护人应对作业人员进行不间断监护，随时纠正其不规范或违章动作。重点关注高处作业人员，使其保持足够的安全距离（符合表1-9-2的规定）。

⑥当耐热导线运行温度高于40℃时，应采用耐高温带电作业工器具，包括耐高温软质绝缘工器具、耐高温硬质绝缘工器具、耐高温隔热屏蔽服等，需采用电位转移棒进入等电位。

⑦隔热防护用具应视为导体，传递和使用过程中应保证相地及极间安全距离满足安全要求。

（3）危险点——高处坠落

预控措施：

①高处作业人员登高前，必须具备符合本项作业要求的身体状况、精神状态和技能素质。

②监护人员应随时纠正其不规范或违章动作，重点关注作业人员在转位的过程中不得失去安全带或绝缘后备保护绳的保护，严禁低挂高用。

（4）危险点——高处坠物伤人

预控措施：

①高处作业人员的个人工具及零星材料应装入工具袋，严禁在高处浮置物件、口中含物。

②地面作业人员必须正确佩戴安全帽，正确使用绳结，与作业点垂直下方距离不得小于坠落半径。

③作业现场设置围栏并挂好警示标示牌。监护人员应随时注意，禁止非工作人员及车辆进入作业区域。

（5）导线过热伤人

预控措施：

①作业前应与调度联系确认运行方式，避免在导线大电流流经时进行带电作业；

②作业人员应穿着专用的耐热性屏蔽服进行带电作业，防止导线过热伤人。

3.2 工器具及材料选择

带电更换±800kV直流接地极线路直线整串绝缘子所需工器具及材料见表1-9-3。工器具出库前，应认真核对工器具的使用电压等级和试验周期，并检查确认外观良好、连接牢固、转动灵活，且符合本次工作任务的要求；工器具出库后，应存放在工具袋或工具箱内进行运输，防止脏污、受潮；金属工具和绝缘工器具应分开装运，防止因混装运输导致工器具变形、损伤等现象发生。

表1-9-3 带电更换±800kV特高压输电线路接地极直线整串绝缘子所需工器具及材料表

序号	名称	规格型号	单位	数量	备注
1	绝缘传递绳	TJS-14×50m	根	2	绝缘工具
2	绝缘传递绳	TJS-10×50m	根	1	绝缘工具
3	绝缘绳套	TJS-14	个	2	绝缘工具
4	绝缘滑车	0.5T	个	2	绝缘工具
5	绝缘拉棒		套	1	绝缘工具
6	导线保护绳	TJS-18	根	1	绝缘工具
7	绝缘软梯	30m	副	2	绝缘工具
8	绝缘操作杆		副	1	绝缘工具
9	横担卡具		套	1	金属工具
10	二分裂提线器		个	1	金属工具
11	跟斗滑车		个	1	金属工具
12	铝合金梯头		个	1	金属工具
13	金属扣环		个	6	金属工具
14	屏蔽服	500kV I型或隔热型	套	3	个人防护用具
15	人体后备保护绳		副	2	个人防护用具
16	安全带		副	3	个人防护用具
17	安全帽		顶	8	个人防护用具
18	个人工具		套	1	其他工具
19	防潮毡布	3m×3m	块	1	其他工具
20	万用表		个	1	其他工具
21	风速湿度仪	HT-8321	只	1	其他工具

续表

序号	名称	规格型号	单位	数量	备注
22	兆欧表及测试电极	5000V	套	1	其他工具
23	激光测距仪		台	1	其他工具
24	红外线测温仪		台	1	其他工具
25	对讲机		台	2	其他工具
26	红马甲	"工作负责人"	件	1	其他工具
27	安全围栏		套	若干	其他工具
28	相同型号绝缘子		片	6	材料

3.3 作业人员分工

本任务作业人员分工如表1-9-4所示。

表1-9-4 带电更换±800kV特高压输电线路接地极直线整串绝缘子人员分工表

序号	工作岗位	数量(人)	工作职责
1	工作负责人	1	负责作业现场的各项工作
2	专责监护人	1	负责塔上监护作业
3	等电位电工	1	负责导线端绝缘子销子拆除、恢复。
4	塔上地电位电工	1	负责布置传递绳、软梯、绝缘子串吊点及工器具的安装等
5	地面电工	2	负责传递工具、材料，配合等电位电工进出等电位

4. 工作程序

本任务工作流程如表1-9-5所示。

表1-9-5 带电更换±800kV特高压输电线路接地极直线整串绝缘子工作流程表

序号	作业内容	作业步骤及标准	安全措施及注意事项	责任人
1	确认线路名称和塔号、杆塔检查	工作负责人负责完成以下工作： (1)现场核对线路名称、杆塔编号、相别无误；基础及杆塔完好无异常；交叉跨越距离符合安全要求；确认缺陷情况及导地线规格型号等； (2)检查地形环境符合作业要求； (3)检查工作票所列安全措施与现场实际情况相符，必要时予以补充	(1)正确穿戴安全帽、工作服、工作鞋、劳保手套； (2)不得在危及作业人员安全的气象条件下作业； (3)严禁非工作人员、车辆进入作业现场	

续表

序号	作业内容	作业步骤及标准			安全措施及注意事项	责任人
2	现场作业环境测量	测量项目	标准值	实测值	风力大于5级,湿度大于80%时一般不宜进行带电作业	
		风速	≤10m/s			
		湿度	≤80%			
		环境温度	0~38℃			
		导线温度	—			
		作业距离	≥0.75m			
		海拔	海拔500m和1000m为界			
3	工作许可	(1)工作负责人负责联系值班调控人员,按工作票内容申请停用线路再启动; (2)经值班调控人员许可后,方可开始带电作业工作			不得未经值班调控人员许可即开始工作	
4	现场布置	正确装设安全围栏并悬挂标示牌: (1)安全围栏范围应充分考虑高处坠物,以及对道路交通的影响; (2)安全围栏出入口设置合理; (3)妥当布置"从此进出""在此工作""从此上下"等标示			对道路交通安全影响不可控时,应及时联系交通管理部门强化现场交通安全管控	
5	召开班前会	(1)全体工作成员列队; (2)工作负责人宣读工作票,明确工作任务及人员分工;讲解工作中的安全措施和技术措施;查(问)全体工作成员精神状态;告知工作中存在的危险点及采取的预控措施; (3)全体工作成员在工作票上签名确认			(1)工作票填写、签发和许可手续规范,签名完整; (2)全体工作成员精神状态良好; (3)全体工作成员明确任务分工、安全措施和技术措施	
6	检查工具	(1)塔上地电位电工和等电位电工正确地穿戴好屏蔽服并检测合格,由负责人监督检查; (2)正确佩戴个人安全用具(大小合适,锁扣自如),由负责人监督检查; (3)检查绝缘工具外观情况并测量其绝缘性能,并做好记录			(1)金属、绝缘工具使用前,应仔细检查其是否损坏、变形、失灵。绝缘工具应使用清洁干燥的毛巾将其擦拭干净并用2500V及以上绝缘电阻表进行分段绝缘检测,阻值应不低于700MΩ; (2)用万用表测量屏蔽服衣裤最远端点之间的电阻值不得大于20Ω。工作负责人认真检查作业电工屏蔽服的连接情况; (3)检查工具组装情况并确认连接可靠; (4)现场所使用的带电作业工具应放置在防潮毡布上	

续表

序号	作业内容	作业步骤及标准	安全措施及注意事项	责任人
7	登塔	(1)核对线路名称、杆塔编号无误后,塔上地电位电工冲击检查安全带、防坠器受力情况; (2)塔上地电位电工携带绝缘传递绳登塔至横担位置,打好安全带及人体后备保护绳,在合适位置布置好传递绳	(1)核对线路名称和杆塔编号无误后,方可登塔作业; (2)登塔过程中应使用塔上安装的防坠装置;杆塔上移动及转位时,不准失去安全保护,作业人员必须攀抓牢固构件; (3)作业电工必须穿全套合格的屏蔽服,且全套屏蔽服必须连接可靠	
8	安装带绝缘绳的跟斗滑车	地面电工将绝缘操作杆及跟斗滑车传至塔上;地电位电工将带有Φ10mm绝缘传递绳的跟斗滑车挂在导线上	(1)地电位电工在杆塔上转位时不得失去安全带或人体后备保护绳的保护; (2)安装跟斗滑车时人体与带电体的距离不得小于0.75m;绝缘操作杆的有效绝缘长度不得小于1.3m; (3)跟斗滑车应可靠安装在导线上且位置合理; (4)专责监护人认真监护	
9	安装软梯	地电位电工配合地面作业人员将软梯挂在导线上,再将软梯拉至安全位置	(1)软梯、梯头应外观完好,无损伤、缺件及变形等情况; (2)软梯与梯头应连接可靠; (3)传递软梯的绑点应合适,布置在梯头上的人体后备保护绳应牢固可靠,其滑车应转动灵活,无卡挂、翻槽等情况; (4)工作负责人认真检查	
10	进入强电场	(1)地面电工对软梯进行冲击检查确认其已可靠安装在导线上后,下压收紧绝缘软梯; (2)工作负责人再次确认等电位电工屏蔽服各部分连接情况; (3)等电位电工在地面电工配合下拴好人体后备保护绳,经工作负责人同意后攀爬软梯进入强电场	(1)地面电工下压收紧软梯时其导线对地及跨越物的最小距离应符合安全规定地面电工控制防坠人体后备保护绳方式合理;攀爬过程中速度平稳,动作规范、熟练; (2)等电位电工电位转移前必须得到工作负责人的许可,电位转移时严禁用头部充放电;人体裸露部分与带电体的距离不得小于0.3m;其组间隙不得小于1.2m; 等电位电工进入电场后应先打好安全带再锁死梯头保险; (3)工作负责人认真监护、提醒	

续表

序号	作业内容	作业步骤及标准	安全措施及注意事项	责任人
11	更换整串绝缘子	(1)地电位电工安装卡具,在等电位电工配合下将拉棒安装到位,打好导线保护绳,将传递绳绑在绝缘子串的横担侧第1、2片之间; (2)地电位电工收紧丝杠使拉棒受力后进行冲击检查,确保拉棒受力正常后,等电位电工拔出导线侧碗头销子; (3)地电位电工继续收紧丝杠使绝缘子串荷载全部转移至拉棒,再次检查承力工具受力无异常后,等电位电工拆开导线侧球头挂环与碗头挂板的连接;地电位电工松出丝杠使导线下降200mm; (4)地电位电工拔出横担侧碗头销子,在地面电工的配合下拆开横担侧球头挂环与绝缘子的连接; (5)地面电工采用新旧绝缘子交替方式将新绝缘子传递横担处,地电位电工恢复横担侧球头挂环与绝缘子的连接,并将销子安装到位; (6)等电位电工在地电位电工的配合下恢复导线侧球头挂环与碗头挂板的连接,并将销子安装到位	(1)卡具安装位置应合理,在安装拉棒时等电位电工和地电位电工不得同时操作,绝缘子串绑点应合适且拴绑牢固; (2)导线保护绳的布置位置和长度应合适,自锁装置应灵活便于操作; (3)摇动丝杠时应匀速用力,避免拉棒扭动过大; (4)拉棒受力后应冲击检查确保受力正常,更换绝缘子串过程中等电位电工和地电位电工不得同时操作; (5)采用新旧绝缘子交替上下传递时必须控制好尾绳,避免发生相互磕碰或磕碰导线、拉棒; (6)更换完毕后应检查销子到位情况,避免销子漏装或安装不到位等情况发生; (7)工作负责人认真监护、提醒	
12	退出电场	(1)等电位电工再次检查导线端绝缘子复位情况及销子到位情况,确认无误后向工作负责人汇报工作结束,申请退出电场; (2)经工作负责人同意后,等电位电工对屏蔽服连接情况进行检查确认连接可靠后,按进电场相反步骤退出电场,沿绝缘软梯下至地面	(1)等电位电工退出电位前必须得到工作负责人的许可; (2)等电位电工进入梯头后应先打好安全带、做好防坠人体后备保护措施后再打开梯头保险; (3)等电位电工在退出电场过程中与接地体和带电体两部分间隙所组成的组合间隙不得小于1.2m; (4)地面电工下压收紧软梯且控制防坠人体后备保护绳方式合理,无未受力情况发生; (5)工作负责人认真监护、提醒	

续表

序号	作业内容	作业步骤及标准	安全措施及注意事项	责任人
13	地电位电工返回地面	地电位电工将工器具传至地面,检查塔上无遗留物后,向工作负责人汇报,得到工作负责人同意后携带绝缘传递绳下塔	下塔过程中应使用塔上安装的防坠装置,杆塔上移动及转位时,不准失去安全保护,作业人员必须攀抓牢固构件	
14	工作结束	(1)工作负责人组织全体工作成员整理工器具和材料,将工器具清洁后放入专用的箱(袋)中;清理现场,做到"工完料尽场地清"。 (2)召开班后会,工作负责人进行工作总结和点评工作。点评本次工作的施工质量;点评全体工作成员的安全措施落实情况。 (3)工作负责人向值班调控人员汇报工作结束,申请恢复线路直流再启动,终结工作票	不得约时恢复线路直流再启动	

二、考核标准

表1-9-6 国网四川省电力公司特高压直流输电线路运检技能考核评分细则

考生填写栏	编号: 姓 名: 所在岗位: 单 位: 日 期: 年 月 日						
考评员填写栏	成绩: 考评员: 考评组长: 开始时间: 结束时间: 操作时长:						
考核模块	带电更换±800kV特高压输电线路接地极直线整串绝缘子	考核对象	±800kV直流接地极输电线路检修人员	考核方式	操作	考核时限	60min
任务描述	沿软梯进入电场对±800kV直流接地极线路直线整串绝缘子进行带电更换。						
工作规范及要求	1.带电作业工作应在良好天气下进行。如遇雷、雨、雪、雾天气不得进行带电作业。风力大于5级、湿度大于80%时,一般不宜进行带电作业。 2.本项作业需工作负责人1名,专责监护人1人,塔上地电工1人,等电位电工1人,地面辅助电工2人,采用沿软梯进入电场对±800kV直流接地极线路直线整串绝缘子进行带电更换。 3.工作负责人职责:负责本次工作任务的人员分工、工作票的宣读、办理线路停用直流再启动、办理工作许可手续、召开工作班前会、工作中突发情况的处理、工作质量的监督、工作后的总结。 4.专责监护人:负责作业现场的安全把控。 5.等电位电工职责:负责导线端绝缘子销子拆除、恢复。						

续表

工作规范及要求	6. 塔上地电位电工职责:负责布置传递绳、软梯、绝缘子串吊点及工器具的安装等。 7. 地面电工职责:负责传递工器具、材料,配合等电位电工进出等电位。 8. 在带电作业中,如遇雷、雨、大风或其他任何情况威胁到工作人员的安全时,工作负责人或监护人可根据情况,临时停止工作。 给定条件: 1. 培训基地:特高压±800kV直流接地极线路铁塔 2. 工作票已办理,安全措施已经完备(直流再启动已停用),工作开始、工作终结时应口头提出申请(调控或考评员)。 3. 安全、正确地使用仪器对绝缘工具进行检测。 4. 必须按工作程序进行操作,工序错误扣除应做项目分值,出现重大人身、器材和操作安全隐患,考评员可下令终止操作(考核)
考核情景准备	1. 线路:特高压±800kV直流接地极线路铁塔,工作内容:带电更换±800kV直流接地极线路直线整串绝缘子,绝缘子型号:XZP-160 2. 所需作业工器具:绝缘传递绳2根(TJS-14);绝缘传递绳1根(TJS-10),导线保护1根(TJS-18),绝缘绳套2个(TJS-14),绝缘滑车2个(JH10-0.5);绝缘拉棒1套,绝缘操作杆1副,绝缘软梯2副,横担卡具1套(带丝杠),二分裂提线器1个,跟斗滑车1个,铝合金梯头1副;屏蔽服(I型或隔热型)3套;绝缘电阻表(5000V型带测试电极),万用表1块,风速湿度仪1块,激光测距仪1台,红外线测温仪1台,苫布1块。 3. 作业现场做好监护工作,作业现场安全措施(围栏等)已全部落实;禁止非作业人员进入现场,工作人员进入作业现场必须戴安全帽。 4. 考生自备工作服,阻燃纯棉内衣,安全帽,线手套,安全带(含二保绳)
备注	1. 各项目得分均扣完为止,出现重大人身、器材和操作安全隐患,考评员可下令终止操作。 2. 设备、作业环境、安全带、安全帽、工器具、屏蔽服等不符合作业条件考评员可下令终止操作

表1-9-7 国网四川省电力公司特高压直流输电线路运检技能考核评分标准

序号	项目名称	质量要求	分值	扣分标准	扣分原因	扣分	得分
1	现场复勘	(1)工作负责人到作业现场核对线路名称和杆塔编号、现场工作条件、缺陷部位等。 (2)检测风速、湿度等现场气象条件符合作业要求。 (3)检查工作票填写完整,无涂改,检查是否所列安全措施与现场实际情况相符,必要时予以补充	5	(1)未进行核对双重称号,扣1分。 (2)未核实现场工作条件(气象)、缺陷部位,扣1分。 (3)工作票填写出现涂改,每项扣0.5分,工作票编号有误,扣1分。工作票填写不完整,扣1.5分			
2	工作许可	(1)工作负责人联系值班调控人员,按工作票内容申请停用线路直流再启动。 (2)汇报内容规范、完整	2	(1)未联系调控部门(裁判)停用直流再启动扣2分。 (2)汇报专业用语不规范或不完整的,各扣0.5分			

第一部分　±800kV 特高压输电线路运检带电作业培训及考核标准

续表

序号	项目名称	质量要求	分值	扣分标准	扣分原因	扣分	得分
3	现场布置	正确装设安全围栏并悬挂标示牌： (1)安全围栏范围应充分考虑高处坠物，以及对道路交通的影响。 (2)安全围栏出入口设置合理。 (3)妥当布置"从此进出""在此工作""从此上下"等标示	3	(1)作业现场未装设围栏，扣0.5分。 (2)未设立警示牌，扣0.5分。 (3)未悬挂登塔作业标志，扣0.5分。			
4	召开班前会	(1)全体工作成员全体人员正确佩戴安全帽、工作服。 (2)工作负责人佩戴红色背心，宣读工作票，明确工作任务及人员分工；讲解工作中的安全措施和技术措施；查(问)全体工作成员精神状态；告知工作中存在的危险点及采取的预控措施。 (3)全体工作成员在工作票上签名确认	3	(1)工作人员着装不整齐，扣0.5分，工作人员着装不整齐，每人次扣0.5分。 (2)未进行分工本项不得分，分工不明扣1分。 (3)现场工作负责人未穿佩安全监护背心，扣0.5分。 (4)工作票上工作班成员未签字或签字不全的，扣1分			
5	工器具检查	(1)工作人员按要求将工器具放在防潮苫布上；防潮苫布应清洁、干燥。 (2)工器具应按定置管理要求分类摆放；绝缘工器具不能与金属工具、材料混放；对工器具进行外观检查。 (3)绝缘工具表面不应磨损、变形损坏，操作应灵活。绝缘工具应使用2500V及以上绝缘电阻表进行分段绝缘检测，阻值应不低于700MΩ，并用清洁干燥的毛巾将其擦拭干净。 (4)塔上地电位和等电位人员按要求正确穿戴全套合格的屏蔽服、导电鞋，且各部分连接应良好；屏蔽服内不得贴身穿着化纤类衣服；并系好安全带；工作负责人应认真检查是否穿戴正确	7	(1)未使用防潮布并定置摆放工器具，扣1分。 (2)未检查工器具试验合格标签及外观检查，每项扣0.5分。 (3)未正确使用检测仪器对工器具进行检测，每项扣1分。 (4)作业人员未正确穿戴屏蔽服且各部位连接良好，每人次扣2分。 (5)现场工作负责人未对登塔作业人员进安全防护装备进行检查，扣1分			

续表

序号	项目名称	质量要求	分值	扣分标准	扣分原因	扣分	得分
6	登塔	(1)核对线路名称、杆塔编号无误后,塔上地电位电工冲击检查安全带、防坠器受力情况。 (2)塔上地电位电工携带绝缘传递绳登塔至横担位置,打好安全带及人体后备保护绳,在合适位置布置好传递绳	5	(1)登塔人员未核对线路双重名称、杆号、相别,每人扣2分。 (2)登塔人员未报告核对结果,每人扣2分。 (3)未系安全带或安全带及后备保护绳未进行冲击试验,各扣2分。 (4)手抓脚钉,扣2分。 (5)滑车传递绳悬挂位置不便工具取用,扣1分。 (6)塔上电工操作不正确,扣2分。 (7)工作负责人监护不到位,扣2分			
7	安装带绝缘绳的跟斗滑车	地面电工将绝缘操作杆及跟斗滑车传至塔上;地电位电工将带有Φ10mm绝缘传递绳的跟斗滑车挂在导线上	4	(1)转移作业位置未采取保护措施,扣1分。 (2)工作负责人未提醒地电位电工转移时采取安全措施,扣2分。 (3)安装位置不合适,扣1分。 (4)传递绳索缠绕,扣1分			
8	安装软梯	地电位电工配合地面作业人员将软梯挂在导线上,再将软梯拉至安全位置	4	(1)高空坠物,扣1分。 (2)软梯安装不牢靠,扣2分;位置不合适,扣2分。 (3)传递绳索缠绕,扣1分			
9	进入强电场	(1)地面电工对软梯进行冲击检查确认其已可靠安装在导线上后,下压收紧绝缘软梯。 (2)工作负责人再次确认等电位电工屏蔽服各部分连接情况; (3)等电位电工在地面电工配合下挂好人体后备保护绳,经工作负责人同意后攀爬软梯进入强电场	15	(1)登未对绝缘软梯做冲击检查,扣2分。 (2)压紧软梯前未检查交叉跨越是否满足安全距离,扣2分。 (3)等电位电工未向工作负责人申请即开始登梯,扣2分;申请了但未同意请即开始登梯,扣1分。 (4)攀登不平稳、登空,每次扣1分			

续表

序号	项目名称	质量要求	分值	扣分标准	扣分原因	扣分	得分
9	进入强电场		15	(5)未有效控制控制后备保护绳,扣1分。 (6)未向工作负责人申请即开始电位转移,扣2分;申请了但未同意请即开始电位转移,扣1分。 (7)申请电位转移位置不合适,扣1分。 (8)转移电位动作不熟练,扣1分。 (9)工作负责人未认真监护,扣2分。			
10	更换整串绝缘子	(1)地电位电工安装卡具,在等电位电工配合下将拉棒安装到位,打好导线保护绳,将传递绳绑在绝缘子串的横担侧第1、2片之间; (2)地电位电工收紧丝杠使拉棒受力后进行冲击检查,确保拉棒受力正常后,等电位电工拔出导线侧碗头销子; (3)地电位电工继续收紧丝杠使绝缘子串荷载全部转移至拉棒,再次检查承力工具受力无异常后,等电位电工拆开导线侧球头挂环与碗头挂板的连接;地电位电工松出丝杠使导线下降200mm。 (4)地电位电工拔出横担侧碗头销子,在地面电工的配合下拆开横担侧球头挂环与绝缘子的连接。 (5)地面电工采用新旧绝缘子交替方式将新绝缘子传递横担处,地电位电工恢复横担侧球头挂环与绝缘子的连接,并将销子安装到位。 (6)等电位电工在地电位电工的配合下恢复导线侧球头挂环与碗头挂板的连接,并将销子安装到位	30	(1)卡具安装不到位,扣2分。 (2)导线保护绳位置布置不合适,扣2分。 (3)传递绳绑在绝缘子串上的位置错误扣2分,未栓牢,扣3分。 (4)丝杠受力后未做冲击检查,扣2分,检查了未向工作负责人汇报,扣1分。 (5)摇动丝杠时拉棒扭动过大,扣0.5分。 (6)未绑好绝缘子即先取横担侧碗头销子,扣2分。 (7)等电位电工和地电位电工同时操作,扣5分。 (8)新、旧绝缘子交替方式传递时发生相互磕碰或磕碰杆塔、导线、拉棒等,扣0.5分/次。 (9)安装绝缘子串顺序不正确,扣3分。 (10)未检查绝缘子串安装情况即开始松丝杠,扣2分;检查了未汇报,扣1分。 (11)销子安装不到位,扣2分/处。 (12)销子每少装1个,扣3分			

续表

序号	项目名称	质量要求	分值	扣分标准	扣分原因	扣分	得分
11	退出电场	(1)等电位电工再次检查导线端绝缘子复位情况及销子到位情况,确认无误后向工作负责人汇报工作结束,申请退出电场。 (2)经工作负责人同意后,等电位电工对屏蔽服连接情况进行检查确认连接可靠后,按进电场相反步骤退出电场,沿绝缘软梯下至地面	8	(1)未汇报工作结束,扣1分。 (2)未向工作负责人申请退出电场,扣2分。 (3)未对屏蔽服连接情况进行检查,扣1分。 (4)未向工作负责人申请即进行电位转移,扣2分;申请了但未得同意即开始,扣1分。 (5)申请电位转移位置不合适,扣1分。 (6)等电位电工退出电场动作不正确,反复放电,扣2分。 (7)攀登不平稳、登空,每次扣1分。 (8)未有效控制控制后备保护绳,扣1分			
12	地电位电工返回地面	地电位电工将工器具传至地面,检查塔上无遗留物后,向工作负责人汇报,得到工作负责人同意后携带绝缘传递绳下塔	4	(1)下塔过程未使用防坠装置,扣2分。 (2)塔上移位失去安全带保护的,扣2分。 (3)下塔抓塔钉,每处扣1分。 (4)塔上有遗留物的,扣2分			
13	工作结束	(1)工作负责人组织全体工作成员整理工器具和材料,将工器具清洁后放入专用的箱(袋)中;清理现场,做到"工完料尽场地清"。 (2)召开班后会,工作负责人进行工作总结和点评工作。点评本次工作的施工质量;点评全体工作成员的安全措施落实情况。 (3)工作负责人向值班调控人员汇报工作结束,申请恢复线路直流再启动,终结工作票	10	(1)工器具未清理,扣2分。 (2)工器具有遗漏,扣2分。 (3)未开班后会,扣3分。 (4)点评不到位,扣1分。 (5)未联系调控汇报工作结束,扣3分。 (6)汇报内容,每缺一项扣0.5分。(单位名称、负责人姓名、线路名称、工作完成情况、设备已恢复正常、人员已撤离) (7)工作票终结填写错误,扣1分			
	合计		100				

模块十　带电修补±800kV特高压输电线路接地极线路导线培训及考核标准

一、培训标准

（一）培训要求

模块名称	带电修补±800kV特高压输电线路接地极线路导线	培训类别	操作类
培训方式	实操培训	培训学时	14学时
培训目标	1.掌握±800kV直流接地极直线塔进、出电场时采用"自爬升装置"作业方式的电学意义。 2.能完成采用"自爬升装置"进入±800kV直流接地极线路等电位作业点。 3.能独立完成±800kV直流接地极线路导线修补的操作（等电位作业法）		
培训场地	特高压直流实训线路		
培训内容	采用从自爬升装置进入电场，采用等电位作业法带电修补±800kV特高压输电线路接地极线路导线		
适用范围	特高压±800kV直流输电线路检修人员		

（二）引用规程规范

（1）《±800kV直流架空输电线路设计规范》（GB/T 50790-2013）

（2）《±800kV直流架空输电线路检修规程》（DL/T 251-2012）

（3）《±800kV直流架空输电线路运行规程》（GB/T 28813-2012）

（4）《±800kV直流线路带电作业技术规范》（DL/T 1242-2013）

（5）《±800kV特高压输电线路金具技术规范》（GB/T 31235-2014）

（6）《国家电网公司带电作业工作管理规定（试行）》（国家电网生〔2007〕751号）

（7）《国家电网公司电力安全工作规程（线路部分）》（Q/GDW1799.2-2013）

（8）《电工术语　架空线路》（GB/T 2900.51-1998）

（9）《电工术语　带电作业》（GB/T 2900.55-2016）

（10）《带电作业工具设备术语》（GB/T 14286-2002）

（11）《带电作业用工具、装置和设备使用的一般要求》（DL/T 877-2004）

(12)《带电作业工具、装置和设备预防性试验规程》(DL/T 976-2005)
(13)《±800kV 特高压输电线路带电作业技术导则》(Q/GDW 302-2009)
(14)《带电作业用屏蔽服装》(GB/T 6568-2008)
(15)《带电作业工具基本技术要求与设计导则》(GB 18037-2008)

（三）培训教学设计

本设计以完成"带电修补±800kV 特高压输电线路接地极线路导线"为工作任务，按工作任务完成的标准化作业流程来设计各个培训阶段，每个阶段包括了具体的培训目标、培训内容、培训学时、培训方法（培训资源）、培训环境和考核评价等内容，如表1-10-1所示。

表1-10-1　带电修补±800kV 特高压输电线路接地极线路导线培训内容设计

培训流程	培训目标	培训内容	培训学时	培训方法与资源	培训环境	考核评价
1.理论教学	1.初步掌握采用自爬升装置进出±800kV 直流接地极线路电场的方法。2.熟悉工器具配置及安全注意事项。3.熟悉±800kV 直流接地极线路导线修补的方法	1.采用自爬升装置进出±800kV 直流接地极线路电场的电学意义。2.工器具配置、危险点分析及安全措施。3.±800kV 直流接地极线路导线修补方法和质量标准	2	培训方法:讲授法。培训资源：PPT、相关规程规范	多媒体教室	考勤、课堂提问和作业
2.准备工作	能完成作业前准备工作	1.作业现场查勘。2.编制培训标准化作业卡。3.填写培训操作工作票。4.完成本操作的工器具及材料准备	1	培训方法：1.现场查勘和工器具及材料清理采用现场实操方法；2.编写作业卡和填写工作票采用讲授方法。培训资源：1.±800kV 接地极实训线路 2.特高压工器具库房 3.空白工作票	1.±800kV 接地极实训线路；2.多媒体教室	

续表

培训流程	培训目标	培训内容	培训学时	培训方法与资源	培训环境	考核评价
3.作业现场准备	能完成作业现场准备工作	1.作业现场复勘。 2.工作申请。 3.作业现场布置。 4.班前会。 5.工器具及材料检查	1	培训方法：演示与角色扮演法。资源：±800kV接地极实训线路	±800kV接地极实训线路	
4.培训师演示	通过现场观摩，使学员初步领会本任务操作流程	1.作业距离测量、导线温度测量。 2.地面电工利用无人机进行牵引绳的展放。 3.地面电工将后备保护绳及导轨绳牵引至工作位置。 4.等电位电工采用自爬升装置进入电场。 5.等电位电工进行导线的修补	2	培训方法：演示法。资源：±800kV接地极实训线路	±800kV接地极实训线路	
5.学员分组训练	通过培训： 1.使学员能完成进出±800kV直流接地极线路电场操作。 2.使学员能完成±800kV直流接地极线路导线修补方法	1.学员分组（6人一组）训练进出±800kV直流接地极线路导线修补技能操作。 2.培训师对学员操作进行指导和安全监护	7	培训方法：角色扮演法。资源：±800kV接地极实训线路	±800kV接地极实训线路	采用技能考核评分细则对学员操作评分
6.工作终结	通过培训： 1.使学员进一步认识操作过程不足处，便于后期提升。 2.培训学员安全文明生产的工作作风	1.作业现场清理。 2.向调度汇报工作。 3.班后会，对本次工作任务进行点评总结	1	培训方法：讲授和归纳法	±800kV接地极实训线路	

（四）作业流程

1. 工作任务

采用自爬升装置进入电场、到达作业点，采用等电位作业法带电修补±800kV特高压输电线路接地极线路导线。

2. 天气及作业现场要求

（1）带电修补±800kV特高压输电线路接地极线路导线应在良好的天气进行。工作现场如遇雷电（听见雷声、看见闪电）、雪、雹、雨、雾等，禁止进行带电作业。风力大于5级（10m/s）时，或空气相对湿度大于80%时，不宜进行本次作业。作业过程中,注意对天气变化的检测,如遇大风、雨雾等紧急情况,按照规程正确采取措施,保证人员和设备安全。在接地极线路上进行带电作业前,应了解线路负荷电流及运行温度等基本情况,在现场作业前及作业过程中应利用适用于输电线路的红外测温仪等监测导线温度。

(2) 作业人员精神状态良好,熟悉工作中保证安全的组织措施和技术措施;应持有在有效期内的带电作业资质证书。

(3) 工作负责人应事先组织相关人员完成现场勘察,根据勘察结果确定本次作业方法和所需工器具,以及应采取的必要措施,并办理带电作业工作票。

(4) 作业现场应合理设置围栏,并妥当布置警示标示牌,禁止非工作人员入内。

(5) 本项目需停用直流再启动装置。

(6) 工作中安全距离及有效绝缘长度如表1-10-2所示。

表1-10-2 带电修补±800kV特高压输电线路接地极线路导线的安全距离(m)

电压等级(海拔高度≤1000)	人身与带电体安全距离	最小有效绝缘长度		最小组合间隙	转移电位时人体裸露部分与带电体的最小距离
		绝缘操作杆	绝缘承力工具、绝缘绳索		
±800kV直流接地极	6.8	6.8	6.8	6.6	0.5

(7) 在±800kV输电线路上作业,应保证作业相良好绝缘子片数不少于32片。

3. 准备工作

3.1 危险点及其预控措施

(1) 危险点——预防系统过电压

预控措施:本次作业为等电位作业,工作负责人在工作开始前,应与值班调控员联系,申请办理电力线路带电作业工作票,并申请停用直流再启动,由调控值班员履行许可手续。带电作业结束后应及时向调控值班员汇报。严禁约时停用和恢复直流再启动。在带电作业过程中如遇设备突然停电,作业人员应视设备仍然带电。工作负责人应尽快与调控联系,调控值班员未与工作负责人取得联系前不得强送电。

(2) 危险点——触电伤害

预控措施:

①绝缘工具及绝缘绳索不得损坏、受潮、变形、失灵,不准使用非绝缘绳索(如棉纱绳、白棕绳、钢丝绳)。

②等电位作业人员应穿着阻燃内衣,衣服外面应穿戴全套屏蔽服(包括帽、衣裤、手套、袜和鞋),且各部分应连接良好,全套屏蔽服电阻不大于20Ω。

③等电位作业人员在电位转移前,应得到工作负责人的许可,人体裸露部分与带电体的最小距离不小于0.5m;地电位作业人员与带电体(等电位作业人员与接地体)的安全距离大于表1-10-2的规定。

④用绝缘绳索传递大件金属物品时,地电位作业人员应将金属物品接地后再接触。

⑤工作负责人和专责监护人应对作业人员进行不间断监护,随时纠正其不规范或违章动作。重点关注高处作业人员,使其保持足够的安全距离(符合表1-10-2的规定)。

⑥当耐热导线运行温度高于40℃时,应采用耐高温带电作业工器具,包括耐高温软质绝缘工器具、耐高温硬质绝缘工器具、耐高温隔热屏蔽服等,需采用电位转移棒进入等电位。

⑦隔热防护用具应视为导体,传递和使用过程中应保证相地及极间安全距离满足安全要求。

(3)危险点——高处坠落

预控措施:

①高处作业人员登高前,必须具备符合本项作业要求的身体状况、精神状态和技能素质。

②监护人员应随时纠正其不规范或违章动作,重点关注作业人员在转位的过程中不得失去安全带或绝缘后备保护绳的保护,严禁低挂高用。

(4)危险点——高处坠物伤人

预控措施:

①高处作业人员的个人工具及零星材料应装入工具袋,严禁在高处浮置物件、口中含物。

②地面作业人员必须正确佩戴安全帽,正确使用绳结,与作业点垂直下方距离不得小于坠落半径。

③作业现场设置围栏并挂好警示标示牌。监护人员应随时注意,禁止非工作人员及车辆进入作业区域。

(5)导线过热伤人

预控措施:

①作业前应与调度联系确认运行方式,避免在导线大电流流经时进行带电作业;

②作业人员应穿着专用的耐热性屏蔽服进行带电作业,防止导线过热伤人。

3.2 工器具及材料选择

带电修补±800kV特高压输电线路接地极线路导线所需工器具及材料见表1-10-3。工器具出库前,应认真核对工器具的使用电压等级和试验周期,并检查确认外观良

好、连接牢固、转动灵活,且符合本次工作任务的要求;工器具出库后,应存放在工具袋或工具箱内进行运输,防止脏污、受潮;金属工具和绝缘工器具应分开装运,防止因混装运输导致工器具变形、损伤等现象发生。

表1-10-3　带电修补±800kV特高压输电线路接地极线路导线所需工器具及材料表

序号	名称	规格型号	单位	数量	备注
1	绝缘传递绳	TJS-14×50m	根	1	
2	绝缘传递绳	TJS-10×50m	根	1	
3	绝缘传递绳	TJS-14×100m	根	1	
4	绝缘传递绳	TJS-4×100m	根	1	
5	绝缘绳套	TJS-14	个	1	
6	绝缘滑车	0.5T	个	1	
7	无人机	六旋翼	架	1	
8	自爬升装置	ACTSAFE-II	台	1	
9	个人工具		套	1	
10	屏蔽服	500kV I型或隔热型	套	1	
11	人体后备保护绳		副	1	
12	安全带		副	1	
13	安全帽		顶	6	
14	防潮毡布	3m×3m	块	1	
15	万用表		个	1	
16	风速湿度仪	HT-8321	只	1	
17	兆欧表及测试电极	5000V	套	1	
18	对讲机		台	2	
19	红马甲	"工作负责人"	件	1	
20	安全围栏		套	若干	
21	预绞丝修补条	标配	套	2	
22	导电膏		盒	1	
23	砂纸		张	1	

3.3　作业人员分工

本任务作业人员分工如表1-10-4所示。

表1-10-4　带电修补±800kV特高压输电线路接地极线路导线人员分工表

序号	工作岗位	数量(人)	工作职责
1	工作负责人	1	负责作业现场的各项工作
2	等电位电工	1	负责安装及使用自爬升装置进入电场修补导线
3	无人机操控人员	1	负责操作无人机展放牵引绳
4	地面电工	3	负责传递工具、材料，配合等电位电工进出等电位

4. 工作程序

本任务工作流程如表1-10-5所示。

表1-10-5　带电修补±800kV特高压输电线路接地极线路导线工作流程表

序号	作业内容	作业步骤及标准	安全措施及注意事项	责任人
1	确认线路名称和塔号、杆塔检查	工作负责人负责完成以下工作： (1)现场核对线路名称、杆塔编号、相别无误；基础及杆塔完好无异常；交叉跨越距离符合安全要求；确认缺陷情况及导地线规格型号等。 (2)检查地形环境符合作业要求。 (3)检查工作票所列安全措施与现场实际情况相符，必要时予以补充	(1)正确穿戴安全帽、工作服、工作鞋、劳保手套。 (2)不得在危及作业人员安全的气象条件下作业。 (3)严禁非工作人员、车辆进入作业现场	
2	现场作业环境测量	测量项目　标准值　实测值 风速　≤10m/s 湿度　≤80% 环境温度　0～38℃ 导线温度　— 作业距离　≥0.5m 海拔高度　海拔500m和1000m为界	风力大于5级，湿度大于80%时一般不宜进行带电作业	
3	工作许可	(1)工作负责人负责联系值班调控人员，按工作票内容申请停用线路直流再启动。 (2)经值班调控人员许可后，方可开始带电作业工作	不得未经值班调控人员许可即开始工作	
4	现场布置	正确装设安全围栏并悬挂标示牌： (1)安全围栏范围应充分考虑高处坠物，以及对道路交通的影响。 (2)安全围栏出入口设置合理。 (3)妥当布置"从此进出""在此工作""从此上下"等标示	对道路交通安全影响不可控时，应及时联系交通管理部门强化现场交通安全管控	

续表

序号	作业内容	作业步骤及标准	安全措施及注意事项	责任人
5	召开班前会	(1)全体工作成员列队。 (2)工作负责人宣读工作票,明确工作任务及人员分工;讲解工作中的安全措施和技术措施;查(问)全体工作成员精神状态;告知工作中存在的危险点及采取的预控措施。 (3)全体工作成员在工作票上签名确认	(1)工作票填写、签发和许可手续规范,签名完整。 (2)全体工作成员精神状态良好。 (3)全体工作成员明确任务分工、安全措施和技术措施	
6	检查工具	(1)等电位电工正确地穿戴好屏蔽服并检测合格,由负责人监督检查。 (2)正确佩戴个人安全用具(大小合适,锁扣自如),由负责人监督检查。 (3)检查绝缘工具外观情况并测量其绝缘性能,并做好记录。 (4)检查好无人机及自爬升装置的运转情况	(1)金属、绝缘工具使用前,应仔细检查其是否损坏、变形、失灵。绝缘工具应使用清洁干燥的毛巾将其擦拭干净并用2500V及以上绝缘电阻表进行分段绝缘检测,阻值应不低于700MΩ。 (2)用万用表测量屏蔽服衣裤最远端点之间的电阻值不得大于20Ω。工作负责人认真检查作业电工屏蔽服的连接情况。 (3)检查工具组装情况并确认连接可靠。 (4)现场所使用的带电作业工具应放置在防潮毡布上	
7	无人机展放导轨绳	(1)无人机操控人员将无人机进行起飞至离地面1.6m左右,悬停好。 (2)地面电工将直径为4mm的绝缘传递绳挂在无人机的脱扣装置上。 (3)无人机操控人员继续操作无人机上升至合适位置时向外牵引迈过导线,继续牵引与导线距离大致相同距离,操控脱扣装置使牵引绳自然下落。 (4)地面电工将牵引绳安装成绝缘无极绳	(1)无人机悬停位置应合适,地面电工悬挂牵引绳时应注意安全。 (2)无人机牵引时应从线路方向内侧起飞向外展放牵引绳	
8	牵引导轨绳及人体后备保护绳	(1)地面电工将牵引绳转移至作业点位置。 (2)地面电工利用牵引绳将自爬升装置的导轨绳进行循环牵引,导轨绳一端稳定的安装在地面牢固位置上,另一端与地面垂直。 (3)地面电工利用牵引绳将人体后备保护绳进行牵引	(1)导轨绳的一端必须固定在地面牢固位置上。 (2)导轨绳及人体后备保护绳应保持一定距离,防止缠绕在一起	

续表

序号	作业内容	作业步骤及标准	安全措施及注意事项	责任人
9	安装自爬升装置	(1)等电位电工再次检查自爬升装置的转动情况及电池电量情况等,确认无误后将自爬升装置安装到导轨绳上。 (2)等电位电工做好人体后备保护措施后,操作自爬升装置起到一定高度(0.8~1m)后,地面电工对自爬升装置及等电位电工进行冲击试验。 (3)冲击试验合格后,等电位电工携带绝缘传递绳进行垂直攀升	(1)自爬升装置应转动良好。 (2)地面电工应对自爬升装置进行冲击试验,冲击时注意自爬升装置、导轨绳固定到地面处的受力情况。 (3)工作负责人认真检查	
10	进入强电场并修补导线	(1)等电位电工乘坐自爬升装置垂直上升至距导线0.3m处,检查屏蔽服各连接处无误,经工作组负责人同意后转移电位。 (2)等电位电工在导线上打好安全带保护,将绝缘传递绳布置在预绞丝缠绕处挂好绝缘传递绳向工作负责人汇报。 (3)地面电工挂好预绞丝向工作负责人汇报。 (4)等电位电工申请平整导线,工作负责人许可。按要求使用预绞丝修补导线向工作负责人汇报	(1)地面电工应控制好人体后备保护绳,确保等电位电工高处作业不失去安全带保护。 (2)等电位电工电位转移前必须得到工作负责人的许可,电位转移时严禁用头部充放电;人体裸露部分与带电体的距离不得小于0.3m;其组合间隙不得小于1.2m。 (3)预绞丝传递过程中应避免碰撞,等单位电工在高处作业时应避免高空落物	
11	退出电场	(1)等电位电工再次检查预绞丝修补导线情况,确认无误后向工作负责人汇报工作结束,申请退出电场。 (2)经工作负责人同意后,等电位电工对屏蔽服连接情况进行检查确认连接可靠后,按进电场相反步骤退出电场,乘坐自爬升装置下至地面	(1)等电位电工退出电位前必须得到工作负责人的许可。 (2)等电位电工在退出电场过程中与接地体和带电体两部分间隙所组成的组合间隙不得小于1.2m。 (3)工作负责人认真监护、提醒	
12	工作结束	(1)工作负责人组织全体工作成员整理工器具和材料,将工器具清洁后放入专用的箱(袋)中;清理现场,做到"工完料尽场地清"。 (2)召开班后会,工作负责人进行工作总结和点评工作。点评本次工作的施工质量;点评全体工作成员的安全措施落实情况。 (3)工作负责人向值班调控人员汇报工作结束,申请恢复线路直流再启动,终结工作票	不得约时恢复直流再启动	

二、考核标准

表1-10-6　国网四川省电力公司特高压直流技能培训考核评分细则

考生填写栏	编号：　　姓　名：　　所在岗位：　　单位：　　日　期：　年　月　日					
考评员填写栏	成绩：　考评员：　考评组长：　开始时间：　结束时间：　操作时长：					
考核模块	带电修补±800kV特高压输电线路接地极线路导线	考核对象	±800kV直流接地极输电线路检修人员	考核方式	操作	考核时限	60min
任务描述	采用"自爬升装置"进入电场对±800kV直流接地极线路受损导线进行带电修补					
工作规范及要求	1. 带电作业工作应在良好天气下进行。如遇雷、雨、雪、雾天气不得进行带电作业。风力大于5级、湿度大于80%时，一般不宜进行带电作业。 2. 本项作业需工作负责人1名，等电位电工1人，无人机操作人员1人，地面辅助电工3人，采用自爬升装置进入电场对±800kV直流接地极线路导线进行带电修补。 3. 工作负责人职责：负责本次工作任务的人员分工、工作票的宣读、办理线路停用直流再启动、办理工作许可手续、召开工作班前会、工作中突发情况的处理、工作质量的监督、工作后的总结。 4. 等电位电工职责：负责安装及使用自爬升装置进入电场修补导线。 5. 无人机操作人员职责：负责操作无人机展放牵引绳。 6. 地面电工职责：负责传递工器具、材料，配合等电位电工进出等电位。 7. 在带电作业中，如遇雷、雨、大风或其他任何情况威胁到工作人员的安全时，工作负责人或监护人可根据情况，临时停止工作。 给定条件： 1. 培训基地：特高压±800kV直流接地极线路 2. 工作票已办理，安全措施已经完备（直流再启动已停用），工作开始、工作终结时应口头提出申请（调控或考评员）。 3. 安全、正确地使用仪器对绝缘工具进行检测。 4. 必须按工作程序进行操作，工序错误扣除应做项目分值，出现重大人身、器材和操作安全隐患，考评员可下令终止操作（考核）					
考核情景准备	1. 线路：特高压±800kV直流接地极线路，工作内容：带电修补±800kV直流接地极线路导线 2. 所需作业工器具：绝缘传递绳2根（TJS-14），绝缘传递绳1根（TJS-10），绝缘传递绳1根（TJS-4），绝缘绳套1个（TJS-14），绝缘滑车1个（JH10-0.5），六旋翼无人机1架，自爬升装置1台，屏蔽服（I型或隔热型）1套，全身式安全带1副，绝缘电阻表（5000V型带测试电极），万用表1块，风速湿度仪1块，苫布1块； 3. 作业现场做好监护工作，作业现场安全措施（围栏等）已全部落实；禁止非作业人员进入现场，工作人员进入作业现场必须戴安全帽。 4. 考生自备工作服，阻燃纯棉内衣，安全帽，线手套，安全带（含二保绳）					
备注	1. 各项目得分均扣完为止，出现重大人身、器材和操作安全隐患，考评员可下令终止操作。 2. 设备、作业环境、安全带、安全帽、工器具、屏蔽服等不符合作业条件考评员可下令终止操作					

表1-10-7　国网四川省电力公司特高压直流技能培训考核评分标准

序号	项目名称	质量要求	分值	扣分标准	扣分原因	扣分	得分
1	现场复勘	(1)工作负责人到作业现场核对线路名称和杆塔编号、现场工作条件、缺陷部位等。 (2)检测风速、湿度等现场气象条件符合作业要求。 (3)检查工作票填写完整，无涂改，检查是否所列安全措施与现场实际情况相符，必要时予以补充	5	(1)未进行核对双重称号，扣1分。 (2)未核实现场工作条件（气象）、缺陷部位，扣1分。 (3)工作票填写出现涂改，每项扣0.5分，工作票编号有误，扣1分。工作票填写不完整，扣1.5分			
2	工作许可	(1)工作负责人联系值班调控人员，按工作票内容申请停用线路直流再启动。 (2)汇报内容规范、完整	2	(1)未联系调控部门（裁判）停用直流再启动，扣2分。 (2)汇报专业用语不规范或不完整的各，扣0.5分			
3	现场布置	正确装设安全围栏并悬挂标示牌： (1)安全围栏范围应充分考虑高处坠物，以及对道路交通的影响。 (2)安全围栏出入口设置合理。 (3)妥当布置"从此进出""在此工作""从此上下"等标示	3	(1)作业现场未装设围栏，扣0.5分。 (2)未设立警示牌，扣0.5分。 (3)未悬挂登塔作业标志，扣0.5分			
4	召开班前会	(1)全体工作成员全体人员正确佩戴安全帽、工作服。 (2)工作负责人佩戴红色背心，宣读工作票，明确工作任务及人员分工；讲解工作中的安全措施和技术措施；查(问)全体工作成员精神状态；告知工作中存在的危险点及采取的预控措施。 (3)全体工作成员在工作票上签名确认	3	(1)工作人员着装不整齐，扣0.5分，工作人员着装不整齐，每人次扣0.5分。 (2)未进行分工本项不得分，分工不明，扣1分。 (3)现场工作负责人未穿佩安全监护背心，扣0.5分。 (4)工作票上工作班成员未签字或签字不全的，扣1分。			
5	工器具检查	(1)工作人员按要求将工器具放在防防潮苫布上；防潮苫布应清洁、干燥。 (2)工器具应按定置管理要求分类摆放；绝缘工器具不能与金属工具、材料混放；对工器具进行外观检查。 (3)绝缘工具表面不应磨损、变形损坏，操作应灵活。绝缘工具应使用2500V及以上绝缘电阻表进行分段绝缘检测，阻值应不低于700MΩ，并用清洁干燥的毛巾将其擦拭干净。	7	(1)未使用防潮布并定置摆放工器具，扣1分。 (2)未检查工器具试验合格标签及外观检查，每项扣0.5分。 (3)未正确使用检测仪器对工器具进行检测，每项扣1分。 (4)作业人员未正确穿戴屏蔽服且各部位连接良好，每人次扣2分。			

续表

序号	项目名称	质量要求	分值	扣分标准	扣分原因	扣分	得分
5	工器具检查	(4)塔上地电位和等电位人员按要求正确穿戴全套合格的屏蔽服、导电鞋,且各部分连接应良好;屏蔽服内不得贴身穿着化纤类衣服;并系好安全带;工作负责人应认真检查是否穿戴正确	7	(5)现场工作负责人未对登塔作业人员进安全防护装备进行检查,扣1分。			
6	无人机展放导轨绳	(1)无人机操控人员将无人机进行起飞至离地面1.6m左右,悬停好; (2)地面电工将直径为4mm的绝缘传递绳挂在无人机的脱扣装置上; (3)无人机操控人员继续操作无人机上升至合适位置时向外牵引迈过导线,继续牵引与导线距离大致相同距离,操控脱扣装置使牵引绳自然下落; (4)地面电工将牵引绳安装成绝缘无极绳	23	(1)无人机悬停位置不当,扣2分,出现人员伤害,扣5分。 (2)展放过程中出现牵引绳缠挂,每次扣1分。 (3)无人机未从内侧起飞,扣3分。 (4)无人机升空位置不足即水平牵引,扣2分。 (5)无人机水平牵引距离不足,扣5分。 (6)未能一次性脱扣,扣2分。 (7)未能一次性展放牵引绳,扣3分。 (8)发生无人机坠落,扣15分			
7	牵引导轨绳及人体后备保护绳	(1)地面电工将牵引绳转移至作业点位置; (2)地面电工利用牵引绳将自爬升装置的导轨绳进行循环牵引,导轨绳一端稳定的安装在地面牢固位置上,另一端与地面垂直; (3)地面电工利用牵引绳将人体后备保护绳进行牵引	5	(1)导轨绳与人体后备保护绳发生缠绕,每次扣2分。 (2)导轨绳与人体后备保护绳未作单独牵引,扣3分。 (3)导轨绳未进行有效固定,扣3分			
8	安装自爬升装置	(1)等电位电工再次检查自爬升装置的转动情况及电池电量情况等,确认无误后将自爬升装置安装到导轨绳上; (2)等电位电工做好人体后备保护措施后,操作自爬升装置起到一定高度(0.8~1m)后,地面电工对自爬升装置及等电位电工进行冲击试验; (3)冲击试验合格后,等电位电工携带绝缘传递绳进行垂直攀升。	5	(1)未检查自爬升装置情况,扣2分。 (2)未做冲击试验,扣3分,冲击试验方法不当,扣2分。 (3)冲击试验方法不当造成自爬升装置损坏,扣5分			

续表

序号	项目名称	质量要求	分值	扣分标准	扣分原因	扣分	得分
9	进入强电场并修补导线	(1)等电位电工乘坐自爬升装置垂直上升至距导线0.3m处,检查屏蔽服各连接处无误,经工作组负责人同意后转移电位。 (2)等电位电工在导线上打好安全带保护,将绝缘传递绳布置在预绞丝缠绕处挂好绝缘传递绳向工作负责人汇报。 (3)地面电工挂好预绞丝向工作负责人汇报。 (4)等电位电工申请平整导线,工作负责人许可。按要求使用预绞丝修补导线向工作负责人汇报	25	(1)等电位电工未向工作负责人申请即开始爬升,扣3分;申请了但未同意请即开始登梯,扣2分。 (2)未有效控制控制后备保护绳,扣2分。 (3)未向工作负责人申请即开始电位转移,扣2分;申请了但未同意请即开始电位转移,扣1分。 (4)申请电位转移位置不合适,扣1分。 (5)转移电位动作不熟练,扣1分。 (6)工作负责人未认真监护,扣2分。 (7)预绞丝修补条传递过程发生磕碰,每次,扣1分。 (8)自爬升装置起降不平稳,每次扣2分			
10	退出电场	(1)等电位电工再次检查预绞丝修补导线情况,确认无误后向工作负责人汇报工作结束,申请退出电场。 (2)经工作负责人同意后,等电位电工对屏蔽服连接情况进行检查确认连接可靠后,按进电场相反步骤退出电场,乘坐自爬升装置下至地面	12	(1)未汇报工作结束,扣1分。 (2)未向工作负责人申请退出电场,扣2分。 (3)未对屏蔽服连接情况进行检查,扣1分。 (4)未向工作负责人申请即进行电位转移,扣2分;申请了但未得同意即开始,扣1分。 (5)申请电位转移位置不合适,扣1分。 (6)等电位电工退出电场动作不正确,反复放电,扣2分。 (7)未有效控制控制后备保护绳,扣1分			

续表

序号	项目名称	质量要求	分值	扣分标准	扣分原因	扣分	得分
11	工作结束	(1)工作负责人组织全体工作成员整理工器具和材料,将工器具清洁后放入专用的箱(袋)中;清理现场,做到"工完料尽场地清"。 (2)召开班后会,工作负责人进行工作总结和点评工作。点评本次工作的施工质量;点评全体工作成员的安全措施落实情况。 (3)工作负责人向值班调控人员汇报工作结束,申请恢复线路直流再启动,终结工作票	10	(1)工器具未清理,扣2分。 (2)工器具有遗漏,扣2分。 (3)未开班后会,扣3分。 (4)点评不到位,扣1分。 (5)未联系调控汇报工作结束,扣3分。 (6)汇报内容,每缺一项扣0.5分。(单位名称、负责人姓名、线路名称、工作完成情况、设备已恢复正常、人员已撤离) (7)工作票终结填写错误,扣1分			
	合计		100				

模块十一　带电更换±800kV特高压输电线路接地极线路间隔棒培训及考核标准

一、培训标准

（一）培训要求

模块名称	带电更换±800kV特高压输电线路接地极线路间隔棒	培训类别	操作类
培训方式	实操培训	培训学时	16学时
培训目标	1.掌握±800kV直流接地极直线塔进、出电场时采用"自爬升装置"作业方式的电学意义。 2.能完成采用"自爬升装置"进入±800kV直流接地极线路等电位作业点。 3.能独立完成±800kV直流接地极线路间隔棒更换的操作（等电位作业法）。		
培训场地	特高压直流实训线路		
培训内容	采用从自爬升装置进入电场，采用等电位作业法带电更换±800kV特高压输电线路接地极线路间隔棒		
适用范围	特高压±800kV直流输电线路检修人员		

（二）引用规程规范

（1）《±800kV直流架空输电线路设计规范》（GB/T 50790-2013）

（2）《±800kV直流架空输电线路检修规程》（DL/T 251-2012）

（3）《±800kV直流架空输电线路运行规程》（GB/T 28813-2012）

（4）《±800kV直流线路带电作业技术规范》（DL/T 1242-2013）

（5）《±800kV直流输电线路金具技术规范》（GB/T 31235-2014）

（6）《国家电网公司带电作业工作管理规定（试行）》（国家电网生〔2007〕751号）

（7）《国家电网公司电力安全工作规程（线路部分）》（Q/GDW1799.2-2013）

（8）《电工术语　架空线路》（GB/T 2900.51-1998）

（9）《电工术语　带电作业》（GB/T 2900.55-2016）

（10）《带电作业工具设备术语》（GB/T 14286-2002）

（11）《带电作业用绝缘滑车》（GB/T 13034-2008）

（12）《带电作业用绝缘绳索》（GB 13035-2008）

（13）《带电作业用工具、装置和设备使用的一般要求》（DL/T 877-2004）

（14）《带电作业工具、装置和设备预防性试验规程》（DL/T 976-2005）

（15）《±800kV直流输电线路带电作业技术导则》（Q/GDW 302-2009）

（16）《带电作业用屏蔽服装》（GB/T 6568-2008）

（17）《带电作业工具基本技术要求与设计导则》（GB 18037-2008）

（18）《带电设备红外线诊断应用规范》（DL/T 664-2016）

（三）培训教学设计

本设计以完成"带电更换±800kV特高压输电线路接地极线路间隔棒"为工作任务，按工作任务完成的标准化作业流程来设计各个培训阶段，每个阶段包括了具体的培训目标、培训内容、培训学时、培训方法（培训资源）、培训环境和考核评价等内容，如表1-11-1所示。

表1-11-1　带电更换±800kV特高压输电线路接地极线路间隔棒培训内容设计

培训流程	培训目标	培训内容	培训学时	培训方法与资源	培训环境	考核评价
1.理论教学	1.初步掌握采用自爬升装置进出±800kV直流接地极线路电场的方法。2.熟悉工器具配置及安全注意事项。3.熟悉±800kV直流接地极线路间隔棒更换的方法	1.采用自爬升装置进出±800kV直流接地极线路电场的电学意义。2.工器具配置、危险点分析及安全措施。3.±800kV直流接地极线路间隔棒更换方法和质量标准	2	培训方法：讲授法 培训资源：PPT、相关规程规范	多媒体教室	考勤、课堂提问和作业
2.准备工作	能完成作业前准备工作	1.作业现场查勘。2.编制培训标准化作业卡。3.填写培训操作工作票。4.完成本操作的工器具及材料准备	1	培训方法：1.现场查勘和工器具及材料清理采用现场实操方法 2.编写作业卡和填写工作票采用讲授方法 培训资源：1.±800kV接地极实训线路 2.特高压工器具库房 3.空白工作票	1.±800kV接地极实训线路；2.多媒体教室	

续表

培训流程	培训目标	培训内容	培训学时	培训方法与资源	培训环境	考核评价
3.作业现场准备	能完成作业现场准备工作	1.作业现场复勘。 2.工作申请。 3.作业现场布置。 4.班前会。 5.工器具检查	1	培训方法：演示与角色扮演法 资源：±800kV接地极实训线路	±800kV接地极实训线路	
4.培训师演示	通过现场观摩，使学员初步领会本任务操作流程	1.作业距离测量、导线温度测量。 2.地面电工利用无人机进行牵引绳的展放。 3.地面电工将后备保护绳及导轨绳牵引至工作位置。 4.等电位电工采用自爬升装置进入电场。 5.等电位电工进行间隔棒的更换	3	培训方法：演示法 资源：±800kV接地极实训线路	±800kV接地极实训线路	
5.学员分组训练	通过培训： 1.使学员能完成进、出±800kV直流接地极线路电场操作。 2.使学员能完成±800kV直流接地极线路间隔棒更换方法	1.学员分组（6人一组）训练进出±800kV直流接地极线路间隔棒更换技能操作。 2.培训师对学员操作进行指导和安全监护	8	培训方法：角色扮演法 资源：±800kV接地极实训线路	±800kV接地极实训线路	采用技能考核评分细则对学员操作评分
6.工作终结	通过培训： 1.使学员进一步认识操作过程不足处，便于后期提升。 2.培训学员安全文明生产的工作作风	1.作业现场清理。 2.向调控汇报工作。 3.班后会，对今天工作任务进行点评总结	1	培训方法：讲授和归纳法	±800kV接地极实训线路	

（四）作业流程

1. 工作任务

采用自爬升装置进入电场、到达作业点，采用等电位作业法带电更换±800kV特高压输电线路接地极线路间隔棒。

2. 天气及作业现场要求

（1）带电更换±800kV直流接地极线路间隔棒应在良好的天气下进行。

工作现场如遇雷电（听见雷声、看见闪电）、雪、雹、雨、雾等情况，应禁止本次作业。风力大于5级，或空气相对湿度大于80%时，不宜进行本次作业；作业过程中，注意天气变化，如遇大风、雨雾等紧急情况，按照规程规定正确采取措施，保证人员和设备安全。

（2）作业人员精神状态良好，熟悉工作中保证安全的组织措施和技术措施；应持有在有效期内的带电作业资质证书。

（3）工作负责人应事先组织相关人员完成现场勘察，根据勘察结果确定本次作业方法和所需工器具，以及应采取的必要措施，并办理带电作业工作票。

（4）作业现场应合理设置围栏，并妥当布置警示标示牌，禁止非工作人员入内。

（5）本项目需停用直流再启动装置。

（6）工作中安全距离及有效绝缘长度如表1-11-2所示。

表1-11-2 带电更换±800kV特高压输电线路接地极线路间隔棒的安全距离（m）

电压等级	人身与带电体安全距离	最小有效绝缘长度		最小组合间隙	转移电位时人体裸露部分与带电体的最小距离
		绝缘操作杆	绝缘承力工具、绝缘绳索		
±800kV直流接地极	0.75	1.3	1.0	1.2	0.3

3. 准备工作

3.1 危险点及其预控措施

（1）危险点——预防系统过电压

预控措施：本次作业为等电位作业，工作负责人在工作开始前，应与值班调控员联系，申请办理电力线路带电作业工作票，并申请停用直流再启动，由调控值班员履行许可手续。带电作业结束后应及时向调控值班员汇报。严禁约时停用和恢复直流再启动。在带电作业过程中如遇设备突然停电，作业人员应视设备仍然带电。工作负责人应尽快与调控联系，调控值班员未与工作负责人取得联系前不得强送电。

(2)危险点——触电伤害

预控措施：

①绝缘工具及绝缘绳索不得损坏、受潮、变形、失灵，不准使用非绝缘绳索（如棉纱绳、白棕绳、钢丝绳）。

②等电位作业人员应穿着阻燃内衣，衣服外面应穿戴全套屏蔽服（包括帽、衣裤、手套、袜和鞋），且各部分应连接良好，全套屏蔽服电阻不大于20Ω。

③等电位作业人员在电位转移前，应得到工作负责人的许可，人体裸露部分与带电体的最小距离不小于0.3m；地电位作业人员与带电体（等电位作业人员与接地体）的安全距离小于表1-11-2的规定。

④用绝缘绳索传递大件金属物品时，地电位作业人员应将金属物品接地后再接触。

⑤工作负责人和专责监护人应对作业人员进行不间断监护，随时纠正其不规范或违章动作。重点关注高处作业人员，使其保持足够的安全距离（符合表1-11-2的规定）。

⑥当耐热导线运行温度高于40℃时，应采用耐高温带电作业工器具，包括耐高温软质绝缘工器具、耐高温硬质绝缘工器具、耐高温隔热屏蔽服等，需采用电位转移棒进入等电位。

⑦隔热防护用具应视为导体，传递和使用过程中应保证相地及极间安全距离满足安全要求。

(3)危险点——高处坠落

预控措施：

①高处作业人员登高前，必须具备符合本项作业要求的身体状况、精神状态和技能素质。

②监护人员应随时纠正其不规范或违章动作，重点关注作业人员在转位的过程中不得失去安全带或绝缘后备保护绳的保护，严禁低挂高用。

(4)危险点——高处坠物伤人

预控措施：

①高处作业人员的个人工具及零星材料应装入工具袋，严禁在高处浮置物件、口中含物。

②地面作业人员必须正确佩戴安全帽，正确使用绳结，与作业点垂直下方距离不得小于坠落半径。

③作业现场设置围栏并挂好警示标示牌。监护人员应随时注意，禁止非工作人员及车辆进入作业区域。

（5）导线过热伤人

预控措施：

①作业前应与调度联系确认运行方式，避免在导线大电流流经时进行带电作业；

②作业人员应穿着专用的耐热性屏蔽服进行带电作业，防止导线过热伤人。

3.2 工器具及材料选择

带电更换±800kV特高压输电线路接地极线路间隔棒所需工器具及材料见表1-11-3。工器具出库前，应认真核对工器具的使用电压等级和试验周期，并检查确认外观良好、连接牢固、转动灵活，且符合本次工作任务的要求；工器具出库后，应存放在工具袋或工具箱内进行运输，防止脏污、受潮；金属工具和绝缘工器具应分开装运，防止因混装运输导致工器具变形、损伤等现象发生。

表1-11-3 带电更换±800kV特高压输电线路接地极线路间隔棒所需工器具及材料表

序号	名称	规格型号	单位	数量	备注
1	绝缘传递绳	TJS-14×50m	根	1	
2	绝缘传递绳	TJS-10×50m	根	1	
3	绝缘传递绳	TJS-14×100m	根	1	
4	绝缘传递绳	TJS-4×100m	根	1	
5	绝缘绳套	TJS-14	个	1	
6	绝缘滑车	0.5T	个	1	
7	无人机	六旋翼	架	1	
8	自爬升装置	ACTSAFE-II	台	1	
9	个人工具		套	1	
10	屏蔽服	500kV I型或隔热型	套	1	
11	人体后备保护绳		副	1	
12	安全带		副	1	
13	安全帽		顶	6	
14	防潮毡布	3m×3m	块	1	
15	万用表		个	1	
16	风速湿度仪	HT-8321	只	1	
17	兆欧表及测试电极	5000V	套	1	
18	对讲机		台	2	
19	红马甲	"工作负责人"	件	1	
20	安全围栏		套	若干	
21	间隔棒	接地极线路用	个	1	

3.3 作业人员分工

本任务作业人员分工如表1-11-4所示。

表1-11-4　带电更换±800kV特高压输电线路接地极线路间隔棒人员分工表

序号	工作岗位	数量（人）	工作职责
1	工作负责人	1	负责作业现场的各项工作
2	等电位电工	1	负责安装及使用自爬升装置进入电场更换间隔棒
3	无人机操控人员	1	负责操作无人机展放牵引绳
4	地面电工	3	负责传递工具、材料，配合等电位电工进出等电位

4. 工作程序

本任务工作流程如表1-11-5所示。

表1-11-5　带电更换±800kV特高压输电线路接地极线路间隔棒工作流程表

序号	作业内容	作业步骤及标准	安全措施及注意事项	责任人
1	确认线路名称和塔号、杆塔检查	工作负责人负责完成以下工作： (1)现场核对线路名称、杆塔编号，相别无误；基础及杆塔完好无异常；交叉跨越距离符合安全要求；确认缺陷情况及导地线规格型号等。 (2)检查地形环境符合作业要求。 (3)检查工作票所列安全措施与现场实际情况相符，必要时予以补充	(1)正确穿戴安全帽、工作服、工作鞋、劳保手套。 (2)不得在危及作业人员安全的气象条件下作业。 (3)严禁非工作人员、车辆进入作业现场	
2	现场作业环境测量	测量项目｜标准值｜实测值 风速｜≤10m/s｜ 湿度｜≤80%｜ 环境温度｜0～38℃｜ 导线温度｜—｜ 作业距离｜≥0.75m｜ 海拔｜海拔500m和1000m为界｜	风力大于5级，湿度大于80%时一般不宜进行带电作业	
3	工作许可	(1)工作负责人负责联系值班调控人员，按工作票内容申请停用线路直流再启动。 (2)经值班调控人员许可后，方可开始带电作业工作	不得未经值班调控人员许可即开始工作	

续表

序号	作业内容	作业步骤及标准	安全措施及注意事项	责任人
4	现场布置	正确装设安全围栏并悬挂标示牌： (1)安全围栏范围应充分考虑高处坠物，以及对道路交通的影响。 (2)安全围栏出入口设置合理。 (3)妥当布置"从此进出""在此工作""从此上下"等标示	对道路交通安全影响不可控时，应及时联系交通管理部门强化现场交通安全管控	
5	召开班前会	(1)全体工作成员列队。 (2)工作负责人宣读工作票，明确工作任务及人员分工；讲解工作中的安全措施和技术措施；查(问)全体工作成员精神状态；告知工作中存在的危险点及采取的预控措施。 (3)全体工作成员在工作票上签名确认	(1)工作票填写、签发和许可手续规范，签名完整。 (2)全体工作成员精神状态良好。 (3)全体工作成员明确任务分工、安全措施和技术措施	
6	检查工具	(1)等电位电工正确地穿戴好屏蔽服并检测合格，由负责人监督检查。 (2)正确佩戴个人安全用具(大小合适，锁扣自如)，由负责人监督检查。 (3)检查绝缘工具外观情况并测量其绝缘性能，并做好记录。 (4)检查好无人机及自爬升装置的运转情况	(1)金属、绝缘工具使用前，应仔细检查其是否损坏、变形、失灵。绝缘工具应使用清洁干燥的毛巾将其擦拭干净并用2500V及以上绝缘电阻表进行分段绝缘检测，阻值应不低于700MΩ。 (2)用万用表测量屏蔽服衣裤最远端点之间的电阻值不得大于20Ω。工作负责人认真检查作业电工屏蔽服的连接情况。 (3)检查工具组装情况并确认连接可靠。 (4)现场所使用的带电作业工具应放置在防潮毡布上	
7	无人机展放导轨绳	(1)无人机操控人员将无人机进行起飞至离地面1.6m左右，悬停好。 (2)地面电工将直径为4mm的绝缘传递绳挂在无人机的脱扣装置上。 (3)无人机操控人员继续操作无人机上升至合适位置时向外牵引迈过导线，继续牵引与导线距离大致相同距离，操控脱扣装置使牵引绳自然下落。 (4)地面电工将牵引绳安装成绝缘无极绳	(1)无人机悬停位置应合适，地面电工悬挂牵引绳时应注意安全。 (2)无人机牵引时应从线路方向内侧起飞向外展放牵引绳	

续表

序号	作业内容	作业步骤及标准	安全措施及注意事项	责任人
8	牵引导轨绳及人体后备保护绳	(1)地面电工将牵引绳转移至作业点位置。(2)地面电工利用牵引绳将自爬升装置的导轨绳进行循环牵引,导轨绳一端稳定的安装在地面牢固位置上,另一端与地面垂直。(3)地面电工利用牵引绳将人体后备保护绳进行牵引	(1)导轨绳的一端必须固定在地面牢固位置上。(2)导轨绳及人体后备保护绳应保持一定距离,防止缠绕在一起	
9	安装自爬升装置	(1)等电位电工再次检查自爬升装置的转动情况及电池电量情况等,确认无误后将自爬升装置安装到导轨绳上。(2)等电位电工做好人体后备保护措施后,操作自爬升装置起到一定高度(0.8~1m)后,地面电工对自爬升装置及等电位电工进行冲击试验。(3)冲击试验合格后,等电位电工携带绝缘传递绳进行垂直攀升	(1)自爬升装置应转动良好。(2)地面电工应对自爬升装置进行冲击试验,冲击时注意自爬升装置、导轨绳固定到地面处的受力情况。(3)工作负责人认真检查	
10	进入强电场并更换间隔棒	(1)等电位电工乘坐自爬升装置垂直上升至距导线0.3m处,检查屏蔽服各连接处无误,经工作组负责人同意后转移电位。(2)等电位电工在导线上打好安全带保护,将绝缘传递绳布置在合适位置。(3)等电位电工将旧间隔棒拆除,用绝缘传递绳将旧间隔棒捆绑好,地面电工采用绝缘传递绳将旧间隔棒传递至地面,同时将新间隔棒传递至等电位电工处。(4)等电位电工在间隔棒原来的位置上安装好新间隔棒	(1)地面电工应控制好人体后备保护绳,确保等电位电工高处作业不失去安全带保护。(2)等电位电工电位转移前必须得到工作负责人的许可,电位转移时严禁用头部充放电;人体裸露部分与带电体的距离不得小于0.3m;其组合间隙不得小于1.2m。(3)间隔棒传递过程中应避免碰撞,等单位电工在高处作业时应避免高空落物	
11	退出电场	(1)等电位电工再次检查间隔棒复位情况及销子到位情况,确认无误后向工作负责人汇报工作结束,申请退出电场。(2)经工作负责人同意后,等电位电工对屏蔽服连接情况进行检查确认连接可靠后,按进电场相反步骤退出电场,乘坐自爬升装置下至地面	(1)等电位电工退出电位前必须得到工作负责人的许可。(2)等电位电工在退出电场过程中与接地体和带电体两部分间隙所组成的组合间隙不得小于1.2m。(3)工作负责人认真监护、提醒。	

续表

序号	作业内容	作业步骤及标准	安全措施及注意事项	责任人
12	工作结束	(1)工作负责人组织全体工作成员整理工器具和材料,将工器具清洁后放入专用的箱(袋)中;清理现场,做到"工完料尽场地清"。 (2)召开班后会,工作负责人进行工作总结和点评工作。点评本次工作的施工质量;点评全体工作成员的安全措施落实情况。 (3)工作负责人向值班调控人员汇报工作结束,申请恢复线路直流再启动,终结工作票	不得约时恢复直流再启动	

二、考核标准

表1-11-6　国网四川省电力公司特高压直流技能培训考核评分细则

考生填写栏	编号:	姓名:	所在岗位:	单位:	日期:	年 月 日
考评员填写栏	成绩:	考评员:	考评组长:	开始时间:	结束时间:	操作时长:
考核模块	带电更换±800kV特高压输电线路接地极线路间隔棒		考核对象	±800kV直流接地极输电线路检修人员	考核方式	操作	考核时限	60min
任务描述	沿软梯进入电场对±800kV直流接地极线路间隔棒进行带电更换。
工作规范及要求	1. 带电作业工作应在良好天气下进行。如遇雷、雨、雪、雾天气不得进行带电作业。风力大于5级、湿度大于80%时,一般不宜进行带电作业。 2. 本项作业需工作负责人1名,等电位电工1人,无人机操作人员1人,地面辅助电工3人,采用自爬升装置进入电场对±800kV直流接地极线路间隔棒进行带电更换。 3. 工作负责人职责:负责本次工作任务的人员分工、工作票的宣读、办理线路停用直流再启动、办理工作许可手续、召开工作班前会、工作中突发情况的处理、工作质量的监督、工作后的总结。 4. 等电位电工职责:负责安装及使用自爬升装置进入电场更换间隔棒。 5. 无人机操作人员职责:负责操作无人机展放牵引绳 6. 地面电工职责:负责传递工器具、材料,配合等电位电工进出等电位。 7. 在带电作业中,如遇雷、雨、大风或其他任何情况威胁到工作人员的安全时,工作负责人或监护人可根据情况,临时停止工作。 给定条件: 1. 培训基地:特高压±800kV直流接地极线路 2. 工作票已办理,安全措施已经完备(直流再启动已停用),工作开始、工作终结时应口头提出申请(调控或考评员)。 3. 安全、正确地使用仪器对绝缘工具进行检测。 4. 必须按工作程序进行操作,工序错误扣除应做项目分值,出现重大人身、器材和操作安全隐患,考评员可下令终止操作(考核)

续表

考核情景准备	1. 线路:特高压±800kV直流接地极线路,工作内容:带电更换±800kV直流接地极线路间隔棒。 2. 所需作业工器具:绝缘传递绳2根(TJS-14);绝缘传递绳1根(TJS-10),绝缘传递绳1根(TJS-4),绝缘绳套1个(TJS-14),绝缘滑车1个(JH10-0.5);六旋翼无人机1架,自爬升装置1台,屏蔽服(I型或隔热型)1套;全身式安全带1副,绝缘电阻表(5000V型带测试电极),万用表1块;风速湿度仪1块,苫布1块。 3. 作业现场做好防护工作,作业现场安全措施(围栏等)已全部落实;禁止非作业人员进入现场,工作人员进入作业现场必须戴安全帽。 4. 考生自备工作服,阻燃纯棉内衣,安全帽,线手套,安全带(含二保绳)
备注	1. 各项目得分均扣完为止,出现重大人身、器材和操作安全隐患,考评员可下令终止操作。 2. 设备、作业环境、安全带、安全帽、工器具、屏蔽服等不符合作业条件考评员可下令终止操作

表1-11-7 国网四川省电力公司特高压直流技能培训考核评分标准

序号	项目名称	质量要求	分值	扣分标准	扣分原因	扣分	得分
1	现场复勘	(1)工作负责人到作业现场核对线路名称和杆塔编号、现场工作条件、缺陷部位等。 (2)检测风速、湿度等现场气象条件符合作业要求。 (3)检查工作票填写完整,无涂改,检查是否所列安全措施与现场实际情况相符,必要时予以补充	5	(1)未进行核对双重称号,扣1分。 (2)未核实现场工作条件(气象)、缺陷部位,扣1分。 (3)工作票填写出现涂改,每项扣0.5分,工作票编号有误,扣1分。工作票填写不完整,扣1.5分			
2	工作许可	(1)工作负责人联系值班调控人员,按工作票内容申请停用线路直流再启动。 (2)汇报内容规范、完整	2	(1)未联系调控部门(裁判)停用直流再启动,扣2分。 (2)汇报专业用语不规范或不完整的各,扣0.5分			
3	现场布置	正确装设安全围栏并悬挂标示牌: (1)安全围栏范围应充分考虑高处坠物,以及对道路交通的影响。 (2)安全围栏出入口设置合理。 (3)妥当布置"从此进出""在此工作""从此上下"等标示	3	(1)作业现场未装设围栏,扣0.5分。 (2)未设立警示牌,扣0.5分。 (3)未悬挂登塔作业标志,扣0.5分			
4	召开班前会	(1)全体工作成员全体人员正确佩戴安全帽、工作服。 (2)工作负责人佩戴红色背心,宣读工作票,明确工作任务及人员分工;讲解工作中的安全措施和技术措施;查(问)全体工作成员精神状态;告知工作中存在的危险点及采取的预控措施。 (3)全体工作成员在工作票上签名确认	3	(1)工作人员着装不整齐,扣0.5分,工作人员着装不整齐,每人次扣0.5分。 (2)未进行分工本项不得分,分工不明,扣1分。 (3)现场工作负责人未穿佩安全监护背心,扣0.5分。 (4)工作票上工作班成员未签字或签字不全的,扣1分			

续表

序号	项目名称	质量要求	分值	扣分标准	扣分原因	扣分	得分
5	工器具检查	(1)工作人员按要求将工器具放在防潮苫布上;防潮苫布应清洁、干燥。(2)工器具应按定置管理要求分类摆放;绝缘工器具不能与金属工具、材料混放;对工器具进行外观检查。(3)绝缘工具表面不应磨损、变形损坏,操作应灵活。绝缘工具应使用2500V及以上绝缘电阻表进行分段绝缘检测,阻值应不低于700MΩ,并用清洁干燥的毛巾将其擦拭干净。(4)塔上地电位和等电位人员按要求正确穿戴全套合格的屏蔽服、导电鞋,且各部分连接应良好;屏蔽服内不得贴身穿着化纤类衣服,并系好安全带;工作负责人应认真检查是否穿戴正确	7	(1)未使用防潮布并定置摆放工器具,扣1分。(2)未检查工器具试验合格标签及外观检查,每项扣0.5分。(3)未正确使用检测仪器对工器具进行检测,每项扣1分。(4)作业人员未正确穿戴屏蔽服且各部位连接良好,每人次扣2分。(5)现场工作负责人未对登塔作业人员进安全防护装备进行检查,扣1分。			
6	无人机展放导轨绳	(1)无人机操控人员将无人机进行起飞至离地面1.6m左右,悬停好。(2)地面电工将直径为4mm的绝缘传递绳挂在无人机的脱扣装置上。(3)无人机操控人员继续操作无人机上升至合适位置时向外牵引迈过导线,继续牵引与导线距离大致相同距离,操控脱扣装置使牵引绳自然下落;(4)地面电工将牵引绳安装成绝缘无极绳。	23	(1)无人机悬停位置不当,扣2分,出现人员伤害,扣5分。(2)展放过程中出现牵引绳缠挂,每次扣1分。(3)无人机未从内侧起飞,扣3分。(4)无人机升空位置不足即水平牵引,扣2分。(5)无人机水平牵引距离不足,扣5分。(6)未能一次性脱扣,扣2分。(7)未能一次性展放牵引绳,扣3分。(8)发生无人机坠落,扣15分			

第一部分 ±800kV 特高压输电线路运检带电作业培训及考核标准

续表

序号	项目名称	质量要求	分值	扣分标准	扣分原因	扣分	得分
7	牵引导轨绳及人体后备保护绳	(1)地面电工将牵引绳转移至作业点位置；(2)地面电工利用牵引绳将自爬升装置的导轨绳进行循环牵引，导轨绳一端稳定的安装在地面牢固位置上，另一端与地面垂直；(3)地面电工利用牵引绳将人体后备保护绳进行牵引。	5	(1)导轨绳与人体后备保护绳发生缠绕，每次扣2分。(2)导轨绳与人体后备保护绳未作单独牵引，扣3分。(3)导轨绳未进行有效固定，扣3分。			
8	安装自爬升装置	(1)等电位电工再次检查自爬升装置的转动情况及电池电量情况等，确认无误后将自爬升装置安装到导轨绳上；(2)等电位电工做好人体后备保护措施后，操作自爬升装置起到一定高度(0.8~1m)后，地面电工对自爬升装置及等电位电工进行冲击试验；(3)冲击试验合格后，等电位电工携带绝缘传递绳进行垂直攀升。	5	(1)未检查自爬升装置情况，扣2分。(2)未做冲击试验，扣3分，冲击试验方法不当，扣2分。(3)冲击试验方法不当造成自爬升装置损坏，扣5分			
9	进入强电场并更换间隔棒	(1)等电位电工乘坐自爬升装置垂直上升至距导线0.3m处，检查屏蔽服各连接处无误，经工作组负责人同意后转移电位；(2)等电位电工在导线上打好安全带保护，将绝缘传递绳布置在合适位置；(3)等电位电工将旧间隔棒拆除，用绝缘传递绳将旧间隔棒捆绑好，地面电工采用绝缘传递绳将旧间隔棒传递至地面，同时将新间隔棒传递至等电位电工处；(4)等电位电工在间隔棒原来的位置上安装好新间隔棒。	25	(1)等电位电工未向工作负责人申请即开始爬升，扣3分；申请了但未同意请即开始登梯，扣2分。(2)未有效控制控制后备保护绳，扣2分。(3)未向工作负责人申请即开始电位转移，扣2分；申请了但未同意请即开始电位转移，扣1分。(4)申请电位转移位置不合适，扣1分。(5)转移电位动作不熟练，扣1分。(6)工作负责人未认真监护，扣2分。(7)间隔棒传递过程发生磕碰，每次扣1分。(8)自爬升装置起降不平稳，每次扣2分。			

续表

序号	项目名称	质量要求	分值	扣分标准	扣分原因	扣分	得分
10	退出电场	(1)等电位电工再次检查间隔棒复位情况及销子到位情况，确认无误后向工作负责人汇报工作结束，申请退出电场；(2)经工作负责人同意后，等电位电工对屏蔽服连接情况进行检查确认连接可靠后，按进电场相反步骤退出电场，乘坐自爬升装置下至地面。	12	(1)未汇报工作结束，扣1分。(2)未向工作负责人申请退出电场，扣2分。(3)未对屏蔽服连接情况进行检查，扣1分。(4)未向工作负责人申请即进行电位转移，扣2分；申请了但未得同意即开始，扣1分。(5)申请电位转移位置不合适，扣1分。(6)等电位电工退出电场动作不正确，反复放电，扣2分。(7)未有效控制控制后备保护绳，扣1分			
11	工作结束	(1)工作负责人组织全体工作成员整理工器具和材料，将工器具清洁后放入专用的箱（袋）中；清理现场，做到"工完料尽场地清"。(2)召开班后会，工作负责人进行工作总结和点评工作。点评本次工作的施工质量；点评全体工作成员的安全措施落实情况。(3)工作负责人向值班调控人员汇报工作结束，申请恢复线路直流再启动，终结工作票	10	(1)工器具未清理，扣2分。(2)工器具有遗漏，扣2分。(3)未开班后会，扣3分。(4)点评不到位，扣1分。(5)未联系调控汇报工作结束，扣3分。(6)汇报内容，每缺一项扣0.5分。(单位名称、负责人姓名、线路名称、工作完成情况、设备已恢复正常、人员已撤离)(7)工作票终结填写错误，扣1分			
	合计		100				

第二部分

±800kV 特高压输电线路运检停电检修培训及考核标准

模块一 停电更换±800kV特高压输电线路直线塔双V型瓷质绝缘子培训及考核标准

一、培训标准

(一)培训要求

模块名称	停电更换±800kV特高压输电线路直线塔双V型瓷质绝缘子	培训类别	操作类
培训方式	实操培训	培训学时	21学时
培训目标	1.掌握各类工器具、机具的使用方案和受力结构,以及更换整串绝缘子技术要点。 2.能熟练掌握停电更换更换±800kV特高压输电线路直线杆塔双V型瓷质绝缘子的操作流程、技术方法和施工作业危险点。 3.作为主要作业人员,能熟练完成更换±800kV特高压输电线路直线杆塔双V型瓷质绝缘子的更换		
培训场地	特高压±800kV直流实训线路		
培训内容	正确使用各类受力工器具的操作方法正确安装各类工器具,采用停电作业法更换±800kV输电线路直线杆塔双V型瓷质绝缘子		
适用范围	特高压直流输电线路检修人员		

(二)引用规程规范

(1)《架空送电线路运行规程》(DL/T 741-2010)

(2)《110~500kV架空送电线路设计技术规程》(DL/T 5092-1999)

(3)《国家电网公司电力安全工作规程(线路部分)》(Q/GDW1799.2-2013)

(4)《±800kV直流架空输电线路设计规范》(GB 50790-2013)

(5)《±800kV直流架空输电线路检修规程》(DL/T 251-2012)

(6)《±800kV直流架空输电线路运行规程》(GB/T 28813-2012)

(7)《110(66)kV~500kV架空输电线路检修规范》(国家电网公司)

(8)《架空输电线路状态检修导则》(DLT 1248-2013)

(9)《输变电设备状态检修管理规定》(国家电网公司)

(10)《输变电设备状态检修试验规程》(国家电网公司)

(三) 培训教学设计

本设计以完成"停电更换±800kV特高压输电线路直线塔双V型瓷质绝缘子"为工作任务，按工作任务完成的标准化作业流程来设计各个培训阶段，每个阶段包括了具体的培训目标、培训内容、培训学时、培训方法（培训资源）、培训环境和考核评价等内容，如表2-1-1所示。

表2-1-1 停电更换±800kV特高压输电线路直线塔双V型瓷质绝缘子培训内容设计

培训流程	培训目标	培训内容	培训学时	培训方法与资源	培训环境	考核评价
1.理论教学	1.掌握各类工器具、机具的使用方案和受力结构，以及更换瓷质绝缘子技术要点。 2.能熟练掌握更换±800kV输电线路直线杆塔双V型瓷质绝缘子的操作流程、技术方法和施工作业危险点。	1.正确使用各类受力工器具，熟悉绞磨等机具的操作方法。 2.正确安装各类工器具。 3.采用停电作业法更换±800kV输电线路直线杆塔双V型瓷质绝缘子。	2	培训方法：讲授法。 培训资源：PPT、相关规程规范。	多媒体教室	考勤、课堂提问和作业
2.准备工作	能完成作业前准备工作	1.作业现场查勘。 2.编制培训标准化作业卡。 3.填写培训操作工作票。 4.完成本操作的工器具及材料准备	1	培训方法： 1.现场查勘和工器具及材料清理采用现场实操方法。 2.编写作业卡和填写工作票采用讲授方法。 培训资源： 1.±800kV实训线路。 2.特高压工器具库房。 3.空白工作票	1.特高压输电实训线路； 2.多媒体教室	
3.作业现场准备	能完成作业现场准备工作	1.作业现场复勘。 2.工作申请。 3.作业现场布置。 4.班前会。 5.工器具及材料检查。	1	培训方法：演示与角色扮演法。 资源：±800kV实训线路	±800kV实训线路	

续表

培训流程	培训目标	培训内容	培训学时	培训方法与资源	培训环境	考核评价
4.培训师演示	通过现场观摩，使学员初步领会本任务操作流程	1.各类工器具使用方法讲解。 2.演示更换瓷质绝缘子的塔上工器具连接方式。 3.高空作业人员配合演示更换瓷质绝缘子的操作流程。 4.利用地面人员配合更换±800kV输电线路直线杆塔双V型瓷质绝缘子。	2	培训方法：演示法。 资源：±800kV实训线路	±800kV实训线路	
5.学员分组训练	1.能掌握各类受力工器具的使用方法和注意事项。 2.掌握更换瓷质绝缘子的全部操作流程。 3.能完成±800kV输电线路直线杆塔双V型瓷质绝缘子的更换	1.学员分组（高空4人、地面配合7人）训练工器具、机具的操作方法和更换双V型瓷质绝缘子的现场实际操作。 2.培训师对学员操作进行指导和安全监护	14	培训方法：角色扮演法。 资源：±800kV实训线路	±800kV实训线路	采用技能考核评分细则对学员操作评分
6.工作终结	1.使学员进一步辨析操作过程不足之处，便于后期提升。 2.培训学员安全文明生产的工作作风	1.作业现场清理。 2.向调度汇报工作。 3.班后会，对本次工作任务进行点评总结。	1	培训方法：讲授和归纳法	±800kV实训线路	

（四）作业流程

1. 工作任务

完成停电更换±800kV特高压输电线路直线塔双V型瓷质绝缘子。

2. 天气及作业现场要求

（1）停电更换±800kV特高压输电线路直线塔双V型瓷质绝缘子应在良好的天气进行。

在5级及以上的大风以及暴雨、雷电、冰雹、大雾、沙尘暴等恶劣天气下，应停止露天高处作业。特殊情况下，确需在恶劣天气进行抢修时，应组织有关人员充分讨论必要的安全措施，经本单位批准后方可进行。

（2）作业人员精神状态良好，工作班成员认真学习工作票和安全技术措施，所有人员做到"四清楚"（作业任务清楚、危险点清楚、作业程序清楚、安全措施清楚）。

（3）作业前停送电联系人必须与调度联系履行工作许可手续，严禁约时停送电。工作负责人必须在得到许可人的许可工作命令后，方可在需检修的线路上验电、挂设接地线和进行检修工作。

（4）停电后，工作负责人应认真做好记录。

（5）登杆前应检查塔上是否有蜂窝，发现蜂窝严禁登塔。

（6）塔上作业人员必须使用双保险安全带，并佩戴护目镜。

3. 准备工作

3.1 危险点及其预控措施

（1）危险点——误登带电线路

预控措施：

①登杆塔作业前，工作负责人、工作班成员应共同认真核查双重名称和识别标记（色标、判别标志等）与停电线路名称相符。

②登杆塔前应检查铁塔根部、基础等，必须牢固可靠。

③登杆塔前应检查登高工器具和设施，如安全带、脚钉、塔材等必须完整牢靠。

④不涉及挂设接地线的中间作业人员，应认真核实线路相序、色标、名称、编号与停电线路相符，确认线路名称无刷错、刷反等情况后，方可登杆。

（2）危险点——登塔时、塔上作业时违反安规进行操作，可能引起高空坠落

预控措施：

①攀爬过程中，为防止登杆人员串落，登杆作业人员间距不得小于1.6m。

②攀爬铁塔前应将脚底泥土清除干净，检查工具包完整，攀爬过程中不得掉落物件伤人。

③作业人员攀登杆塔时应戴好安全帽，穿软底鞋，动作不能过大，匀步攀登。

④攀爬过程中，安全带应收拾妥当，长尾绳放置在工具包内，主带应挂在肩上，防止攀爬过程中安全带勾挂脚钉和塔材，致使作业人员高空坠落。

⑤杆塔上移位时，不得失去安全带保护，做到踩稳抓牢。

⑥到达作业点位置，系好安全带（绳），应牢固可靠，不得低挂高用。

⑦未验电前，人体、绳索等与导线的安全距离必须不小于10.1m，工作中应设专人监护。

（3）危险点——高处坠物伤人

预控措施：

①地面人员不得站在作业点垂直下方。塔上人员应防止落物伤人，使用的工具、材料应用绳索传递。

②在高处作业应使用工具袋，较大的工具应固定在牢固的构件上，不准随便乱放。

③使用绞磨起吊过程中，应设专人指挥，统一配合，瓷质绝缘子串刚离地后应进行冲击检查。

（4）危险点——防止感应电伤人

①为防感应电伤人，塔上作业人员应穿全套屏蔽服。

②如需接触架空地线，在架空地线接触前应进行可靠接地。

（5）危险点——现场作业安全监护

①自作业开始至作业终结，安全监护人必须始终在现场对作业人员进行不间断的安全监护。

②工作负责人，监护人必须穿安全监护背心。

（6）危险点——交通安全

①出车时应注意车辆行驶安全，谨慎驾驶车辆，禁止违法行车。

3.2 工器具及材料选择

停电更换±800kV特高压输电线路直线塔双V型瓷质绝缘子所需工器具及材料见表2-1-2。工器具出库前，应认真核对工器具的使用电压等级和试验周期，并检查确认外观良好、连接牢固、转动灵活，且符合本次工作任务的要求；工器具出库后，应防止脏污、受潮；金属工具和绝缘工器具应分开装运，防止因混装运输导致工器具变形、损伤等现象发生。

表2-1-2 停电更换±800kV特高压输电线路直线塔双V型瓷质绝缘子所需工器具及材料表

序号	名称	规格型号	单位	数量	备注
1	接地线	±800kV	组	2	绝缘工具
2	绝缘手套	10kV	付	2	绝缘工具
3	验电器	±800kV专用	支	2	绝缘工具
4	全身式安全带	含带缓冲包长20M的后保绳	套	3	个人防护用具
5	绞磨	5T	台	2	机械工具
6	六勾卡	适用于六分裂导线	套	2	金属工具
7	卸扣	10T	个	20	金属工具
8	手扳葫芦	6T	付	2	金属工具
9	手扳葫芦	12T	付	2	金属工具
10	个人手动工具		套	4	金属工具

续表

序号	名称	规格型号	单位	数量	备注
11	个人保安线	（直径不小于16mm²）	根	1	其他工具
12	钢丝套	Φ22	根	6	其他工具
13	磨绳	Φ6	米	2圈每圈200米	其他工具
14	对讲机		台	5	其他工具
15	吊绳滑车	1T	个	2	其他工具
16	传递绳	Φ16	套	3	其他工具
17	拔销器		把	2	其他工具
18	安全背心		件	2	其他工具
19	安全围栏		卷	4	其他工具
20	垫木		块	若干	其他工具
21	防潮苫布		张	1	其他工具
22	瓷质绝缘子		串	1	材料

3.3 作业人员分工

本任务作业人员分工如表2-1-3所示。

表2-1-3 停电更换±800kV特高压输电线路直线塔双V型瓷质绝缘子人员分工表

序号	工作岗位	数量（人）	工作职责
1	工作负责人	1	负责本次工作任务的人员分工、工作前的现场查勘、作业方案的制定、工作票的填写、现场复勘、办理工作许可手续、召开工作班前会、落实现场安全措施、负责作业过程中的安全监督、工作中突发情况的处理、工作质量的监督、工作后的总结
2	安全监护人	2	负责本次工作过程中的安全监护工作
3	高空作业人员	4	负责本次停电更换±800kV双V型瓷质绝缘子操作
4	地面辅助人员	7	负责本次作业过程的地面辅助工作，包括2名绞磨操作人员

4. 工作程序

本任务工作流程如表2-1-4所示。

表2-1-4 停电更换±800kV特高压输电线路直线塔双V型瓷质绝缘子工作流程表

序号	作业内容	作业标准	安全注意事项	责任人
1	现场复勘	工作负责人负责完成以下工作： (1)现场核对线路名称无误；基础及铁塔完好无异常；交叉跨越距离符合安全要求；确认缺陷情况及导地线规格型号等。 (2)检查地形环境符合作业要求。 (3)检查工作票所列安全措施与现场实际情况相符，必要时予以补充	(1)正确穿戴安全帽、工作服、工作鞋、劳保手套。 (2)不得在危及作业人员安全的气象条件下作业。 (3)严禁非工作人员、车辆进入作业现场	
2	工作许可	作业前停送电联系人必须与调度联系履行工作许可手续。	(1)不得未经工作许可人许可即开始工作。 (2)严禁约时停送电	
3	现场布置	正确装设安全围栏并悬挂标示牌： (1)安全围栏范围应充分考虑高处坠物，以及对道路交通的影响。 (2)安全围栏出入口设置合理。 (3)妥当布置"从此进出""在此工作""车辆慢行"或"车辆绕行"等标示	对道路交通安全影响不可控时，应及时联系交通管理部门强化现场交通安全管控	
4	召开班前会	(1)全体工作成员列队。 (2)工作负责人宣读工作票，明确工作任务及人员分工；讲解工作中的安全措施和技术措施；查(问)全体工作成员精神状态；告知工作中存在的危险点及采取的预控措施。 (3)全体工作成员在工作票上签名确认	(1)工作票填写、签发和许可手续规范，签名完整。 (2)全体工作成员精神状态良好。 (3)全体工作成员明确任务分工、安全措施和技术措施	
5	检查工器具	(1)在防潮苫布上，将工器具按作业要求准备齐备，并分类定置摆放整齐。检查工器具外观和试验合格证，无遗漏。 (2)检查人员向工作负责人汇报各项检查结果符合作业要求	(1)防潮苫布数量足够，设置位置合理，保持清洁、干燥。 (2)工器具外观检查合格，无损伤、受潮、变形、失灵现象，合格证在有效期内	
6	登杆塔	(1)登杆塔作业前，必须先核对线路名称及编号。对同塔多回线路，工作负责人、工作班成员应共同认真核查双重名称和识别标记(色标、判别标志等)。 (2)登杆塔前应检查铁塔根部、基础等，必须牢固可靠。 (3)攀爬过程，为防止登杆人员串落，登杆作业人员间距不得小于1.6m，安全带收拾妥当，后保绳放置在工具包内，主带应挂在肩	(1)作业人员攀登杆塔时应戴好安全帽，穿软底鞋，动作不能过大，匀步攀登。 (2)攀爬过程中，安全带应收拾妥当，长尾绳放置在工具包内，主带应挂在肩上，防止攀爬过程中安全带勾挂脚钉和塔材，致使作业人员高空坠落。	

续表

序号	作业内容	作业标准	安全注意事项	责任人
		上,防止攀爬过程中安全带勾攀脚钉和塔材,致使作业人员高空坠落。 (4)登杆塔至横担处时,监护人和作业人员应再次核对停电线路的识别标记和双重名称,确实无误后方可进入作业点位	(3)杆塔上移位时,不得失去安全带保护,做到踩稳抓牢。 (4)到达作业点位置,系好安全带(绳),应牢固可靠,不得低挂高用。 (5)未验电前,人体、无头绳等与导线的安全距离必须不小于10.1m,工作中应设专人监护	
7	验电、装设接地线	验电杆就位后,将安全带系在牢固可靠构件或电杆上,必须检查扣环是否正确就位。验电器等工器具必须使用传递绳传递。 (2)检查接地线完好,按程序(先接接地端,后导线端)装设好接地线。 (3)装设接地线时,必须使用绝缘绳或绝缘手柄进行操作,禁止直接用手操作接地线的金属部分的方式装设接地线。确认接地线的夹头与导线连接紧密可靠	(1)验电器在领用时和使用前应检查是否正常。 (2)禁止以缠绕导线的方式装设接地线	
8	更换双V型瓷质绝缘子	高空作业人员到达作业点位后,2名高空人员利用通过瓷质绝缘子进入导线,塔上作业人员利用传递绳将两套六勾卡与两把12T手板葫芦传递至塔上,并固定在垂直于线路方向的瓷质绝缘子及金具两侧后,架设牢固,并保留足够的人员操作空间。 确认所有连接部位已固定后,将1把12T手板葫芦和绝缘子后备保护绳在绝缘子两端装设完毕,2名高空人员同时收紧两把12T手板葫芦,待12T手板葫芦完全承受瓷质绝缘子串的拉力后,应再次检查横担有无变形,连接部位有无异常,确认一切正常后,使用传递绳将磨绳传递瓷质绝缘子接地端并绑扎牢固后,再利用两把6T手板葫芦收紧导线端绝缘子,并将磨绳在瓷质绝缘子导线端绑扎牢固,导线端高空人员取下瓷质绝缘子与导线端的连接金具的连接部分,地面人员绞磨收紧接地端磨绳,铁塔端高空人员配合取下瓷质绝缘子铁塔端连接金具的连接部分。 待瓷质绝缘子的连接部分均已取下后,方可利用两台绞磨同时缓慢将瓷质绝缘子下放,地面人员取下瓷质绝缘子,更换完好瓷质绝缘子,并将瓷质绝缘子绑扎牢固到位后,方可传递至塔上作业人员处	(1)使用手扳葫芦、六勾卡更换绝缘子串过程中,在手扳葫芦开始承受导线荷载后,必须检查手扳葫芦、钢丝绳套(吊装带)、卸扣的连接和受力情况,并做冲击试验,确认完全可靠后方可继续收紧手扳葫芦。 (2)在使用手扳葫芦过程中,应防止与瓷质绝缘子碰撞,避免损伤瓷质绝缘子。 (3)在使用绞磨吊放瓷质绝缘子串时,地面工作人员和塔上的工作人员必须密切配合,防止磨绳缠绕,绝缘子串碰撞损伤导线。吊放绝缘子串时,地面人员不得站在垂直下方	

续表

序号	作业内容	作业标准	安全注意事项	责任人
		将瓷质绝缘子拉至作业点位后,塔上作业人员应先连接导线端连接金具的连接部分,并确保销子到位后方可继续进行连接瓷质绝缘子接地端的连接工作。 安装好瓷质绝缘子后,应检查连接金具是否安装到位、销子是否安装到位,并进行冲击试验,确认安装正确,连接可靠后方可松出绝缘子后备保护绳再缓慢放松2把12T手扳葫芦		
9	拆除工器具	拆除六勾卡、手扳葫芦、钢丝套等工器具	上下传递工具过程中不得碰撞,绑扎绳应正确可靠,防止高处坠物	
10	工作结束	(1)工作负责人组织全体工作成员整理工器具和材料,清理现场,做到"工完料尽场地清"。 (2)召开班后会,工作负责人进行工作总结和点评工作。点评本次工作的施工质量;点评全体工作成员的安全措施落实情况。 (3)工作负责人向工作许可人汇报工作结束,恢复停电线路送电,终结工作票		

二、考核标准

表2-1-5 国网四川省电力公司特高压直流输电线路运检技能考核评分细则

考生填写栏	编号:	姓名:	所在岗位:	单位:	日期:	年 月 日		
考评员填写栏	成绩:	考评员:	考评组长:	开始时间:	结束时间:	操作时长:		
考核模块	停电更换±800kV特高压输电线路直线塔双V型瓷质绝缘子		考核对象	特高压直流输电线路检修人员	考核方式	操作	考核时限	150min
任务描述	停电更换±800kV直流输电线路级ⅠV型瓷质绝缘子							
工作规范及要求	1. 给定条件:±800kV实训线路C相V型瓷质绝缘子损坏,需要更换。线路已经停电、验电、挂接地线,所使用绝缘子已经过测试,工作票已办理,安全措施已经完备。 2. 整个过程主要操作流程由工作负责人1名、专责监护人2人,塔上高空人员4人配合完成,地面辅助7人,协助参考人员完成工器具、材料的上、下传递以及其他非技术性工作。 3. 操作前参考人员应作必要的安全检查。 4. 工作开始应口头提出申请,工作结束时应口头汇报							

续表

考核情景准备	1. 工器具:六勾卡2套、个人保安线1根、Φ22钢丝套6根、10T卸扣20个、12T手扳葫芦2把、6T手扳葫芦1把、传递绳3根、1T滑车2个、绳套2根、对讲机5个、双保险安全带4套、防潮苫布1张。 2. 材料:同型号瓷质绝缘子一串。 3. 在培训线路上操作
备注	1. 个人工器具由参考人员自备。 2. 各项目得分均扣完为止

表2-1-6　国网四川省电力公司特高压直流输电线路运检技能考核评分标准

序号	项目名称	质量要求	分值	扣分标准	扣分原因	扣分	得分
1	工具材料准备						
1.1	个人工具检查	活动扳手、平口钳、拔销钳、工具包符合质量要	2	错漏1项,扣1分			
1.2	受力工具检查	六勾卡、手扳葫芦、钢丝绳检查,在试验合格期内	3	错漏1项,扣1分			
1.3	安全工器具检查	双保险安全带、个人保安线、绝缘手套符合质量要求,并在试验合格周期内	3	错漏1项,扣1分			
1.4	材料检查	核对瓷质绝缘子串型号,外观检查符合要求	2	错漏1项,扣1分			
2	场地布置						
2.1	场地围栏	场地围栏布置	2	未布置,扣2分			
3	登塔及横担上的操作						
3.1	登塔	(1)检查杆塔基础无异常。 (2)正确携带传递绳(吊绳头折双、打死结、斜挎身上)。 (3)沿脚钉侧主材正确登塔	6	(1)未检查,每项扣1分。 (2)未携带传递绳,扣2分,传递绳携带方式不规范,扣1分。 (3)手抓脚钉,每次扣1分。 (4)未沿脚钉侧主材登塔,扣2分			
3.2	进入横担上工作点	由塔身到横担上工作点不得失去安全带保护	3	未正确使用安全带,扣3分			
3.3	安装滑车、个人保安线	传递滑车安装位置正确,方便操作,并加挂个人保安线	3	滑车安装不规范,扣1分;个人保安线漏挂,扣2分			
4	绝缘子串上操作						

续表

序号	项目名称	质量要求	分值	扣分标准	扣分原因	扣分	得分
4.1	进入工作点	(1)将双保险安全带的安全绳系在横担适当位置。 (2)沿绝缘子串进入作业点,将围杆带系在绝缘子串上。 (3)检查绝缘子串锁紧销,连接金具	6	(1)未使用双保险安全带,扣4分。 (2)未正确使用双保险安全带,每次扣2分。 (3)未检查,扣2分			
4.2	安装工器具	(1)正确安装钢丝绳套、6T手扳葫芦、12T手扳葫芦。 (2)正确安装六勾卡。 (3)钢丝绳套、卸扣型号选用正确、安装可靠 (4)安装的手扳葫芦、个人保安线不影响人员操作	9	(1)未对塔材采取保护,每处扣1分。 (2)手扳葫芦、六勾卡安装不对,每处扣1分。 (3)手扳葫芦受力后有碰撞、缠绕,每处扣1分。 (4)钢丝绳套、卸扣型号选用不正确,每处扣1分。 (5)手扳葫芦漏,每套扣4分			
4.3	收紧瓷质绝缘子串	(1)同时收紧2把12T手扳葫芦,确保手扳葫芦受力均衡。 (2)手扳葫芦串受力后,检查滑车、卸扣、钢丝绳等连接部位,确认连接可靠,并对绝缘子串做冲击。 (3)确认受力无误后继续收紧手扳葫芦,直至瓷质绝缘子串松弛。	6	(1)碰响绝缘子,每次扣1分。 (2)未冲击试验,扣3分。 (3)使用手扳葫芦不正确,扣2分			
4.4	更换直线杆塔双V型瓷质绝缘子	(1)将传递绳绑扎在瓷质绝缘子的合适位置,高空人员先把两把6T手扳葫芦再取脱瓷质绝缘子导线端,并将导线端磨绳在瓷质绝缘子导线端绑扎牢固,地面绞磨操作人员收紧接地端磨绳后再取脱瓷质绝缘子接地端。 (2)两名地面绞磨操作人员同时操作将瓷质绝缘子缓慢传送至地面。 (3)将新瓷质绝缘子通过磨绳传递到杆塔上,塔上作业先连接导线端,并将销子安装到位。 (4)继续安装瓷质绝缘子接地端,安装金具R销子,并检查是否安装到位。 (5)确认连接无误后,缓慢松出手扳葫芦,使瓷质绝缘子受力,并做冲击试验	16	(1)未冲击试验,扣2分。 (2)绳索缠绕,扣2分。 (3)传递物有撞击现象,每次扣2分。 (4)掉落物件,扣5分。 (5)未装锁紧销,每个扣1分。 (6)拆装导线端和铁塔端顺序错误,扣4分			

续表

序号	项目名称	质量要求	分值	扣分标准	扣分原因	扣分	得分
4.5	撤除工器具	(1)检查锁紧销、球头是否齐全到位,碗口朝向正确,清洁绝缘子串表面污垢。(2)拆除手扳葫芦、六勾卡及钢丝绳套,传递至地面	7	(1)绳索缠绕,扣2分。(2)传递物有撞击现象,每次,扣2分。(3)锁紧销位置不正确,扣1分。(4)碗口朝向不正确,扣1分。(5)未清洁绝缘子串,扣1分			
4.6	清理杆塔上工器具	确认无遗留物	4	有遗留物扣4分			
4.7	从导线到塔身	(1)沿瓷质绝缘子到进入到铁塔横担。(2)攀爬软梯过程中不得失去安全带保护	8	(1)失去安全带保护,扣4分。(2)未正确使用软梯,扣4分			
5	下塔	(1)必须沿脚钉侧主材正确下塔。(2)正确携带传递绳(吊绳头折双、打死结,斜挎肩上)	8	(1)未携带传递绳,扣4分,传递绳携带方式不规范,扣2分。(2)手抓脚钉,每次扣1分。(3)未沿脚钉侧主材登塔,扣4分			
6	其他要求						
6.1	塔上作业	(1)严禁高处坠物。(2)在操作过程中应双手协调配合操作。(3)严禁浮置物品。(4)严禁口中含物	5	(1)高处坠物,每件扣5分。(2)动作不协调,扣2分。(3)浮置物品,扣2分。(4)口中含物,扣2分			
6.2	着装	工作服、工作胶鞋、安全帽、劳保手套穿戴正确	2	漏一项,扣2分			
6.3	清理现场	完工后清理作业现场,符合文明生产要求	2	未清理作业现场扣2分			
6.4	完成时间	在规定时间内按要求完成	3	超过时间10min,扣1分,达到480min即终止操作,只记完成部分得分			
	合计		100				

模块二 停电更换±800kV特高压输电线路耐张整串绝缘子培训及考核标准

一、培训标准

(一) 培训要求

模块名称	停电更换±800kV特高压输电线路耐张整串绝缘子	培训类别	操作类
培训方式	实操培训	培训学时	32学时
培训目标	1.掌握各类工器具、机具的使用方案和受力结构,以及更换整串绝缘子技术要点。 2.能熟练掌握停电更换±800kV特高压输电线路耐张整串绝缘子的操作流程、技术方法和施工作业危险点。 3.作为高空主要作业人员,能熟练完成±800kV特高压输电线路耐张整串绝缘子的更换换。		
培训场地	特高压直流实训线路		
培训内容	正确使用各类受力工器具,熟练绞磨等机具的操作方法,采用"滑轮组"作业方式正确安装各类工器具,采用停电作业法更换±800kV特高压输电线路耐张整串绝缘子。		
适用范围	特高压直流输电线路检修人员		

(二) 引用规程规范

(1)《架空送电线路运行规程》(DL/T 741-2010)

(2)《110~500kV架空送电线路设计技术规程》(DL/T 5092-1999)

(3)《国家电网公司电力安全工作规程(线路部分)》(Q/GDW1799.2-2013)

(4)《±800kV直流架空输电线路设计规范》(GB 50790-2013)

(5)《±800kV直流架空输电线路检修规程》(DL/T 251-2012)

(6)《±800kV直流架空输电线路运行规程》(GB/T 28813-2012)

(7)《110(66)kV~500kV架空输电线路检修规范》(国家电网公司)

(8)《架空输电线路状态检修导则》(DLT 1248-2013)

(9)《输变电设备状态检修管理规定》(国家电网公司)

(10)《输变电设备状态检修试验规程》(国家电网公司)

（三）培训教学设计

本设计以完成"停电更换±800kV特高压输电线路耐张整串绝缘子"为工作任务，按工作任务完成的标准化作业流程来设计各个培训阶段，每个阶段包括了具体的培训目标、培训内容、培训学时、培训方法（培训资源）、培训环境和考核评价等内容，如表2-2-1所示。

表2-2-1 停电更换±800kV特高压输电线路耐张整串绝缘子培训内容设计

培训流程	培训目标	培训内容	培训学时	培训方法与资源	培训环境	考核评价
1.理论教学	1.掌握各类工器具、机具的使用方案和受力结构，以及更换整串绝缘子技术要点。 2.能熟练掌握更换±800kV输电线路耐张整串绝缘子的操作流程、技术方法和施工作业危险点。	1.正确使用各类受力工器具，熟悉绞磨等机具的操作方法。 2.采用"滑轮组"作业方式正确安装各类工器具。 3.正确利用绞磨配合避开跳线串传递绝缘子。 4.采用停电作业法更换±800kV输电线路耐张整串绝缘子。	2	培训方法：讲授法。 培训资源：PPT、相关规程规范、技术标准。	多媒体教室	考勤、课堂提问和作业
2.准备工作	能完成作业前准备工作	1.作业现场查勘。 2.编制培训标准化作业卡。 3.填写培训操作工作票。 4.完成本操作的工器具及材料准备	1	培训方法： 1.现场查勘和工器具及材料清理采用现场实操方法。 2.编写作业卡和填写工作票采用讲授方法。 培训资源： 1.±800kV实训线路。 2.特高压工器具库房。 3.空白工作票。	1.特高压输电实训线路 2.多媒体教室	
3.作业现场准备	能完成作业现场准备工作	1.作业现场复勘。 2.工作申请。 3.作业现场布置。 4.班前会。 5.工器具及材料检查。	3	培训方法：演示与角色扮演法。 资源：±800kV实训线路。	±800kV实训线路	

续表

培训流程	培训目标	培训内容	培训学时	培训方法与资源	培训环境	考核评价
4.培训师演示	通过现场观摩,使学员初步领会本任务操作流程。	1.各类工器具使用方法讲解。 2.演示更换整串绝缘子的塔上工器具连接方式。 3.高空作业人员配合演示更换整串绝缘子的操作流程。 4.利用绞磨配合更换±800kV线路耐张整串绝缘子	7	培训方法:演示法。资源:±800kV实训线路。	±800kV实训线路	
5.学员分组训练	1.能掌握各类受力工器具、绞磨机具的使用方法和注意事项。 2.掌握更换整串绝缘子的全部操作流程。 3.能协助负责完成±800kV输电线路耐张整串绝缘子的更换。	1.学员分组(高空6人、地面配合9人一组)训练工器具、机具的操作方法和更换整串绝缘子的现场实际操作。 2.培训师对学员操作进行指导和安全监护。	14	培训方法:角色扮演法。资源:±800kV实训线路	±800kV实训线路	采用技能考核评分细则对学员操作评分
6.工作终结	1.使学员进一步辨析操作过程不足之处,便于后期提升。 2.培训学员安全文明生产的工作作风。	1.作业现场清理。 2.向调度汇报工作。 3.班后会,对本次工作任务进行点评总结。	1	培训方法:讲授和归纳法	±800kV实训线路	

(四)作业流程

1. 工作任务

完成停电更换±800kV特高压输电线路耐张整串绝缘子。

2. 天气及作业现场要求

(1)停电更换±800kV特高压输电线路耐张整串绝缘子应在良好的天气进行。

在5级及以上的大风以及暴雨、雷电、冰雹、大雾、沙尘暴等恶劣天气下,应停止露天高处作业。特殊情况下,确需在恶劣天气进行抢修时,应组织有关人员充分讨论必要的安全措施,经本单位批准后方可进行。

(2)作业人员精神状态良好,工作班成员认真学习工作票和安全技术措施,所有人员做到"四清楚"（作业任务精楚、危险点清楚、作业程序清楚、安全措施清楚）。

（3）作业前停送电联系人必须与调度联系履行工作许可手续，严禁约时停送电。工作负责人必须在得到许可人的许可工作命令后，方可在需检修的线路上验电、挂设接地线和进行检修工作。

（4）停电后，工作负责人应认真做好记录。

（5）登杆前应检查塔上是否有蜂窝，发现蜂窝严禁登塔。

（6）塔上作业人员必须使用双保险安全带，并佩戴护目镜。

3. 准备工作

3.1 危险点及其预控措施

（1）危险点——误登带电线路

预控措施：

①登杆塔作业前，工作负责人、工作班成员应共同认真核查双重名称和识别标记（色标、判别标志等）与停电线路名称相符。

②登杆塔前应检查铁塔根部、基础等，必须牢固可靠。

③登杆塔前应检查登高工器具和设施，如安全带、脚钉、塔材等必须完整牢靠。

④不涉及挂设接地线的中间作业人员，应认真核实线路相序、色标、名称、编号与停电线路相符，确认线路名称无刷错、刷反等情况后，方可登杆。

（2）危险点——登塔时、塔上作业时违反安规进行操作，可能引起高空坠落

预控措施：

①攀爬过程中，为防止登杆人员串落，登杆作业人员间距不得小于1.6m。

②攀爬铁塔前应将脚底泥土清除干净，检查工具包完整，攀爬过程中不得掉落物件伤人。

③作业人员攀登杆塔时应戴好安全帽，穿软底鞋，动作不能过大，匀步攀登。

④攀爬过程中，安全带应收拾妥当，长尾绳放置在工具包内，主带应挂在肩上，防止攀爬过程中安全带勾挂脚钉和塔材，致使作业人员高空坠落。

⑤杆塔上移位时，不得失去安全带保护，做到踩稳抓牢。

⑥到达作业点位置，系好安全带（绳），应牢固可靠，不得低挂高用。

⑦未验电前，人体、无头绳等与导线的安全距离必须不小于10.1m，工作中应设专人监护。

（3）危险点——高处坠物伤人

预控措施：

①地面人员不得站在作业点垂直下方。塔上人员应防止落物伤人，使用的工具、

材料应用绳索传递。

②在高处作业应使用工具袋,较大的工具应固定在牢固的构件上,不准随便乱放。

③使用绞磨起吊过程中,应设专人指挥,统一配合,绝缘子串刚离地后应进行冲击检查。

(4)危险点——防止感应电伤人

①为防感应电伤人,塔上作业人员应穿全套屏蔽服。

②如需接触架空地线,在架空地线接触前应进行可靠接地。

(5)危险点——现场作业安全监护

①自作业开始至作业终结,安全监护人必须始终在现场对作业人员进行不间断的安全监护。

②工作负责人,监护人必须穿安全监护背心。

(6)危险点——交通安全

①出车时应注意车辆行驶安全,谨慎驾驶车辆,禁止违法行车。

3.2 工器具及材料选择

停电更换±800kV特高压输电线路耐张整串绝缘子所需工器具及材料见表2-2-2。工器具出库前,应认真核对工器具的使用电压等级和试验周期,并检查确认外观良好、连接牢固、转动灵活,且符合本次工作任务的要求;工器具出库后,应防止脏污、受潮;金属工具和绝缘工器具应分开装运,防止因混装运输导致工器具变形、损伤等现象发生。

表2-2-2 停电更换±800kV特高压输电线路耐张整串绝缘子所需工器具及材料表

序号	名称	规格型号	单位	数量	备注
1	接地线	±800kV专用	组	2	绝缘工具
2	绝缘手套	10kV	付	2	绝缘工具
3	验电器	±800kV专用	支	2	其他工具
4	全身式安全带	含带缓冲包长24M的后保绳	套	8	个人防护用具
5	个人保安线	(直径不小于16mm²)	根	2	其他工具
6	钢丝套	Φ24	根	20	其他工具
7	铁滑车	15T	个	12	金属工具
8	卸扣	18T	个	12	金属工具
9	铁滑车	5T	个	8	金属工具
10	钢丝绳	Φ24	米	150	其他工具
11	磨绳	Φ17.5	米	600	其他工具

续表

序号	名称	规格型号	单位	数量	备注
12	绞磨	5T	台	3	机动工具
13	手扳葫芦	9T	付	2	金属工具
14	手扳葫芦	6T	付	2	金属工具
15	个人手动工具		套	8	其他工具
16	对讲机		台	10	其他工具
17	吊绳滑车	1T	个	2	金属工具
18	传递绳	Φ16	套	2	其他工具
19	拔销器		把	3	金属工具
20	安全背心		件	3	其他工具
21	护目镜		付	6	其他工具
22	安全围栏		卷	5	其他工具
23	垫木		块	若干	其他工具
24	防潮苫布		张	1	其他工具
25	钢钎		根	2	其他工具
26	铁锤		把	2	其他工具
27	玻璃绝缘子	U550BP/240T	片	81	材料

3.3 作业人员分工

本任务作业人员分工如表2-2-3所示。

表2-2-3 停电更换±800kV特高压输电线路耐张整串绝缘子人员分工表

序号	工作岗位	数量(人)	工作职责
1	工作负责人	1	负责本次工作任务的人员分工、工作前的现场查勘、作业方案的制定、工作票的填写、现场复勘、办理工作许可手续、召开工作班前会、落实现场安全措施、负责作业过程中的安全监督、工作中突发情况的处理、工作质量的监督、工作后的总结
2	安全监护人	2	负责本次工作过程中的安全监护工作
3	高空作业人员	6	负责本次停电更换±800kV耐张整串绝缘子操作
4	地面辅助人员	5	负责本次作业过程的地面辅助工作
5	绞磨操作人员	2	负责本次作业过程的绞磨操作工作
6	信号指挥人员	2	负责2台绞磨启停的指挥工作

4. 工作程序

本任务工作流程如表2-2-4所示。

表2-2-4 停电更换±800kV特高压输电线路耐张整串绝缘子工作流程表

序号	作业内容	作业标准	安全注意事项	责任人
1	工作许可	作业前停送电联系人必须与调度联系履行工作许可手续	(1)不得未经工作许可人许可即开始工作。(2)严禁约时停送电	
2	现场布置	正确装设安全围栏并悬挂标示牌:(1)安全围栏范围应充分考虑高处坠物,以及对道路交通的影响。(2)安全围栏出入口设置合理。(3)妥当布置"从此进出""在此工作""车辆慢行"或"车辆绕行"等标示	对道路交通安全影响不可控时,应及时联系交通管理部门强化现场交通安全管控	
3	召开班前会	(1)全体工作成员列队。(2)工作负责人宣读工作票,明确工作任务及人员分工;讲解工作中的安全措施和技术措施;查(问)全体工作成员精神状态;告知工作中存在的危险点及采取的预控措施。(3)全体工作成员在工作票上签名确认	(1)工作票填写、签发和许可手续规范,签名完整。(2)全体工作成员精神状态良好。(3)全体工作成员明确任务分工、安全措施和技术措施	
4	检查工器具	(1)在防潮苫布上,将工器具按作业要求准备齐备,并分类定置摆放整齐。检查工器具外观和试验合格证,无遗漏。(2)检查人员向工作负责人汇报各项检查结果符合作业要求	(1)防潮苫布数量足够,设置位置合理,保持清洁、干燥。(2)工器具外观检查合格,无损伤、受潮、变形、失灵现象,合格证在有效期内	
5	登杆塔	(1)登杆塔作业前,必须先核对线路名称及编号。对同塔多回线路,工作负责人、工作班成员应共同认真核查双重名称和识别标记(色标、判别标志等)。(2)登杆塔前应检查铁塔根部、基础等,必须牢固可靠。(3)攀爬过程,为防止登杆人员串落,登杆作业人员间距不得小于1.6m,安全带收拾妥当,长尾绳放置在工具包内,主带应挂在肩上,防止攀爬过程中安全带勾攀脚钉和塔材,致使作业人员高空坠落。(4)登杆塔至横担处时,监护人和作业人员应再次核对停电线路的识别标记和双重名称,确实无误后方可进入停电线路侧的横担	(1)作业人员攀登杆塔时应戴好安全帽,穿软底鞋,动作不能过大,匀步攀登。(2)攀爬过程中,安全带应收拾妥当,长尾绳放置在工具包内,主带应挂在肩上,防止攀爬过程中安全带勾挂脚钉和塔材,致使作业人员高空坠落。(3)杆塔上移位时,不得失去安全带保护,做到踩稳抓牢。(4)到达作业点位置,系好安全带(绳),应牢固可靠,不得低挂高用。(5)未验电前,人体、无头绳等与导线的安全距离必须不小于10.1m,工作中应设专人监护	

续表

序号	作业内容	作业标准	安全注意事项	责任人
6	验电、装设接地线	杆就位后,将安全带系在牢固可靠构件 或电杆上,必须检查扣环是否正确就位。验电杆(笔)等工器具必须使用绳索传递。查接地线完好,按程序(先接接地端,后接导线端)装设好接地线。设接地线时,必须使用绝缘绳或绝缘手柄进行操作,禁止直接用手操作接地线的金属部分的方式装设设接地线。确认接地线的夹头与导线连接紧密可靠	(1)验电杆在领用时和使用前应检查是否正常。(2)禁止以缠绕导线的方式装设接地线	
7	更换耐张整串绝缘子	高空作业人员到达作业点位后,3名高空人员在横担端的2串耐张绝缘子之间的牢固塔材上安装3个15T的铁滑车,3名高空人员在对应的带电侧联板上安装3个15T的铁滑车,将钢丝套与手扳葫芦勾卡固定于横担侧的3个铁滑车的一边,再将手扳葫芦链条连接的钢丝绳依次穿入两边的6个滑车中,构成一套滑轮组,并保留足够的人员操作空间,确认手扳葫芦与滑轮组连接部位已连接牢固后,再收紧9T手扳葫芦,当9T手板葫芦已完全承受绝缘子串的拉力后,应再次检查横担有无变形,连接部位有无异常,确认一切正常。3名高空人员将1号绞磨上的磨绳穿过设置于横担上5T滑轮后,再将其固定于绝缘子串的横担端的第三片处,固定稳固后,再将2号绞磨上的磨绳依次穿过设置于横担端与带电端连接金具上的2个5T滑轮后,再将2号绞磨的磨绳固定于带电端的第3片绝缘子处,待所有连接部位都已连接稳固后,方可开动绞磨。地面绞磨操作人员开动2号绞磨,应将绝缘子串带电端拉紧后,待高空人员取下带电端绝缘子串的连接部分后,再缓慢将带电端的绝缘子串向下松出,待绝缘子串垂直于地面后,再开动1号绞磨,将绝缘子拉紧提起一端,高空人员取下横担端绝缘子串的连接部分,然后两端同时缓慢松出,松至地面,在传递至地面过程中应使用绞磨配合控制绝缘子串,防止与跳线串相互碰撞,放下绝缘子串后,地面人员配合取下绝缘子。待地面人员取下绝缘子,更换完好绝缘子,先开动1号绞磨,再开动2号,安装顺序与拆除顺序相反,先将横担端安装完毕后,再紧固带电端。绝缘子到位后,塔山作业人员应先连接铁塔端连接金具,并确保子到位后方可继续进行金具与铁塔、导线的连接工作。安装好绝缘串后,应检查连接金具是否安装到位、销子是否安装到位,并进行冲击试验,确认安装正确,连接可靠后方可缓慢松出9T手扳葫芦	(1)使用手扳葫芦卡更换绝缘子串过程中,在手扳葫芦开始承受导线荷载后,必须检查手扳葫芦、钢丝绳套(吊装带)、卸扣的连接和受力情况,并做冲击试验,确认完全可靠后方可继续收紧手扳葫芦。(2)在使用手扳葫芦时,应防止与绝缘子碰撞,避免损伤绝缘子。(3)在使用传递绳吊放整串绝缘子串时,地面工作人员和塔上的工作人员必须密切配合,防止起吊绳缠绕,绝缘子串碰撞损伤导线。吊放绝缘子串时,地面人员不得站在垂直下方。(4)绞磨起吊整串绝缘子时,应有专人指挥,2台绞磨应密切配合,绝缘子串一起地,应再次检查绝缘子串是否绑扎牢固,并做冲击试验,确认无误后方可继续起吊	

续表

序号	作业内容	作业标准	安全注意事项	责任人
8	拆除工器具	依序拆除磨绳、手扳葫芦、钢丝套、滑车等工器具		
9	工作结束	(1)工作负责人组织全体工作成员整理工器具和材料,清理现场,做到"工完料尽场地清"。 (2)召开班后会,工作负责人进行工作总结和点评工作。点评本次工作的施工质量;点评全体工作成员的安全措施落实情况。 (3)工作负责人向工作许可人汇报工作结束,恢复停电线路送电,终结工作票		

二、考核标准

表2-2-5 国网四川省电力公司特高压直流输电线路运检技能考核评分细则

考生填写栏	编号: 姓名: 所在岗位: 单位: 日期: 年 月 日				
考评员填写栏	成绩: 考评员: 考评组长: 开始时间: 结束时间: 操作时长:				
考核模块	停电更换±800kV特高压线路耐张整串绝缘子	考核对象	特高压直流输电线路检修人员	考核方式	操作	考核时限	360min
任务描述	更换±800kV特高压输电线路C相右串整串玻璃绝缘子				
工作规范及要求	1. 给定条件:±800kV直流实训线路C相右串整串玻璃绝缘子老化,需要更换。线路已经停电、验电、挂接地线,所使用绝缘子已经过测试,工作票已办理,安全措施已经完备。 2. 整个过程主要操作流程由工作负责人1名、专责监护人1人,塔上电工6人配合完成,地面辅助工7人,绞磨操作人员2人,协助参考人员完成工器具、材料的上、下传递工作,以及其他非技术性工作。 3. 操作前参考人员应作必要的安全检查。 4. 更换整串绝缘子时采用的工具应满足受力要求。 5. 工作开始应口头提出申请,工作结束时应口头汇报				
考核情景准备	1. 工器具:Φ24钢丝套150m、15T铁滑车12个、5T铁滑车8个、18T卸扣12个、Φ24钢丝绳150m、Φ16磨绳一捆、9T手扳葫芦2把、6T手扳葫芦2把、5T绞磨2台、吊绳滑车2套、绳套2根、对讲机若干、双保险安全带6套。 2. 材料:同型号玻璃绝缘子一串。 3. 在培训线路上操作				
备注	1. 个人工器具由参考人员自备。 2. 各项得分均扣完为止。				

表2-2-6 国网四川省电力公司特高压直流输电线路运检技能考核评分标准

序号	项目名称	质量要求	分值	扣分标准	扣分原因	扣分	得分
1	工具材料准备						
1.1	个人工具检查	活动扳手、平口钳、拔销钳、工具包符合质量要求	2	错漏1项扣1分			
1.2	机具检查	绞磨试机,检查档位是否正常;手扳葫芦检查	2	错漏1项扣1分			
1.3	钢丝绳、滑车检查	对钢丝绳、滑车进行检查,确认连接可靠,受力满足要求	2	错漏1项扣1分			
1.4	安全带	双保险安全带符合质量要求,在试验周期内	1	未检查扣1分			
1.5	专用工具检查	个人保安线、绝缘手套外观检查符合要求,在试验周期内	1	错漏1项扣0.5分			
1.6	材料检查	清洁绝缘子,核对绝缘子串数量,外观检查符合要求	2	错漏1项扣1分			
2	场地布置						
2.1	绞磨场地布置	绞磨位置布置、转角滑车位置布置	2	错漏1项扣1分			
2.2	场地围栏	场地围栏布置	1	未布置扣1分			
3	登塔及横担上的操作						
3.1	登塔	(1)检查杆塔基础无异常。 (2)正确携带传递绳(吊绳头折双、打死结、斜挎肩上)。 (3)沿脚钉侧主材正确登塔	6	(1)未检查,每项扣1分。 (2)未携带传递绳,扣2分,传递绳携带方式不规范,扣1分。 (3)手抓脚钉,每次扣1分。 (4)未沿脚钉侧主材登塔,扣2分			
3.2	进入横担上工作点	由塔身到横担上工作点不得失去安全带保护。	3	未正确使用安全带,扣3分。			
3.3	传递滑车安装	传递滑车安装位置正确,方便操作	1	安装不规范,扣1分			
4	绝缘子串上操作						
4.1	进入工作点	(1)将双保险安全带的安全绳系在横担适当位置。 (2)沿绝缘子串进入作业点,将围杆带系在绝缘子串上。 (3)检查绝缘子串锁紧销	5	(1)未使用双保险安全带,扣5分。 (2)未正确使用双保险安全带,每次扣3分。 (3)未检查,扣3分			

续表

序号	项目名称	质量要求	分值	扣分标准	扣分原因	扣分	得分
4.2	安装3-3滑车组	(1)导线端和铁塔端3-3滑车组位置安装正确。 (2)钢丝绳穿向正确,不缠绕。 (3)钢丝绳套、卸扣型号选用正确、安装可靠	6	(1)滑车组安装不对,每处扣1分。 (2)钢丝绳穿向位置不对,每处扣1分。 (3)钢丝绳受力后有碰撞、缠绕,每处扣1分。 (4)钢丝绳套、卸扣型号选用不正确,每处扣1分			
4.3	安装手扳葫芦	(1)将手扳葫芦传递至塔上,并正确安装。 (2)将手扳葫芦与3-3滑车组连接,使其略为受力	5	(1)绳索缠绕,扣1分。 (2)位置不正确,扣2分。 (3)碰响绝缘子,每次扣1分			
4.4	收紧绝缘子串	(1)收紧手扳葫芦,手扳葫芦受力后检查滑车、卸扣、钢丝绳等连接部位,确认连接可靠,并对绝缘子串做冲击。 (2)确认受力无误后继续收紧手扳葫芦,直至绝缘子串松弛。	5	(1)未冲击试验,扣3分。 (2)使用手扳葫芦不正确,扣2分			
4.5	更换绝缘子串	(1)在已松弛的绝缘子串上正确的安装6T辅助手扳葫芦。 (2)在绝缘子串上适当位置连接起吊磨绳。 (3)收紧9T手扳葫芦,受力后进行冲击试验,无异常,收紧6T手扳葫芦,取出R销。 (4)取下绝缘子串两端连接金具,收紧起吊磨绳。 (5)磨绳受力后,检查磨绳连接部位,并冲击试验。 (6)拆除6T手扳葫芦,通过起吊磨绳将绝缘子串传送至地面。 (7)将新绝缘子串传递到杆塔上,安装9T手扳葫芦。 (8)先收紧9T手扳葫芦,再收紧6T手扳葫芦,安装两端R销子,并检查是否安装到位。	18	(1)手扳葫芦安装位置不对、使用操作不对,每次扣2分。 (2)磨绳起吊滑车位置不正确,扣2分。 (3)未冲击试验,扣3分。 (4)传递绳索缠绕,扣2分。 (5)传递物有撞击现象,每次扣2分。 (6)掉落物件,扣5分。 (7)未装锁紧销1个,扣2分。			

续表

序号	项目名称	质量要求	分值	扣分标准	扣分原因	扣分	得分
4.6	撤除工器具	(1)检查锁紧销、球头是否齐全到位,碗口朝正确,清洁绝缘子串。(2)松出3-3滑车组绝缘子串受力后,对绝缘子串做冲击试验。(3)拆除手扳葫芦、滑车组及钢丝绳,传递至地面。	10	(1)绳索缠绕,扣2分。(2)传递物有撞击现象,每次扣2分。(3)未转移围杆带到绝缘子串上,扣5分。(4)锁紧销位置不正确,扣2分。(5)绝缘子大口朝向不正确,扣2分。(6)未作冲击试验,扣3分。(7)未清洁绝缘子串,扣1分。			
4.7	清理杆塔上工器具	确认无遗留物。	4	有遗留物,扣4分			
4.8	从绝缘子串到塔身	由绝缘子串到塔身上不得失去安全带保护。	5	失去安全带保护,扣5分			
5	下塔	(1)必须沿脚钉侧主材正确下塔。(2)正确携带传递绳(吊绳头折双、打死结,斜挎肩上)。	8	(1)未携带传递绳扣4分;吊绳携带方式不规范扣2分。(2)手抓脚钉,每次扣1分。(3)未沿脚钉侧主材登塔,扣4分			
6	其他要求						
6.1	塔上作业	(1)严禁高处坠物。(2)在操作过程中应双手协调配合操作。(3)严禁浮置物品。(4)严禁口中含物。	5	(1)高处坠物,每件扣5分。(2)动作不协调,扣2分。(3)浮置物品,扣2分。(4)口中含物,扣2分			
6.2	着装	工作服、工作胶鞋、安全帽、劳保手套穿戴正确	2	漏一项扣2分			
6.3	清理现场	完工后清理作业现场,符合文明生产要求	2	未清理作业现场,扣2分			
6.4	完成时间	在规定时间内按要求完成	2	超过时间10min,扣1分;达到480min即终止操作,只记完成部分得分			
	合计		100				

模块三 停电修补±800kV特高压输电线路架空地线培训及考核标准

一、培训标准

(一) 培训要求

模块名称	停电修补±800kV特高压输电线路架空地线	培训类别	操作类
培训方式	实操培训	培训学时	14学时
培训目标	1.能使用飞车沿±800kV特高压输电线路架空地线到达指定作业位置； 2.能独立完成用预绞丝补修条修补架空地线的操作		
培训场地	特高压直流实训线路		
培训内容	使用飞车沿±800kV特高压输电线路架空地线到达指定作业位置，用预绞丝补修条修补±800kV特高压输电线路架空地线		
适用范围	特高压直流输电线路检修人员		

(二) 引用规程规范

(1)《架空送电线路运行规程》(DL/T 741-2010)

(2)《110~500kV架空送电线路设计技术规程》(DL/T 5092-1999)

(3)《国家电网公司电力安全工作规程（线路部分）》(Q/GDW1799.2-2013)

(4)《±800kV直流架空输电线路设计规范》(GB 50790-2013)

(5)《±800kV直流架空输电线路检修规程》(DL/T 251-2012)

(6)《±800kV直流架空输电线路运行规程》(GB/T 28813-2012)

(7)《110 (66) kV~500kV架空输电线路检修规范》(国家电网公司)

(8)《架空输电线路状态检修导则》(DLT 1248-2013)

(9)《输变电设备状态检修管理规定》(国家电网公司)

(10)《输变电设备状态检修试验规程》(国家电网公司)

(三) 培训教学设计

本设计以完成"停电修补±800kV特高压输电线路架空地线"为工作任务，按工作任务完成的标准化作业流程来设计各个培训阶段,每个阶段包括了具体的培训目标、培训内容、培训学时、培训方法(培训资源)、培训环境和考核评价等内容,如表2-3-1所示。

表2-3-1 停电修补±800kV特高压输电线路架空地线培训内容设计

培训流程	培训目标	培训内容	培训学时	培训方法与资源	培训环境	考核评价
1.理论教学	1.初步掌握使用飞车沿±800kV特高压输电线路架空地线到达指定作业位置基本方法。2.熟悉架空地线受损的修补方法	1.飞车的分类、结构及其使用注意事项。2.使用飞车沿±800kV特高压输电线路架空地线到达指定作业位置方法。3.输电线路架空地线修补方法和质量标准	2	培训方法:讲授法。培训资源:PPT、相关规程规范	多媒体教室	考勤、课堂提问和作业
2.准备工作	能完成作业前准备工作	1.作业现场查勘。2.编制培训标准化作业卡。3.填写输电线路第一种工作票。4.完成本操作的工器具及材料准备	1	培训方法:1.现场查勘和工器具及材料清理采用现场实操方法。2.编写作业卡和填写工作票采用讲授方法。培训资源:1.±800kV实训线路。2.特高压工器具库房。3.空白工作票	1.特高压输电实训线路。2.多媒体教室	
3.作业现场准备	能完成作业现场准备工作	1.作业现场复勘。2.工作申请。3.作业现场布置。4.班前会。5.工器具及材料检查	1	培训方法:演示与角色扮演法。资源:±800kV实训线路	±800kV实训线路	
4.培训师演示	通过现场观摩,使学员初步领会本任务操作流程	1.装设接地线。2.安装作业飞车。3.使用飞车沿±800kV特高压输电线路架空地线到达指定作业位置。4.用预绞丝补修条完成架空地线修补	2	培训方法:演示法。资源:±800kV实训线路	±800kV实训线路	

续表

培训流程	培训目标	培训内容	培训学时	培训方法与资源	培训环境	考核评价
5.学员分组训练	1.能完成飞车的正确安装。2.能使用飞车沿±800kV直流输电线路架空地线到达指定作业位置。3.能完成±800kV输电线路架空地线修补作业	1.学员分组(6人一组)训练飞车的使用和修补架空地线技能操作。2.培训师对学员操作进行指导和安全监护	7	培训方法：角色扮演法。资源：±800kV实训线路	±800kV实训线路	采用技能考核评分细则对学员操作评分
6.工作终结	1.使学员进一步辨析操作过程不足之处，便于后期提升。2.培养学员安全文明生产的工作意识	1.作业现场清理。2.向调度汇报工作。3.班后会，对本次工作任务进行点评总结	1	培训方法：讲授和归纳法	±800kV实训线路	

（四）作业流程

1. 工作任务

完成停电修补±800kV特高压输电线路架空地线任务。

2. 天气及作业现场要求

（1）停电修补±800kV特高压输电线路架空地线应在良好的天气进行。

在5级及以上的大风以及暴雨、雷电、冰雹、大雾、沙尘暴等恶劣天气下，应停止露天高处作业。特殊情况下，确需在恶劣天气进行抢修时，应组织有关人员充分讨论必要的安全措施，经本单位批准后方可进行。

（2）作业人员精神状态良好，工作班成员认真学习工作票和安全技术措施，所有人员做到"四清楚"（作业任务清楚、危险点清楚、作业程序清楚、安全措施清楚）。

（3）作业前工作负责人必须与调度联系履行工作许可手续，严禁约时停送电。工作负责人必须在得到许可人的许可工作命令后，方可在需检修的线路上验电、装设接地线和进行检修工作。

（4）停电后，工作负责人应认真做好记录。

（5）登塔前应检查塔上是否有蜂窝，发现蜂窝严禁登塔。

（6）塔上作业人员必须使用双保险安全带，并佩戴护目镜。

3. 准备工作

3.1 危险点及其预控措施

（1）危险点——误登带电线路

预控措施：

①登塔作业前，工作负责人、工作班成员应认真核对双重称号和识别标记（色标、判别标志等）与停电线路名称相符。

②登塔前应检查铁塔根部、基础等，必须牢固可靠。

③登杆塔前应检查登高工器具和设施，如安全带、脚钉、塔材等必须完整牢靠。

④不涉及挂设接地线的中间作业人员，应认真核实线路相序、色标、名称、编号与停电线路相符，确认线路名称无刷错、刷反等情况后，方可登杆。

（2）危险点——登塔和塔上作业时违反安规进行操作，可能引起高处坠落

预控措施：

①登塔过程中，为防止登塔人员相互碰撞，登塔作业人员间距不得小于1.6m。

②登塔前应将脚底泥土清除干净，检查工具包完整，登塔过程中不得掉负重登杆。

③作业人员登塔时应戴好安全帽，穿软底工作鞋，动作不宜过大，匀步攀登。

④登塔过程中，安全带应收拾妥当，后背保护绳放置在工具包内，主带挂在肩上，防止登塔过程中安全带勾挂脚钉和塔材，致使作业人员高处坠落。

⑤塔上移位时，不得失去安全带保护，做到踩稳抓牢。

⑥到达作业点位置，系好安全带，将安全带、后备保护绳应分别系在牢固构件上，不得低挂高用。

⑦未验电前，人体、传递绳等与导线的安全距离不小于10.1m，工作中应设专人监护。

（3）危险点——高处坠物伤人

预控措施：

①地面人员不得站在作业点正下方。塔上人员应防止坠物伤人，使用的工具、材料应用绳索传递。

②高处作业应使用工具袋，较大的工器具应固定在牢固的构件上，不准随便乱放。

（4）危险点——防止感应电伤人

①接触架空地线前，应将地线接地端可靠接地。

②挂接地线时，作业人员戴绝缘手套，手握绝缘部位。

（5）危险点——现场作业安全监护

①作业过程中，安全监护人对作业人员进行不间断监护。

②工作负责人，监护人必须有符合身份的明显标识。

（6）危险点——交通安全

①出车时应注意车辆行驶安全，谨慎驾驶车辆，禁止违法行车。

3.2 工器具及材料选择

停电修补±800kV特高压输电线路架空地线所需工器具及材料见表2-3-2。工器具出库前,应认真核对工器具的使用电压等级和试验周期,并检查确认外观良好、连接牢固、转动灵活,且符合本次工作任务的要求;工器具出库后,应防止脏污、受潮;金属工具和绝缘工器具应分开装运,防止因混装运输导致工器具变形、损伤等现象发生。

表2-3-2 停电修补±800kV直流输电线路架空地线所需工器具及材料表

序号	名称	规格型号	单位	数量	备注
1	作业飞车		付	1	金属工具
2	安全带	含带缓冲包长20M的后保绳	根	2	个人防护用具
3	绝缘手套		双	1	绝缘工具
4	传递绳	Φ16	根	1	其他工具
5	铁滑车	0.5T	个	1	金属工具
6	钢丝套	Φ8	根	1	其他工具
7	个人工具		套	2	其他工具
8	砂纸		张	1	其他工具
9	钢卷尺		把	1	其他工具
10	记号笔		根	1	其他工具
11	木榔头		把	1	其他工具
12	对讲机		台	3	其他工具
13	安全背心		件	2	其他工具
14	安全围栏		卷	4	其他工具
15	防潮苫布		张	1	其他工具
16	抛挂式接地线	±800kV	组	3	其他工具
17	地线接地线	25mm²	组	1	其他工具
18	验电器	±800kV	个	1	其他工具
19	防坠器	T型	个	1	其他工具
20	预绞丝	与地线型号对应	组	1	材料

3.3 作业人员分工

本任务作业人员分工如表2-3-3所示。

表2-3-3 停电修补±800kV直流输电线路架空地线人员分工表

序号	工作岗位	数量(人)	工作职责
1	工作负责人	1	负责本次工作任务的人员分工、工作前的现场查勘、作业方案的制定、工作票的填写、现场复勘、办理工作许可手续、召开工作班前会、落实现场安全措施,负责作业过程中的安全监督、工作中突发情况的处理、工作质量的监督、工作后的总结
2	安全监护人	1	工作前,向被监护人员交代监护范围内的安全措施、告知危险点和安全注意事项;监督被监护人员遵守本规程并严格执行现场安全措施,及时纠正被监护人员不安全动作和行为。
3	高处作业人员	2	负责本次停电修补±800kV直流输电线路架空地线操作
4	地面辅助人员	2	负责本次作业过程的地面辅助工作

4. 工作程序

本任务工作流程如表2-3-4所示。

表2-3-4 停电修补±800kV直流输电线路架空地线工作流程表

序号	作业内容	作业步骤及标准	安全措施及注意事项	责任人
1	现场复勘	工作负责人负责完成以下工作: (1)现场核对线路名称、铁塔编号无误;基础及塔身完好无异常;交叉跨越距离符合安全要求;确认缺陷情况及地线规格型号等。 (2)检查地形环境符合作业要求。 (3)检查工作票所列安全措施与现场实际情况相符,必要时予以补充	(1)正确穿戴安全帽、工作服、工作鞋、劳保手套。 (2)严禁非工作人员、车辆进入作业现场	
2	工作许可	作业前工作负责人必须与调度联系履行工作许可手续	(1)不得未经工作许可人许可即开始工作。 (2)严禁约时停送电	
3	现场布置	正确设置安全围栏并悬挂标示牌: (1)安全围栏范围应充分考虑高处坠物,以及对道路交通的影响。 (2)安全围栏出入口设置合理。 (3)妥当布置"从此进出""在此工作""车辆慢行"或"车辆绕行"等标示	对道路交通安全影响不可控时,应及时联系交通管理部门强化现场交通安全管控	
4	召开班前会	(1)全体工作成员列队。 (2)工作负责人宣读工作票,明确工作任务及人员分工;讲解工作中的安全措施和技术措施;查问全体工作成员精神状态;告知工作中存在的危险点及采取的预控措施。 (3)全体工作成员在工作票上签名确认	(1)工作票填写、签发和许可手续规范,签名完整。 (2)全体工作成员精神状态良好。 (3)全体工作成员明确任务分工、安全措施和技术措施	

续表

序号	作业内容	作业步骤及标准	安全措施及注意事项	责任人
5	检查工器具	(1)在防潮垫布上,将工器具按作业要求准备齐备,并分类定置摆放整齐。检查工器具外观和试验合格证,无遗漏。 (2)检查人员向工作负责人汇报各项检查结果符合作业要求	(1)防潮垫布设置位置合理,保持清洁、干燥。 (2)工器具外观检查合格,无损伤、受潮、变形、失灵现象,合格证在有效期内	
6	登塔	(1)登塔作业前,必须先核对线路双重名称及编号。对同塔多回线路,工作负责人、工作班成员应共同认真核查双重名称和识别标记(色标、判别标志等)。 (2)登塔前应检查塔身、基础等,必须牢固可靠。 (3)登塔过程,为防止登塔人员相互碰撞,登塔作业人员相互间距不得小于1.6m,安全带收拾妥当,后备保护绳放置在工具包内,主带应挂在肩上。 (4)登塔至横担处时,看清楚行走通道,与导线的安全距离不小于10.1m	(1)作业人员登塔时应戴好安全帽,穿软底工作鞋,匀步攀登。 (2)登塔过程中,安全带应收拾妥当,后背保护绳放置在工具包内,主带应挂在肩上,防止登塔过程中安全带勾挂脚钉和塔材,致使作业人员高处坠落。 (3)塔上移位时,不得失去安全带保护,做到踩稳抓牢。 (4)到达作业点位置,将安全带、后备保护绳应分别系在牢固构件上,不得低挂高用。 (5)未验电前,人体、传递绳等与导线的安全距离不小于10.1m,工作中应设专人监护	
7	验电、装设导线接地线	(1)登塔至指定位置后,将安全带系在牢固的构件上,检查扣环是否扣牢;滑车安装位置正确,方便操作;工器具必须使用绳索传递。 (2)使用验电器验电前,戴好绝缘手套自检其声光信号正常。使用伸缩式验电器时,应将其各段绝缘杆全部拉出到位,以保证绝缘杆的有效绝缘长度;验电时,作业人员应手持验电器绝缘手柄,保证人体与导线间的足够安全距离;先验下层后验上层,先验近侧后验远侧,线路验电应逐相进行。 (3)验明确无电压后立即装设导线接地线,先用砂纸打磨接地端安装位置;先安装接地端,后装导线端。 (4)装设导线接地线时,必须使用绝缘手套采用抛挂方式进行操作,禁止直接用手操作接地线的金属部分的方式装设导线接地线。确认接地线的挂钩与导线连接紧密可靠	(1)验电器在领用时和使用前应检查是否正常。 (2)接地线与导线及塔材接触良好,禁止以缠绕导线的方式装设接地线。 (3)人体与导线保持足够安全距离	

续表

序号	作业内容	作业步骤及标准	安全措施及注意事项	责任人
8	装设地线接地线	(1)登塔至指定位置后,将安全带系在牢固的构件上,检查扣环是否扣牢。使用绳索传递工器具。 (2)检查接地线完好,先用砂纸打磨塔材接地端,并先安装接端后装导线端。 (3)装设地线接地线时,必须使用绝缘手套进行操作,禁止直接用手操作接地线的金属部分的方式装设地线接地线。确认地线接地线的挂钩与地线连接紧密可靠	(1)禁止以缠绕导线的方式装设地线接地线。 (2)挂接地线时,作业人员戴绝缘手套,手握绝缘棒	
9	安装作业飞车	(1)作业人员进入地线前应检查连接金具锈蚀情况。 (2)检查地线绝缘子是否完好,销子是否齐全。 (3)对地线做冲击试验。 (4)塔上辅助人员配合在地线上安装作业飞车	(1)业人员进入地线前应检查连接金具锈蚀情况。 (2)安装作业飞车前,应对地线做冲击试验	
10	进入作业点	正确使用作业飞车,移动速度均匀,无危险动作;在地线上移动时,不得失去安全带的保护	(1)整个过程不能失去安全带保护。 (2)移动飞车速度不能过快	
11	修补断股地线	(1)到达作业点位后固定好作业飞车。 (2)作业人员对地线损伤处进行打磨处埋。 (3)量出预绞丝长度,用钢卷尺在导线损伤处的一端量出1/2预绞丝长的位置画印。 (4)预绞丝中心应安装在导线损伤部位严重处,不得有缝隙。 (5)用木榔头轻敲预绞丝端头,端头应平整	(1)正确使用作业飞车,移动速度均匀,无危险动作;在地线上移动时,不得失去安全带的保护。 (2)断股地线修补前应对损伤点进行打磨处理。 (3)预绞丝应与断股地线绞制方向一致,缠绕应平滑、紧密,损伤部位应位于预绞丝的中间位置。 (4)缠绕预绞丝时,不得用力强扭、撬动,防止其变形,缠绕时两端应保持平整	
12	进入横担	(1)作业人员移动飞车沿地线返回横担。 (2)进入横担过程中不得失去安全带的保护。 (3)传递绳将作业飞车传至地面	(1)整个过程不能失去安全带保护。 (2)移动飞车速度不能过快	
13	清理铁塔上工器具	(1)依次拆作业飞车、地线接地线、导线接地线、钢丝套、滑车等工器具。 (2)拆除地线上的接地线和导线上的接地线,先拆导(地)线端,后拆接地端。 (3)确认无遗留物	(1)拆除地线上的接地线和导线上的接地线,作业人员戴绝缘手套,手握绝缘棒。 (2)人体与导线保持足够安全距离。 (3)防止高处坠物	

续表

序号	作业内容	作业步骤及标准	安全措施及注意事项	责任人
14	下塔	(1)沿脚钉侧主材正确下塔。 (2)正确携带传递绳(传递绳头折双、打死结,斜挎肩上)	防止高处坠落	
15	工作结束	(1)工作负责人组织全体工作成员整理工器具和材料,清理现场,做到"工完料尽场地清"。 (2)召开班后会,工作负责人进行工作总结和点评工作。点评本次工作的施工质量;点评全体工作成员的安全措施落实情况。 (3)作负责人向工作许可人汇报工作结束,恢复停电线路送电,终结工作票	现场不能有遗留物	

二、考核标准

表2-3-5 特高压直流输电线路运检技能考核评分细则

考生填写栏	编号:	姓 名:	所在岗位:	单 位:	日 期:	年 月 日		
考评员填写栏	成绩:	考评员:	考评组长:	开始时间:	结束时间:	操作时长:		
考核模块	停电修补±800kV特高压输电线路架空地线		考核对象	特高压直流输电线路检修人员	考核方式	操作	考核时限	100min
任务描述	停电修补±800kV输电线路架空地线							
工作规范及要求	1. 给定条件:±800kV直流实训线路左架空地线需要修补。线路已经停电、验电、装设接地线,所使用工器具已经过测试,工作票已办理,安全措施已经完备。 2. 操作前参考人员应作必要的安全检查。 3. 作业飞车应满足受力要求。 4. 整个过程主要操作流程由工作负责人1名、专责监护人1人、塔上电工1人、塔上辅助工1人、地面辅助工2人完成。工作负责人职责:负责本次工作任务的人员分工、工作前的现场查勘、作业方案的制定、工作票的填写、现场复勘、办理工作许可手续、召开工作班前会、落实现场安全措施、负责作业过程中的安全监督、工作中突发情况的处理、工作质量的监督、工作后的总结。安全监护人责任:工作前,对被监护人员交代监护范围内的安全措施、告知危险点和安全注意事项;监督被监护人员遵守本规程和执行现场安全措施,及时纠正被监护人员的不安全动作和行为。塔上作业人员:负责停电修补±800kV直流输电线路架空地线。塔上辅助工:负责传递工具、材料配合作业人员安装飞车。地面电工职责:协助参考人员完成工器具、材料的上、下传递工作,以及其他辅助工作。 给定条件: 1. 培训线路:特高压±800kV直流实训线路左架空地线,架空地线型号:LBGJ-150-20AC。 2. 必须按工作程序进行操作,工序错误扣除相应项目分值,出现重大人身、器材和操作安全隐患,考评员可下令终止考核							

续表

考核情景准备	1. 线路：特高压±800kV直流实训线路左架空地线，工作内容：停电修补±800kV直流输电线路左架空地线，架空地线型号：LBGJ-150-20AC。 2. 所需作业工器具：作业飞车1付、安全带2套、对讲机3个、导线接地线1根、地线接地线1根、绝缘手套1双、传递绳1根、0.5T滑车1个、钢丝套1根、砂纸1张、个人工具2套、木榔头1个。 3. 材料：与地线型号相对应的预绞丝一组
备注	1. 个人工器具由考生自备。 2. 各项目得分均扣完为止

表2-3-6 特高压直流输电线路运检技能考核评分标准

序号	项目名称	质量要求	分值	扣分标准	扣分原因	扣分	得分
1	着装	工作服、工作鞋、安全帽、劳保手套穿戴正确	4	漏一项扣1分			
2	个人工具检查	活动扳手、平口钳、工具包符合质量要求	3	未检查1项扣1分			
3	安全带检查	安全带符合质量要求，在试验周期内	2	(1)未进行外观检查，扣1分。 (2)未检查出厂合格证和试验合格证，扣1分			
4	作业工具检查	作业飞车检查符合要求，在试验周期内	2	(1)未进行外观检查，扣1分。 (2)未检查出厂合格证和试验合格证，扣1分			
5	材料检查	确认预绞丝型号与地线相对应，外观检查符合要求无损伤	2	(1)错选预绞丝型号，扣1分。 (2)未进行检查，扣1分			
6	登塔	(1)现场核对线路名称、铁塔编号无误；基础及塔身完好无异常；交叉跨越距离符合安全要求；确认缺陷情况及地线规格型号等。 (2)正确携带传递绳（传递绳头折双、打死结、斜挎肩上）。 (3)沿脚钉侧主材正确登塔。 (4)登塔至指定位置后，将安全带系在牢固的构件上，检查扣环是否扣牢	9	(1)未确认线路双重称号，扣2分；未检查基础、塔身、交叉跨越，每项扣1分；未确认地线缺陷，扣1分。 (2)未携带传递绳扣2分；携带方式不规范扣1分。 (3)手抓脚钉，每次扣1分。 (4)未沿脚钉侧主材登塔，扣2分。 (5)安全带使用不规范，扣2分。 (6)踏空、踩滑，每次扣1分			

续表

序号	项目名称	质量要求	分值	扣分标准	扣分原因	扣分	得分
7	验电、挂设导线接地线	(1)登塔至指定位置后，将安全带系在牢固的构件上，检查扣环是否扣牢；滑车安装位置正确，方便操作；工器具必须使用绳索传递。 (2)使用验电器前，再次检查其声光信号正常。使用伸缩式验电器时，应将其各段绝缘杆全部拉出到位，以保证绝缘杆的有效绝缘长度；验电时，作业人员应手持验电器绝缘手柄，保证人体与导线间的足够安全距离；先验下层后验上层，先验近侧后验远侧，线路验电应逐相进行。 (3)验明确无电压后立即装设导线接地线，先用砂纸打磨接地端安装的塔材位置；先安装接地端，后装导线端；装设顺序为先中相，后两边相。 (4)装设导线接地线时，必须使用绝缘手套进行操作，禁止直接用手操作接地线的金属部分的方式装设导线接地线。确认接地线的挂钩与导线连接紧密可靠	8	(1)滑车安装不规范，扣2分。 (2)安全带使用不规范，扣2分。 (3)验电、安装接地线未戴绝缘手套，扣2分。 (4)验电器使用不规范，扣2分；验电顺序错误，扣2分。 (5)接地线两端安装顺序错误，扣2分；连接处不牢固，扣1分；未对安装好的端部进行检查，扣1分。 (6)接地线缠绕，扣2分			
8	挂设地线接地线	(1)登塔至指定位置后，将安全带系在牢固的构件上，检查扣环是否扣牢。使用绳索传递工器具。 (2)检查接地线完好，先用砂纸打磨塔材接地端，并先安装接地端后装导线端。 (3)装设地线接地线时，必须使用绝缘手套进行操作，禁止直接用手操作接地线的金属部分的方式装设地线接地线。确认地线接地线的挂钩与地线连接紧密可靠	6	(1)安全带使用不规范，扣2分。 (2)安装接地线未戴绝缘手套，扣2分。 (3)接地线两端安装顺序错误，扣2分；连接处不牢固，扣1分；未对安装好的端部进行检查，扣1分。 (4)接地线缠绕，扣2分			

续表

序号	项目名称	质量要求	分值	扣分标准	扣分原因	扣分	得分
9	安装作业飞车	(1)作业人员进入地线前应检查连接金具锈蚀情况。 (2)检查地线绝缘子是否完好,销子是否齐全。 (3)对地线做冲击试验。 (4)塔上辅助人员配合安装作业飞车	9	(1)未检查连接金具锈蚀情况,扣2分。 (2)未检查地线绝缘子是否完好、销子是否齐全,每项扣1分。 (3)未对地线做冲击试验,扣2分。 (4)作业飞车安装不规范,扣2分			
10	进入作业点	正确使用作业飞车,移动速度均匀,无危险动作;在地线上移动时,不得失去安全带的保护	9	(1)失去安全带保护,每次扣3分; (2)移动飞车速度过快,扣2分; (3)移动过程中有危险动作,每次扣3分			
11	修补断股地线	(1)到达作业点位后固定好作业飞车; (2)损伤地线修复、打磨处理平整; (3)量出预绞丝长度,用钢卷尺在导线损伤处的一端量出1/2预绞丝长的位置画印; (4)预绞丝中心应安装在导线损伤部位严重处,不得有缝隙; (5)用木榔头轻敲预绞丝端头,端头应平整	16	(1)未在合适位置固定作业飞车,扣3分。 (2)地线未打磨处理平整,扣3分。 (3)未正确画印,扣2分。 (4)预绞丝安装顺序不规范,扣4分。 (5)预绞丝末端若不平整,扣2分。 (6)预绞丝安装有缝隙,每处扣1分			
12	进入横担	(1)作业人员移动飞车沿地线返回横担; (2)进入横担过程中不得失去安全带的保护; (3)用传递绳将作业飞车传至地面	8	(1)作业飞车移动速度控制不当,扣2分。 (2)进入横担过程中失去安全带的保护,每次扣2分。 (3)传递飞车时,传递不规范,扣2分			
13	清理铁塔上工器具	(1)依次拆作业飞车、地线接地线、导线接地线、钢丝套、滑车等工器具; (2)拆除地线上的接地线和导线上的接地线,先拆导(地)线端、后拆接地端; (3)确认无遗留物	7	(1)拆除顺序错误,扣3分。 (2)有遗留物,扣4分			

续表

序号	项目名称	质量要求	分值	扣分标准	扣分原因	扣分	得分
14	下塔	(1)携带传递绳沿脚钉匀步下塔。 (2)正确携带传递绳(传递绳头折双、打死结,斜挎肩上)	8	(1)未携带传递绳,扣4分;传递绳携带方式不规范,扣2分。 (2)手抓脚钉,每次扣1分。 (3)未沿脚钉侧主材登塔,扣4分			
15	摆放工器具	完工后清理作业现场,按要求摆放工器具整齐	2	未按要求摆放工器具,扣2分			
16	文明施工	(1)严禁高处坠物。 (2)严禁浮置物品。 (3)严禁口中含物	5	(1)高处坠物,扣5分。 (2)浮置物品,扣2分。 (3)口中含物,扣2分			
17	完成时间	在规定时间内按要求完成。		在规定时间内按要求完成,超过时间10分钟即终止操作,只记完成部分得分			
18	合计		100				

模块四 停电更换±800kV特高压输电线路架空地线培训及考核标准

一、培训标准

(一) 培训要求

模块名称	停电更换±800kV特高压输电线路架空地线	培训类别	操作类
培训方式	实操培训	培训学时	21学时
培训目标	1.了解架空地线的型号、结构及安装方式。 2.能在±800kV特高压输电线路的架空地线上验电、装设接地线。 3.能完成停电更换±800kV特高压输电线路架空地线		
培训场地	特高压直流实训线路		
培训内容	停电更换±800kV特高压直流输电线路架空地线		
适用范围	特高压直流输电线路检修人员		

(二) 引用规程规范

(1)《±800kV直流线路带电作业技术规范》(DL/T 1242-2013)

(2)《±800kV直流架空输电线路运行规程》(GB/T 28813-2012)

(3)《±800kV直流架空输电线路检修规程》(DL/T 251-2012)

(4)《±800kV直流输电线路带电作业技术导则》(Q/GDW 302-2009)

(5)《带电作业用屏蔽服装》(GB/T 6568-2008)

(6)《带电作业用绝缘配合导则》(DL/T 867-2004)

(7)《带电作业用绝缘工具试验导则》(DL/T 878-2004)

(8)《国家电网公司带电作业工作管理规定(试行)》(国家电网生〔2007〕751号)

(9)《国家电网公司电力安全工作规程(线路部分)》(Q/GDW 1799.2-2013)

(10)《电工术语 架空线路》(GB/T 2900.51-1998)

(11)《电工术语 带电作业》(GB/T 2900.55-2016)

(12)《带电作业工具设备术语》(GB/T 14286-2008)

(13)《带电作业用工具、装置和设备使用的一般要求》(DL/T 877-2004)

(14)《带电作业工具、装置和设备预防性试验规程》(DL/T 976-2005)

(15)《带电作业用绝缘滑车》(GB/T 13034-2008)

(16)《带电作业用绝缘绳索》(GB 13035-2008)

(三)培训教学设计

本设计以完成"停电更换±800kV特高压输电线路架空地线"为工作任务,按工作任务完成的标准化作业流程来设计各个培训阶段,每个阶段包括了具体的培训目标、培训内容、培训学时、培训方法(培训资源)、培训环境和考核评价等内容,如表2-4-1所示。

表2-4-1 停电更换±800kV特高压直流输电线路架空地线培训内容设计

培训流程	培训目标	培训内容	培训学时	培训方法与资源	培训环境	考核评价
1.理论教学	1.熟悉架空地线的型号、结构及安装方式。2.熟悉验电、装设接地线的操作流程。3.熟悉停电更换±800kV特高压直流输电线路架空地线的工作流程	1.架空地线的型号、结构及安装方式。2.验电、装设接地线的操作流程。3.停电更换±800kV特高压直流输电线路架空地线的工作流程	2	培训方法:讲授法。培训资源:PPT、相关规程规范	多媒体教室	考勤、课堂提问和作业
2.准备工作	能完成作业前准备工作	1.作业现场查勘。2.编制培训标准化作业卡。3.填写培训操作工作票。4.完成本操作的工器具及材料准备	1	培训方法:1.现场查勘和工器具及材料清理采用现场实操方法。2.编写作业卡和填写工作票采用讲授方法。培训资源:1.±800kV实训线路。2.特高压工器具库房。3.空白工作票	1.特高压输电实训线路;2.多媒体教室	

续表

培训流程	培训目标	培训内容	培训学时	培训方法与资源	培训环境	考核评价
3.作业现场准备	能完成作业现场准备工作	1.作业现场复勘。 2.工作许可。 3.作业现场布置。 4.班前会。 5.工器具及材料检查	2	培训方法：演示与角色扮演法。 培训资源：±800kV实训线路	±800kV实训线路	
4.培训师演示	通过现场观摩，使学员初步领会本任务操作流程	1.验电、装设接地线。 2.直线塔翻线。 3.耐张塔松线及连接新旧地线。 4.渡线及附件安装	8	培训方法：演示法。 培训资源：±800kV特高压直流实训线路	±800kV实训线路	
5.学员分组训练	1.能完成验电、装设接地线操作。 2.能完成停电更换±800kV特高压直流输电线路架空地线作业	1.学员分组（22人一组）训练验电、装设接地线和更换架空地线技能操作。 2.培训师对学员操作进行指导和安全监护	7	培训方法：角色扮演法。 培训资源：实训线路	±800kV实训线路	采用技能考核评分细则对学员操作评分
6.工作终结	1.使学员进一步辨析操作过程不足之处，便于后期提升。 2.培训学员安全文明生产的工作作风	1.作业现场清理。 2.向调度汇报工作。 3.班后会，对本次工作任务进行点评总结	1	培训方法：讲授和归纳法	±800kV实训线路	

（四）作业流程

1.工作任务

完成停电更换±800kV特高压输电线路架空地线。

2.天气及作业现场要求

（1）停电更换±800kV特高压输电线路架空地线应在良好的天气进行。

如遇雷电（听见雷声、看见闪电）、雪、雹、雨、雾等，风力大于5级，不得进行作业；恶劣天气下必须开展停电抢修时，应组织有关人员充分讨论并编制必要的安全措施，经本单位批准后方可进行。

（2）作业人员精神状态良好，工作班成员认真学习工作票和安全技术措施，所有人员做到"四清楚"（作业任务清楚、危险点清楚、作业程序清楚、安全措施清楚）。

（3）作业前停送电联系人必须与调度联系履行工作许可手续，严禁约时停送电。工作负责人必须在得到许可人的许可工作命令后，方可在需检修的线路上验电、挂设接地线和进行检修工作。

(4) 停电后，工作负责人应认真做好记录。

(5) 登杆前应检查塔上是否有蜂窝，发现蜂窝严禁登塔。

(6) 塔上作业人员必须使用双保险安全带。

3. 准备工作

3.1 危险点及其预控措施

(1) 危险点——误登带电线路

预控措施：

①登杆塔作业前，工作负责人、工作班成员应共同认真核查双重名称和识别标记（色标、判别标志等）与停电线路名称相符。

②登杆塔前应检查铁塔根部、基础等，必须牢固可靠。

③登杆塔前应检查登高工器具和设施，如安全带、脚钉、塔材等必须完整牢靠。

④不涉及挂设接地线的中间作业人员，应认真核实线路相序、色标、名称、编号与停电线路相符，确认线路名称无刷错、刷反等情况后，方可登杆。

(2) 危险点——高处坠落

预控措施：

①攀爬过程中，为防止登杆人员串落，登杆作业人员间距不得小于1.6m。

②攀爬铁塔前应将脚底泥土清除干净，检查工具包完整，攀爬过程中不得掉落物件伤人。

③作业人员攀登杆塔时应戴好安全帽，穿软底鞋，动作不能过大，匀步攀登。

④攀爬过程中，安全带应收拾妥当，长尾绳放置在工具包内，主带应挂在肩上，防止攀爬过程中安全带勾挂脚钉和塔材，致使作业人员高空坠落。

⑤杆塔上移位时，不得失去安全带保护，做到踩稳抓牢。

⑥到达作业点位置，系好安全带（绳），应牢固可靠，不得低挂高用。

⑦未验电前，人体、无头绳等与导线的安全距离必须不小于10.1m，工作中应设专人监护。

(3) 危险点——高处坠物伤人

预控措施：

①地面人员不得站在作业点垂直下方。塔上人员应防止落物伤人，使用的工具、材料应用绳索传递。

②在高处作业应使用工具袋，较大的工具应固定在牢固的构件上，不准随便乱放。

③使用绞磨起吊过程中，应设专人指挥，统一配合，绝缘子串刚离地后应进行冲

击检查。

（4）危险点——防止感应电伤人

①为防感应电伤人，需在检修的线路上验电、挂设接地线。

②如需接触架空地线，在架空地线接触前应进行可靠接地。

（5）危险点——现场作业安全监护

①自作业开始至作业终结，安全监护人必须始终在现场对作业人员进行不间断的安全监护。

②工作负责人，监护人必须穿安全监护背心。

（6）危险点——交通安全

①出车时应注意车辆行驶安全，谨慎驾驶车辆，禁止违法行车。

3.2 工器具及材料选择

停电更换±800kV特高压输电线路架空地线所需工器具及材料见表2-4-2。工器具出库前，应认真核对工器具的使用电压等级和试验周期，并检查确认外观良好、连接牢固、转动灵活，且符合本次工作任务的要求；工器具出库后，应存放在工具袋或工具箱内进行运输，防止脏污、受潮；金属工具和绝缘工器具应分开装运，防止因混装运输导致工器具变形、损伤等现象发生。

表2-4-2　停电更换±800kV特高压输电线路架空地线所需工器具及材料表

序号	名称	规格型号	单位	数量	备注
1	作业飞车		付	1	其他工具
2	磨绳	Φ16	盘	1	其他工具
3	手板葫芦	6T	把	3	金属工具
4	绞磨	5T	台	3	机动工具
5	吊绳滑车	1T	套	3	金属工具
6	绳套		根	3	其他工具
7	卡线器		个	2	金属工具
8	提线器		套	1	金属工具
9	单轮滑车		个	3	金属工具
10	放线架		套	2	金属工具
11	放线盘		套	2	金属工具
12	钢丝套	Φ18	根	8	其他工具

续表

序 号	名 称	规格型号	单位	数量	备 注
13	铁滑车	10T	个	8	金属工具
14	卸扣	10T	个	8	金属工具
15	卸扣	8T	个	6	金属工具
16	液压机(含液压钳)		套	2	机动工具
17	对讲机		个	10	其他工具
18	全身式安全带	含带缓冲包长20M的后保绳	套	7	个人防护用具
19	安全背心		件	3	其他工具
20	安全围栏		卷	5	其他工具
21	垫木		块	若干	其他工具
22	地线	铝包钢绞线 JLB20A,150	盘	3	材料
23	耐张线夹	同型号	套	2	材料

3.3 作业人员分工

本任务作业人员分工如表2-4-3所示。

表2-4-3　停电更换±800kV特高压输电线路架空地线人员分工表

序号	工作岗位	数量(人)	工作职责
1	工作负责人	1	负责本次工作任务的人员分工、工作前的现场查勘、作业方案的制定、工作票的填写、现场复勘、办理工作许可手续、召开工作班前会、落实现场安全措施、负责作业过程中的安全监督、工作中突发情况的处理、工作质量的监督、工作后的总结
2	安全监护人	1	负责本次工作过程中的安全监护工作
3	高空作业人员	7	负责本次停电更换±800kV直流输电线路架空地线操作
4	地面辅助人员	10	负责本次作业过程的地面辅助工作
5	绞磨操作人员	2	负责本次作业过程的绞磨操作工作
6	绞磨信号指挥人员	2	负责2台绞磨启停的指挥工作

4. 工作程序

本任务工作流程如表2-4-4所示。

表2-4-4 停电更换±800kV特高压输电线路架空地线工作流程表

序号	作业内容	作业步骤及标准	安全措施及注意事项	责任人
1	现场复勘	工作负责人负责完成以下工作： (1)现场核对线路名称、杆塔编号、相别无误；基础及杆塔完好无异常；交叉跨越距离符合安全要求；确认缺陷情况及导地线规格型号等。 (2)检测风速等现场气象条件符合作业要求。 (3)检查地形环境符合作业要求。 (4)检查工作票所列安全措施与现场实际情况相符，必要时予以补充	(1)正确穿戴安全帽、工作服、工作鞋、劳保手套。 (2)不得在危及作业人员安全的气象条件下作业。 (3)严禁非工作人员、车辆进入作业现场	
2	工作许可	作业前停送电联系人必须与调度联系履行工作许可手续	不得未经工作许可人许可即开始工作	
3	现场布置	正确装设安全围栏并悬挂标示牌： (1)安全围栏范围应充分考虑高处坠物，以及对道路交通的影响。 (2)安全围栏出入口设置合理。 (3)妥当布置"从此进出""在此工作""车辆慢行"或"车辆绕行"等标示	对道路交通安全影响不可控时，应及时联系交通管理部门强化现场交通安全管控	
4	召开班前会	(1)全体工作成员列队。 (2)工作负责人宣读工作票，明确工作任务及人员分工；讲解工作中的安全措施和技术措施；查(问)全体工作成员精神状态；告知工作中存在的危险点及采取的预控措施。 (3)全体工作成员在工作票上签名确认	(1)工作票填写、签发和许可手续规范，签名完整。 (2)全体工作成员精神状态良好。 (3)全体工作成员明确任务分工、安全措施和技术措施	
5	检查工具	(1)在防潮苫布上，将工器具按作业要求准备齐备，并分类定置摆放整齐。检查工器具外观和试验合格证，无遗漏。 (2)检查人员向工作负责人汇报各项检查结果符合作业要求	(1)防潮苫布数量足够，设置位置合理，保持清洁、干燥。 (2)工器具外观检查合格，无损伤、受潮、变形、失灵现象，合格证在有效期内	

续表

序号	作业内容	作业步骤及标准	安全措施及注意事项	责任人
6	登塔	(1)登杆塔作业前,必须先核对线路名称及编号。对同塔多回线路,工作负责人、工作班成员应共同认真核查双重名称和识别标记(色标、判别标志等)。 (2)登杆塔前应检查铁塔根部、基础等,必须牢固可靠。 (3)攀爬过程,为防止登杆人员串落,登杆作业人员间距不得小于1.6m,安全带收拾妥当,长尾绳放置在工具包内,主带应挂在肩上,防止攀爬过程中安全带勾攀脚钉和塔材,致使作业人员高空坠落。 (4)登杆塔至横担处时,监护人和作业人员应再次核对停电线路的识别标记和双重名称,确实无误后方可进入停电线路侧的横担	(1)作业人员攀登杆塔时应戴好安全帽,穿软底鞋,动作不能过大,匀步攀登。 (2)攀爬过程中,安全带应收拾妥当,长尾绳放置在工具包内,主带应挂在肩上,防止攀爬过程中安全带勾挂脚钉和塔材,致使作业人员高空坠落。 (3)杆塔上移位时,不得失去安全带保护,做到踩稳抓牢。 (4)到达作业点位置,系好安全带(绳),应牢固可靠,不得低挂高用。 (5)未验电前,人体、无头绳等与导线的安全距离必须不小于9.5m,工作中应设专人监护	
7	验电、装设接地线	(1)登杆就位后,将安全带系在牢固可靠构件或电杆上,必须检查扣环是否正确就位。验电杆(笔)等工器具必须使用绳索传递。 (2)验电时站位正确:宜工作且无触电危险的位置;验电时必须带绝缘手套。对线路的验电应逐相进行,按照先近后远,先验下层、后验上层的顺序。 (3)检查接地线完好,按程序(先接接地端,后接导线端)装设好接地线。 (4)装设接地线时,必须使用绝缘绳或绝缘手柄进行操作,禁止直接用手操作接地线的金属部分的方式装设设接地线。确认接地线的夹头与导线连接紧密可靠	(1)验电前,应先在有电设备上进行试验,验证验电器良好,无法在有电设备上试验时可用高压发生器等确证验电器良好。 (2)高压验电必须戴绝缘手套、验电器的伸缩式绝缘杆长度应拉足,验电时手应握在手柄处不得超过护环,人体与验电设备保持安全距离,雨雪天气时不得进行室外直接验电。 (3)装设接地线必须先接接地端,后接导体端,必须接触良好,拆除时顺序相反。禁止以缠绕导线的方式装设设接地线	
8	直线塔翻线	(1)作业人员在直线塔地线横担合适位置安装单轮滑车,并拆除地线上的防震锤。 (2)作业人员使用提线器和6T手板葫芦提升地线。 (3)将连接部分金具取下后,再将地线翻进固定于地线横担上的单轮滑车中		

续表

序号	作业内容	作业步骤及标准	安全措施及注意事项	责任人
9	安装卡线器	(1)作业人员在地线上量出防震锤位置并记录后,拆除防震锤。 (2)作业人员乘坐飞车将卡线器卡至地线上合适位置后,并收紧磨绳。 (3)在地线两端通过6T手板葫芦收紧地线后,取脱地线连接金具	收紧地线后,应做冲击试验,确认无误后方可进行下一步操作	
10	耐张塔松线及连接新旧地线	(1)作业人员松出两端的6T手扳葫芦,使磨绳受力后,拆除手扳葫芦。 (2)启动绞磨,利用绞磨将两端的地线松至地面。 (3)松至地面后,地面人员开断地线端的耐张线夹,打磨并压接新旧地线	(1)放线过程中,收线绞磨操作应平稳,保持地线牵引平衡,预防地线跳槽。 (2)使用绞磨时,绞磨盘上缠绕圈数不得少于5圈,且从下方卷入,并排列整齐,尾绳控制人员不得少于2人。 (3)放、紧线时,人员不得站在或跨在已受力的牵引绳、地线的内角侧和展放的地线圈内以及牵引绳或架空线的垂直下方,防止意外跑线时伤人。 (4)地线压接时人员禁止将身体任何部位放在液压机正上方,压接机具应有固定设施,操作时放置平稳,两侧扶线人员对准位置,手指不得伸入压模内	
11	渡线及附件安装	(1)牵引场与张力场同时2台绞磨同时启动,并有专人统一指挥。 (2)待新地线到达牵引场后张力场压接耐张线夹,压接后利用绞磨收线将压接后的地线连接到绝缘子金具上。 (3)牵引场收紧地线,待地线弧垂与原弧垂一致后作记号。 (4)地线松至地面压接另一端耐张线夹并收紧磨绳,将地线连接到绝缘子串金具上,直线塔将新地线连接到线夹里,紧固螺栓,按照之前记号位置安装防震锤	(1)放线施工过程中,临时拉线、交叉跨越、直线塔、受力工器具应有专人看守,并检查受力情况良好。 (2)放线时,应保持通讯畅通,统一信号、统一指挥。 (3)施工过程中,严禁任何人在地线下方穿越或逗留。 (4)渡线过程中,应有专人随时检查地锚、转向滑车、磨绳的受力情况,并适时进行调整。 (5)放线过程中,收线绞磨操作应平稳,保持地线牵引平衡,预防地线跳槽	
12	拆除工器具	(1)依序拆磨绳、手扳葫芦、卡线器、后备保护绳等工器具。 (2)拆除接地线	拆除接地线的顺序为先导体端,后接地端	

续表

序号	作业内容	作业步骤及标准	安全措施及注意事项	责任人
13	工作结束	(1)工作负责人组织全体工作成员整理工器具和材料,清理现场,做到"工完料尽场地清"。 (2)召开班后会,工作负责人进行工作总结和点评工作。点评本次工作的施工质量;点评全体工作成员的安全措施落实情况。 (3)工作负责人向工作许可人汇报工作结束,恢复停电线路送电,终结工作票		

二、考核标准

表2-4-5　特高压直流输电线路运检技能考核评分细则

考生填写栏	编号:	姓名:	所在岗位:	单位:	日期:	年　月　日
考评员填写栏	成绩:	考评员:	考评组长:	开始时间:	结束时间:	操作时长:
考核模块	停电更换±800kV特高压输电线路架空地线		考核对象	特高压直流输电线路检修人员	考核方式	操作	考核时限	360min
任务描述	停电更换±800kV特高压输电线路耐张段左架空地线							
工作规范及要求	1. 给定条件:特高压±800kV实训线路001#-003#塔耐张段左架空线需要更换。线路已经停电、验电、挂接地线,所使用工器具、材料已经过测试,工作票已办理,安全措施已经完备。 2. 本作业工作应在良好天气下进行。如遇雷电(听见雷声、看见闪电)、雪、雹、雨、雾等,风力大于5级,不得进行作业。 3. 本项作业需工作负责人1名,安全监护人1人、塔上电工1人、塔上辅助工6人、地面辅助工10人,绞磨操作人员2人,协助人员完成工器具、材料的上、下传递工作,以及其他非技术性工作。 4. 工作负责人职责:负责本次工作任务的人员分工、工作前的现场查勘、作业方案的制定、工作票的填写、现场复勘、办理工作许可手续、召开工作班前会、落实现场安全措施、负责作业过程中的安全监督、工作中突发情况的处理、工作质量的监督、工作后的总结。 5. 专责监护人:负责作业现场的安全把控。 6. 高空作业人员职责:负责停电更换±800kV特高压输电线路架空地线。 7. 地面辅助人员职责:负责传递工具、材料配合塔上电工。 8. 绞磨操作人员职责:负责本次作业过程的绞磨操作工作。 9. 在作业过程中,如遇雷、雨、大风或其他任何情况威胁到工作人员的安全时,工作负责人或监护人可根据情况,临时停止工作。 给定条件: 1. 培训基地:特高压±800kV特高压实训线路001#～003#塔耐张段左架空线,地线型号:铝包钢绞线JLB20A,150。 2. 工作票已办理,安全措施已经完备,工作开始、工作终结时应口头提出申请(调度或考评员)。 3. 安全、正确地使用仪器对绝缘工具进行检测。 4. 必须按工作程序进行操作,工序错误扣除应做项目分值,出现重大人身、器材和操作安全隐患,考评员可下令终止操作(考核)							

续表

考核情景准备	1. 线路:特高压±800kV线路001#～003#塔耐张段左架空线,工作内容:停电更换±800kV特高压输电线路架空地线,地线型号:JLB20A,150。 2. 所需作业工器具:作业飞车1付、Φ16磨绳一捆、6T手扳葫芦3把、5T绞磨3台、吊绳滑车3套、绳套3根、卡线器2个、提线器1套、单轮滑车3个、放线架2套、放线盘2套、Φ18钢丝套8根、10T铁滑车8个、10T卸扣8个、8T卸扣6个、液压机(含液压钳)2套、对讲机若干、双保险安全带7套、垫木若干。 3. 材料:同型号地线1盘、接续管1套、耐张线夹2套。 4. 作业现场做好监护工作,作业现场安全措施(围栏等)已全部落实;禁止非作业人员进入现场,工作人员进入作业现场必须戴安全帽。 5. 考生自备工作服,安全帽,线手套,安全带(含二保绳)
备注	1. 各项目得分均扣完为止,出现重大人身、器材和操作安全隐患,考评员可下令终止操作。 2. 设备、作业环境、安全带、安全帽、工器具等不符合作业条件考评员可下令终止操作

表2-4-6 特高压直流输电线路运检技能考核评分标准

序号	项目名称	质量要求	分值	扣分标准	扣分原因	扣分	得分
1	着装	工作服、工作鞋、安全帽、劳保手套穿戴正确	5	错漏1项扣1分			
2	现场布置	正确装设安全围栏并悬挂标示牌: (1)安全围栏范围应充分考虑高处坠物,以及对道路交通的影响。 (2)安全围栏出入口设置合理。 (3)妥当布置"从此进出""在此工作""从此上下"等标示	3	(1)作业现场未装设围栏,扣0.5分。 (2)未设立警示牌,扣0.5分。 (3)未悬挂登塔作业标志,扣0.5分。			
3	召开班前会	(1)工作负责人佩戴红色背心,宣读工作票,明确工作任务及人员分工;讲解工作中的安全措施和技术措施;查(问)全体工作成员精神状态;告知工作中存在的危险点及采取的预控措施。 (2)全体工作成员在工作票上签名确认	3	(1)未进行分工本项不得分,分工不明,扣1分。 (2)现场工作负责人未穿佩安全监护背心,扣0.5分。 (3)工作票上工作班成员未签字或签字不全的,扣1分			
4	工器具检查	(1)工作人员按要求将工器具放在防防潮苫布上;防潮苫布应清洁、干燥。 (2)工器具应按定置管理要求分类摆放;检查工器具外观和试验合格证,无遗漏。 (3)活动扳手、平口钳、拔销钳、工具包符合质量要求。 (4)绞磨试机,检查档位是否正常;手扳葫芦检查。	10	(1)未使用防潮布并定置摆放工器具,扣1分。 (2)未检查工器具试验合格标签及外观检查,每项扣0.5分。 (3)未正确对工器具进行检测,每项扣1分。 (4)未正确绞磨试机,扣2分;未正确检查手扳葫芦,扣2分。			

续表

序号	项目名称	质量要求	分值	扣分标准	扣分原因	扣分	得分
4	工器具检查	(5)对钢丝绳、滑车进行检查,确认连接可靠,受力满足要求。 (6)安全带、个人保安线符合质量要求,在试验周期内。 (7)提线器、卡线器检查符合要求,在试验周期内。 (8)确认地线、接续管型号、长度,外观检查符合要求。 (9)登塔人员再次核对双重名称、杆号、相别并报告	10	(5)错漏,每项扣1分。 (6)未正确检查,每项扣1分。 (7)错漏,每项扣0.5分。 (8)错漏,每项扣1分。 (9)未核对双重名称,扣3分			
5	登塔	(1)检查杆塔基础无异常。 (2)塔上电工穿好将安全带做冲击试验后,系好安全带后正确携带吊绳(吊绳头折双、打死结、斜挎肩上)相继登塔。 (3)登塔过程中应系好防坠落保护装置,匀速登塔,手抓主材;将安全带挂在肩上,登塔作业人员间距不得小于1.6m,工作负责人加强作业监护。 (4)由塔身到横担上工作点不得失去安全带保护。 (5)传递滑车安装位置正确,方便操作。	10	(1)未检查,每项扣1分。 (2)未系安全带或安全带及后备保护绳未进行冲击试验,扣2分;未携带吊绳,扣1分;吊绳携带方式不规范,扣1分。 (3)手抓脚钉,扣2分;未沿脚钉侧主材登塔扣1分。 (4)未正确使用安全带,扣2分。 (5)滑车传递绳悬挂位置不便工具取用,扣1分。 (6)传递时金属工具难以保证安全距离,扣2分;工具绑扎不牢,扣2分。 (7)传递时高空落物,扣2分。 (8)传递过程工具与塔身磕碰,扣2分。 (9)传递工具绳索打结混乱,扣1分			
6	验电、装设接地线	(1)登杆就位后,将安全带系在牢固可靠构件或电杆上,必须检查扣环是否正确就位。验电杆(笔)等工器具必须使用绳索传递。 (2)检查接地线完好,按程序(先接接地端,后接导线端)装设好接地线,接地夹头在横担上连接牢固。	11	(1)未进行验电、装设接地线,扣5分。 (2)装设接地线顺序错误,扣3分。			

续表

序号	项目名称	质量要求	分值	扣分标准	扣分原因	扣分	得分
6	验电、装设接地线	(3)装设接地线时,必须使用绝缘绳或绝缘手柄进行操作,禁止直接用手操作接地线的金属部分的方式装设接地线。确认接地线的夹头与导线连接紧密可靠	11	(3)接地线连接不牢固,每处扣3分。(4)未使用绝缘手套,每次扣2分。			
7	直线塔翻线	(1)作业人员在直线塔地线横担合适位置安装单轮滑车,并拆除地线上的防震锤。(2)作业人员使用提线器和6T手扳葫芦提升地线。(3)将连接部分金具取下后,再将地线翻进固定于地线横担上的单轮滑车中	6	(1)提升地线时未进行保护地线和塔材,扣2分。(2)操作手扳葫芦不当,扣2分。(3)上下过程中软梯使用不规范,扣2分。			
8	安装卡线器	(1)作业人员在地线上量出防震锤位置并记录后,拆除防震锤。(2)作业人员乘坐飞车将卡线器卡至地线上合适位置后,并收紧磨绳。(3)在地线两端通过6T手扳葫芦收紧地线,做冲击试验,确认无误后取脱地线连接金具	8	(1)未记录防震锤位置,扣3分。(2)未正确使用作业飞车,扣3分。(3)取脱连接金具前未冲击试验,扣2分			
9	耐张塔松线及连接新旧地线	(1)作业人员松出两端的6T手扳葫芦,使磨绳受力后,冲击磨绳,拆除手扳葫芦。(2)启动绞磨,利用绞磨将两端的地线松至地面。(3)布置张力场塔上新地线的转角滑车及地线走向。(4)地面人员开断地线端的耐张线夹,打磨并压接新旧地线,接续管压接工艺满足要求。	15	(1)未冲击试验,扣2分。(2)转向滑车安装未对塔材保护,每处扣1分。(3)新地线走向布置不规范,扣3分。(4)地线未进行打磨处理,每处扣2分。(5)压接工艺不满足要求,扣3分			

续表

序号	项目名称	质量要求	分值	扣分标准	扣分原因	扣分	得分
10	渡线及附件安装	(1)牵引场与张力场同时2台绞磨同时启动,并有专人统一指挥。 (2)渡线过程中档中央应派专人看守,防止地线落地。 (3)塔上作业人员看好接头,防止卡线。 (4)绞磨操作人员控制好渡线速度,防止张力过大。 (5)待新地线到达牵引场后张力场压接耐张线夹,压接后利用绞磨收线将压接后的地线连接到绝缘子金具上。 (6)牵引场收紧地线,待地线弧垂与原弧垂一致后作记号。 (7)地线松至地面压接另一端耐张线夹并收紧磨绳,将地线连接到绝缘子串金具上,直线塔将新地线连接到线夹里,紧固螺栓,按照之前记号位置安装防震锤。 (8)直线塔将新地线连接到线夹里,紧固螺栓,按照原要求安装防震锤。	12	(1)渡线时地线掉落地面,每次扣2分。 (2)渡线过程中张力过大,扣1分。 (3)未派专人指挥导致卡线,扣2分。 (4)旧地线未及时整理回收,扣2分。 (5)未指派专人测量弧垂扣2分;弧垂与原弧垂不一致,扣2分。 (6)耐张线夹型号与地线不符,扣3分。 (7)连接完毕后未进行冲击试验,扣2分。 (8)压接工艺不满足要求,扣2分。 (9)未检查螺栓是否紧固到位,扣2分。 (10)间隔棒安装位置不正确,每处扣1分;间隔棒安装不规范,安装方向不正确,扣2分			
11	返回地面	(1)检查金具、附件是否齐全到位,连接部位是否牢固,松出磨绳、检查塔材是否损伤,检查地线弧垂,连接金具是否符合要求。 (2)依序拆磨绳、手扳葫芦、接地线、滑车、卡线器、后备保护绳等工器具,传递至地面。拆除接地线时,应先拆导线端,后拆接地端。先拆上层,后拆下层,先拆远端,后拆近端。 (3)塔上电工检查塔上无遗留物后,向工作负责人汇报,得到工作负责人同意后携带绝缘传递绳下塔 (4)必须沿脚钉侧主材正确下塔,正确携带吊绳(吊绳头折双、打死结,斜挎肩上)。 (5)确认无遗留物	7	(1)未检查金具连接情况,扣3分;未检查地线弧垂,扣2分。 (2)传递绳缠绕,每次扣2分;塔材损伤,每处扣2分。 (3)接地线拆除顺序不正确,扣2分。 (4)未携带吊绳,扣4分;吊绳携带方式不规范,扣2分。 (5)手抓脚钉,每次扣1分;未沿脚钉侧主材登塔,扣4分。 (6)有遗留物,扣4分			

续表

序号	项目名称	质量要求	分值	扣分标准	扣分原因	扣分	得分
12	工作结束	(1)工作负责人组织全体工作成员整理工器具和材料,将工器具清洁后放入专用的箱(袋)中;清理现场,做到"工完料尽场地清"。 (2)召开班后会,工作负责人进行工作总结和点评工作。点评本次工作的施工质量;点评全体工作成员的安全措施落实情况。 (3)工作负责人向工作许可人汇报工作结束,恢复停电线路送电,终结工作票。 (4)在规定时间内按要求完成	10	(1)未清理作业现场,漏一项扣3分。 (2)未开班后会,扣2分。 (3)未向工作许可人汇报,扣2分。 (4)每超过1分钟,扣1分;超过5分钟后,未完成项不计分			
	合计		100				

模块五 停电更换±800kV特高压输电线路接地极耐张整串绝缘子培训及考核标准

一、培训标准

（一）培训要求

模块名称	停电更换±800kV特高压输电线路接地极耐张整串绝缘子	培训类别	操作类
培训方式	实操培训	培训学时	14学时
培训目标	1.会检查和使用停电更换±800kV特高压输电线路接地极耐张整串绝缘子的工器具 2.掌握停电更换±800kV特高压输电线路接地极耐张整串绝缘子工作的操作流程和工艺要求 3.掌握停电更换±800kV特高压输电线路接地极耐张整串绝缘子危险点		
培训场地	特高压直流实训线路		
培训内容	1.停电更换±800kV特高压输电线路接地极耐张整串绝缘子工器选择和检查 2.停电更换±800kV特高压输电线路接地极耐张整串绝缘子工作的操作流程和工艺要求 3.停电更换±800kV接地极耐张整串绝缘子工作的危险点		
适用范围	特高压直流输电线路检修人员		

（二）引用规程规范

（1）《架空送电线路运行规程》（DL/T 741-2010）

（2）《110~500kV架空送电线路设计技术规程》（DL/T 5092-1999）

（3）《国家电网公司电力安全工作规程（线路部分）》（Q/GDW1799.2-2013）

（4）《±800kV直流架空输电线路设计规范》（GB 50790-2013）

（5）《±800kV直流架空输电线路检修规程》（DL/T 251-2012）

（6）《±800kV直流架空输电线路运行规程》（GB/T 28813-2012）

（7）《110（66）kV~500kV架空输电线路检修规范》（国家电网公司）

（8）《架空输电线路状态检修导则》（DLT 1248-2013）

（9）《输变电设备状态检修管理规定》（国家电网公司）

(10)《输变电设备状态检修试验规程》(国家电网公司)

(三)培训教学设计

本设计以完成"停电更换±800kV特高压输电线路接地极耐张整串绝缘子"为工作任务,按工作任务完成的标准化作业流程来设计各个培训阶段,每个阶段包括了具体的培训目标、培训内容、培训学时、培训方法(培训资源)、培训环境和考核评价等内容,如表2-5-1所示。

表2-5-1　停电更换±800kV特高压输电线路接地极耐张整串绝缘子培训内容设计

培训流程	培训目标	培训内容	培训学时	培训方法与资源	培训环境	考核评价
1.理论教学	1.会检查和使用停电更换±800kV特高压输电线路接地极耐张整串绝缘子的工器具 2.掌握停电更换±800kV特高压输电线路接地极耐张整串绝缘子工作的操作流程和工艺要求 3.掌握停电更换±800kV特高压输电线路接地极耐张整串绝缘子危险点	1.停电更换±800kV特高压输电线路接地极耐张整串绝缘子工器具选择和检查 2.停电更换±800kV特高压输电线路接地极耐张整串绝缘子工作的操作流程和工艺要求 3.停电更换±800kV接地极耐张整串绝缘子工作的危险点	2	培训方法:讲授法。 培训资源:PPT、相关规程规范	多媒体教室	考勤、课堂提问和作业
2.准备工作	能完成作业前准备工作	1.作业现场查勘。 2.编制培训标准化作业卡。 3.填写输电线路第一种工作票。 4.完成本操作的工器具及材料准备	1	培训方法: 1.现场查勘和工器具及材料清理采用现场实操方法。 2.编写作业卡和填写工作票采用讲授方法。 培训资源: 1.±800kV实训线路; 2.特高压工器具库房; 3.空白工作票	1.特高压输电实训线路; 2.多媒体教室	

续表

培训流程	培训目标	培训内容	培训学时	培训方法与资源	培训环境	考核评价
3.作业现场准备	能完成作业现场准备工作	1. 作业现场复勘。 2. 工作申请。 3. 作业现场布置。 4. 班前会。 5. 工器具及材料准备	1	培训方法：演示与角色扮演法。 资源：±800kV实训线路	±800kV实训线路	
4.培训师演示	通过现场观摩，使学员初步领会本任务操作流程	1. 各类工器具检查和使用方法讲解。 2. 装设接地线。 3. 安装作业架，作业人员到达工作位置。 4. 利用双沟更换±800kV接地极耐张整串绝缘子。 5. 拆除工器具，下塔。	1	培训方法：演示法。 资源：±800kV实训线路	±800kV实训线路	
5.学员分组训练	能够分组完成停电更换±800kV接地极耐张整串绝缘子工作	1. 学员分组（5人一组，工作负责人1人，塔上作业人员2人，地面辅助人员2人）训练停电更换±800kV接地极耐张整串绝缘子。 2. 培训师对学员操作进行指导和安全监护	8	培训方法：角色扮演法。 资源：±800kV实训线路	±800kV实训线路	采用技能考核评分细则对学员操作评分
6.工作终结	1. 使学员进一步辨析操作过程不足之处，便于后期提升。 2. 培训学员安全文明生产的工作作风	1. 作业现场清理。 2. 向调度汇报工作。 3. 班后会，对本次工作任务进行点评总结	1	培训方法：讲授和归纳法。	±800kV实训线路	

（四）作业流程

1. 工作任务

完成停电更换±800kV特高压输电线路接地极耐张整串绝缘子。

2. 天气及作业现场要求

（1）停电更换±800kV特高压输电线路接地极耐张整串绝缘子应在良好的天气进行。

在5级及以上的大风以及暴雨、雷电、冰雹、大雾、沙尘暴等恶劣天气下，应停止露天高处作业。特殊情况下，确需在恶劣天气进行抢修时，应组织有关人员充分讨论必要的安全措施，经本单位批准后方可进行。

(2）作业人员精神状态良好，工作班成员认真学习工作票和安全技术措施，所有人员做到"四清楚"（作业任务清楚、危险点清楚、作业程序清楚、安全措施清楚）。

(3）作业前停送电联系人必须与调度联系履行工作许可手续，严禁约时停送电。工作负责人必须在得到许可人的许可工作命令后，方可在需检修的线路上验电、挂设接地线和进行检修工作。

(4）停电后，工作负责人应认真做好记录。

(5）登塔前应检查塔上是否有蜂窝，发现蜂窝严禁登塔。

(6）塔上作业人员必须使用双保险安全带，并佩戴护目镜。

3. 准备工作

3.1 危险点及其预控措施

(1）危险点——误登带电线路

预控措施：

①登杆塔作业前，工作负责人、工作班成员应共同认真核查双重名称和识别标记（色标、判别标志等）与停电线路名称相符。

②登杆塔前应检查铁塔根部、基础等，必须牢固可靠。

③登杆塔前应检查登高工器具和设施，如安全带、脚钉、塔材等必须完整牢靠。

④不涉及挂设接地线的中间作业人员，应认真核实线路相序、色标、名称、编号与停电线路相符，确认线路名称无刷错、刷反等情况后，方可登杆。

(2）危险点——登塔时、塔上作业时违反安规进行操作，可能引起高空坠落

预控措施：

①攀爬过程中，为防止登杆人员串落，登杆作业人员间距不得小于1.6m。

②攀爬铁塔前应将脚底泥土清除干净，检查工具包完整，攀爬过程中不得掉落物件伤人。

③作业人员攀登杆塔时应戴好安全帽，穿软底鞋，动作不能过大，匀步攀登。

④攀爬过程中，安全带应收拾妥当，长尾绳放置在工具包内，主带应挂在肩上，防止攀爬过程中安全带勾挂脚钉和塔材，致使作业人员高空坠落。

⑤杆塔上移位时，不得失去安全带保护，做到踩稳抓牢。

⑥到达作业点位置，系好安全带（绳），应牢固可靠，不得低挂高用。

⑦未验电前，人体、无头绳等与导线的安全距离必须不小于1.6m，工作中应设专人监护。

(3）危险点——高处坠物伤人

预控措施：

①地面人员不得站在作业点垂直下方。塔上人员应防止落物伤人，使用的工具、材料应用绳索传递。

②在高处作业应使用工具袋，较大的工具应固定在牢固的构件上，不准随便乱放。

（4）危险点——防止感应电伤人

①为防感应电伤人，塔上作业人员应使用个人保安线。

（5）危险点——现场作业安全监护

①自作业开始至作业终结，安全监护人必须始终在现场对作业人员进行不间断的安全监护。

②工作负责人，监护人必须穿安全监护背心。

（6）危险点——交通安全

出车时应注意车辆行驶安全，谨慎驾驶车辆，禁止违法行车。

3.2 工器具及材料选择

停电更换±800kV特高压输电线路接地极耐张整串绝缘子所需工器具及材料见表2-5-2。工器具出库前，应认真核对工器具的使用电压等级和试验周期，并检查确认外观良好、连接牢固、转动灵活，且符合本次工作任务的要求；工器具出库后，应防止脏污、受潮；金属工具和绝缘工器具应分开装运，防止因混装运输导致工器具变形、损伤等现象发生。

表2-5-2 停电更换±800kV特高压输电线路接地极耐张整串绝缘子所需工器具及材料表

序号	名称	规格型号	单位	数量	备注
1	安全帽		顶	5	
2	风速风向仪		支	1	
3	绝缘电阻测试仪		支	1	
4	防坠器	与杆塔防坠器装置型号对应	个	2	
5	安全带		副	2	
6	传递绳		根	1	
7	绳套		根	1	
8	滑车		个	2	
9	导线后背保护绳		根	1	
10	卡线器	与导线相匹配	个	2	
11	双沟	5T	把	1	
12	作业架		架	1	

续表

序号	名称	规格型号	单位	数量	备注
13	绝缘子	与原绝缘子一致	串	1	
14	防潮帆布	2m×4m	块	2	
15	安全围栏		套	若干	
16	警示标示牌	"在此工作""从此进出""从此上下"	套	1	
17	红马甲	"工作负责人"	件	1	
18	清洁毛巾		条	1	

3.3 作业人员分工

本任务作业人员分工如表2-5-3所示。

表2-5-3 停电更换±800kV特高压输电线路接地极耐张整串绝缘子人员分工表

序号	工作岗位	数量(人)	工作职责
1	工作负责人（安全监护人）	1	负责本次工作任务的人员分工、工作前的现场查勘、作业方案的制定、工作票的填写、现场复勘、办理工作许可手续、召开工作班前会、落实现场安全措施，负责作业过程中的安全监督、工作中突发情况的处理、工作质量的监督、工作后的总结。 负责本次工作过程中的安全监护工作
2	塔上作业人员	2	负责本次停电更换±800kV特高压输电线路接地极耐张整串绝缘子操作
3	地面辅助人员	2	负责本次作业过程的地面辅助工作

4. 工作程序

本任务工作流程如表2-5-4所示。

表2-5-4 停电更换±800kV特高压输电线路接地极耐张整串绝缘子工作流程表

序号	作业内容	作业标准	安全注意事项	责任人
1	现场复勘	工作负责人负责完成以下工作： (1)现场核对线路名称、铁塔编号无误；基础及塔身完好无异常；交叉跨越距离符合安全要求；确认缺陷情况及地线规格型号等。 (2)检查地形环境符合作业要求。 (3)检查工作票所列安全措施与现场实际情况相符，必要时予以补充	(1)正确穿戴安全帽、工作服、工作鞋、劳保手套。 (2)严禁非工作人员、车辆进入作业现场	
2	工作许可	作业前停送电联系人必须与调度联系履行工作许可手续。	(1)不得未经工作许可人许可即开始工作。 (2)严禁约时停送电	

续表

序号	作业内容	作业标准	安全注意事项	责任人
3	现场布置	正确装设安全围栏并悬挂标示牌： (1)安全围栏范围应充分考虑高处坠物,以及对道路交通的影响。 (2)安全围栏出入口设置合理。 (3)妥当布置"从此进出""在此工作""车辆慢行"或"车辆绕行"等标示	对道路交通安全影响不可控时,应及时联系交通管理部门强化现场交通安全管控	
4	召开班前会	(1)全体工作成员列队。 (2)工作负责人宣读工作票,明确工作任务及人员分工；讲解工作中的安全措施和技术措施；查(问)全体工作成员精神状态；告知工作中存在的危险点及采取的预控措施。 (3)全体工作成员在工作票上签名确认	(1)工作票填写、签发和许可手续规范,签名完整。 (2)全体工作成员精神状态良好。 (3)全体工作成员明确任务分工、安全措施和技术措施	
5	检查工器具	(1)在防潮苫布上,将工器具按作业要求准备齐备,并分类定置摆放整齐。检查工器具外观和试验合格证,无遗漏。 (2)检查人员向工作负责人汇报各项检查结果符合作业要求	(1)防潮苫布数量足够,设置位置合理,保持清洁、干燥。 (2)工器具外观检查合格,无损伤、受潮、变形、失灵现象,合格证在有效期内。	
6	登杆塔	(1)登杆塔作业前,必须先核对线路名称及编号。工作负责人、工作班成员应共同认真核查双重名称和识别标记(色标、判别标志等)。 (2)登杆塔前应检查铁塔根部、基础等,必须牢固可靠。 (3)攀爬过程,为防止登杆人员串落,登杆作业人员间距不得小于1.6m,安全带收拾妥当,长尾绳放置在工具包内,主带应挂在肩上,防止攀爬过程中安全带勾攀脚钉和塔材,致使作业人员高空坠落。 (4)登杆塔至横担处时,监护人和作业人员应再次核对停电线路的识别标记和双重名称,确实无误后方可进入停电线路侧的横担	(1)作业人员攀登杆塔时应戴好安全帽,穿软底鞋,动作不能过大,匀步攀登。 (2)攀爬过程中,安全带应收拾妥当,长尾绳放置在工具包内,主带应挂在肩上,防止攀爬过程中安全带勾挂脚钉和塔材,致使作业人员高空坠落。 (3)杆塔上移位时,不得失去安全带保护,做到踩稳抓牢。 (4)到达作业点位置,系好安全带(绳),应牢固可靠,不得低挂高用。 (5)未接地前,人体、无头绳等与导线的安全距离必须不小于1.6m,工作中应设专人监护。	

续表

序号	作业内容	作业标准	安全注意事项	责任人
7	验电、装设接地线	杆就位后,将安全带系在牢固可靠构件上,必须检查扣环是否正确就位。验电器等工器具必须使用绳索传递。 查接地线完好,按程序(先接接地端,后接导线端)装设好接地线。 设接地线时,必须使用绝缘绳或绝缘手柄进行操作,禁止直接用手操作接地线的金属部分的方式装设接地线。确认接地线的夹头与导线连接紧密可靠	(1)验电杆在领用时和使用前应检查是否正常。 (2)禁止以缠绕导线的方式装设接地线	
8	更换耐张整串绝缘子	(1)将双保险安全带的安全绳系在横担适当位置。 (2)设置好传递绳吊点。 (3)将作业架传递至横担。 (4)塔上作业人员将作业架前段挂在导线上,把作业架后端放在横担上并系牢固。 (5)塔上一名作业人员沿作业架平缓进入作业点,将安全带主带系在绝缘子串上。 (6)将卡线器、双钩、钢丝套传递至杆塔上。 (7)安装钢丝套、双钩紧线器,将卡线器安装在导线适当位置。 (8)收紧双钩紧线器,使其略为受力。 (9)将导线后备保护钢丝套一端安装在横担上,固定牢固;另一端用卡线器固定在导线上。 (10)导线后备保护卡线器应安装在连接双钩紧线器的卡线器前面适当位置。 (11)收紧双钩紧线器,进行冲击试验,将安全带转移到双钩上。 (12)在被更换绝缘子前、后,用短绳将绝缘子串固定在双钩上。 (13)取出 M 销,取下需更换的绝缘子并传递至地面。 (14)将新绝缘子传递到杆塔上,安装绝缘子及 M 销。 (15)检查 M 销、球头是否齐全到位,碗口朝向正确,清洁绝缘子串。 (16)松出双钩使绝缘子受力后,对绝缘子串做冲击试验		
9	拆除工器具	(1)拆除双钩紧线器及后备保护钢丝套,传递至地面。 (2)拆除作业架,传递至地面。 (3)确认工作点无遗留物。		

续表

序号	作业内容	作业标准	安全注意事项	责任人
10	工作结束	(1)工作负责人组织全体工作成员整理工器具和材料,清理现场,做到"工完料尽场地清"。 (2)召开班后会,工作负责人进行工作总结和点评工作。点评本次工作的施工质量;点评全体工作成员的安全措施落实情况。 (3)工作负责人向工作许可人汇报工作结束,恢复停电线路送电,终结工作票		

二、考核标准

表2-5-5　国网四川省电力公司特高压直流技能培训考核评分细则

考生填写栏	编号：	姓名：	所在岗位：	单位：	日期：	年 月 日			
考评员填写栏	成绩：	考评员：	考评组长：	开始时间：	结束时间：	操作时长：			
考核模块	停电更换±800kV特高压输电线路接地极耐张整串绝缘子		考核对象	特高压±800kV直流输电线路检修人员		考核方式	操作	考核时限	60min
任务描述	停电更换±800kV特高压输电线路接地极耐张整串绝缘子								
工作规范及要求	给定条件： 1. 培训基地：特高压直流±800kV输电线路接地极耐张塔 2. 所使用绝缘子已经过测试,工作票已办理,安全措施已经完备,工作开始、工作终结时应口头提出申请(调度或考评员)。 3. 必须按工作程序进行操作,工序错误扣除应做项目分值,出现重大人身、器材和操作安全隐患,考评员可下令终止操作(考核)。 4. 整个过程由参考人员独立完成;杆塔上辅助工1人、地面辅助工2人,协助参考人员完成工器具、材料的上、下传递工作,以及其他非技术性工作。 5. 工作开始应口头提出申请,工作结束时应口头汇报。								
考核情景准备	1. 工器具：作业架1付、5T双钩紧线器1把、卡线器2个、无极绳滑车(1T)1套、钢丝套1根、导线后备保护钢丝套1根、5T卸扣3个、吊绳2根、短绳1根(3m)、肩背双控式安全带、登杆工具。 2. 材料：同型号悬式绝缘子。 3. 在培训线路上操作。								
备注	1. 各项目得分均扣完为止,出现重大人身、器材和操作安全隐患,考评员可下令终止操作。 2. 设备、作业环境、安全带、安全帽、工器具等不符合作业条件考评员可下令终止操作								

表2-5-6　国网四川省电力公司特高压直流技能培训考核评分标准

序号	项目名称	质量要求	分值	扣分标准	扣分原因	扣分	得分
1	着装	工作服、工作鞋、安全帽、劳保手套穿戴正确	4	漏一项扣1分			
2	工具材料准备	(1)个人工具检查:活动扳手、平口钳、拔销钳、工具包符合质量要求。(2)登杆工具、安全带:升降板/脚扣、双保险安全带进行外观、试验周期检查,并进行冲击试验。(3)专用工具检查:外观检查符合要求,双钩紧线器在试验周期内,调整好双钩紧线器	8	(1)未进行外观检查扣1分。(2)未检查出厂合格证和试验合格证,扣1分。(3)漏一项,扣1分			
3	材料检查	清洁绝缘子,外观检查符合要求	2	(1)错选绝缘子型号,扣2分。(2)未进行检查,扣1分			
4	登塔	(1)现场核对线路名称、铁塔编号无误;基础及塔身完好无异常;交叉跨越距离符合安全要求;确认缺陷情况及地线规格型号等。(2)登杆(塔)动作规范、无危险动作。(3)正确携带吊绳(吊绳头折双、打死结,斜挎肩上)。(4)由杆塔身到横担上工作点不得失去安全带保护。(5)将无极绳安装在横担上方适当位置	8	(1)未确认线路双重称号,扣4分;未检查基础、塔身、交叉跨越,每项扣1分;未确认地线缺陷,扣1分。(2)未携带吊绳,扣4分;吊绳携带方式不规范,扣2分。(3)登杆(塔)动作不规范,每次扣1分。(4)未正确使用安全带,扣5分。(5)无极绳固定位置不当,扣2分			
5	验电、挂设导线接地线	(1)登塔至指定位置后,将安全带系在牢固的构件上,检查扣环是否扣牢;滑车安装位置正确,方便操作;工器具必须使用绳索传递。(2)使用验电器前,再次检查其声光信号正常。使用伸缩式验电器时,应将其各段绝缘杆全部拉出到位,以保证绝缘杆的有效绝缘长度;验电时,作业人员应手持验电器绝缘手柄,保证人体与导线间的足够安全距离;先验下层后验上层,先验近侧后验远侧,线路验电应逐相进行。(3)验明确无电压后立即装设导线接地线,先用砂纸打磨接地端安装的塔材位置;先安装接地端,后装导线端;装设顺序为先中相,后两边相	10	(1)滑车安装不规范,扣2分。(2)安全带使用不规范,扣2分。(3)验电、安装接地线未戴绝缘手套,扣2分。(4)验电器使用不规范,扣2分;验电顺序错误,扣2分			

续表

序号	项目名称	质量要求	分值	扣分标准	扣分原因	扣分	得分
5	验电、挂设导线接地线	(4)装设导线接地线时,必须使用绝缘手套进行操作,禁止直接用手操作接地线的金属部分的方式装设导线接地线。确认接地线的挂钩与导线连接紧密可靠		(5)接地线两端安装顺序错误,扣2分;连接处不牢固,扣1分;未对安装好的端部进行检查,扣1分。(6)接地线缠绕,扣2分			
6	安装作业架并进入工作位置	(1)检查绝缘子串连接可靠。(2)将作业架传递至横担上,安装位置正确、固定牢固,将双保险安全带的安全绳系在横担适当位置。(3)沿作业架平缓进入作业点,将安全带的安全绳系在横担适当位置,围杆带系在绝缘子串上	8	(1)未检查绝缘子串,扣2分。(2)作业架安装位置不正确,扣3分;不牢固,扣5分。(3)未使用双保险安全带的安全绳系,扣5分;位置系错,扣3分			
7	安装紧线工具	(1)检查绝缘子串M销。(2)将卡线器、双钩、钢丝套传递至杆塔上。(3)安装钢丝套、双钩紧线器,将卡线器安装在导线适当位置。(4)收紧双钩紧线器,使其略为受力。(5)将导线后备保护钢丝套一端安装在横担上,固定牢固;另一端用卡线器固定在导线上。(6)导线后备保护卡线器应安装在连接双钩紧线器的卡线器前面适当位置	16	(1)未检查绝缘子串M销,扣3分。(2)传递时绳索缠绕,每次扣1分。(3)卡线器安装位置不正确,扣2分。(4)卡线器固定位置不正确,扣2分。(5)导线后备保护钢丝套松紧不适,扣2分。(6)传递物有撞击现象,每次扣2分。(7)掉落物件,每件扣5分			
8	更换绝缘子	(1)收紧双钩紧线器,进行冲击试验,将安全带转移到双钩上。(2)在被更换绝缘子前、后,用短绳将绝缘子串固定在双钩上。(3)取出M销,取下需更换的绝缘子并传递至地面。(4)将新绝缘子传递到杆塔上,安装绝缘子及M销。(5)检查M销、球头是否齐全到位,碗口朝向正确,清洁绝缘子串。(6)松出双钩使绝缘子受力后,对绝缘子串做冲击试验	18	(1)未转移围杆带到受力物件上,每次扣5分。(2)失去安全带保护,扣5分。(3)未清洁绝缘子串,扣2分。(4)碰响绝缘子一次,扣1分。(5)未装M销,每个扣5分;M销位置不正确,扣2分;绝缘子大口朝向不正确,扣2分。(6)未冲击试验,每次扣5分			

续表

序号	项目名称	质量要求	分值	扣分标准	扣分原因	扣分	得分
9	拆除工器	(1)拆除双钩紧线器及后备保护钢丝套,传递至地面。(2)拆除作业架,传递至地面。(3)确认工作点无遗留物。(4)由横担到塔身上不得失去安全带保护	10	(1)失去安全带保护,扣5分。(2)传递物有撞击现象,每次扣2分,传递时绳索缠绕,每次扣1分。(3)有遗留物,扣5分			
10	下杆塔	(1)下杆(塔)动作规范、无危险动作。(2)正确携带吊绳(吊绳头折双、打死结,斜挎肩上)。(3)必须沿脚钉侧主材正确下塔	8	(1)未携带吊绳,扣4分,吊绳携带方式不规范,扣2分。(2)下杆(塔)动作不规范,每次扣1分。(3)脚扣下杆未拴安全带,扣4分。(4)未沿脚钉侧主材下塔,扣4分			
11	其他要求	(1)严禁高处坠物。(2)在操作过程中应双手协调配合操作。(3)严禁浮置物品。(4)严禁口中含物。(5)上、下杆塔过程中不得出现危险动作。(6)工作服、工作胶鞋、安全帽、劳保手套穿戴正确。(7)完工后清理作业现场,符合文明生产要求	10	(1)高处坠物,每件扣5分。(2)动作不协调,扣2分。(3)浮置物品,扣2分。(4)口中含物,扣2分。(5)打滑,每次扣2次。(6)出现失稳悬空,本模块不合格,记零分。(7)劳保用品穿戴,错漏一项扣1分。(8)未清理作业现场,扣2分			
12	合计		100				

模块六　停电更换±800kV特高压输电线路接地极导线培训及考核标准

一、培训标准

（一）培训要求

模块名称	停电更换±800kV特高压输电线路接地极导线	培训类别	操作类
培训方式	实操培训	培训学时	14学时
培训目标	1.掌握各类工器具、机具的使用方案和受力结构，以及更换接地极级导线的技术要点。 2.能熟练掌握停电更换±800kV特高压输电线路接地极导线的操作流程、技术方法和施工作业危险点。 3.能完成停电更换±800kV特高压输电线路接地极导线工作		
培训场地	特高压直流实训线路		
培训内容	正确使用各类受力工器具的操作方法正确安装各类工器具，采用停电作业法更换±800kV特高压输电线路接地极导线		
适用范围	特高压±800kV直流输电线路检修人员		

（二）引用规程规范

（1）《架空送电线路运行规程》（DL/T 741-2010）

（2）《110～500kV架空送电线路设计技术规程》（DL/T 5092-1999）

（3）《国家电网公司电力安全工作规程（线路部分）》（Q/GDW1799.2-2013）

（4）《±800kV直流架空输电线路设计规范》（GB 50790-2013）

（5）《±800kV直流架空输电线路检修规程》（DL/T 251-2012）

（6）《±800kV直流架空输电线路运行规程》（GB/T 28813-2012）

（7）《110（66）kV～500kV架空输电线路检修规范》（国家电网公司）

（8）《架空输电线路状态检修导则》（DLT 1248-2013）

（9）《输变电设备状态检修管理规定》（国家电网公司）

（10）《输变电设备状态检修试验规程》（国家电网公司）

(三)培训教学设计

本设计以完成"停电更换±800kV特高压输电线路接地极导线"为工作任务,按工作任务完成的标准化作业流程来设计各个培训阶段,每个阶段包括了具体的培训目标、培训内容、培训学时、培训方法(培训资源)、培训环境和考核评价等内容,如表2-6-1所示。

表2-6-1 停电更换±800kV特高压输电线路接地极导线培训内容设计

培训流程	培训目标	培训内容	培训学时	培训方法与资源	培训环境	考核评价
1.理论教学	熟悉停电更换±800kV输电线路接地极导线工作的操作步骤	讲授停电更换±800kV输电线路接地极导线工作的操作步骤	2	培训方法:讲授法。培训资源:PPT、相关规程规范	多媒体教室	考勤、课堂提问和作业
2.准备工作	能完成作业前准备工作	1.作业现场查勘。2.编制培训标准化作业卡。3.填写培训操作工作票。4.完成本操作的工器具及材料准备	1	培训方法:1.现场查勘和工器具及材料清理采用现场实操方法。2.编写作业卡和填写工作票采用讲授方法。培训资源:1.±800kV实训线路。2.特高压工器具库房。3.空白工作票	1.特高压输电实训线路;2.多媒体教室	
3.作业现场准备	能完成作业现场准备工作	1.作业现场复勘。2.工作申请。3.作业现场布置。4.班前会。5.工器具及材料检查	1	培训方法:演示与角色扮演法。资源:±800kV实训线路	±800kV实训线路	
4.培训师演示	通过现场观摩,使学员初步领会本任务操作流程	1.登塔。2.验电、挂接地线。3.导线上的操作。4.渡线及附件安装。5.拆除工器具	1	培训方法:演示法。资源:±800kV实训线路	±800kV实训线路	

续表

培训流程	培训目标	培训内容	培训学时	培训方法与资源	培训环境	考核评价
5.学员分组训练	能够分组完成停电更换±800kV输电线路接地极导线工作	1.学员分组（22人一组）训练停电更换±800kV输电线路接地极导线。 2.培训师对学员操作进行指导和安全监护	8	培训方法：角色扮演法。 资源：±800kV实训线路	±800kV实训线路	采用技能考核评分细则对学员操作评分
6.工作终结	1.使学员进一步辨析操作过程不足之处，便于后期提升。 2.培训学员安全文明生产的工作作风	1.作业现场清理。 2.向调度汇报工作。 3.班后会，对本次工作任务进行点评总结	1	培训方法：讲授和归纳法	±800kV实训线路	

（四）作业流程

1. 工作任务

完成停电更换±800kV特高压输电线路接地极导线。

2. 天气及作业现场要求

（1）停电更换±800kV特高压输电线路接地极导线应在良好的天气进行。

在5级及以上的大风以及暴雨、雷电、冰雹、大雾、沙尘暴等恶劣天气下，应停止露天高处作业。特殊情况下，确需在恶劣天气进行抢修时，应组织有关人员充分讨论必要的安全措施，经本单位批准后方可进行。

（2）作业人员精神状态良好，工作班成员认真学习工作票和安全技术措施，所有人员做到"四清楚"（作业任务精楚、危险点清楚、作业程序清楚、安全措施清楚）。

（3）作业前停送电联系人必须与调度联系履行工作许可手续，严禁约时停送电。工作负责人必须在得到许可人的许可工作命令后，方可在需检修的线路上验电、挂设接地线和进行检修工作。

（4）停电后，工作负责人应认真做好记录。

（5）登杆前应检查塔上是否有蜂窝，发现蜂窝严禁登塔。

（6）塔上作业人员必须使用双保险安全带，并佩戴护目镜。

3. 准备工作

3.1 危险点及其预控措施

（1）危险点——误登带电线路

预控措施：

①登塔作业前，工作负责人、工作班成员应共同认真核查双重名称和识别标记

（色标、判别标志等）与停电线路名称相符。

②登塔前应检查铁塔根部、基础等，必须牢固可靠。

③登塔前应检查登高工器具和设施，如安全带、脚钉、塔材等必须完整牢靠。

④不涉及挂设接地线的中间作业人员，应认真核实线路相序、色标、名称、编号与停电线路相符，确认线路名称无刷错、刷反等情况后，方可登杆。

（2）危险点——登塔时、塔上作业时违反安规进行操作，可能引起高空坠落

预控措施：

①攀爬过程中，为防止登杆人员串落，登杆作业人员间距不得小于1.6m。

②攀爬铁塔前应将脚底泥土清除干净，检查工具包完整，攀爬过程中不得掉落物件伤人。

③作业人员攀登塔时应戴好安全帽，穿软底鞋，动作不能过大，匀步攀登。

④攀爬过程中，安全带应收拾妥当，长尾绳放置在工具包内，主带应挂在肩上，防止攀爬过程中安全带勾挂脚钉和塔材，致使作业人员高空坠落。

⑤塔上移位时，不得失去安全带保护，做到踩稳抓牢。

⑥到达作业点位置，系好安全带（绳），应牢固可靠，不得低挂高用。

⑦未验电前，人体、无头绳等与导线的安全距离必须不小于9.5m，工作中应设专人监护。

（3）危险点——高处坠物伤人

预控措施：

①地面人员不得站在作业点垂直下方。塔上人员应防止落物伤人，使用的工具、材料应用绳索传递。

②在高处作业应使用工具袋，较大的工具应固定在牢固的构件上，不准随便乱放。

③使用绞磨起吊过程中，应设专人指挥，统一配合，绝缘子串刚离地后应进行冲击检查。

（4）危险点——防止感应电伤人

①为防感应电伤人，塔上作业人员应穿全套屏蔽服。

②如需接触架空地线，在架空地线接触前应进行可靠接地。

（5）危险点——现场作业安全监护

①自作业开始至作业终结，安全监护人必须始终在现场对作业人员进行不间断的安全监护。

②工作负责人，监护人必须穿安全监护背心。

(6)危险点——交通安全

出车时应注意车辆行驶安全,谨慎驾驶车辆,禁止违法行车。

3.2 工器具及材料选择

停电更换±800kV特高压输电线路接地极导线所需工器具及材料见表2-6-2。工器具出库前,应认真核对工器具的使用电压等级和试验周期,并检查确认外观良好、连接牢固、转动灵活,且符合本次工作任务的要求;工器具出库后,应防止脏污、受潮;金属工具和绝缘工器具应分开装运,防止因混装运输导致工器具变形、损伤等现象发生。

表2-6-2 停电更换±800kV特高压输电线路接地极导线所需工器具及材料表

序号	名称	规格型号	单位	数量	备注
1	传递绳		根	2	
2	保护绳		根	2	
3	滑车		个	2	
4	安全帽		顶	6	
5	风速风向仪		块	1	
6	温湿度仪		块	1	
7	万用表		块	1	
8	防潮帆布	2m×4m	块	2	
9	绳套		根	4	
10	防坠器	与塔防坠器装置型号对应	只	2	
11	安全带		副	2	
12	安全围栏		套	若干	
13	警示标示牌	"在此工作""从此进出""从此上下"	套	1	
14	红马甲	"工作负责人"	件	1	
15	砂纸		张	1	
16	清洁毛巾		条	1	
17	对讲机		台	4	
18	操作杆		根	1	

3.3 作业人员分工

本任务作业人员分工如表2-6-3所示。

表2-6-3　停电更换±800kV特高压输电线路接地极导线人员分工表

序号	工作岗位	数量(人)	工作职责
1	工作负责人	1	负责本次工作任务的人员分工、工作前的现场查勘、作业方案的制定、工作票的填写、现场复勘、办理工作许可手续、召开工作班前会、落实现场安全措施、负责作业过程中的安全监督、工作中突发情况的处理、工作质量的监督、工作后的总结
2	安全监护人	1	负责本次工作过程中的安全监护工作
3	高空作业人员	7	负责本次停电更换±800kV特高压输电线路接地极导线操作
4	地面辅助人员	10	负责本次作业过程的地面辅助工作
5	绞磨操作人员	2	负责本次作业过程的绞磨操作工作
6	信号指挥人员	2	负责2台绞磨启停的指挥工作

4. 工作程序

本任务工作流程如表2-6-4所示。

表2-6-4　停电更换±800kV特高压输电线路接地极导线工作流程表

序号	作业内容	作业标准	安全注意事项	责任人
1	现场复勘	工作负责人负责完成以下工作： (1)现场核对线路名称、铁塔编号无误；基础及塔身完好无异常；交叉跨越距离符合安全要求；确认缺陷情况及地线规格型号等； (2)检查地形环境符合作业要求； (3)检查工作票所列安全措施与现场实际情况相符，必要时予以补充	(1)正确穿戴安全帽、工作服、工作鞋、劳保手套。 (2)严禁非工作人员、车辆进入作业现场	
2	工作许可	作业前停送电联系人必须与调度联系履行工作许可手续	(1)不得未经工作许可人许可即开始工作 (2)严禁约时停送电	
3	现场布置	正确装设安全围栏并悬挂标示牌： (1)安全围栏范围应充分考虑高处坠物，以及对道路交通的影响。 (2)安全围栏出入口设置合理。 (3)妥当布置"从此进出""在此工作""车辆慢行"或"车辆绕行"等标示	对道路交通安全影响不可控时，应及时联系交通管理部门强化现场交通安全管控	
4	召开班前会	(1)全体工作成员列队。 (2)工作负责人宣读工作票，明确工作任务及人员分工；讲解工作中的安全措施和技术措施；查(问)全体工作成员精神状态；告知工作中存在的危险点及采取的预控措施。 (3)全体工作成员在工作票上签名确认	(1)工作票填写、签发和许可手续规范，签名完整。 (2)全体工作成员精神状态良好。 (3)全体工作成员明确任务分工、安全措施和技术措施	

续表

序号	作业内容	作业标准	安全注意事项	责任人
5	检查工器具	(1)在防潮苫布上,将工器具按作业要求准备齐备,并分类定置摆放整齐。检查工器具外观和试验合格证,无遗漏。 (2)检查人员向工作负责人汇报各项检查结果符合作业要求	(1)防潮苫布数量足够,设置位置合理,保持清洁、干燥。 (2)工器具外观检查合格,无损伤、受潮、变形、失灵现象,合格证在有效期内	
6	登塔	(1)登塔作业前,必须先核对线路名称及编号。对同塔多回线路,工作负责人、工作班成员应共同认真核查双重名称和识别标记(色标、判别标志等)。 (2)登塔前应检查铁塔根部、基础等,必须牢固可靠。 (3)攀爬过程,为防止登杆人员串落,登杆作业人员间距不得小于1.6m,安全带收拾妥当,长尾绳放置在工具包内,主带应挂在肩上,防止攀爬过程中安全带勾攀脚钉和塔材,致使作业人员高空坠落。 (4)登塔至横担处时,监护人和作业人员应再次核对停电线路的识别标记和双重名称,确实无误后方可进入停电线路侧的横担	(1)作业人员攀登塔时应戴好安全帽,穿软底鞋,动作不能过大,匀步攀登。 (2)攀爬过程中,安全带应收拾妥当,长尾绳放置在工具包内,主带应挂在肩上,防止攀爬过程中安全带勾挂脚钉和塔材,致使作业人员高空坠落。 (3)塔上移位时,不得失去安全带保护,做到踩稳抓牢。 (4)到达作业点位置,系好安全带(绳),应牢固可靠,不得低挂高用。 (5)未验电前,人体、无头绳等与导线的安全距离必须不小于9.5m,工作中应设专人监护	
7	验电、装设接地线	(1)登杆就位后,将安全带系在牢固可靠构件上,必须检查扣环是否正确就位。验电杆(笔)等工器具必须使用绳索传递。 (2)检查接地线完好,按程序(先接接地端,后接导线端)装设好接地线。 (3)装设接地线时,必须使用绝缘绳或绝缘手柄进行操作,禁止直接用手操作接地线的金属部分的方式装设接地线。确认接地线的夹头与导线连接紧密可靠	(1)验电杆在领用时和使用前应检查是否正常。 (2)禁止以缠绕导线的方式装设接地线	

序号	作业内容	作业标准	安全注意事项	责任人
8	导线上的操作	(1)作业人员进入接地极导线,并拆除导线上的间隔棒,并在拆除位置做好标记。 (2)作业人员在直线塔沿软梯进入导线,并在安装单轮滑车,然后拆除防震锤,使用提线器和6T手扳葫芦提升接地极导线,取脱金具,将接地极导线翻进滑车。 (3)将连接部分金具取下后,再将接地极导线翻进固定于地线横担上的单轮滑车中。 (4)作业人员利用卡线器将5T绞磨连接接地极导线,再利用9T手扳葫芦收紧接地极导线,取脱导线,两端耐张塔利用绞磨将两端导线松至地面。 (5)作业人员拆除导线端的耐张线夹,新旧导线安装网套,并绑扎牢固,使用旋转接头连接新旧导线		
9	渡线及附件安装	(1)牵引场与张力场同时2台绞磨同时启动,并有专人统一指挥。 (2)待新导线到达牵引场后张力场压接耐张线夹,压接后利用绞磨收线将压接后的导线连接到绝缘子金具上。 (3)牵引场收紧地线,待接地极导线弧垂与原弧垂一致后作记号。 (4)接地极导线松至地面压接另一端耐张线夹并收紧磨绳,将地线连接到绝缘子串金具上,直线塔将新导线连接到线夹里,紧固螺栓,按照之前记号位置安装防震锤	(1)放线施工过程中,临时拉线、交叉跨越、直线塔、受力工器具应有专人看守,并检查受力情况良好。 (2)放线时,应保持通讯畅通,统一信号、统一指挥。 (3)施工过程中,严禁任何人在地线下方穿越或逗留。 (4)渡线过程中,应有专人随时检查地锚、转向滑车、磨绳的受力情况,并适时进行调整。 (5)放线过程中,收线绞磨操作应平稳,保持地线牵引平衡,预防地线跳槽	
10	拆除工器具	依序拆磨绳、手扳葫芦、滑车、卡线器、后备保护绳等工器具		
11	工作结束	(1)工作负责人组织全体工作成员整理工器具和材料,清理现场,做到"工完料尽场地清"。 (2)召开班后会,工作负责人进行工作总结和点评工作。点评本次工作的施工质量;点评全体工作成员的安全措施落实情况。 (3)工作负责人向工作许可人汇报工作结束,恢复停电线路送电,终结工作票		

二、考核标准

表2-6-5　国网四川省电力公司特高压直流技能培训考核评分细则

考生填写栏	编号：　　姓名：　　所在岗位：　　单位：　　日期：　　年　月　日				
考评员填写栏	成绩：　考评员：　考评组长：　开始时间：　结束时间：　操作时长：				
考核模块	停电更换±800kV特高压输电线路接地极导线	考核对象	特高压±800kV直流输电线路检修人员	考核方式	操作	考核时限	480min
任务描述	停电更换±800kV输电线路XX#-XX#塔耐张段接地极导线				
工作规范及要求	1. 整个过程主要操作流程由参考人员1人独立完成；塔上辅助工6人、地面辅助工10人，绞磨操作人员2人，协助参考人员完成工器具、材料的上、下传递工作，以及其他非技术性工作。 2. 操作前参考人员应作必要的安全检查。 3. 更换接地极导线时采用的工具应满足受力要求。 4. 工作开始应口头提出申请，工作结束时应口头汇报。 给定条件： 1. 培训基地：特高压直流±800kV输电线路耐张段接地极导线 2. 工作票已办理，安全措施已经完备，工作开始、工作终结时应口头提出申请（调度或考评员）。 3. 必须按工作程序进行操作，工序错误扣除应做项目分值，出现重大人身、器材和操作安全隐患，考评员可下令终止操作（考核）				
考核情景准备	1. 工器具：Φ16磨绳一捆、9T手扳葫芦2把、6T手扳葫芦1把、5T绞磨3台、吊绳滑车3套、绳套3根、旋转接头2套、网套2付、卡线器2个、提线器1套、翻线器1套、间隔棒专用工具2套、单轮滑车3个、放线架2套、放线盘2套、Φ20钢丝绳8根、15T铁滑车8个、18T卸扣8个、10T卸扣6个、液压机（含液压钳）2套、对讲机若干、软梯1付、双保险安全带7套、垫木若干 2. 材料：同型号导线3盘、耐张线夹2套。 3. 在培训线路上操作。				
备注	1. 各项目得分均扣完为止，出现重大人身、器材和操作安全隐患，考评员可下令终止操作。 2. 设备、作业环境、安全带、安全帽、工器具、屏蔽服等不符合作业条件考评员可下令终止操作				

表2-6-6　国网四川省电力公司特高压直流技能培训考核评分标准

序号	项目名称	质量要求	分值	扣分标准	扣分原因	扣分	得分
1	着装	工作服、工作鞋、安全帽、劳保手套穿戴正确	4	错漏1项扣1分			
2	工具材料准备						
2.1	个人工具检查	检查个人工具，活动扳手、平口钳、拔销钳、工具包符合质量要求。	2	错漏1项扣1分			
2.2	机具检查	绞磨试机，检查档位是否正常；手扳葫芦是否符合质量要求。	2	错漏1项扣1分			

续表

序号	项目名称	质量要求	分值	扣分标准	扣分原因	扣分	得分
2.3	钢丝绳、滑车检查	对钢丝绳、滑车进行检查,确认连接可靠,受力满足要求。	2	错漏1项扣1分			
2.4	安全带检查	检查安全带、个人保安线是否符合质量要求,在试验周期内。	2	错漏1项扣1分			
2.5	专用工具检查	检查提线器、卡线器、翻线器是否符合要求,在试验周期内。	2	错漏1项扣1分			
2.6	材料检查	确认更换的导线型号,长度,外观检查符合要求。	2	错漏1项扣1分			
3	现场布置	正确装设安全围栏并悬挂标示牌: (1)安全围栏范围应充分考虑高处坠物,以及对道路交通的影响。 (2)安全围栏出入口设置合理。 (3)妥当布置"从此进出""在此工作""从此上下"等标示 (4)牵引场绞磨布置,正确完成绞磨位置布置、转角滑车位置布置。 (5)张力场绞磨布置,正确完成绞磨位置布置、转角滑车位置布置、线盘布置	10	(1)作业现场未装设围栏,扣0.5分。 (2)未设立警示牌,扣0.5分。 (3)未悬挂登塔作业标志,扣0.5分。 (4)错漏1项,扣2分。			
4	登塔	(1)塔上电工核对线路双重名称、杆号、相别,检查铁塔是否满足登塔条件,并将结果向工作负责人汇报。 (2)塔上电工系好安全带,并正确对安全带、后备保护绳以及防坠器进行做冲击试验,向工作负责人汇报以后,方可登塔。 (3)登塔过程中系好防坠落保护装置,匀速登塔、脚踩脚钉,手抓主材。到达横担工作点后,将安全带系在牢固可靠构件上,必须检查扣环是否正确就位,选择合适位置布置滑车传递绳。	5	(1)未系安全带或安全带及后备保护绳未进行冲击试验,各扣2分。 (2)手抓脚钉,扣2分。 (3)滑车传递绳悬挂位置不便工具取用,扣1分。 (4)传递时高空落物,扣2分。 (5)传递过程工具与塔身磕碰,扣2分。 (6)传递工具绳索打结混乱,扣1分。 (7)工作负责人监护不到位,扣2分。 (8)塔上电工操作不正确,扣2分			

续表

序号	项目名称	质量要求	分值	扣分标准	扣分原因	扣分	得分
5	验电、装设接地线	(1)地面电工将验电杆(笔)等工器具使用绳索传递给塔上电工。 (2)验电、并装设接地线。装设接地线时，必须使用绝缘绳或绝缘手柄进行操作，禁止直接用手操作接地线的金属部分的方式装设接地线。确认接地线的夹头与导线连接紧密可靠	4	(1)验电、装设接地线未佩戴绝缘手套，每项，扣4分。 (2)使用缠绕导线的方式装设接地线，扣4分			
6	导线上操作	(1)塔上电工检查金具锈蚀情况，对绝缘子串进行冲击，经工作负责人许可后绝缘子串进入导线。 (2)塔上电工进入接地极导线，并拆除导线上的间隔棒，并在拆除位置做好标记。 (3)塔上电工在直线塔沿软梯进入导线，并在安装单轮滑车，然后拆除防震锤，使用提线器和6T手扳葫芦提升接地极导线导线，取脱金具，将接地极导线翻进单轮滑车。 (4)将连接部分金具取下后，再将地线翻进固定于地线横担上的单轮滑车中。 (5)作业人员利用卡线器将5T绞磨连接地极导线，再利用9T手扳葫芦收紧接地极导线，取脱导线，两端耐张塔利用绞磨将两端导线松至地面。 (6)作业人员拆除导线端的耐张线夹，新旧导线安装网套，并绑扎牢固，使用旋转接头连接新旧导线。	15	(1)未使用双保险安全带或失去安全保护，扣3分。 (2)未对绝缘子串进行冲击，扣2分。 (3)未检查金具，扣1分。 (4)未使用传递绳传递间隔棒，扣2分。 (5)未做记号，扣1分。 (6)提升导线时未进行保护导线和塔材，扣2分。 (7)上下过程中软梯使用不规范，扣2分。 (8)未先将绞磨与导线连接，扣2分。 (9)手扳葫芦安装位置不正确，扣2分。 (10)转向滑车安装未对塔材保护，每处扣1分。 (11)新导线走向布置不规范，扣3分。 (12)网套未进行绑扎处理，扣3分。 (13)网套长度不够，扣2分。 (14)未正确使用旋转接头，扣1分			

续表

序号	项目名称	质量要求	分值	扣分标准	扣分原因	扣分	得分
7	渡线及附件安装	(1)牵引场与张力场同时2台绞磨同时启动,并有专人统一指挥。 (2)待新导线到达牵引场后张力场压接耐张线夹,压接后利用绞磨收线将压接后的导线连接到绝缘子金具上。 (3)牵引场收紧地线,待导线弧垂与原弧垂一致后作记号。 (4)导线松至地面压接另一端耐张线夹并收紧磨绳,将地线连接到绝缘子串金具上,直线塔将新导线连接到线夹里,紧固螺栓,按照之前记号位置安装防震锤。 (5)依序拆磨绳、手扳葫芦、滑车、卡线器、后备保护绳等工器具。 (6)拆除接地线	40	(1)渡线是导线掉落地面,每次扣2分。 (2)渡线过程中张力过大,扣1分。 (3)未派专人指挥导致卡线,扣2分。 (4)旧导线未及时整理回收,扣2分。 (5)未指派专人测量弧垂,扣2分。 (6)弧垂与其他导线弧垂不一致,扣2分。 (7)耐张线夹型号与导线不符,扣3分。 (8)连接完毕后未进行冲击试验,扣2分。 (9)压接工艺不满足要求,扣2分。 (10)未检查螺栓是否紧固到位,扣2分。 (11)间隔棒安装位置不正确,每处扣1分。 (12)间隔棒安装不规范,安装方向不正确,扣2分。 (13)未检查金具连接情况,扣3分。 (14)传递绳缠绕,每次扣2分。 (15)未检查导线弧垂,扣2分。 (16)塔材损伤,每处扣2分。			
8	返回地面	塔上电工检查塔上无遗留物后,向工作负责人汇报,得到工作负责人同意后传递绳下塔	5	(1)下塔过程未使用防坠装置,扣2分。 (2)塔上移位失去安全带保护的,扣2分。 (3)下塔抓塔钉,每处扣1分。 (4)塔上有遗留物的,扣2分			

续表

序号	项目名称	质量要求	分值	扣分标准	扣分原因	扣分	得分
9	工作结束	(1)工作负责人组织全体工作成员整理工器具和材料,将工器具清洁后放入专用的箱(袋)中;清理现场,做到"工完料尽场地清"。 (2)召开班后会,工作负责人进行工作总结和点评工作。点评本次工作的施工质量;点评全体工作成员的安全措施落实情况	5	(1)工器具未清理,扣2分。 (2)工器具有遗漏,扣2分。 (3)未开班后会,扣2分。 (4)未拆除围栏扣2分。 (5)未向调度汇报,扣2分			
	合计		100				

Part I
Standard for Training and Assessment on Live Working for Operation Maintenance of ±800kV UHV Power Transmission Line

Module 1 Standard for Training and Assessment on Live Replacement of Single-V Composite Insulator for ±800kV UHV Power Transmission Line Tangent Tower

I. Standard for Training

(I) Training Requirements

Designation of module	Live replacement of single-V composite insulator for ±800kV UHV power transmission line tangent tower	Type of training	Operation
Training method	Practical operation training	Hours of training	21 training hours
Training objectives	1. Master the electrical significance of "basket method" operation mode when the tangent tower enters and exits ±800kV intense electric field. 2. Be able to enter ±800kV equipotential operation point by adopting the "basket method". 3. Be able to independently complete the live replacement of single-V composite insulator for ±800kV UHV transmission line tangent tower (equipotential operation method)		
Training venue	UHV ±800kV practical training line		
Training content	With the cooperation of equipotential with ground potential, enter the equipotential for operation through the "basket method", and carry out live replacement of the single-V composite insulator for ±800kV UHV transmission line tangent tower by equipotential operation method		
Scope of application	Maintenance personnel for ±800kV U HV DC power transmission line		

(II) Referenced Procedures and Specifications

(1) Technical Specification for Live Working of ±800kV DC Line (DL/T1242-2013)

(2) Operating Code for ±800kV DC Overhead Transmission Line (GB/T28813-2012)

(3) Maintenance Specification for ±800kV DC Overhead Transmission Line (DL/

T251-2012)

(4) Technical Guide for Live Working of ± 800kV DC Transmission Line (Q/GDW302-2009)

(5) Calculation Method of Live Working Minimum Approach Distance on AC Transmission Line (GB/T 19185-2008)

(6) Guidelines of Insulation Coordination for Live Working (DL/T867-2004)

(7) Test Guide of the Insulating Tool for Live Working (DL/T878—2004)

(8) State Grid Corporation of China on the Management Regulations of Live Working (Trial Implementation) (SGCC [2007] No.751)

(9) State Grid Corporation of China Working Regulations of Power Safety (Transmission Line Section) (Q/GDW1799.2-2013)

(10) Electrotechnical Terminology - Overhead Line (GB/T 2900.51-1998)

(11) Electrotechnical Terminology - Live Working (GB/T2900.55-2016)

(12) Live Working - Terminology for Tools, Equipment and Devices (GB/T 14286-2008)

(13) Minimum Requirements for Utilization of Tools, Devices and Equipment for Live Working (DL/T 877-2004)

(14) Preventive Test Code of Tools, Devices and Equipment for Live Working (DL/T 976-2005)

(15) Insulated Tackles for Live Working (GB/T13034-2008)

(16) Live Working-Insulating Ropes (GB 13035-2008)

(17) Shielding Clothes for Live Working (GB/T6568—2008)

(18) Screen Clothes for Live Working on 1000kV AC (GB/T25726-2010)

(III) Teaching Design for Training

To complete the work task of "live replacement of single-V composite insulator for ± 800kV UHV transmission line tangent tower", each training stage shall be designed according to the standard operation procedure for work task completion. Each stage includes specific training objectives, training content, hours of training, training methods (training resources), training environment, assessment and evaluation, etc, as shown in the Table 1-1-1.

Table 1-1-1 Live Replacement of Single-V Composite Insulator for ±800kV UHV Power Transmission Line Tangent Tower

Training schedule	Training objectives	Training contents	Hours of training	Training methods and resources	Preparation of training conditions	Assessment and evaluation
1. Theoretical teaching	1. Preliminarily master the basic method for entering and exiting ±800kV electric field in basket method. 2. Be familiar with the method to replace the single-V composite insulator for ±800kV transmission line tangent tower	1. The electrical significance of operation mode of entering and exiting the intense electric field in "basket method". 2. Method and quality standard for the replacement of single-V composite insulator for ±800kV transmission line tangent tower	2	Training methods: Lecture. Training resources: PPT, relevant regulations and specifications	Multimedia classroom	Attendance, classroom questions and assignments
2. Preparations	Be able to complete the preparation before operation	1. Work site survey. 2. Preparation of the standardized operation card. 3. Filling of the work order. 4. Preparation of tools and materials for this operation	2	Training methods: 1. Site survey and cleaning of tools and materials shall be practiced at site. 2. Preparation of operation card and the filling of work order shall adopt lecture method. Training resources: 1. UHV practical training line (±800kV practical training line). 2. UHV tools warehouse. 3. Blank work order	1. UHV transmission line for practical training 2. Multimedia classroom	

Part I Standard for Training and Assessment on Live Working for Operation Maintenance of ±800kV UHV Power Transmission Line

Table (Cont'd)

Training schedule	Training objectives	Training contents	Hours of training	Training methods and resources	Preparation of training conditions	Assessment and evaluation
3. Work site preparation	Complete the preparations of work site	1. Work site re-survey. 2. Work application. 3. Work site layout. 4. Pre-shift meeting. 5. Inspection of tools	1	Training methods: demonstration and role play. Resources: UHV practical training line (±800kV practical training line)	±800kV practical training line	
4. Trainer's demonstration	The trainees can preliminarily understand the operation process of the task through inspecting and learning from each other's work	1. The arrangement and installation of the tools on tower. 2. The equipotential electrician enters and exits the intense electric field by adopting the basket method. 3. The ground potential electrician and the equipotential electrician cooperate with each other to complete the replacement of single-V composite insulator by using the load transfer device	2	Training methods: Demonstration. Resource: UHV practical training line (±800kV practical training line)	±800kV practical training line	

Table (Cont'd)

Training schedule	Training objectives	Training contents	Hours of training	Training methods and resources	Preparation of training conditions	Assessment and evaluation
5. Group training of trainees	Through training: 1. Be able to complete the operation of entering and exiting ±800kV intense electric field. 2. Be able to complete the replacement of single-V composite insulator for ±800kV transmission line tangent tower	1. Trainees are grouped (12 in a group) to train the skill of entering and exiting ±800kV intense electric field and replacing single-V composite insulator for tangent tower. 2. Trainers guide the operation of trainees and conduct safety supervision Custody	12	Training methods: Role play. Resources; UHV practical training line (±800kV practical training line)	±800kV practical training line	Score the operation of trainees according to the detailed rules for skill assessment and scoring
6. End of the work	Through training: 1. Trainees can further understand the shortcomings during the operation process for later improvement; 2. Train the trainees in the working style of safe and civilized production	1. Clean up the work site. 2. Report to dispatcher. 3. Comment and summarize the current work task at post-shift meeting	1	Training method: Lecture and inductive method	±800kV practical training line	

(IV) Work Flow

1. Work task

With the cooperation between equipotential electrician and ground potential electrician, enter the equipotential field for operation through the "basket method", and carry out live replacement of the single-V composite insulator for ±800kV UHV transmission line tangent tower.

2. Requirements for Weather and Work Site

(1) The live replacement of single-V composite insulator for ±800kV UHV transmission line tangent tower shall be carried out in good weather.

In case of lightning (hearing thunder or seeing lightning), snow, hail, rain, fog and so on, live working is prohibited. When the wind force is greater than level 5, or the relative humidity of the air is greater than 80%, it is unsuitable for live working; when emergency live repair is required in bad weather, relevant personnel shall be organized to fully discuss and prepare necessary safety measures, which can be implemented after being approved by the unit.

(2) The operating personnel should be in good mental states and be familiar with the organizational and technical measures to ensure safety in work; they should hold the qualification certificate for live working within the validity period.

(3) Responsible Person should organize the relevant personnel to complete field investigation in advance, determine the operating methods, required working apparatus and necessary measures according to the results, and handle the tickets for live working.

(4) The work site should be reasonably set up with fence and warning signs. Non-operating personnel is forbidden to enter.

(5) DC restart device of the line shall be deactivated for the Project.

(6) Safe working distance and effective insulation length during operation are shown in Table 1-1-2.

Table 1-1-2 Safe Distance (m) for Live Replacement of Single-V Composite Insulator for ±800kV UHV Power Transmission Line Tangent Tower

Altitude	Minimum safe distance between equipotential electrician and grounding frame	Minimum effective insulation length of insulating tools	Minimum combination gap
$H \leqslant 1000$	6.8	6.8	6.7
$1000 < H \leqslant 2000$	7.3	7.3	7.3
$2000 < H \leqslant 2500$	7.9	7.8	7.8

Note: The minimum safe distance and minimum combined gap in the table include 0.5m occupying gap of human body.

3. Preparations

3.1 Hazards and precontrol measures

(1) Hazard —— electric shock

Precontrol measures:

① Before work, the Responsible Person shall contact the control personnel on duty, deactivate the line DC restart device, and perform the licensing procedures.

② Before climbing the tower, ground potential operators on tower must carefully check the name of the line, the number of the pole and tower, and phase, and then the tower can be climbed after all have been confirmed correct.

③ If lines lose power suddenly during work, operators shall consider it as still charged. The Responsible Person shall contact the control personnel as soon as possible, and no forced energization is allowed before the on-duty control personnel getting in touch with the Responsible Person.

④ Insulating tool and insulating ropes shall be free of damage, moisture, deformation, and failure. It is not allowed to use non-insulating ropes (such as cotton rope, manila rope, and steel wire rope).

⑤ The equipotential operator shall wear flame-retardant underwear and full set of screen clothes outside (including hat, mask, dresses & trousers, gloves, socks and shoes). All parts shall be in excellent connection conditions.

⑥ Before potential transfer, equipotential operator shall obtain the approval of Responsible Person, and the minimum distance between exposed part of human body and electrified body shall not be lower than 0.5m. During potential transfer, the operation shall be fast, and end shall not be used for power charging and discharging; During delivering tools and materials to ground potential operators, the effective length of insulating tools or insulating ropes shall not be lower than the requirements as specified in Table 1-1-2.

⑦ When transmitting large metal objects by using insulating rope, the ground potential electricians and ground operators shall ground the metal objects before contacting.

⑧ The special Supervisor shall continuously monitor the operators and correct their nonstandard operation or actions in violation at any time. Special attention shall be paid to operators working at heights to ensure that there is enough safe distance (meeting the requirements in Table 1-1-2). It is forbidden to contact two non-connected electrified bodies or make contact with electrified body and grounding body at the same time.

(2) Hazard——falling accident

Precontrol measures:

① Before climbing, operators working at heights must satisfy the requirements of this operation, such as physical condition, mental state, and skill and quality.

② After climbing the tower to the operation point, the ground electrician shall fasten the safety rope, check and confirm that it is firm. The track rope of basket and the body backup protection rope for equipotential electrician shall be reasonably arranged and reliably anchored. The basket shall be firmly hung by four insulating ropes on its four sides. The hauling length of insulating ropes shall be calculated accurately or be measured on site, to ensure the height of equipotential operator's head not to exceed the grading ring on conductor side. The equipotential electrician shall fasten the insulated and main protection ropes, and pass the impact test on the basket before entering the basket.

③ Supervisors shall correct the nonstandard or illegal actions at any time. Special attention shall be paid to operators to prevent them from losing the protection of safety belt or insulated backup protection rope during transposition, and it is forbidden to fasten the safety belt or insulated backup protection rope in a position lower than the operating personnel.

④ Personnel shall inspect the shackles and tower materials for fastening condition before climbing the tower, and shall grasp the main materials by hand but not grasp the shackles by hand when climbing the tower.

(3) Hazard —— injury caused by objects falling from high place

Precontrol measures:

① Operator working at heights should put personal tools and fragmentary materials into the tools bag. It is strictly forbidden to hang objects in high place or keep in the mouth.

② Ground operator should correctly wear a helmet and use the knots. The vertical distance from the operation point should not be less than the falling radius.

③ The work site shall be set up with fence and warning signs. It shall be noted at any time that Supervisor shall prohibit irrelevant personnel and vehicles from entering operation area.

3.2 Selection of tools, instruments and materials

The tools and materials required for live replacement of single-V composite insulator for ±800kV transmission line tangent tower are shown as Table 1-1-3. Before being taken out of the warehouse, the tools and instruments shall be carefully inspected for their service voltage class and test cycle, and shall be inspected to confirm that they are in intact appearance, firm connection and flexible rotation, and meet the requirements of the work task. After being taken out of the warehouse, the tools and instruments shall be stored in the tool bag or tool kit for transportation to avoid contamination and damp. Metal tools and insulated tools shall be separately loaded and transported to avoid deformation and damage caused by mixed loading and transportation.

Table 1-1-3　Tools and Materials Required for Live Replacement of Single-V Composite Insulator for ±800kV Transmission Line Tangent Tower

S/N	Name	Specification and Model	Unit	Qty.	Remarks
1	Insulated transmission rope	TJS-10	Nos.	1	Insulating tool
2	Insulated transmission rope	TJS-16	Nos.	3	Insulating tool
3	Insulated protection rope	TJS-16	Nos.	2	Insulating tool
4	Basket track rope	TJS-16	Nos.	1	Insulating tool
5	Insulated tackle	JH10-1	Nos.	5	Insulating tool
6	Insulated suspender		Set	2	Insulating tool
7	Insulated tightener	0.5T	pcs	1	Insulating tool
8	Insulated noose		Nos.	Several	Insulating tool
9	Cradle		Set	1	Insulating tool
10	Hydraulic leading screw		Nos.	2	Metal tool
11	Wire lifting clamps		Nos.	2	Metal tool
12	Engine driven winching		Set	1	Metal tool
13	Shielding clothes	Shielding efficiency ≥ 60 dB(shield efficiency of shielding mask ≥ 20 dB)	Set	4	PPE
14	Safety belt		Nos.	4	PPE
15	Safety helmet		pcs	12	PPE

Table (Cont'd)

S/N	Name	Specification and Model	Unit	Qty.	Remarks
16	Insulation resistance tester	5000V, with electrode width 2cm and interelectrode width 2cm	pcs	1	Other tools
17	Anemometer		pcs	1	Other tools
18	Temperature and humidity meter		pcs	1	Other tools
19	Multimeter		pcs	1	Other tools
20	Moisture-proof canvas	2m×4m	pcs	4	Other tools
21	Red waistcoat	"Responsible Person"	pcs	1	Other tools
22	Interphone		Set	4	Other tools
23	Falling protector	Corresponding to the type of pole and tower falling protector	pcs	4	Other tools
24	Security fence		Set	Several	Other tools
25	Personal tools		Set	2	Other tools
26	Warning sign	"Work Here", "Access from Here", "Access from Here"	Set	1	Other tools
27	Composite insulators		Set	1	Material

3.3 Division of labor for operators

Division of labor for operators of the task is shown in Table 1-1-4.

Table 1-1-4 Division of Labor for Live Replacement of Single-V Composite Insulator for ±800kV UHV Transmission Line Tangent Tower

S/N	Post	Qty. (person)	Responsibilities
1	Responsible Person	1	Be responsible for various work on the work site
2	Special Supervisor	1	Be responsible for the safety control of the work site
3	Equipotential electrician	2	Be responsible for entering the equipotential field and completing the replacement of insulator on live terminal
4	Ground electrician on the tower	2	Be responsible for completing the placement of insulator on ground terminal and assisting the equipotential electrician to enter and exit the electric field
5	Ground electrician	6	Be responsible for transferring tools and materials and cooperating with equipotential electrician in entering and exiting the equipotential

4. Working Procedures

The workflow of this task is shown in Table 1-1-5.

Table 1-1-5 Workflow for Live Replacement of Single-V Composite Insulator for ±800kV UHV Transmission Line Tangent Tower

S/N	Work Content	Operation Steps and Standards	Safety Measures and Precautions	Responsible Person
1	Site re-survey	The Responsible Person shall complete the following work: (1) Check the line name, the number of the pole and tower and ensure the phases are correct; guarantee that the foundation and the pole and tower are intact and in normal condition; ensure that the cross and span distance meets the safety requirements; confirm the defect conditions, etc. (2) Check that the site meteorological conditions such as wind speed and humidity should meet the operation requirements. (3) Check that the terrain and environment should meet the operation requirements. (4) Check that the safety measures listed in the work order are in line with the actual situations on site. The measures will be added if necessary.	(1) Correctly wear helmet, working clothes, work shoes and protective gloves. (2) Operation under meteorological conditions that may endanger the safety of operators is forbidden. (3) Non-operation personnel and vehicles are strictly prohibited from entering the working site	
2	Work Permit	(1) The Responsible Person shall contact the on-duty control personnel and apply for stopping the transmission line DC restart device as per the contents of the work order. (2) Live working could be started only after being approved by control personnel on duty.	Live working shall not be started without the permission of the on-duty control personnel	
3	Site layout	Install the security fence and hang the signboards correctly: (1) The security fence should take full account of falling objects from the high place and the influence on road traffic. (2) The entrance and exit of the security fence shall be set reasonably. (3) Signs such as "Access from Here", "Work Here", "Access from Here" shall be properly arranged	When the influence on road traffic safety is uncontrollable, the traffic management department should be contacted in time to strengthen the on-site control of traffic safety	

Table (Cont'd)

S/N	Work Content	Operation Steps and Standards	Safety Measures and Precautions	Responsible Person
4	Holding a pre-shift meeting	(1) All working personnel shall line up. (2) Responsible Person will read out the work order and be clear with work task and division of personnel; explain safety measures and technical measures in work; check (inquire after) mental state of all working personnel; inform of hazards in work and precontrol measures. (3) All working personnel shall sign on the work order for confirmation.	(1) Work order shall be filled and issued with standardized licensing procedure and complete signature. (2) All working personnel shall be in good mental states. (3) All working personnel shall be clear with task division of works, safety measures and technical measures	
5	Inspection of tools	(1) The ground electrician on tower and equipotential electrician shall wear the shielding clothes in a right way and pass the inspection, which shall be supervised and inspected by the Responsible Person. (2) Wear personal safety equipment correctly (proper size and easy lock), and the Responsible Person shall supervise and inspect it. (3) Measure the wind speed, wind direction and humidity, check the insulation performance of insulating tools, and make records	(1) Check carefully for damage, deformation and failure before using metal and insulating tools. Carry out segment insulation detection with such insulated tools with insulation resistance meter of 2,500V or above and with the resistance no less than $700 M\Omega$, and wipe it off with a clean dry towel. (2) Use a multimeter to measure the resistance between the farthest ends of the shielding clothes and trousers, which shall not be greater than 20Ω. The Responsible Person shall check the connection of the electrician's shielding clothing. (3) Check the tool assembly and make sure the connection is reliable. (4) The tools and instruments for live working shall be placed on the moisture-proof canvas	

S/N	Work Content	Operation Steps and Standards	Safety Measures and Precautions	Responsible Person
6	Climbing the tower	(1) After checking the line name and pole and tower number, the ground potential electrician on tower and equipotential electrician shall impact to check the stress of safety belt and falling protector. (2) The ground potential electrician on tower carries the insulated transmission rope to climb the tower, and the equipotential electrician then climbs the tower. When they reach the operation point of the cross arm, they choose the appropriate position to fasten the safety belt, and the ground potential electrician on tower installs the insulated tackle and the insulated transmission rope in the appropriate position of the cross arm. Then he shall cooperate with the ground electrician to separate the insulated transmission rope for lifting preparation	(1) After checking the correct line name and pole and tower number, he can climb for operation. (2) The anti-falling device installed on tower shall be used during climbing the tower; when moving and transposing on the pole and tower, the safety protection shall not be lost, and the operators must climb and grasp the components securely. (3) The working electrician must wear full set of qualified shielding clothes which must be connected reliably. Before the cross arm enters the equipotential, the equipotential electrician shall check and confirm that each part of the shielding clothes are connected reliably before the next operation	
7	Installation of tackle block and basket	(1) The ground electrician transfers the basket, basket track rope, and 2-2 insulated tackle block to the cross arm by means of insulated transmission rope. (2) The electrician on tower installs the 2-2 insulated tackle block and the basket in the appropriate position of the upper plane of the cross arm, and reliably connects the basket with the group II tackle block; then the electrician installs one end of the basket track rope in the appropriate position of the cross arm, and reliably connects the other end with the basket	(1) The lifting of insulating sling shall be smooth, free of collision and winding in transfer. (2) After the basket is installed, the electrician on tower shall carry out the careful inspection and check to the basket situation. (3) 2-2 tackle block and basket shall be installed reliably in a suitable position on the cross arm	

Part I Standard for Training and Assessment on Live Working for Operation Maintenance of ±800kV UHV Power Transmission Line

Table (Cont'd)

S/N	Work Content	Operation Steps and Standards	Safety Measures and Precautions	Responsible Person
8	Entering the intense electric field	(1) An equipotential electrician shall fasten the insulated protection rope to enter the basket, and the ground electrician slowly releases the control rope of 2-2 insulated tackle block and slows down when the basket is about 2 m away from the live conductor. (2) During the continuous movement of the basket to the conductor, the equipotential electrician faces the live conductor, and meanwhile applies for potential transfer to the Responsible Person. After obtaining the approval, the equipotential electrician quickly grasp the nearest sub-conductor by hand for potential transfer when the basket is 0.5 m away from the conductor. (3) After entering the intense electric field, the equipotential electrician shall inspect the insulated backup protection rope, and meanwhile control his head not to exceed the grading ring on the conductor side. (4) The ground electrician tightens the control rope of 2-2 insulated tackle block, and transfers the basket up to the cross-arm. No. 2 equipotential electrician fastens the insulated protection rope and enters the basket, and then enter the intense electric field by same method	(1) Before entering the equipotential, the equipotential electrician shall re-check and confirm that each part of the shielding clothes, the potential transfer rod and the insulated shielding clothes are connected reliably before the next operation. (2) The equipotential electrician must obtain the permission of the Responsible Person before entering the potential. (3) The equipotential electrician must fasten the insulated protection rope before entering the basket. (4) When the ground electrician cooperates with the equipotential electrician to enter the equipotential, the control rope of the tackle block shall be pulled and released stably. (5) The combined gap composed of the gaps between the equipotential electrician and the grounding body and the electrified body in the process of entering the potential shall not be less than the requirement as shown in Table 1-1-2. (6) The Special Supervisor shall be responsible for monitoring the safety precautions for the equipotential electrician entering the intense electric field. The Special Supervisor shall remind the dangerous and irregular actions of the operators on tower in a timely manner and stop them if necessary. (7) The equipotential electrician shall not release the insulated protection rope after entering the electric field, and the safety belt shall not be fastened on the sub-conductor.	

Table (Cont'd)

S/N	Work Content	Operation Steps and Standards	Safety Measures and Precautions	Responsible Person
9	Install tools and transfer conductor load	(1) The ground electrician transfers the insulated suspender, wire lifting clamps, and hydraulic leading screw to the working position, and the equipotential electrician and ground potential electrician cooperate and properly install the tools for replacement of composite insulator. (2) After checking the reliable connection of each component and obtaining the consent from the Responsible Person, the equipotential electrician firstly tightens the mechanical leading screw, then tightens the hydraulic leading screw after the mechanical leading screw is properly stressed, allows it being stressed slightly and inspect the force bearing point. (3) The ground electrician transfers the control rope of the composite insulator string to the equipotential electrician who installs it at the tail of the composite insulator string. The ground electrician tightens the control rope of composite insulator string. (4) After checking that the load-bearing tools are stressed normally and obtaining the consent of the Responsible Person, the equipotential electrician dismantles the socket hanging plate bolts on the conductor side, and the ground electrician slowly releases the control rope of composite insulator string to the extent that it is naturally vertical. (5) The ground potential electrician ties the insulated transmission rope to the upper end of the composite insulator string, and then takes out the fitting pin connected between the composite insulator string and the ball head hanging ring. The ground electrician cooperates with the ground potential electrician to disconnect the composite insulator string with the ball head hanging ring	(1) The upper and lower working electricians shall cooperate closely, and all working electricians shall obey the unified command of the Responsible Person. (2) The minimum safe distance of the ground potential electrician to the electrified body and the equipotential electrician to the grounding body shall not be less than the requirements in Table 1-1-2. (3) The binding rope buckles of the upper and lower transfer tools of the pole and tower shall be correct and reliable, and the electrician on tower shall not drop objects from high place. (4) After the tool is stressed and passes the impulse inspection, it shall be reported to the Responsible Person, and then the operation can be continued only with the permission of the Responsible Person. (5) The Special Supervisor shall remind the dangerous and irregular actions of the operators on tower in a timely manner and stop them if necessary. (6) In the process of load transfer, the operators shall always pay attention to the stress condition of tools and instruments, and report to the Responsible Person immediately in any abnormal case	

Table (Cont'd)

S/N	Work Content	Operation Steps and Standards	Safety Measures and Precautions	Responsible Person
10	Replacement of insulator string	(1) The ground electrician controls the control rope of the composite insulator string, and slowly puts the composite insulator string to the ground by using the motor winching. Attention shall be paid to the control rope of composite insulator string, which shall not collide with the load-bearing tools, conductor and pole and tower. (2) The ground electrician transfers the insulated transmission rope and composite insulator control rope to new composite insulator string respectively, and then transmits the new composite insulator string to the working position on tower by using the motor winching. Then the ground potential electrician restores the connection between the new composite insulator string and the ball head hanging ring, and restores the fitting pin. (3) The ground electrician slowly releases the motor winching to the extent that the composite insulator string is naturally vertical, and then tightens the control rope of composite insulator string in order to pull the tail of the composite insulator string to the working position of the equipotential electrician on conductor side. The equipotential electrician restores the connection between the socket hanging plate and the yoke plate by using the insulated tightener, and installs the cotter pin.	(1) When the insulator string is withdrawn from operation, the receiving parts shall be checked in detail for good condition, and shall not be removed until the inspection is completed without any problem and the consent from Responsible Person is obtained. (2) The insulator string shall not collide with the pole and tower in the process of hoisting, and the tail rope of the insulator string shall be controlled well. (3) The arrangement shall be steady when the insulator string is being hoisted by motor winching. The tail rope shall be controlled by experienced operator and shall not be negligently loosened. (4) The stress conditions of the winching and steering tackle must be checked before operation. (5) The binding rope buckles of the upper and lower transfer tools of the pole and tower shall be correct and reliable	

Table (Cont'd)

S/N	Work Content	Operation Steps and Standards	Safety Measures and Precautions	Responsible Person
11	Remove tools	(1) After checking the reliable connection of the composite insulator string and obtaining the consent of the Responsible Person, the ground potential electrician veers away the hydraulic screw rod. (2) After checking that the stress of the composite insulator string is normal and obtaining the consent of the Responsible Person, the ground potential electrician shall cooperate with the equipotential electrician to remove the insulated suspender, hydraulic screw rod, etc., and transfer them to the ground.	(1) After the composite insulator is installed and reset, the connection of each part shall be checked in detail for correctness, and the wire lifting tools shall not be removed until the consent of the Responsible Person is obtained. (2) The tools shall not collide with each other in the process of transfer, and the rope buckle shall be bound correctly and reliably.	
12	Leaving the electric field	(1) An equipotential electrician shall fasten the insulated protective rope, enter the basket, and then keep the arm straight so that the basket is 0.5m away from the sub-conductor. (2) The equipotential electrician applies to the Responsible Person for exiting the potential. After obtaining the consent, the equipotential electrician quickly removes the sub-conductor. (3) At the same time, the ground electrician quickly tightens the control rope of the 2-2 insulated tackle block, pulls the basket upward to the cross arm and stops it. Then, the equipotential electrician climbs the cross arm and fastens the safety belt. (4) The ground electrician transfers the basket to the another equipotential electrician by means of insulated transmission rope. Then, the equipotential electrician enters the basket after checking that there is no object left on the conductor, and exits the potential by the same method.	(1) The upper and lower working electricians shall cooperate closely and follow the command of the Responsible Person. (2) The equipotential electrician must obtain the permission of the Responsible Person before leaving the potential. (3) The equipotential electrician must fasten the insulated protection rope before entering the basket. (4) When the ground electrician cooperates with the equipotential electrician to enter the equipotential, the control rope of the tackle block shall be pulled and released stably. (5) The combined gap composed of the gaps between the equipotential electrician and the grounding body and the electrified body in the process of exiting the potential shall not be less than the provision as shown in Table 1-1-2. (6) The Special Supervisor shall be responsible for supervising the safety precautions when the equipotential electrician exits intense electric field, reminding the dangerous and irregular actions of the operators on tower in a timely manner, and stopping them if necessary.	

S/N	Work Content	Operation Steps and Standards	Safety Measures and Precautions	Responsible Person
13	Remove the basket and return to the ground	(1) The electrician on tower shall co-operate the removal of the basket track rope, the insulated protection rope, the 2-2 insulated tackle block and the basket and transfer of them to the ground. (2) After checking that there is no object left on the tower, the equipotential electricians on the tower shall report it to the Responsible Person and then climb down the tower with insulated transmission rope after obtaining the consent of the Responsible Person	(1) The tools shall not collide with each other in the process of transfer, and the rope buckle shall be bound correctly and reliably. (2) The anti-falling device installed on tower shall be used in the process of climbing the tower; when moving and transposition on the pole and tower, the safety protection shall not be lost, and the operators must grasp the components securely.	
14	End of the work	(1) The site and tools shall be cleaned up, and any left objects on the pole (tower) shall be carefully checked. The Responsible Person shall comprehensively inspect the completion of the work, count the number of people, declare the end of the work if no error is found and evacuate from the construction site. (2) The dispatcher shall be notified of the end of work, the completion formalities for the work order shall be handledIt is forbidden to restore the line restart device at the appointed time	It is forbidden to restore the line restart device at the appointed time	

II. Assessment Standard

Table 1-1-6 Detailed Rules for Assessment and Scoring of Operation and Inspection Skills of UHVDC Transmission Line of State Grid Sichuan Electric Power Corporation

Fill-in Column of Examinee	No.:	Name:	Position:	Unit:	Date: YYYY MM DD
Fill-in Column of Assessor	Grade:	Assessor:	Assessment Team Leader:	Starting time:	Closing time: Operation Duration:

Assessment module	Assessee	Maintenance personnel for ±800kV UHV DC power transmission line	Assessment method	Operation	Assessment Time Limit
					120min

Job Description	Live replacement of single-V composite insulator for ±800kV UHV power transmission line tangent tower

Work Specifications and Requirements	1. Live working shall be carried out in good weather. In case of thunder, rain, snow or fog, no live working shall be carried out. When the wind force is greater than Level 5 and the humidity is greater than 80%, live working should not be carried out. 2. Persons required for this operation include 1 Responsible Person, 1 special Supervisor, 2 ground electricians on tower, 2 equipotential electricians and 6 ground auxiliary electricians. Basket transfer method shall be adopted to enter the intense electric field for replacement of insulator. 3. Responsibilities of Responsible Person: Be responsible for division of operating personnel of the task, work order reading, handling formalities for stopping the DC restart device for the line, getting work permits, holding pre-shift meeting, dealing with emergency situations in work, quality surveillance, and the summary after work. 4. Special Supervisor: Be responsible for safety control of the work site. 5. Responsibilities of ground electrician: Cooperate with ground potential electrician in installing wire lifting system, operate hydraulic leading screw to transfer wire load, and disassemble and assemble composite insulator string.

Part I Standard for Training and Assessment on Live Working for Operation Maintenance of ±800kV UHV Power Transmission Line

Table (Cont'd)

Work Specifications and Requirements	6. Responsibilities of ground potential electrician: Install basket, wire lifting system, insulating grinding rope and cooperating with equal potential electrician in entering and exiting the potential, disassemble and assemble composite insulator string 7. Responsibilities of ground electrician: Be responsible for transferring tools and materials and cooperating with equipotential electrician in entering and exiting the equipotential. 8. During the live working, if thunder, rain, strong wind or any other circumstance threaten the safety of the staff, the Responsible Person or Supervisor may stop working temporarily according to the circumstances.
Work Specifications and Requirements	Given conditions: 1. Training base: UHV AC ±800k grounding line tower. 2. Work orders have been handled, safety measures have been completed (DC restart device has been disabled), and oral application (dispatcher or assessor) shall be made at the beginning and end of the work. 3. The instrument shall be used safely and correctly to test the insulating tool. 4. The operation must be carried out according to the working procedures. The relevant item scores shall be deducted for the process error. In case of major hidden dangers of personal, equipment and operational safety, the assessor may order the termination of the operation (assessment)
Assessment scenario preparation	1. Line: 003# tower of UHV AC ±800kV line; work content: single-V composite insulator for ±800kV transmission line tangent pole and tower. 2. Required operation tools and instruments: 1 insulated transmission rope (TJS-12), 3 insulated transmission rope (TJS-16), 2 insulated protection rope (TJS-16), 1 basket track rope (TJS-16), 1 basket, 2 hydraulic leading screws, 2 groups of insulated suspender, 2 sets of wire lifting clamps, 1 motor winching, 6 insulated tackles (JH10-1), 2 2-2 insulated tackles (JH20-2), 4 sets of shielding clothes (shielding efficiency ≥ 60dB), 1 insulation tester, 1 multimeter, 1 thermohygrometer, 1 speed meter, 2 pieces of moisture-proof canvas, and 2 sets of personal tools. 3. The work site shall be monitored, and the safety measures (fence, etc.) on the work site have been fully implemented; non-operation personnel are prohibited from entering the site, and the staff must wear safety helmets when entering the work site. 4. Examinees shall bring their own work clothes, flame retardant cotton underwear, safety helmets, gloves, and safety belts (including double-protective ropes)
Remarks	1. The deduction shall be done until the scores of each item are deducted completely. In case of major hidden dangers of personal, equipment and operational safety, the assessor may order the termination of the operation. 2. When equipment, working environment, safety belt, safety helmet, tool, shielding clothes, etc., do not conform to the operation condition, the assessor may order the termination of the operation

Table 1-1-7 Standards for Assessment and Scoring of Operation and Inspection Skills of UHVDC Transmission Line of State Grid Sichuan Electric Power Corporation

S/N	Project name	Quality requirements	Score	Deduction standard	Reasons for deduction	Deduction	Scoring
1	Site re-survey	(1) The Responsible Person shall go to the work site to check the line name, pole and tower number, on-site working conditions, defective parts and so on. (2) Check that the site meteorological conditions such as wind speed and humidity should meet the operation requirements. (3) Check whether the work order is complete and unmodified, check whether the safety measures listed are consistent with the actual situation on site, and supplement it if necessary	5	(1) Deduct 1 point for failure to check the line title. (2) Deduct 1 point for failure to check on-site working conditions (meteorology) and defective parts. (3) Deduct 0.5 points/item for any alteration in the work order filling; deduct 1 point for incorrect work order number; deduct 1.5 points for incomplete work order filling			
2	Work Permit	(1) The Responsible Person shall contact the on-duty control personnel and apply for stopping the transmission line DC restart device as per the contents of the work order. (2) Reporting content is standardized and complete	2	(1) Deduct 2 points for failure to contact the dispatching department (referee) for disabling the DC restart device. (2) Deduct 0.5 points for non-standard or incomplete terminology reporting respectively			

Part I Standard for Training and Assessment on Live Working for Operation Maintenance of ±800kV UHV Power Transmission Line

Table (Cont'd)

S/N	Project name	Quality requirements	Score	Deduction standard	Reasons for deduction	Deduction	Scoring
3	Site layout	Install the security fence and hang the signboards correctly: (1) The security fence should take full account of falling objects from the high place and the influence on road traffic. (2) The entrance and exit of the security fence shall be set reasonably. (3) Signs such as "Access from Here", "Work Here", "Access from Here" shall be properly arranged	3	(1) Deduct 0.5 points for failure to arrange the fence at the work site. (2) Deduct 0.5 points for failure to arrange the warning board. (3) Deduct 0.5 points for failure to hang the tower climbing operation sign			
4	Holding a pre-shift meeting	(1) All staff and personnel shall wear safety helmets and work clothes correctly. (2) Responsible Person shall wear red vest and read out the work order and be clear with work task and division of personnel; explain safety measures and technical measures in work; check (inquire after) mental state of all working personnel; inform of hazards in work and precontrol measures. (3) All working personnel shall sign on the work order for confirmation.	3	(1) Deduct 0.5 points/person for the staff not dressing uniformly. Deduct 0.5 points/person for the staff not dressing uniformly. (2) Give no points to this item for no division of labor, and deduct 1 point for unclear division of labor. (3) Deduct 0.5 points for the on-site Responsible Person not wearing a safety monitoring vest. (4) Deduct 1 point for the work shift member failing to sign or signing incompletely on the work order.			

Table (Cont'd)

S/N	Project name	Quality requirements	Score	Deduction standard	Reasons for deduction	Deduction	Scoring
5	Inspection of tools	(1) The staff shall place the tools on the moisture-proof canvas as required; the moisture-proof canvas shall be clean and dry. (2) The tools shall be placed in category according to the requirements of the fixed management; the insulated tools shall not be mixed with metal tools and materials; and the appearance inspection shall be done on the tools. (3) The surface of insulating tools shall not be worn, deformed and damaged, and the operation shall be flexible. Carry out segment insulation detection with such insulated tools with insulation resistance meter of 2,500V or above and with the resistance no less than 700M Ω, and wipe it off with a clean dry towel. (4) The ground potential and equipotential personnel on tower shall correctly wear a whole suit of qualified shield clothes and conductive shoes as required, with each part connected well, shall not wear chemical fiber clothes next to the skin in the shielding clothes and shall fasten safety belts; the Responsible Person shall carefully check whether they wears it correctly. (5) Tower climbing personnel shall check the line name, pole number and phase again and report them	7	(1) Deduct 1 point for failure to use moisture-proof cloth and place tools to designed positions. (2) Deduct 0.5 points/item for failure to check qualified label of tool test and appearance inspection. (3) Deduct 1 point/item for failure to use testing instrument for testing the tools. (4) Deduct 2 points/person time for the operator failing to wear the shielding clothes correctly and each part connected well. (5) Deduct 1 point for the on-site Responsible Person failing to check the safety protective equipment of the tower climbing operators. (6) Deduct 2 points/person for the tower climbing personnel failing to check the line name, pole number and phase. (7) Deduct 2 points/person for the tower climbing personnel failing to report the check results			

Part I Standard for Training and Assessment on Live Working for Operation Maintenance of ±800kV UHV Power Transmission Line

Table (Cont'd)

S/N	Project name	Quality requirements	Score	Deduction standard	Reasons for deduction	Deduction	Scoring
6	Climbing the tower	(1) The ground potential electrician and the equipotential electrician on tower shall wear a whole suit of qualified shielding clothes, fasten the safety belt after performing the impulse test on the safety belt, and carry the insulated transmission rope to climb the tower one after another. (2) During the tower climbing process, they shall fasten the anti-fall protection device, climb the tower to an appropriate position, fasten the safety belt, arrange the insulated transmission rope, and then cooperate with the ground electrician to make lifting preparation of the insulated transmission rope separately. (3) During tower climbing, the electrician shall fasten the anti-fall protection device, climb the tower at a uniform speed, grasp the main material by hand, hang the safety belt on the shoulder and keep the safety distance of more than 6.8m away from the electrified body, and the Responsible Person shall strengthen the operation monitoring.	5	(1) Deduct 2 points respectively for failure to fasten the safety belt or for failure to perform the impulse test on the safety belt and the backup protection rope. (2) Deduct 2 points for grasping the shackles with hands. (3) Deduct 1 point for inconvenient suspension position of tackle transmission rope for taking tools. (4) Deduct 2 points for metal tools that are difficult to ensure safe distance during transfer; deduct 2 points for tools that are not bound securely. (5) Deduct 2 points for falling object at high place. (6) Deduct 2 points for tools colliding with the tower body during the transfer process. (7) Deduct 1 point for knotting and disordered rope in tool transfer. (8) Deduct 2 points for the Responsible Person failing to monitor the operation in place. (9) Deduct 2 points for incorrect operation of electrician on tower			

Table (Cont'd)

S/N	Project name	Quality requirements	Score	Deduction standard	Reasons for deduction	Deduction	Scoring
7	Installation of tackle block and basket	(1) The lifting of insulating sling shall be smooth, free of collision and winding in transfer. (2) After the basket is installed, the electrician on tower shall carry out the careful inspection and check to the basket situation. (3) 2-2 tackle block and basket shall be installed reliably in a suitable position on the cross arm	5	(1) Deduct 0.5 points for rope winding of 2-2 tackle block. (2) Deduct 1 point for the unreasonable installation position of the track rope. (3) Deduct 1 point for unsuitable length of insulated protection rope. (4) Deduct 1 point for unsteadiness and collision of transfer tools.			
8	Enter the intense electric field	(1) After re-checking and confirming that each part of the shielding clothes is connected reliably, the equal potential electrician carries out the impulse test on the basket, and fastens the protection rope to climb the basket after reporting it to the Responsible Person. (2) The ground electrician slowly loosens the control rope of the 2-2 insulated tackle block, and slows down when it is about 2m away from the conductor. The equipotential electrician applies to the Responsible Person for potential transfer at a distance of 0.5m from the conductor, and quickly grasps the nearest sub-conductor by hand for potential transfer after obtaining the consent.	13	(1) Deduct 1 point for the equipotential electrician failing to conduct the impulse on the basket. (2) Deduct 1 point for failure to fasten the insulated protection rope. (3) Deduct 2 points for the ground potential electrician failing to check the safety measures of equipotential electrician. (4) Deduct 1 point for the ground electrician failing to control the tail rope of the tackle steadily. (5) Deduct 2 points for the equipotential electrician failing to apply to the Responsible Person before entering the intense electric field; deduct 1 point for starting to enter the intense electric field without consent after application.			

Part I Standard for Training and Assessment on Live Working for Operation Maintenance of ±800kV UHV Power Transmission Line

Table (Cont'd)

S/N	Project name	Quality requirements	Score	Deduction standard	Reasons for deduction	Deduction	Scoring
8	Enter the intense electric field	(3) After entering the intense electric field, the equipotential electrician shall protect the human body, and control the head not to exceed the grading ring on the conductor side. (4) The combined gap composed of the gaps between the equipotential electrician and the grounding body and the electrified body in the process of entering the potential shall not be less than 6.8m	13	(6) Deduct 1 point for being unskilled in potential transfer. (7) Deduct 2 points for equipotential electrician unfastening the body backup protection rope after entering the intense electric field. (8) Deduct 2 points for the equipotential electrician with head exceeding the grading ring on the conductor side after entering the intense electric field.			
9	Install tools and transfer conductor load	(1) The ground electrician transfers the insulated suspender, wire lifting clamps, and hydraulic leading screw to the working position, and the equipotential electrician and ground potential electrician cooperate and properly install the tools for replacement of composite insulator. (2) After checking the reliable connection of each component and obtaining the consent from the Responsible Person, the equipotential electrician firstly tightens the mechanical leading screw, then tightens the hydraulic leading screw after the mechanical leading screw is properly stressed, allows it being stressed slightly and inspect the force bearing point.	15	(1) Deduct 1 point for unsteady transfer of tools; deduct 1 point for each collision. (2) Deduct 2 points for failure to check the reliable installation and good stress of load-bearing tools, and 2 points for failure to report it and obtain the consent of the Responsible Person. (3) Deduct 1 point for ineffective communication between the ground potential electrician and the equipotential electrician.			

Table (Cont'd)

S/N	Project name	Quality requirements	Score	Deduction standard	Reasons for deduction	Deduction	Scoring
9	Install tools and transfer conductor load	(3) The ground electrician transfers the control rope of the composite insulator string to the equipotential electrician who installs it at the tail of the composite insulator string. The ground electrician tightens the control rope of composite insulator string. (4) After checking that the load-bearing tools are stressed normally and obtaining the consent of the Responsible Person, the equipotential electrician dismantles the connection on the conductor side, and the ground electrician slowly releases the control rope of composite insulator string to the extent that it is naturally vertical. (5) The ground potential electrician ties the insulated transmission rope to the upper end of the composite insulator string, and then the ground potential electrician cooperates with the ground potential electrician to disconnect the ground terminal of composite insulator string	15	(4) Deduct 5 points for failure to check the load-bearing tools before removing the insulator string, and 2 points for failure to report the check results. (5) Deduct 2 points for inappropriate position of the insulating rope fastening the composite insulator. (6) Deduct 2 points for each collision during the transfer of the insulator string. (7) Deduct 1 point for the ground electrician failing to control the motor winching in place.			

Table (Cont'd)

S/N	Project name	Quality requirements	Score	Deduction standard	Reasons for deduction	Deduction	Scoring
10	Replacement of insulator string	(1) The ground electrician controls the control rope of the composite insulator string, and slowly puts the composite insulator string to the ground by using the motor winching. Attention shall be paid to the control rope of composite insulator string, which shall not collide with the load-bearing tools, conductor and pole and tower. (2) The ground electrician transfers the insulated transmission rope and the control rope of the composite insulator string to the new composite insulator respectively. (3) The ground electrician starts the motor winching to transfer the new composite insulator string to the working position on tower. The ground potential electrician restores the connection between the new composite insulator string and ball head hanging ring, and restore the fitting pin. (4) The ground electrician slowly loosens the motor winching to make the composite insulator string naturally vertical, and the equipotential electrician restores the connection between the socket hanging plate and the yoke plate, and installs the cotter pin.	17	(1) Deduct 1 point for the ground electrician failing to control the tail rope of the insulator well. (2) Deduct 1 point for unreasonable rope buckle binding. (3) Deduct 2 points for failure to check the steering of winching and the stress of tackle. (4) Deduct 2 points for each collision during the transfer of the insulator string. (5) Deduct 5 points for failing to install the insulator string in place. (6) Deduct 5 points for the operator failing to check the connection of insulator string. (7) Deduct 2 points for the operator failing to check the installation of pin in place. (8) Deduct 2 points for the Special Supervisor failing to fulfill the supervision responsibility.			

Table (Cont'd)

S/N	Project name	Quality requirements	Score	Deduction standard	Reasons for deduction	Deduction	Scoring
11	Exiting the potential	(1) An equipotential electrician shall fasten the insulated protective rope, enter the basket, and then keep the arm straight so that the basket is 0.5m away from the sub-conductor. After obtaining the consent from the Responsible Person, the equipotential electrician quickly disconnect the sub-conductor. (2) At the same time, the ground electrician quickly tightens the control rope of the 2-2 insulated tackle block, pulls the basket upward to the cross arm and stops it. Then, the equipotential electrician climbs the cross arm and fastens the safety belt. (3) The ground electrician transfers the basket to the another equipotential electrician by means of insulated transmission rope. Then, the equipotential electrician enters the basket after checking that there is no object left on the conductor, and exits the potential by the same method.	10	(1) Deduct 2 points for failing to report the end of work, and deduct 1 point for each item left in the intense electric field. (2) Deduct 1 point for the equipotential electrician failing to fasten the insulated protection rope. (3) Deduct 1 point for the ground electrician failing to control the tail rope of the tackle steadily. (4) Deduct 2 points for the equipotential electrician failing to apply to the Responsible Person before exiting the intense electric field; deduct 1 point for starting to enter the intense electric field without consent after application. (5) Deduct 1 point for being unskilled in potential transfer.			

Part I Standard for Training and Assessment on Live Working for Operation Maintenance of ±800kV UHV Power Transmission Line

Table (Cont'd)

S/N	Project name	Quality requirements	Score	Deduction standard	Reasons for deduction	Deduction	Scoring
12	Remove the basket and return to the ground	(1)The electrician on tower shall cooperate the removal of the basket track rope, the insulated protection rope, the 2-2 insulated tackle block and the basket and transfer of them to the ground. (2)After checking that there is no object left on the tower, the equipotential electrician on the tower shall report it to the Responsible Person and then climb down the tower with insulated transmission rope after obtaining the consent of the Responsible Person	5	(1)Deduct 2 points for failure to use the falling protector when climbing down the tower. (2)Deduct 2 points for loosing the protection of the safety belt when moving on the tower. (3)Deduct 1 point for grasping the tower nail when climbing down the tower. (4)Deduct 2 points for any objects left on the tower			
13	End of the work	(1)The Responsible Person shall organize all working personnel to put working apparatus and materials in order and put them in a special kit (bag) after cleaning; clean the site to ensure that "the materials are removed and the site is cleaned after construction". (2)After a post-shift meeting is held, the Responsible Person shall make work summaries and comments. Comments include the construction quality of this work and the implementation of safety measures from all working personnel. (3)Responsible Person shall report to the on-duty control personnel that the work is over, apply for the restoration of circuit re-closing and terminate the work order.	10	(1) Deduct 2 points for failure to clean the tools. (2) Deduct 2 points for missing tools. (3)Deduct 2 points for failure to hold the post-shift meeting. (4)Deduct 2 points for failure to remove the fence. (5)Deduct 2 points for failure to report to dispatcher			
	Total		100				

Module 2 Standards for Training and Assessment on Live Replacement of 1-3 Glass Insulators on the Cross Arm Side of ±800kV UHV Transmission Line Resisting-tensile Tower

I. Training Standards

(I) Training Requirements

Designation of module	Live replacement of 1-3 glass insulators on the cross arm side of ±800kV UHV transmission line resisting-tensile tower	Type of training	Operation
Training method	Practical operation training	Hours of training	21 training hours
Training objectives	1. Master the electrical significance of electromagnetic protection in ground potential operation method. 2. Be able to independently complete the live replacement of 1-3 glass insulators on the cross arm side of ±800kV UHV transmission line resisting-tensile tower (intermediate potential operation method)		
Training venue	UHV ±800kV DC practical training line		
Training content	Operation of live replacement of 1-3 glass insulators on the cross arm side of ±800kV UHV transmission line resisting-tensile tower by adopting the ground potential operation method		
Scope of application	Maintenance personnel for ±800kV UHV DC power transmission line		

(II) Referenced Procedures and Specifications

(1) Technical Specification for Live Working of ±800kV DC Line (DL/T1242-2013)

(2) Operating Code for ±800kv DC Overhead Transmission Line (GB/T28813-2012)

(3) Maintenance Specification for ±800kV DC Overhead Transmission Line (DL/T251-2012)

(4) Technical Guide for Live Working of ± 800kV DC Transmission Line (Q/GDW302-2009)

(5) Calculation Method of Live Working Minimum Approach Distance on AC Transmission Line (GB/T 19185-2008)

(6) Guidelines of Insulation Coordination for Live Working (DL/T867-2004)

(7) Test Guide of the Insulating Tool for Live Working (DL/T878—2004)

(8) State Grid Corporation of China on the Management Regulations of Live Working (Trial Implementation) (SGCC [2007] No.751)

(9) State Grid Corporation of China Working Regulations of Power Safety (Transmission Line Section) (Q/GDW1799.2-2013)

(10) Electrotechnical Terminology - Overhead Line (GB/T 2900.51-1998)

(11) Electrotechnical Terminology - Live Working (GB/T2900.55-2016)

(12) Live Working - Terminology for Tools, Equipment and Devices (GB/T 14286-2008)

(13) Minimum Requirements for Utilization of Tools, Devices and Equipment for Live Working (DL/T 877-2004)

(14) Preventive Test Code of Tools, Devices and Equipment for Live Working (DL/T 976-2005)

(15) Insulated Tackles for Live Working (GB/T13034-2008)

(16) Live Working-Insulating Ropes (GB 13035-2008)

(17) Shielding Clothes for Live Working (GB/T6568-2008)

(18) Screen Clothes for Live Working on 1000kV AC (GB/T25726-2010)

(III) Teaching Design for Training

To complete the work task of "live replacement of 1-3 glass insulators on the cross arm side of ±800kV UHV transmission line resisting-tensile tower", each training stage shall be designed according to the standard operation procedure for work task completion. Each stage includes specific training objectives, training content, hours of training, training methods (training resources), training environment, assessment and evaluation, etc, as shown in the Table 1-2-1.

Standard for Professional Training and Assessment for Operation Maintenance of UHV DC Power Transmission Line

Table 1-2-1 Design of Training Content for Live Replacement of 1-3 Glass Insulators on the Cross Arm Side of ±800kV UHV Transmission Line Resisting-tensile Tower

Training schedule	Training objectives	Training content	Hours of training	Training methods and resources	Preparation of training conditions	Assessment and evaluation
1. Theoretical teaching	1. Preliminarily master the basic method for electromagnetic protection in ground potential operation method. 2. Be familiar with the method for the replacement of 1-3 glass insulators on the cross arm side of ±800kV UHV transmission line resisting-tensile tower	1. The electrical significance of electromagnetic protection in ground potential operation method. 2. Method and quality standard for the replacement of 1-3 glass insulators on the cross arm side of ±800kV UHV transmission line resisting-tensile tower	2	Training method: Lecture. Training resources: PPT, relevant regulations and specifications	Multimedia classroom	Attendance, classroom questions and assignments
2. Preparations	Be able to complete the preparations before operation	1. Work site survey. 2. Preparation of the standardized operation card. 3. Filling of the work order. 4. Preparation of tools and materials for this operation	2	Training methods: 1. Site survey and cleaning of tools and materials shall be practiced at site. 2. Preparation of operation card and the filling of work order shall adopt lecture method. Training resources: 1.UHV practical training line (±800kV practical training line). 2.UHV tools warehouse. 3.Blank work order	1. UHV transmission line for practical training; 2. Multimedia classroom	

· 282 ·

Part I Standard for Training and Assessment on Live Working for Operation Maintenance of ±800kV UHV Power Transmission Line

Table (Cont'd)

Training schedule	Training objectives	Training content	Hours of training	Training methods and resources	Preparation of training conditions	Assessment and evaluation
3. Work site preparation	Be able to complete the preparations of work site	1. Work site re-survey. 2. Work application. 3. Work site layout. 4. Pre-shift meeting. 5. Inspection of tools	1	Training methods: Demonstration and role play. Resource: UHV practical training line (±800kV practical training line)	±800kV Practical training line	
4. Trainer's demonstration	The trainees can preliminarily understand the operation process of the task through inspecting and learning from each other's work	1. The ground potential electrician assembles tools. 2. The ground potential electrician completes the replacement of single glass insulator	2	Training method: Demonstration. Resource: UHV practical training line (±800kV practical training line)	±800kV Practical training line	
5. Group training of trainees	Be able to complete the replacement of 1-3 glass insulators on the cross arm side of ±800kV UHV transmission line resisting-tensile tower	1. The trainees are grouped (7 in a group) to train the skill operation of replacing insulator. 2. Trainers guide the operation of trainees and conduct safety supervision	12	Training methods: Role play. Resource: UHV practical training line (±800kV practical training line)	±800kV Practical training line	Score the operation of trainees according to the detailed rules for skill assessment and scoring Score
6. End of the work	Through training: 1. Trainees can further understand the shortcomings during the operation process for later improvement. 2. Train the trainees in the working style of safe and civilized production	1. Clean up the work site. 2. Report to dispatcher. 3. Comment and summarize the current work task at post-shift meeting	1	Training methods: Lecture and inductive method	±800kV Practical training line	

(IV) Work Flow

1. Work task

Live replacement of 1-3 glass insulators on the cross arm side of ±800kV UHV transmission line resisting-tensile tower.

2. Requirements for Weather and Work Site

(1) The live replacement of 1-3 glass insulators on the cross arm side of ±800kV UHV transmission line resisting-tensile tower shall be carried out in good weather. In case of lightning (hearing thunder or seeing lightning), snow, hail, rain, fog and so on, live working is prohibited. When the wind force is greater than level 5, or the relative humidity of the air is greater than 80%, it is unsuitable for live working; when emergency live repair is required in bad weather, relevant personnel shall be organized to fully discuss and prepare necessary safety measures, which can be implemented after being approved by the unit.

(2) The operating personnel should be in good mental states and be familiar with the organizational and technical measures to ensure safety in work; they should hold the qualification certificate for live working within the validity period.

(3) Responsible Person should organize the relevant personnel to complete field investigation in advance, determine the operating methods, required working apparatus and necessary measures according to the results, and handle the tickets for live working.

(4) The work site should be reasonably set up with fence and warning signs. Non operating personnel is forbidden to enter.

(5) DC restart device of the line shall be disabled for the Project.

(6) Safe working distance and effective insulation length during operation are shown in Table 1-2-2.

Table 1-2-2　Safe Distance (m) for Live Replacement of 1-3 Glass Insulators on the Cross Arm Side of ±800kV UHV Transmission Line Resisting-tensile Tower

Altitude	Minimum safe distance between equipotential electrician and grounding frame	Minimum effective insulation length of insulating tools	Minimum combination gap
H≤1000	6.8	6.8	6.7
1000<H≤2000	7.3	7.3	7.3
2000<H≤2500	7.9	7.8	7.8

Note: The minimum safe distance and minimum combined gap in the table include 0.5m occupying gap of human body.

(7) When the ground potential electrician enters the cross arm side of strain insulator string, the number of insulators shorted by the human body shall not be more than 4. The minimum number of good insulators shall meet the requirements of Table 1-2-3 after deducting the number of insulators shorted by human body and defective insulators from the strain insulator string.

Table 1-2-3　Minimum Number of Good Insulators of Strain Insulator String

Altitude (m)	Structural height of single glass insulator (mm)	Minimum total length of good insulator string (m)	Minimum number of good insulators
$H \leqslant 1000$	170	6.2	37
	195		32
	205		31
	240		26
$1000 < H \leqslant 2000$	170	7.1	42
	195		37
	205		35
	240		30
$2000 < H \leqslant 2500$	170	7.55	45
	195		39
	205		37
	240		32

3. Preparations

3.1　Hazards and precontrol measures

(1) Hazard —— electric shock

Precontrol measures:

① Before work, the Responsible Person shall contact the on-duty control personnel, deactivate the line DC restart device, and perform the licensing procedures.

② Before climbing the tower, ground potential operators on tower must carefully check the name of the line, the number of the pole and tower, and phase, and then the tower can be climbed after all have been confirmed correct.

③ If lines lose power suddenly during work, operators shall consider it as still charged. The Responsible Person shall contact the control personnel as soon as possible, and no forced energization is allowed before the on-duty control personnel getting in touch with the Responsible Person.

④ Insulating tool and insulating ropes shall be free of damage, moisture, deformation, and failure. It is not allowed to use non-insulating ropes (such as cotton rope, manila rope, and steel wire rope).

⑤ The ground electrician shall wear clean and dry gloves when operating the insulating tools. When entering the work site, the live working tools shall be placed on damp-proof canvas or insulating mat to prevent dirt and dampness of the insulating tools in use.

⑥ The ground potential electrician shall wear flame-retardant underwear and full set of qualified shielding clothes outside (including hat, mask, dresses & trousers, gloves, socks and shoes). All parts shall be in excellent connection conditions.

⑦ When the ground potential electrician enters the cross arm side of the strain insulator string, the positions of the hands and feet shall be correspondingly consistent, and the number of insulators shorted by the human body and the tools shall be no greater than 4.

⑧ When transmitting large metal objects by using insulating rope, the ground potential electricians and ground electricians shall ground the metal objects before contacting.

⑨ During the live working, the Responsible Person (Supervisor) shall continuously monitor the operators and correct their nonstandard operation or actions in violation at any time. Special attention shall be paid to operators work at heights to ensure that there is enough safe distance (meeting the requirements in Table 1-2-2). It is forbidden to contact two non-connected electrified bodies or make contact with electrified body and grounding body at the same time.

(2) Hazard —— falling from high places

Precontrol measures:

① Before climbing, operators working at heights must satisfy the requirements of this operation, such as physical condition, mental state, and skill and quality.

② Supervisor shall correct irregularities and violations at any time. Special attention shall be paid to operators to prevent them from losing the protection of the safety belt or insulated backup protection rope during transposition. The safety belt or insulated backup protection rope shall not be fastened lower than operating personnel.

(3) Hazard —— injury caused by objects falling from high place

Precontrol measures:

① High-place operators should put personal tools and fragmentary materials into the tool bag. It is strictly forbidden to hang objects in high place or keep in the mouth.

② Ground operator should correctly wear a helmet and use the knots. The vertical distance from the operation point should not be less than the falling radius.

③ The work site shall be set up with fence and warning signs. It shall be noted at any time that Supervisor shall prohibit irrelevant personnel and vehicles from entering operation area.

3.2 Selection of tools, instruments and materials

Tools and materials required for live replacement of 1-3 glass insulators on the cross arm side of ±800kV UHV transmission line resisting-tensile tower are shown as Table 1-2-4. Before delivering tools and instruments out of warehouse, application voltage class and test period shall be carefully checked and they shall be inspected to ensure that appearance is intact, connection is firm, rotation is flexible and meet the working task requirements. After delivering tools and instruments out of warehouse, they shall be stored in tools bag or tool kit for transportation to avoid contamination and damp. Metal tools and insulated tools shall be separately loaded and transported to avoid deformation, damage or other defects caused by mixed loading and transportation.

Table 1-2-4 Tools and Materials Required for Live Replacement of 1-3 Glass Insulators on the Cross Arm Side of ±800kV UHV Transmission Line Resisting-tensile Tower

S/N	Name	Specification and Model	Unit	Qty.	Remarks
1	Insulated transmission rope	TJS-12	Nos.	2	Insulating tool
2	Insulated protection rope	TJS-16	Nos.	2	Insulating tool
3	Insulated noose	Φ14mm	Nos.	2	Insulating tool
4	Insulated tackle	JH10-1	Nos.	2	Insulating tool
5	Strain end clamp		Nos.	1	Metal tool
6	Hydraulic leading screw		Nos.	2	Metal tool
7	Closed clamp (rear clamp)		Nos.	1	Metal tool
8	Shielding clothes	Shielding efficiency ≥ 60dB (shield efficiency of shielding mask ≥20dB)	Set	2	PPE
9	Conductive shoes	The size depends on the wearer	Pair	2	PPE
10	Flame retardant underwear	Pure mulberry silk	Set	2	PPE
11	Safety belt of double insurance	Suspender	Nos.	2	PPE
12	Safety helmet		pcs	7	PPE
13	Goggles		Nos.	2	PPE

S/N	Name	Specification and Model	Unit	Qty.	Remarks
14	Insulation resistance tester	5000V, with electrode width 2cm and interelectrode width 2cm	Set	1	Other tools
15	Anemometer		pcs	1	Other tools
16	Temperature and humidity meter		pcs	1	Other tools
17	Multimeter		pcs	1	Other tools
18	Moisture-proof canvas	2m×4m	pcs	2	Other tools
S/N	Name	Specification and Model	Unit	Qty.	Remarks
19	Falling protector	Corresponding to the type of pole and tower falling protector	pcs	2	Other tools
20	Security fence		Set	Several	Other tools
21	Warning sign	"Work Here", "Access from Here""Access from Here"	Set	1	Other tools
22	Red waistcoat	"Responsible Person"	pcs	1	Other tools
23	Personal tools	Wrench, vice	Set	1	Other tools
24	Pin puller		pcs	1	Other tools
25	Falling protector	Corresponding to the type of pole and tower anti-fall device	pcs	2	Other tools
26	Clean towel	Cotton	pcs	1	Other tools
27	Interphone		Set	3	Other tools
28	Insulator		pcs	1	Material

3.3 Division of labor for operators

Division of labor for operators of the task is shown in Table 1-2-5.

Table 1-2-5 Personnel Allocation for Live Replacement of 1-3 Glass Insulators on the Cross Arm Side of ±800kV UHV Transmission Line Resisting-tensile Tower

S/N	Post	Qty. (person)	Responsibilities
1	Responsible Person	1	Be responsible for various work on the work site
2	Specific responsible supervisor	1	Be responsible for the safety control of the work site
3	Ground potential electrician	2	Be responsible for tool installation and insulator replacement
4	Ground electrician	3	Be responsible for passing on tools and materials, and cooperating with the ground potential electrician to replace insulators

4. Working Procedures

The workflow of this task is shown in Table 1-2-6.

Table 1-2-6   Workflow for Live Replacement of 1-3 Glass Insulators on the Cross Arm Side of ± 800kV UHV Transmission Line Resisting-tensile Tower

S/N	Work Content	Operation Steps and Standards	Safety Measures and Precautions	Responsible Person
1	Site re-survey	The Responsible Person shall complete the following work: (1) Check the line name, the number of the pole and tower and ensure the phases are correct; guarantee that the foundation and the pole and tower are intact and in normal condition; ensure that the cross and span distance meets the safety requirements; confirm the defect conditions, etc. (2) Check that the site meteorological conditions such as wind speed and humidity should meet the operation requirements. (3) Check that the terrain and environment should meet the operation requirements. (4) Check that the safety measures listed in the work order are in line with the actual situations on site. The measures will be added if necessary.	(1) Correctly wear helmet, working clothes, work shoes and protective gloves. (2) Operation under meteorological conditions that may endanger the safety of operators is forbidden. (3) Non-operation personnel and vehicles are strictly prohibited from entering the working site	
2	Work Permit	(1)The Responsible Person shall contact the on-duty control personnel and apply for stopping the transmission line DC restart device as per the contents of the work order. (2)Live working could be started only after being approved by control personnel on duty.	Live working shall not be started without the permission of the on-duty control personnel	
3	Site layout	Install the security fence and hang the signboards correctly: (1)The security fence should take full account of falling objects from the high place and the influence on road traffic. (2)The entrance and exit of the security fence shall be set reasonably. (3)Signs such as "Access from Here", "Work Here", "Access from Here" shall be properly arranged	When the influence on road traffic safety is uncontrollable, the traffic management department should be contacted in time to strengthen the on-site control of traffic safety	

Table (Cont'd)

S/N	Work Content	Operation Steps and Standards	Safety Measures and Precautions	Responsible Person
4	Holding a pre-shift meeting	(1)All working personnel shall line up. (2)Responsible Person will read out the work order and be clear with work task and division of personnel; explain safety measures and technical measures in work; check (inquire after) mental state of all working personnel; inform of hazards in work and precontrol measures. (3)All working personnel shall sign on the work order for confirmation.	(1)Work order shall be filled and issued with standardized licensing procedure and complete signature. (2)All working personnel shall be in good mental states. (3)All working personnel shall be clear with task division of works, safety measures and technical measures	
5	Inspection tool	(1)The ground potential electrician shall wear the shielding clothes in a right way and pass the inspection, which shall be supervised and inspected by the Responsible Person. (2)Wear personal safety appliance correctly (proper size and easy lock), and the Responsible Person shall supervise and inspect it. (3)Measure the wind speed, wind direction and humidity, check the insulation performance of insulating tools, and make records	(1)Check carefully for damage, deformation and dysfunction before using metal and insulating tools. Carry out segment insulation detection with such insulated tools with insulation resistance meter of 2,500V or above and with the resistance no less than $700M\Omega$, and wipe it off with a clean dry towel. (2)Use a multimeter to measure the resistance between the farthest ends of the shielding clothes and trousers, which shall not be greater than 20Ω. The Responsible Person shall check the connection of the electrician's shielding clothes. (3)Check the tool assembly and make sure the connection is reliable. (4)The tools and instruments for live working shall be placed on the moisture-proof canvas	

Table (Cont'd)

S/N	Work Content	Operation Steps and Standards	Safety Measures and Precautions	Responsible Person
6	Climbing the tower	(1)After checking the line name and pole and tower number, the ground potential electrician on tower shall check the stress of safety belt and falling protector. (2)The ground potential electrician carries the insulated transmission rope to climb the tower. When two electricians reach the operation point of the cross arm, they choose the appropriate position to fasten the safety belt, and the ground potential electrician on the tower installs the insulated tackle and the insulated transmission rope in the appropriate position of the cross arm. Then he shall cooperate with the ground electrician to separate the insulated transmission rope for lifting preparation	(1)After checking the correct line name and pole and tower number, he can climb the tower for operation. (2)The anti-falling device installed on the tower shall be used during climbing the tower; when moving and transposing on the pole and tower, the safety protection shall not be lost, and the operators must climb and grasp the components securely. (3)The ground potential electrician must wear full set of qualified shielding clothes which must be connected reliably. Before the cross arm enters the insulator string, the ground potential electrician shall check and confirm that each part of the shielding clothes are connected reliably before the next operation	
7	Install tools and transfer conductor tension	(1)The ground electrician transfers the closed clamp (rear clamp), hydraulic leading screw, strain end clamp and so on to the ground potential electrician by using the insulated transmission rope. (2) The ground potential electrician first installs the strain end clamp on the towing plate and then the closed clamp (rear clamp) on the fourth insulator on the cross arm side, and connects the hydraulic leading screw. (3) Equipotential electricians shall check and confirm that all parts of the load-bearing tools are installed in good condition. With the permission of the Responsible Person, they can operate the hydraulic leading screw so that it is gradually stressed and the insulators to be replaced are relaxed.	(1)The lifting process shall be smooth, free from collision and winding, and the knot shall be used correctly to avoid falling object from high place. (2) After being installed, the tools and instruments shall be checked for firm and reliable installation on each part; in the process of operation, the number of human body and tools shall be no greater than 4, and the minimum safe distance between human body and electrified body shall be no less than the requirements in Table 1-2-2. (3) In the process of tightening the hydraulic leading screw, it shall keep both sides being stressed uniformly and synchronously, and the leading screw shall be stable and strong when being shaken.	

Table (Cont'd)

S/N	Work Content	Operation Steps and Standards	Safety Measures and Precautions	Responsible Person
8	Replacement of insulator	(1) The ground potential electrician shall perform impulse test, check and confirm that the load-bearing tools are stressed normally. With the permission of the Responsible Person, they shall tie the old insulator with the insulated transmission rope, take out the fitting pins at both ends of the old insulator, continue to operate and tighten the hydraulic leading screw until the old insulator is removed. The two hydraulic leading screws shall be stressed uniformly, and the operating handle shall not knock on the insulator. (2) The ground electrician uses the other end of the insulated transmission rope to fasten the new insulator, and lifts the new insulator to the ground potential electrician by means of the old lower and the new upper. (3) The ground potential electrician installs the new insulator, resets the fitting pin at its two ends, and confirms the installation in place	(1) In the process of regulating the hydraulic leading screw, it shall keep both sides being stressed uniformly and synchronously, and the handle shall not knock on the glass insulator. (2) The lifting process shall be smooth, free from collision and winding, and the knot shall be used correctly. The ground electricians shall be cooperate with each other to prevent the insulators from colliding. (3) In the process of operation, attention shall be always paid to the stress condition of each part.	
9	Remove tools	(1) The ground potential electricians check and confirm that the connection of the new insulator is reliable and, with the permission of the Responsible Person, they operate and loosen the hydraulic leading screw so that the replaced insulator is gradually stressed. (2) After load transfer, the ground potential electricians shall perform the impulse test, check and confirm that the new insulator is in good condition. With the permission of the Responsible Person, they can remove the insulated transmission rope tied to the insulator, fasten it to the proper position of the load-bearing tools, remove the closed clamp (rear clamp), hydraulic leading screw, strain end clamp and other load-bearing tools, and transfer them to the ground with the cooperation of the ground electrician.	(1) Before the tools and instruments are removed, the fitting pins of insulators shall be inspected for installation in place. (2) The transfer process shall be smooth, free from collision and winding, and the knot shall be used correctly, to protect people against objects falling from high places	

Table (Cont'd)

S/N	Work Content	Operation Steps and Standards	Safety Measures and Precautions	Responsible Person
10	Return to the ground	After checking that there is no object left on the tower, the electrician on the tower shall report it to the Responsible Person and then climb down the tower with insulated transmission rope after obtaining the consent of the Responsible Person	The anti-falling device installed on the tower shall be used in the process of climbing down the tower; when moving and transposition on the pole and tower, the safety protection shall not be lost, and the operators must grasp the components securely	
11	End of the work	(1) The Responsible Person shall organize all working personnel to put working apparatus and materials in order and put them in a special kit (bag) after cleaning; clean the site to ensure that "the materials are removed and the site is cleaned after construction". (2) Hold a post-shift meeting, and the Responsible Person shall make work summaries and comments on the construction quality of this work and the implementation of safety measures from all working personnel. (3) Responsible Person shall report to the on-duty control personnel the end of the work, apply for the restoration of line DC restarting device and terminate the work order	It is forbidden to restore the DC transmission line restarting device at the appointed time	

Standard for Professional Training and Assessment for Operation Maintenance of UHV DC Power Transmission Line

II. Assessment standard

Table 1-2-7 Detailed Rules for Assessment and Scoring of Operation and Inspection Skills of UHVDC Transmission Line of State Grid Sichuan Electric Power Corporation

Fill-in Column of Examinee	No.:	Name:	Position:	Unit:	Date:	MM/DD/YYYY		
Fill-in Column of Assessor	Grade:	Assessor:	Assessment Team Leader:	Starting time:	Closing time:	Operation Duration:		
Assessment module	Live replacement of 1-3 glass insulators on the cross arm side of ±800kV UHV transmission line resisting-tensile tower		Appraisal Object	Maintenance personnel for ±800kV UHV DC power transmission line		Assessment method	Operation	Assessment Time Limit
								90min
Job Description	Live replacement of 1-3 glass insulators on the cross arm side of ±800kV UHV transmission line resisting-tensile tower.							
Work Specifications and Requirements	1. Live working shall be carried out in good weather. In case of thunder, rain, snow or fog, no live working shall be carried out. When the wind force is greater than Level 5 and the humidity is greater than 80%, live working should not be carried out. 2.1 Responsible Person, 1 Special Supervisor, 2 ground electricians on tower and 3 ground support electricians are required for this operation. 3. Responsibilities of Responsible Person: Be responsible for division of operating personnel of the task, work order reading, stopping the DC restart device for the transmission line, getting work permits, holding pre-shift meeting, dealing with emergency situations in work, quality surveillance, and the summary after work. 4. Responsibilities of Special Supervisor: Be responsible for safety control of the work site. 5. Responsibilities of ground potential electricians: Be responsible for tool installation and insulator replacement. 6. Responsibilities of ground electrician: Be responsible for transferring tools and materials and cooperating with ground electrician to replace insulators.							

Part I Standard for Training and Assessment on Live Working for Operation Maintenance of ±800kV UHV Power Transmission Line

Table (Cont'd)

Work Specifications and Requirements	7. During the live working, if thunder, rain, strong wind or any other circumstance threaten the safety of the staff, the Responsible Person or Supervisor may stop working temporarily according to the circumstances. Given conditions: 1. Training base: Tower of ±800kV UHV DC line; model of insulator: U500BP. 2. Work orders have been handled, safety measures have been completed (DC restart device has been disabled), and oral application (dispatcher or assessor) shall be made at the beginning and end of the work. 3. The instrument shall be used safely and correctly to test the insulating tool. 4. The operation must be carried out according to the working procedures. The scores of the items to be carried out shall be deducted for the process error. In case of major hidden dangers of personal, equipment and operational safety, the assessor may order the termination of the operation (assessment)
Assessment scenario preparation	1. Line: Tower of ±800kV UHV DC line; work content: replacement of 1-3 glass insulators on cross arm side of ±800kV transmission line resisting-tensile pole and tower; model of insulator: U500BP. 2. Tools and instruments required for operation: 2 insulated transmission ropes (TJS-12), 2 insulated protection ropes (TJS-16), 2 insulated tackles (JH10-1), 1 set of closed clamps, 2 hydraulic leading screws, 2 sets of shielding clothes, 1 insulation resistance tester, 1 multimeter, 1 thermohygrometer, 1 anemometer, 2 pieces of moisture-proof canvas, and 1 set of personal tools. 3. The work site shall be monitored, and the safety measures (fence, etc.) on the work site have been fully implemented; non-operation personnel are prohibited from entering the site, and the staff must wear safety helmet when entering the work site. 4. Examinees shall bring their own work clothes, flame retardant cotton underwear, safety helmets, gloves, and safety belts (including double-protective ropes)
Remarks	1. The deduction shall be done until the scores of each item are deducted completely. In case of major hidden dangers of personal, equipment and operational safety, the assessor may order the termination of the operation. 2. When equipment, working environment, safety belt, safety helmet, tool, shielding clothes, etc., do not conform to the operation condition, the assessor may order the termination of the operation

Table 1-2-8 Standards for Assessment and Scoring of Operation and Inspection Skills of UHVDC Transmission Line of State Grid Sichuan Electric Power Corporation

S/N	Project name	Quality requirements	Score	Deduction standard	Reasons for deduction	Deduction	Scoring
1	Site re-survey	(1) The Responsible Person shall go to the work site to check the line name, pole and tower number, on-site working conditions, defective parts and so on. (2) Check that the site meteorological conditions such as wind speed and humidity should meet the operation requirements. (3) Check whether the work order is complete and unmodified, check whether the safety measures listed are consistent with the actual situation on site, and supplement it if necessary	5	(1) Deduct 1 point for failure to check the line title. (2) Deduct 1 point for failure to check on-site working conditions (meteorology) and defective parts. (3) Deduct 0.5 points/item for any alteration in the work order filling; deduct 1 point for incorrect work order number; deduct 1.5 points for incomplete work order filling			
2	Work Permit	(1) The Responsible Person shall contact the on-duty control personnel and apply for stopping the transmission line DC restart device as per the contents of the work order. (2) Reporting content shall be standardized and complete	2	(1) Deduct 2 points for failure to contact the dispatching department (referee) for disabling the restart device. (2) Deduct 0.5 points for non-standard or incomplete terminology reporting respectively			

Table (Cont'd)

S/N	Project name	Quality requirements	Score	Deduction standard	Reasons for deduction	Deduction	Scoring
3	Site layout	Install the security fence and hang the signboards correctly: (1)The security fence should take full account of falling objects from the high place and the influence on road traffic. (2) The entrance and exit of the security fence shall be set reasonably. (3) Signs such as "Access from Here", "Work Here", "Access from Here" shall be properly arranged;	3	(1)Deduct 0.5 points for failure to arrange the fence at the work site. (2)Deduct 0.5 points for failure to arrange the warning board. (3) Deduct 0.5 points for failure to hang the tower climbing operation sign			
4	Holding a pre-shift meeting	(1)All staff and personnel shall wear safety helmets and work clothes correctly. (2) Responsible Person shall wear red waistcoat and read out the work order and be clear with work task and division of personnel; explain safety measures and technical measures in work; check (inquire after) mental state of all working personnel; inform of hazards in work and precontrol measures. (3) All working personnel shall sign on the work order for confirmation.	3	(1) Deduct 0.5 points/person for the staff not dressing uniformly; deduct 0.5 points/person for the staff not dressing uniformly (2) Give no points to this item for no division of labor, and deduct 1 point for unclear division of labor. (3) Deduct 0.5 points for the on-site Responsible Person not wearing a safety monitoring vest. (4) Deduct 1 point for the work shift member failing to sign or signing incompletely on the work order.			

Standard for Professional Training and Assessment for Operation Maintenance of UHV DC Power Transmission Line

Table (Cont'd)

S/N	Project name	Quality requirements	Score	Deduction standard	Reasons for deduction	Deduction	Scoring
5	Inspection of tools	(1) The staff shall place the tools on the moisture-proof canvas as required; the moisture-proof canvas shall be clean and dry. (2) The tools shall be placed in category according to the requirements of the fixed management; the insulated tools shall not be mixed with metal tools and materials; and the appearance inspection shall be done on the tools. (3) The surface of insulating tools shall not be worn, deformed and damaged, and the operation shall be flexible. Carry out segment insulation detection with such insulated tools with insulation resistance meter of 2,500V or above and with the resistance no less than 700MΩ, and wipe it off with a clean dry towel. (4) The ground potential personnel on tower shall correctly wear a whole suit of qualified shield clothes and conductive shoes as required, with each part connected well, shall not wear chemical fiber clothes next to the skin in the shielding clothes and shall fasten safety belts; the Responsible Person shall carefully check whether they wears it correctly. (5) Tower climbing personnel shall check the line name, pole number and phase again and report them	7	(1) Deduct 1 point for failure to use moisture-proof tarpaulin and place tools to designed positions. (2) Deduct 0.5 points/item for failure to check qualified label of tool test and appearance inspection. (3) Deduct 1 point/item for failure to use testing instrument for testing the tools. (4) Deduct 2 points/person/time for the operator failing to wear the shielding clothes correctly and each part connected well. (5) Deduct 1 point for the on-site Responsible Person failing to check the safety protective equipment of the tower climbing operators. (6) Deduct 2 points/person for the tower climbing personnel failing to check the line name, pole number and phase. (7) Deduct 2 points/person for the tower climbing personnel failing to report the check results			

Table (Cont'd)

S/N	Project name	Quality requirements	Score	Deduction standard	Reasons for deduction	Deduction	Scoring
6	Climbing the tower	(1) The ground potential electrician on tower shall wear a whole suit of qualified shielding clothes, fasten the safety belt after performing the impulse test on the safety belt, and carry the insulated transmission rope to climb the tower one after another. (2) During the tower climbing process, they shall fasten the anti-fall protection device, climb the tower to an appropriate position, fasten the safety belt, arrange the insulated transmission rope, and then cooperate with the ground electrician to make lifting preparation of the insulated transmission rope separately. (3) During tower climbing, the electrician shall fasten the anti-fall protection device, climb the tower at a uniform speed, grasp the main material by hand, hang the safety belt on the shoulder and keep the safety distance of more than 6.8m away from the electrified body, and the Responsible Person shall strengthen the operation monitoring.	5	(1) Deduct 2 points respectively for failure to fasten the safety belt or to perform the impulse test on the safety belt and the backup protection rope. (2) Deduct 2 points for grasping the shackles with hands. (3) Deduct 1 point for inconvenient suspension position of tackle transmission rope for taking tools. (4) Deduct 2 points for metal tools that are difficult to ensure safe distance during transfer; deduct 2 points for tools that are not bound securely. (5) Deduct 2 points for falling object at heights. (6) Deduct 2 points for tools colliding with the tower body during the transfer process. (7) Deduct 1 point for knotting and disordered rope in tool transfer. (8) Deduct 2 points for the Responsible Person failing to monitor the operation in place. (9) Deduct 2 points for incorrect operation of electrician on tower			

Table (Cont'd)

S/N	Project name	Quality requirements	Score	Deduction standard	Reasons for deduction	Deduction	Scoring
7	Installation tool	(1) The ground electrician hoists the closed clamp (rear clamp), hydraulic leading screw, strain end clamp and so on to the ground electrician by using the insulated transmission rope. The lifting process shall be smooth, free from collision and winding, and the knot shall be used correctly. (2) The ground potential electrician first installs the strain end clamp on the towing plate and then the closed clamp (rear clamp) on the fourth insulator on the cross arm side, and connects the hydraulic leading screw. Each part of the load-bearing tools is securely and reliably installed. (3) Equipotential electricians shall check and confirm that all parts of the load-bearing tools are installed in good condition. With the permission of the Responsible Person, they can operate the hydraulic leading screw so that it is gradually stressed and the insulators to be replaced are relaxed. The two hydraulic leading screws shall be uniformly stressed.	20	(1) Deduct 1 point/time for unstable lifting process, colliding and winding. (2) Deduct 2 points/time for falling objects from heights. (3) Deduct 2 points for improper installation and fixing of clamp. (4) Deduct 3 points for failure to check the installation of load-bearing tools; deduct 1 point for failure to report after checking; deduct 1 point for starting tightening the leading screw without the consent of Responsible Person after reporting. (5) Deduct 3 points/time for the number of shorted insulators exceeding 4 during the operation. (6) Deduct 2 points for collision and breakage of insulators in the clamp installation. (7) Deduct 2 points for failure to tighten leading screw in a balanced way			

Part I   Standard for Training and Assessment on Live Working for Operation Maintenance of ±800kV UHV Power Transmission Line

Table (Cont'd)

S/N	Project name	Quality requirements	Score	Deduction standard	Reasons for deduction	Deduction	Scoring
8	Replacement of insulator	(1) The ground potential electrician shall perform impulse test, check and confirm that the load-bearing tools are stressed normally. With the permission of the Responsible Person, they shall tie the old insulator with the insulated transmission rope, take out the fitting pins at both ends of the old insulator, continue to operate and tighten the hydraulic leading screw until the old insulator is removed. The two hydraulic leading screws shall be stressed uniformly, and the operating handle shall not knock on the insulator. (2) The ground electrician uses the other end of the insulated transmission rope to fasten the new insulator, and lifts the new insulator to the ground potential electrician by means of the old lower and the new upper. The lifting process shall be smooth, free from collision and winding, and the knot shall be used correctly. (3) The ground potential electrician installs the new insulator, resets the fitting pin at its two ends	20	(1) Deduct 3 points for failure to check the stress of load-bearing tools; deduct 2 points for failure to report after checking; deduct 1 point for removing fitting pins at both ends of the old insulator without the consent of Responsible Person after reporting. (2) Deduct 2 points for failure to tighten leading screw in a balanced way. (3) Deduct 1 point/time for the operating handle hitting the insulator. (4) Deduct 1 point for wrong knot. (5) Deduct 2 points/time for falling objects from heights. (6) Deduct 1 point for new and old insulators colliding with each other. (7) Deduct 1 point for the transmission insulator and tower body colliding with each other. (8) Deduct 2 points for winding insulating ropes			

Table (Cont'd)

S/N	Project name	Quality requirements	Score	Deduction standard	Reasons for deduction	Deduction	Scoring
9	Remove tools	(1) The ground potential electricians check that the connection of the new insulator is reliable and, with the permission of the Responsible Person, they operate and loosen the hydraulic leading screw so that the replaced insulator is gradually stressed. (2) After load transfer, the equipotential electricians shall perform the impulse test, check and confirm that the new insulator is in good condition. With the permission of the Responsible Person, they can remove the insulated transmission rope tied to the insulator, fasten it to the proper position of the load-bearing tools, remove the closed clamp (rear clamp), hydraulic leading screw, strain end clamp and other load-bearing tools, and transfer them to the ground with the co-operation of the ground electrician. The transfer process shall be smooth, free from collision and winding, and the knot shall be used correctly.	20	(1) Deduct 3 points for failure to check the connection of new insulator; deduct 2 points for failure to report after checking; deduct 1 point for starting loosening the hydraulic leading screw without the consent of Responsible Person after reporting. (2) Deduct 3 points for failure to check the stress of new insulator; deduct 2 points for failure to report after checking; deduct 1 point for starting removing the hydraulic leading screw without the consent of Responsible Person after reporting. (3) Deduct 1 point for incorrect use of knot when binding the tools. (4) Deduct 2 points/time for falling objects from heights. (5) Deduct 1 point/time for the tools colliding with each other; deduct 1 point/time for the tool colliding with the electrified body or tower body; deduct 2 points for winding insulating ropes			

Part I Standard for Training and Assessment on Live Working for Operation Maintenance of ±800kV UHV Power Transmission Line

Table (Cont'd)

S/N	Project name	Quality requirements	Score	Deduction standard	Reasons for deduction	Deduction	Scoring
10	Return to the ground	The ground potential electricians shall remove the insulated transmission rope after checking that there is no object left on the tower. After being approved by the responsible person, the falling protector shall be properly hung and the safety belt shall be unfastened and put straight, the electrician shall carry the insulated transmission rope correctly and step down to the ground from the tower in uniform steps by treading on shackles and grasping the main materials	5	(1) Deduct 5 points for failure to use the falling protector when climbing down the tower. (2) Deduct 5 points for loosing the protection of the safety belt when moving on the tower. (3) Deduct 1 point/time for hands grasping the shackles when climbing down the tower. (4) Deduct 2 points for any objects left on the tower			
11	End of the work	(1) The Responsible Person shall organize all working personnel to put working apparatus and materials in order and put them in a special kit (bag) after cleaning; clean the site to ensure that "the materials are removed and the site is cleaned after construction". (2) After a post-shift meeting is held, the Responsible Person shall make work summaries and comments. Comments include the construction quality of this work and the implementation of safety measures from all working personnel. (3) Responsible Person shall report to the on-duty control personnel the end of the work, apply for the restoration of line restarting and terminate the work order.	10	(1) Deduct 2 points for failure to clean the tools. (2) Deduct 2 points for missing tools. (3) Deduct 10 points for failure to hold the post-shift meeting. (4) Deduct 2 points for failure to remove the fence. (5) Deduct 2 points for failure to report to dispatcher			
	Total		100				

Module 3 Standards for Training and Assessment on Live Replacement of 1-3 Glass Insulators on the Conductor Side of ±800kV UHV Transmission Line Resisting-tensile Tower

I. Training Standards

(I) Training Requirements

Designation of module	Live replacement of 1-3 glass insulators on the conductor side of ±800kV UHV transmission line resisting-tensile tower	Type of training	Operation
Training method	Practical operation training	Hours of training	21 training hours
Training objectives	1. Master the "two-span and three-short-circuit" operation mode in entering and exiting ±800kV intense electric field along strain insulator string and electrical significance of potential transfer. 2. Be able to complete the entry of ±800kV equipotential operation point along the strain insulator string. 3. Can independently complete the live replacement of 1-3 glass insulators on the conductor side of ±800kV UHV transmission line resisting-tensile tower (equipotential operation method)		
Training venue	UHV DC practical training line		
Training content	The "two-span and three-short-circuit" operation mode is adopted to enter the equipotential along the tensile glass insulator string, and the equipotential operation method is adopted for live replacement of 1-3 glass insulators on the conductor side of ±800kV UHV transmission line resisting-tensile tower		
Scope of application	Maintenance personnel for UHV DC Transmission Line		

(II) Referenced procedures and specifications

(1) Technical Specification for Live Working of ±800kV DC Line (DL/T1242-2013)

(2) Operating Code for ±800kV DC Overhead Transmission Line (GB/T28813-2012)

(3) Maintenance Specification for ± 800kV DC Overhead Transmission Line (DL/T251-2012)

(4) Technical Guide for Live Working of ± 800kV DC Transmission Line (Q/GDW302-2009)

(5) Calculation Method of Live Working Minimum Approach Distance on AC Transmission Line (GB/T 19185-2008)

(6) Guidelines of Insulation Coordination for Live Working (DL/T867-2004)

(7) Test Guide of the Insulating Tool for Live Working (DL/T878—2004)

(8) State Grid Corporation of China on the Management Regulations of Live Working (Trial Implementation) (SGCC [2007] No.751)

(9) State Grid Corporation of China Working Regulations of Power Safety (Transmission Line Section) (Q/GDW1799.2-2013)

(10) Electrotechnical Terminology - Overhead Line (GB/T 2900.51-1998)

(11) Electrotechnical Terminology - Live Working (GB/T2900.55-2016)

(12) Live Working - Terminology for Tools, Equipment and Devices (GB/T 14286-2008)

(13) Minimum Requirements for Utilization of Tools, Devices and Equipment for Live Working (DL/T 877-2004)

(14) Preventive Test Code of Tools, Devices and Equipment for Live Working (DL/T 976-2005)

(15) Insulated Tackles for Live Working (GB/T13034-2008)

(16) Live Working-Insulating Ropes (GB 13035-2008)

(17) Shielding Clothes for Live Working (GB/T6568-2008)

(18) Screen Clothes for Live Working on 1000kV AC (GB/T25726-2010)

(III) Teaching Design for Training

To complete the work task of "live replacement of 1-3 glass insulators on the conductor side of ±800kV UHV transmission line resisting-tensile tower", each training stage shall be designed according to the standard operation procedure for work task completion. Each stage includes specific training objectives, training content, hours of training, training methods (training resources), training environment, assessment and evaluation, etc, as shown in the Table 1-3-1.

Table 1-3-1 Live Replacement of 1-3 Glass Insulators on the Conductor Side of ±800kV UHV Transmission Line Resisting-tensile Tower

Training schedule	Training objectives	Training content	Hours of training	Training methods and resources	Preparation of training conditions	Assessment and evaluation
1. Theoretical teaching	1. Preliminarily master the basic method for entering and exiting ±800kV intense electric field along the insulator string. 2. Be familiar with the methods for potential transfer. 3. Be familiar with the replacement method of strain single insulator of transmission line	1. The electrical significance of the "two-span and three-short-circuit" operation mode when entering and leaving the intense electric field along the insulator. 2. The use methods for potential transfer rod. 3. Method and quality standard for replacement of strain single insulator of transmission line	2	Training method: Lecture. Training resources: PPT, relevant regulations and specifications	Multimedia classroom	Attendance, classroom questions and assignments
2. Preparations	Be able to complete the preparations before operation	1. Work site survey. 2. Preparation of the standardized operation card. 3. Filling of the work order. 4. Preparation of tools and materials for this operation	1	Training methods: 1. Site survey and cleaning of tools and materials shall be practiced at site. 2. Preparation of operation card and the filling of work order shall adopt lecture method. Training resources: 1. ±800kV practical training line. 2. UHV tools warehouse. 3. Blank work order	1. UHV transmission line for practical training. 2. Multimedia classroom	
3. Work site preparation	Be able to complete the preparations of work site	1. Work site re-survey. 2. Work application. 3. Work site layout. 4. Pre-shift meeting. 5. Tools and materials inspection	1	Training methods: Demonstration and role play. Resources: ±800kV practical training line	±800kV practical training line	

· 306 ·

Part I Standard for Training and Assessment on Live Working for Operation Maintenance of ±800kV UHV Power Transmission Line

Table (Cont'd)

Training schedule	Training objectives	Training content	Hours of training	Training methods and resources	Preparation of training conditions	Assessment and evaluation
4. Trainer's demonstration	The trainees can preliminarily understand the operation process of the task through inspecting and learning from each other's work	1. The equipotential electrician enters and exits the intense electric field along the strain insulator string. 2. The equipotential electrician assembles tools. 3. The equipotential electrician completes the replacement of single glass insulator	2	Training method: Demonstration. Resources: ±800kV practical training line	±800kV practical training line	
5. Group training of trainees	1. Be able to complete the operation of entering and exiting ±800kV intense electric field. 3. Be able to complete the replacement of single glass insulator	1. The trainees are grouped (6 in a group) to train the skill operation of entering and exiting ±800kV intense electric field and replacing insulator. 2. Trainers guide the operation of trainees and conduct safety supervision	14	Training methods: Role play. Resources: ±800kV practical training line	±800kV practical training line	Score the operation of trainees according to the detailed rules for skill assessment and scoring
6. End of the work	1. Enable the trainees to further distinguish the shortcomings of the operation process and facilitate the promotion in the later stage. 2. Train the trainees in the working style of safe and civilized production	1. Clean up the work site. 2. Report to dispatcher. 3. Comment and summarize the work task this time at post-shift meeting	1	Training methods: Lecture and inductive method	±800kV training line	

(IV) Work Flow

1. Work task

The live replacement of 1-3 glass insulators on the conductor side of ±800kV UHV transmission line resisting-tensile tower.

2. Requirements for Weather and Work Site

(1) The live replacement of 1-3 glass insulators on the conductor side of ±800kV UHV transmission line resisting-tensile tower should be carried out in good weather. In case of lightning (hearing thunder or seeing lightning), snow, hail, rain, fog and so on, live working is prohibited. When the wind force is greater than level 5, it is unsuitable for live working; when the relative humidity of the air is greater than 80%, insulating tools with moisture-proof properties shall be used if live working needs to be carried out. When emergency live repair is required in bad weather, relevant personnel shall be organized to fully discuss and prepare necessary safety measures, which can be implemented after being approved by the unit. In case of sudden change of weather during live working, which may endanger the safety of person or equipment, the work shall be stopped immediately; in the case of ensuring personal safety, the normal state of equipment shall be restored as soon as possible, or other measures shall be taken.

(2) The operating personnel should be in good mental states, be familiar with the organizational and technical measures to ensure safety in work and master the methods for emergency rescue and first aid for electric shock in heights; they should hold the qualification certificate for live working within the validity period.

(3) Responsible Person should organize the relevant personnel to complete field investigation in advance, determine the operating methods, required working apparatus and necessary measures according to the results, and handle the tickets for live working.

(4) The work site should be reasonably set up with fence and warning signs. Non operating personnel is forbidden to enter.

(5) DC restart protection device shall be deactivated for the Project.

(6) Mode of operation: Equipotential operation.

(7) Safe working distance and effective insulation length during operation are shown in Table 1-3-2.

Table 1-3-2 Safe Distance for Live Replacement of 1-3 Glass Insulators on the Conductor Side of ± 800kV UHV Transmission Line Resisting-tensile Tower (m)

Altitude	Minimum safe distance between equipotential electrician and grounding frame	Minimum effective insulation length of insulating tools	Minimum combination gap
$H \leqslant 1000$	6.8	6.8	6.7
$1000 < H \leqslant 2000$	7.3	7.3	7.3
$2000 < H \leqslant 2500$	7.9	7.8	7.8

Note: The minimum safe distance and minimum combined gap in the table include 0.5m occupying gap of human body.

(8) When the equipotential electrician enters the equipotential along the strain insulator string, the number of insulators shorted by the human body shall not be more than 4. The minimum number of good insulators shall meet the requirements of Table 1-4-3 after deducting the number of insulators shorted by human body and defective insulators from the strain insulator string.

Table 1-3-3 Minimum Combined Gap and Minimum Number of Good Insulators

Altitude (m)	Structural height of single glass insulator (mm)	Minimum total length of good insulator string (m)	Minimum number of good insulators
$H \leqslant 1000$	170	6.2	37
	195		32
	205		31
	240		26
$1000 < H \leqslant 2000$	170	7.1	42
	195		37
	205		35
	240		30
$2000 < H \leqslant 2500$	170	7.55	45
	195		39
	205		37
	240		32

Note: The values in the table do not include the human body occupying gap which shall not be less than 0.5 m during operation.

3. Preparations

3.1 Hazards and precontrol Measures

(1) Hazard —— electric shock

Precontrol measures:

① Prior to work, the Responsible Person shall contact the on-duty control personnel to disable the DC restart protection device, and perform the licensing procedures. It is strictly prohibited to disable or resume the DC restart protection device at about the time. It shall be reported to the dispatcher in time after the end of the work.

② Before climbing the tower, operators on the tower must carefully check the dual names of the line and the number of the pole and tower, and then the tower can be climbed after all have been confirmed correct.

③ If lines lose power suddenly during work, operators shall consider it as still charged. The Responsible Person shall contact the control personnel as soon as possible, and no forced energization is allowed before the on-duty control personnel getting in touch with the Responsible Person.

④ Insulated tool and insulated ropes shall be free of damage, moisture, deformation, and failure, with the effective length no less than that specified in Table1-3-2 4. It is not allowed to use non-insulated ropes (such as cotton rope, manila rope, and steel wire rope).

⑤ The ground electrician shall wear clean and dry gloves when operating the insulating tools. When entering the work site, the live working tools shall be placed on damp-proof canvas or insulating mat to prevent dirt and dampness of the insulating tools in use.

⑥ The equipotential electrician shall wear flame-retardant underwear and full set of qualified shielding clothes outside (including hat, dresses & trousers, gloves, socks and shoes). All parts shall be in excellent connection conditions.

⑦ When the equipotential electrician moves along the insulator string, the positions of the hands and feet shall be correspondingly consistent, and the number of intact insulators shorted by the human body and the tools shall conform to the provisions of Table 1-3-3.

⑧ The equipotential electrician enters the intense electric field by means of "two-span and three-short-circuit" operation mode (also called free operation method). When the operator moves in parallel to the three insulators of the grading ring on the conductor side, the movement shall be stopped, and the potential transfer shall be carried out by using the potential transfer rod which shall be 0.4 m long. At the time of potential transfer, the distance between the human face and the charged body shall not be less than 0.5 m.

⑨ When transmitting large metal objects by using insulating ropes, ground operators shall not touch them before ground connection.

⑩ During the live working, the Responsible Person (Supervisor) shall continuously monitor the operators and correct their nonstandard operation or actions in violation at any time. Special attention shall be paid to high-place operators to ensure that there is enough safety distance and combined gap (meeting the requirements in Table 1-3-2). It is forbidden to touch two non-connected electrified bodies or electrified body and grounding body.

(2) Hazard —— falling from high places

Precontrol measures:

① Before climbing, operators work at heights must satisfy the requirements of this operation, such as physical condition, mental state, and skill and quality.

② Double safety belts shall be used by the personnel for Work at Heights. When climbing up and down the tower, the main material shall be grasped with hands, with shackles trodden and moving at a uniform speed.

③ Before the operation, the equipotential electrician shall carefully inspect hydraulic lead screw, closed clamp, wire end clamp and so on to ensure that the load-bearing tools are qualified; when moving along the insulator string, the positions of hands and feet must be correspondingly consistent, and the safety belt shall be fastened to the insulator string supported by hands and moved synchronously; when replacing insulators, the load-bearing tools shall be installed reliably. Before and after load transfer, impulse test shall be conducted to determine their reliability, and it shall be reported to the Responsible Person in a timely manner. The implementation can only be carried out with the permission of the Responsible Person.

④ Supervisor shall correct irregularities and violations at any time. Special attention shall be paid to operators for work at heights to prevent them from losing the protection of the safety belt or insulated backup protection rope during transposition. The safety belt or insulated backup protection rope shall not be fastened lower than operating personnel.

(3) Hazard —— injury caused by objects falling from high place

Precontrol measures:

① Operators working at heights should put personal tools and fragmentary materials into the tools bag. It is strictly forbidden to hang objects in high place or keep in the mouth.

② When the heavy equipment is transferred by the way of up and down circulation exchange, the control rope should be fastened to prevent the transferred articles from colliding with each other and accidentally touching the bearing tools in the working state.

③ Ground operator should correctly wear a helmet and use the knots. The vertical distance from the work site should not be less than the falling radius.

④ The operation site shall be set up with fence and warning signs. It shall be noted at any time that Supervisor shall prohibit irrelevant personnel and vehicles from entering operation area.

3.2 Selection of tools, instruments and materials

The tools and materials required for live replacement of 1-3 glass insulators on the conductor side of ±800kV UHV transmission line resisting-tensile tower are shown in Table 1-3-4. Before delivering the tools and instruments out of the warehouse, application voltage class and test period of the tools and instruments shall be carefully checked and they shall be inspected to ensure that appearance is intact, connection is firm, rotation is flexible and the requirements of the work task are met. After delivering the tools and instruments out of the warehouse, they shall be stored in the tool bag or tool kit for transportation to avoid contamination and damp. Metal tools and insulated tools shall be separately loaded and transported to avoid deformation and damage caused by mixed loading and transportation.

Table 1-3-4 Tools and Materials Required for Live Replacement of 1-3 Glass Insulators on the Conductor Side of ±800kV UHV Transmission Line Resisting-tensile Tower

S/N	Name	Specification and Model	Unit	Qty.	Remarks
1	Shielding clothes	Shielding efficiency ≥ 60dB (shield efficiency of shielding mask ≥ 20dB)	Set	2	PPE
2	Conductive shoes	The size depends on the wearer	Pair	2	PPE
3	Flame retardant underwear	Pure mulberry silk	Set	2	PPE
4	Safety belt of double insurance	Suspender	Nos.	2	PPE
5	Safety helmet		pcs	6	PPE
6	Goggles		Nos.	2	PPE
7	Insulated transmission rope	TJS-12	Nos.	1	Insulating tool
8	Insulated backup protection rope	ϕ16mm	Nos.	2	Insulating tool
9	Insulated noose	ϕ14mm	Nos.	1	Insulating tool

Table (Cont'd)

S/N	Name	Specification and Model	Unit	Qty.	Remarks
10	Insulated tackle	JH10-1B	Nos.	1	Insulating tool
11	Conductor end clamp		Nos.	1	Metal tool
12	Hydraulic leading screw	8T	Nos.	2	Metal tool
13	Closed clamp (front clamp)	Tc4	Nos.	1	Metal tool
14	Potential transfer rod		Nos.	2	Other tools
15	Pin puller		pcs	1	Other tools
16	Insulation resistance tester	5000V, with electrode width 2cm and interelectrode width 2cm	Set	1	Other tools
17	Multimeter		Set	1	Other tools
18	Wind speed, temperature and humidity tester		pcs	1	Other tools
19	Security fence		Set	Several	Other tools
20	Warning sign	"Work Here", "Access from Here", "Slow Down", "Blocking"	Set	1	Other tools
21	Red waistcoat	"Responsible Person" "Special Supervisor"	pcs	1	Other tools
22	Moisture-proof tarpaulin	3m×3m	pcs	2	Other tools
23	Personal tools	Wrench, vice	Set	1	Other tools
24	Falling protector	Corresponding to the model of pole and tower anti-fall device. Division of labor for operators of the task is shown in Table 1-3-5.	pcs	2	Other tools
25	Towel	Cotton	pcs	1	Other tools
26	Insulator		pcs	1	Material

3.3 Division of labor for operators

Division of labor for operators of the task is shown in Table 1-3-5.

Table 1-3-5 Personnel Allocation for Live Replacement of 1-3 Glass Insulators on the Conductor Side of ±800kV UHV Transmission Line Resisting-tensile Tower

S/N	Post	Qty. (person)	Responsibilities
1	Responsible Person	1	Be responsible for division of operating personnel of the task, work order reading, handling re-closing deactivation of the line and the formalities of work permit, getting work permits, holding pre-shift meeting, dealing with emergency situations in work, quality surveillance, and the summary after work
2	Special Supervisor	1	Be responsible for the safety control of the work site
3	Equipotential electrician	2	Be responsible for tool installation and insulator replacement
4	Ground electrician	2	Be responsible for ground auxiliary works during the operation.

4. Working Procedures

The workflow of this task is shown in Table 1-3-6.

Table 1-3-6 Workflow for Live Replacement of 1-3 Glass Insulators on the Conductor Side of ±800kV UHV Transmission Line Resisting-tensile Tower

S/N	Work Content	Operation Steps and Standards	Safety Measures and Precautions	Responsible Person
1	Site re-survey	The Responsible Person shall complete the following work: (1) Check the line name, tower number and dualism number to ensure they are correct; guarantee that the base and power are intact and in normal operation, and that the crossed crossing distance meets the safety requirements; confirm the defects conditions and specifications and models of wires. (2) Check that the site meteorological conditions such as wind speed and humidity should meet the operation requirements. (3) Check that the terrain and environment should meet the operation requirements. (4) Check that the safety measures listed in the work order are in line with the actual situations on site, and the measures will be supplemented if necessary.	(1) Correctly wear helmet, working clothes, work shoes and protective gloves. (2) Operation under meteorological conditions that may endanger the safety of operators is forbidden. (3) Non-operation personnel and vehicles are strictly prohibited from entering the working site.	

S/N	Work Content	Operation Steps and Standards	Safety Measures and Precautions	Responsible Person
2	Work Permit	(1) The Responsible Person is responsible for contacting the on-duty control personnel and applying for stopping the DC restart protection device as per the contents of the work order. (2) Live working could be started only after being approved by the on-duty control personnel.	Live operation shall not be started without the permission of the on-duty control personnel.	
3	Site layout	Install the security fence and hang the signboards correctly: (1) The security fence should take full account of falling objects from the high place and the influence on road traffic. (2) The entrance and exit of the security fence shall be set reasonably. (3) Signs such as "Access from Here", "Work Here", "Slow Down" or "Blocking" shall be properly arranged.	When the influence on road traffic safety is uncontrollable, the traffic management department should be contacted in time to strengthen the on-site control of traffic safety.	
4	Holding a pre-shift meeting	(1) All working personnel shall line up. (2) Responsible Person will read out the work order and be clear with work task and division of personnel; explain safety measures and technical measures in work; check (inquire after) mental state of all working personnel; inform of hazards in work and precontrol measures. (3) All working personnel shall sign on the work order for confirmation.	(1) Work order shall be filled and issued with standardized licensing procedure and complete signature. (2) All working personnel shall be in good mental states. (3) All working personnel shall be clear with task division of works, safety measures and technical measures.	

Table (Cont'd)

S/N	Work Content	Operation Steps and Standards	Safety Measures and Precautions	Responsible Person
5	Inspection of tools and instruments	(1) All necessary tools and instruments shall be prepared as per the operation requirements and placed on the waterproof tarpaulin regularly according to the category and location. The appearance and test certificate of tools and instruments shall be checked to ensure there is no omission. (2) The surface insulation resistance of insulating tools and insulating ropes shall be tested with insulation resistance tester in correct methods, and the value shall be not less than 700MΩ. (3) The new insulator is wiped up, and it is intact in appearance inspection, without rust, cracks and breakage. The surface insulation resistance shall be tested with insulation resistance tester in correct methods, and the value shall be not less than 700MΩ. (4) The internal resistance of full shielding clothes shall be tested with a multimeter in correct methods, and the value shall be not more than 20Ω. (5) The inspector shall report to the Responsible Person that all inspection results are in conformity with the operation requirements.	(1) The waterproof tarpaulin shall be enough in quantity and reasonable in position, and be clean and dry. (2) Before using metal and insulating tools, they shall be carefully checked for damage, dampness, deformation and failure, and the certificate of conformity is within the validity period. (3) Insulated tools and insulated ropes shall be tested as acceptable. (4) The shielding clothes are in good appearance and pass the test.	
6	Climbing the tower	(1) No.1 and No.2 equipotential electricians shall check the double name and phase of the line again, check and confirm that the shackles are complete and firm; fasten safety belts and attach falling protectors; perform impact inspection on safety belt, backup protection rope and falling protector in correct method; the Responsible Person shall check and confirm that the connection conditions of each connection point of the double safety belt and shielding clothes worn by the equipotential electricians are in good condition, with the appropriate size and the flexible locking and buckling.	(1) Safety belt, backup protection ropes and falling protector shall pass the impact inspection. (2) The safety belt and insulated transmission rope shall be prevented from hooking up tower materials. (3) The minimum safe distance between the human body and the conductor shall conform to the regulations shown in Table 1-3-2. (4) It is forbidden to grasp shackles with hands.	

Part I Standard for Training and Assessment on Live Working for Operation Maintenance of ±800kV UHV Power Transmission Line

Table (Cont'd)

S/N	Work Content	Operation Steps and Standards	Safety Measures and Precautions	Responsible Person
6	Climbing the tower	(2) The electrician shall carry toolkit and insulated transmission rope (including insulated tackle) in correct methods. (3) Main band of safety belt and backup protection rope shall obliquely across the shoulder. (4) The soles shall be cleaned, and climbing the tower shall be done with the permission of the Responsible Person. (5) The electrician shall tread on shackles, grasp the main materials, and climb the tower uniformly to the proper position of the cross arm with safety belt well fastened and then break away from the falling protector. The ground electricians shall control the tail rope to avoid stranding the transmission rope with the shackles and tower materials.	(5) Falling protector shall be properly used. (6) Protection of safety belt shall not be lost during transposition.	
7	Entering the intense electric field	(1) The No. 1 equipotential electrician carries the insulated transmission rope, transposes it to the hanging point of the strain insulator string, hangs the main safety belt on the insulator string connection fittings, and leaves the backup protection rope at the proper position of the cross arm. (2) The good connection of all parts of the shielding clothes and the insulator string and the location of the faulty insulators shall be re-checked and confirmed. With the permission of the Responsible Person, the electrician shall grasp one string and tread the other one, and enter the intense electric field along the insulator string in the operation mode of "two-span and three-short-circuit", with the shorted insulators no less than 4 pieces. When reaching the three pieces of insulators outside the grading ring on the conductor side, the operator shall stop the movement, and hook the connection fittings at the conductor end with the potential transfer rod for potential transfer, so as to realize the equipotential between the operator and the conductor. (3) No.2 equipotential electrician enters the electric field in the same way.	(1) The safety belt and insulated transmission rope shall be prevented from hooking up tower materials. (2) Protection of safety belt shall not be lost during transposition. (3) The safe distance between the human body and the grounding body, and the combined gap between the human body and the grounding body and the electrified body shall not be less than that specified in Table 1-3-2. (4) Prior to potential transfer, permission shall be obtained from the Responsible Person. The potential transfer rod shall be electrically connected with the shielding clothes, with smooth, accurate and fast actions. (5) When the potential is transferred, the distance between the human face and the electrified body shall not be less than 0.5m.	

Table (Cont'd)

S/N	Work Content	Operation Steps and Standards	Safety Measures and Precautions	Responsible Person
8	Installation of Load-bearing Tools	(1) After entering the electric field, No. 1 and No. 2 equipotential electricians shall fasten their safety belts and arrange insulated transmission ropes firmly and reliably at appropriate positions, facilitating operation. The ground electrician lifts the closed clamp (front clamp), hydraulic leading screw, conductor end clamp and so on to the equipotential electrician by using the insulated transmission rope. The lifting process shall be smooth, free from collision and winding, and the knot shall be used correctly. (2) No.1 and No.2 equipotential electricians shall cooperate with each other to install the conductor end clamp at the appropriate position on the conductor side, install the closed clamp (front clamp) on the third insulator on the conductor side, and connect the hydraulic leading screw. Each part of the load-bearing tools is securely and reliably installed. (3) Equipotential electricians shall check and confirm that all parts of the load-bearing tools are installed in good condition. With the permission of the Responsible Person, they can operate the hydraulic leading screw so that it is gradually stressed and the insulators to be replaced are relaxed. The two hydraulic leading screws shall be uniformly stressed.	(1) The safe distance between the human body and the grounding body, and the effective insulation length of the insulated transmission rope shall not be less than that specified in Table 1-3-2. (2) The upper and lower conveying objects shall be securely tied with ropes for conveying, so as to prevent objects falling from heights. (3) After deducting inferior insulators, insulators shorted by human body and tools, the number of good insulators shall conform to that specified in Table 1-3-3.	
9	Replacement of insulator	(1) No.1 and No.2 equipotential electricians shall perform impulse test, check and confirm that the load-bearing tools are stressed normally. With the permission of the Responsible Person, they shall tie the old insulator with the insulated transmission rope, take out the fitting pins at both ends of the old insulator, continue to operate and tighten the hydraulic leading screw until the old insulator is removed. The two hydraulic leading screws shall be stressed uniformly, and the operating handle shall not knock on the insulator.	(1) The safe distance between the human body and the grounding body shall not be less than that specified in Table 1-3-2. (2) The upper and lower transferred tools shall not collide with each other. The binding rope shall be correct and reliable, so as to prevent objects falling from heights.	

Table (Cont'd)

S/N	Work Content	Operation Steps and Standards	Safety Measures and Precautions	Responsible Person
9	Replacement of insulator	(2) The ground electrician uses the other end of the insulated transmission rope to fasten the new insulator, and transfers the new insulator to the intermediate potential electrician by means of the old lower and the new upper. The lifting process shall be smooth, free from collision and winding, and the knot shall be used correctly. (3) They shall install new insulator and reset fitting pins at both ends.		
10	Remove tools	(1) The equipotential electricians check that the connection of the new insulator is reliable and, with the permission of the Responsible Person, they operate and loosen the hydraulic leading screw so that the replaced insulator is gradually stressed. (2) After load transfer, the equipotential electricians shall perform the impulse test, check and confirm that the new insulator is in good condition. With the permission of the Responsible Person, they can remove the insulated transmission rope tied to the insulator, fasten it to the proper position of the load-bearing tools, remove the hydraulic leading screw, closed clamp, conductor end clamp and other load-bearing tools, and transfer them to the ground with the cooperation of the ground electrician. The transfer process shall be smooth, free from collision and winding, and the knot shall be used correctly.	(1) The safe distance between the human body and the grounding body shall not be less than that specified in Table 1-3-2. (2) The upper and lower transferred tools shall not collide with each other. The binding rope shall be correct and reliable, so as to prevent objects falling from heights.	
11	Exit the intense electric field	(1) The equipotential electrician shall remove and put the insulated transmission rope in order after checking that there is no object left in the operation area. (2) The main safety belt shall be transferred and fastened at an appropriate position of the insulator string, returned to the insulator string through the grading ring, hooked at the appropriate position of the grading ring by the potential transfer rod, and stopped when the main belt is moved along the insulator string toward the later	(1) The combined gap between the human body and the grounding body and the charged body shall not be less than that specified in Table 1-3-2. (2) Protection of safety belt shall not be lost during transposition.	

S/N	Work Content	Operation Steps and Standards	Safety Measures and Precautions	Responsible Person
11	Exit the intense electric field	al side of the cross arm to three pieces of insulators outside the grading ring; the equipotential electrician shall grasps the insulator with one hand and the potential transfer rod with the other, so as to quickly disengage from the equipotential by using the potential transfer rod. (3) They can reach the cross arm along the insulator string in the "two-span and three-short-circuit" operation mode.		
12	Evacuation from the pole and tower	The equipotential electrician shall check that there is no object left on the tower. After being approved by the Responsible Person, the falling protector shall be properly hung and the safety belt shall be unfastened and put straight, the electrician shall carry the insulated transmission rope correctly and step down to the ground from the tower in uniform steps by treading on shackles and grasping the main materials	(1) Protection of safety belt shall not be lost during transposition. (2) Hand slippage and step missing shall be avoided and it's forbidden to grasp the shackles with hands. (3) Falling protector shall be properly used. (4) The insulated transmission rope and safety belt shall be prevented from hooking up the tower materials or shackles	
13	End of the work	(1) The Responsible Person shall organize all working personnel to put working apparatus and materials in order and put them in a special kit (bag) after cleaning; clean the site to ensure that "the materials are removed and the site is cleaned after construction". (2) After a post-shift meeting is held, the Responsible Person shall make work summaries and comments. Comments include the construction quality of this work and the implementation of safety measures from all working personnel. (3) Responsible Person shall report to the on-duty control personnel the end of the work and terminate the work order.		

II. Assessment Standard

Table 1-3-7 Detailed Rules for Assessment and Scoring of Operation and Inspection Skills of UHVDC Transmission Line of State Grid Sichuan Electric Power Corporation

Fill-in Column of Examinee	No.: Name: Position: Unit: Date: MM/DD/YYYY						
Fill-in Column of Assessor	Grade: Assessor: Assessment Team Leader: Starting time: Closing time: Operation Duration:						
Assessment Module	The live replacement of 1-3 glass insulators on the conductor side of ±800kV UHV transmission line resisting-tensile tower	Assessee	Maintenance Personnel of UHV DC Transmission Line	Assessment method	Operation	Assessment Time Limit	90min
Job Description	The live replacement of 1-3 glass insulators on the conductor side of ±800kV transmission line resisting-tensile tower						
Work specifications and requirements	1.Live working shall be carried out in good weather. In case of thunder, rain, snow or fog, no live working shall be carried out. When the wind force is greater than Level 5, live working should not be carried out. When the humidity is greater than 80%, insulating tools with moisture-proof properties shall be used if live working needs to be carried out. 2.Six persons are required for this operation, including 1 Responsible Person, 1 special Supervisor, 2 equipotential electricians and 2 ground electricians. 3.Responsibilities of Responsible Person (Supervisor): Be responsible for division of operating personnel of the task, work order reading, handling re-closing deactivation of the line and the formalities of work permit, getting work permits, holding pre-shift meeting, safety supervision in the operation process, dealing with emergency situations in work, quality surveillance, and the summary after work						

	Table (Cont'd)
Work specifications and requirements	4. During the live working, if thunder, rain, strong wind or any other circumstance threaten the safety of the staff, the Responsible Person or Supervisor may stop working temporarily according to the circumstances Given conditions: 1. Work orders have been handled, safety measures have been completed (re-closing has been deactivated), and oral application (dispatcher or assessor) shall be made at the beginning and end of the work. 2. The instrument shall be used safely and correctly to test the insulating tool. 3. The operation must be carried out according to the working procedures. The scores of the items to be carried out shall be deducted for the process error. In case of major hidden dangers of personal, equipment and operational safety, the assessor may order the termination of the operation (assessment)
Assessment scenario preparation	1. Tower shape: ±800kV resisting-tensile tower. 2. Required operation tools: 2 safety belts (including double-protective ropes), 2 sets of shielding clothes, 2 sheets of moisture-proof cloth, 1 multimeter, 1 insulation resistance detector, 1 anemometer, 1 2-in-one temperature and humidity detector, 2 hydraulic leading screws, 1 insulating rope, 1 insulating rope socket, 1 conductor end clamp, 1 closed clamp (front clamp), 1 tackle, 2 potential transfer rods, 1 goggles, 1 pin puller and 1 set of hand-operated tools. 3. The work site shall be monitored, and the safety measures (fence, etc.) on the work site have been fully implemented; non-operation personnel are prohibited from entering the site, and the staff must wear safety helmets when entering the work site. 4. Examinees shall bring their own work clothes, helmets and gloves
Remarks	1. The deduction shall be done until the scores of each item are deducted completely. In case of major hidden dangers of personal, equipment and operational safety, the assessor may order the termination of the operation. 2. When equipment, working environment, safety belt, safety helmet, tool, shielding clothes, etc., do not conform to the operation condition, the assessor may order the termination of the operation

Table 1-3-8 Standards for Assessment and Scoring of Operation and Inspection Skills of UHVDC Transmission Line of State Grid Sichuan Electric Power Corporation

S/N	Project name	Quality requirements	Score	Deduction standard	Reasons for deduction	Deduction	Scoring
1	Site re-survey	(1) The Responsible Person shall go to the work site to check the line name, pole and tower number, on-site working conditions, defective parts and so on. (2) Check that the site meteorological conditions such as wind speed and humidity should meet the operation requirements. (3) Check whether the work order is complete and unmodified, check whether the safety measures listed are consistent with the actual situation on site, and supplement it if necessary	5	(1) Deduct 1 point/item for failure to check the line name, pole and tower number, on-site working conditions and defective parts. (2) Deduct 1 point/item for failure to detect wind speed, humidity and other on-site meteorological conditions. (3) Deduct 0.5 points/part for any alteration in the work order filling; deduct 1 point for incorrect work order number; deduct 1.5 points for incomplete work order filling			
2	Work Permit	(1) The Responsible Person is responsible for contacting the on-duty control personnel (referee) and applying for stopping the line re-closing as per the contents of the work order. (2) Reporting content is standardized and complete	2	(1) Deduct 2 points for failure to contact the dispatching department (referee) for deactivating the re-closing. (2) Deduct 1 point for non-standard or incomplete terminology reporting			

Table (Cont'd)

S/N	Project name	Quality requirements	Score	Deduction standard	Reasons for deduction	Deduction	Scoring
3	Site layout	Install the security fence and hang the signboards correctly: (1) The security fence should take full account of falling objects from the high place and the influence on road traffic. (2) The entrance and exit of the security fence shall be set reasonably. (3) Signs such as "Access from Here", "Work Here", "Access from Here" shall be properly arranged	3	(1) Deduct 1 point for failure to arrange the fence at the work site. (2) Deduct 1 point for failure to arrange the warning board. (3) Deduct 1 point for failure to hang the tower climbing operation sign			
4	Holding a pre-shift meeting	(1) All staff and personnel shall wear safety helmets and work clothes correctly. (2) Responsible Person shall wear red vest and read out the work order and be clear with work task and division of personnel; explain safety measures and technical measures in work; check (inquire after) mental state of all working personnel; inform of hazards in work and precontrol measures. (3) All working personnel shall sign on the work order for confirmation.	3	(1) Deduct 0.5 points/person for the staff not dressing uniformly. (2) Deduct 3 points for no division of labor; deduct 1 point for unclear division of labor. (3) Deduct 1 point for the on-site Responsible Person not wearing a safety monitoring vest. (4) Deduct 1 point for the work shift member failing to sign or signing incompletely on the work order.			

Part I Standard for Training and Assessment on Live Working for Operation Maintenance of ±800kV UHV Power Transmission Line

Table (Cont'd)

S/N	Project name	Quality requirements	Score	Deduction standard	Reasons for deduction	Deduction	Scoring
5	Inspection of tools and instruments	(1) All necessary tools and instruments shall be prepared as per the operation requirements and placed on the moisture-proof tarpaulin regularly according to the category and location. The appearance and test certificate of tools and instruments shall be checked to ensure there is no omission. (2) The surface insulation resistance of insulating tools and insulating ropes shall be tested with insulation resistance tester in correct methods, and the value shall be not less than 700MΩ. (3) The new insulator is wiped up, and it is intact in appearance inspection, without rust, cracks and breakage. The surface insulation resistance shall be tested with insulation resistance tester in correct methods, and the value shall be not less than 700MΩ. (4) The internal resistance of full shielding clothes shall be tested with a multimeter in correct methods, and the value shall be not more than 20Ω. (5) The inspector shall report to the Responsible Person that all inspection results are in conformity with the operation requirements.	7	(1) Deduct 1 point for failure to use moisture-proof tarpaulin and place tools to designed positions. (2) Deduct 0.5 points/item for failure to check the appearance of tools and pass the test certificate. (3) Deduct 1 point/item for failure to use testing instrument for testing the tools. (4) Deduct 1 point for non-standard reporting of test results; deduct 0.5 points/item for incomplete reporting			

Table (Cont'd)

S/N	Project name	Quality requirements	Score	Deduction standard	Reasons for deduction	Deduction	Scoring
6	Climbing the tower	(1) No.1 and No.2 equipotential electricians shall check the double name and phase of the line again, check and confirm that the shackles are complete and firm; fasten safety belts and attach falling protectors; perform impact inspection on safety belt, backup protection rope and falling protector in correct method; the Responsible Person shall check and confirm that the connection conditions of each connection point of the double safety belt and shielding clothes worn by the equipotential electricians are in good condition, with the appropriate size and the flexible locking and buckling. (2) The electrician shall carry toolkit and insulated transmission rope (including insulated tackle) in correct methods. (3) Main band of safety belt and backup protection rope shall obliquely across the shoulder. (4) The soles shall be cleaned, and climbing the tower shall be done with the permission of the Responsible Person. (5) The electrician shall tread on shackles, grasp the main materials, and climb the tower uniformly to the proper position of the cross arm with safety belt well fastened and then break away from the falling protector. The ground electricians shall control the tail rope to avoid stranding the transmission rope with the shackles and tower materials.	5	(1) Deduct 1 point/item for the equipotential electrician failing to check the double name of the line, pole number, phase and tower material; deduct 1 point for non-report after checking. (2) Deduct 2 points/item for failure to perform the impulse test on double safety belts and falling protector. (3) Deduct 1 point for the on-site Responsible Person failing to check the safety protection equipment of the equipotential electrician. (4) Deduct 0.5 points/time for hands grasping the shackles. (5) Deduct 1 point for the unreasonable suspension position of the tackle transmission rope. (6) Deduct 5 points for loosing the protection of the safety belt during transposition			

Part I Standard for Training and Assessment on Live Working for Operation Maintenance of ±800kV UHV Power Transmission Line

Table (Cont'd)

S/N	Project name	Quality requirements	Score	Deduction standard	Reasons for deduction	Deduction	Scoring
7	Entering the intense electric field	(1) The No. 1 equipotential electrician carries the insulated transmission rope, transposes it to the hanging point of the strain insulator string, hangs the main safety belt on the insulator string connection fittings, and leaves the backup protection rope at the proper position of the cross arm. (2) The good connection of all parts of the shielding clothes and the insulator string and the location of the faulty insulators shall be re-checked and confirmed. With the permission of the Responsible Person, the electrician shall grasp one string and tread the other one, and enter the intense electric field along the insulator string in the operation mode of "two-span and three-short-circuit", with the shorted insulators no less than 4 pieces. When reaching the three pieces of insulators outside the grading ring on the conductor side, the operator shall stop the movement, and hook the connection fittings at the conductor end with the potential transfer rod for potential transfer, so as to realize the equipotential between the operator and the conductor. (3) No.2 equipotential electrician enters the electric field in the same way	5	(1) Deduct 2 points for irrational fastening position and irregular use of safety belt back protection rope. (2) Deduct 1 point/item for the equipotential electrician failing to check the connection of the shielding clothes and insulator string and the location of the faulty insulator. (3) Deduct 5 points for entering the intense electric field without permission of the Responsible Person. (4) Deduct 2 points/time for the incorrect action of equipotential electrician entering the equipotential and repeat discharging. (5) Deduct 3 points for the incorrect action of potential transfer; deduct 2 points for failure to report and 1 point for potential transfer without the consent of the Responsible Person after reporting. (6) Deduct 1 point for the unreasonable installation position of the insulated transmission rope. (7) Deduct 2 points/time for falling objects from heights. (8) Deduct 5 points for loosing the protection of the safety belt during transposition			

Table (Cont'd)

S/N	Project name	Quality requirements	Score	Deduction standard	Reasons for deduction	Deduction	Scoring
8	Installation tool	(1) After entering the electric field, No. 1 and No. 2 equipotential electricians shall fasten their safety belts and arrange insulated transmission ropes firmly and reliably at appropriate positions, facilitating operation. The ground electrician lifts the closed clamp (front clamp), hydraulic leading screw, conductor end clamp and so on to the equipotential electrician by using the insulated transmission rope. The lifting process shall be smooth, free from collision and winding, and the knot shall be used correctly. (2) No.1 and No.2 equipotential electricians shall cooperate with each other to install the conductor end clamp at the appropriate position on the conductor side, install the closed clamp (front clamp) on the third insulator on the conductor side, and connect the hydraulic leading screw. Each part of the load-bearing tools is securely and reliably installed. (3) Equipotential electricians shall check and confirm that all parts of the load-bearing tools are installed in good condition. With the permission of the Responsible Person, they can operate the hydraulic leading screw so that it is gradually stressed and the insulators to be replaced are relaxed. The two hydraulic leading screws shall be uniformly stressed.	15	(1) Deduct 1 point/time for unstable lifting process, colliding and winding. (2) Deduct 2 points/time for falling objects from heights. (3) Deduct 2 points for improper installation and fixing of clamp. (4) Deduct 3 points for failure to check the installation of load-bearing tools; deduct 1 point for failure to report after checking; deduct 1 point for starting tightening the leading screw without the consent of Responsible Person after reporting. (5) Deduct 3 points/time for the number of shorted insulators exceeding 4 during the operation. (6) Deduct 2 points for collision and breakage of insulators in the clamp installation. (7) Deduct 2 points for failure to tighten leading screw in a balanced way.			47

Part I Standard for Training and Assessment on Live Working for Operation Maintenance of ±800kV UHV Power Transmission Line

S/N	Project name	Quality requirements	Score	Deduction standard	Reasons for deduction	Deduction	Scoring
9	Replacement of insulator	(1) No.1 and No.2 equipotential electricians shall perform impulse test, check and confirm that the load-bearing tools are stressed normally. With the permission of the Responsible Person, they shall tie the old insulator with the insulated transmission rope, take out the fitting pins at both ends of the old insulator, continue to operate and tighten the hydraulic leading screw until the old insulator is removed. The two hydraulic leading screws shall be stressed uniformly, and the operating handle shall not knock on the insulator. (2) The ground electrician uses the other end of the insulated transmission rope to fasten the new insulator, and transfers the new insulator to the intermediate potential electrician by means of the old lower and the new upper. The lifting process shall be smooth, free from collision and winding, and the knot shall be used correctly. (3) They shall install new insulator and reset fitting pins at both ends.	20	(1) Deduct 3 points for failure to check the stress of load-bearing tools; deduct 2 points for failure to report after checking; deduct 1 point for removing fitting pins at both ends of the old insulator without the consent of Responsible Person after reporting. (2) Deduct 2 points for failure to tighten leading screw in a balanced way. (3) Deduct 1 point/time for the operating handle hitting the insulator. (4) Deduct 1 point for wrong knot. (5) Deduct 2 points/time for falling objects from heights. (6) Deduct 1 point for new and old insulators colliding with each other. (7) Deduct 1 point for the transmission insulator and tower body colliding with each other. (8) Deduct 2 points for winding insulating ropes			

Table (Cont'd)

S/N	Project name	Quality requirements	Score	Deduction standard	Reasons for deduction	Deduction	Scoring
10	Remove tools	(1) The equipotential electricians check that the connection of the new insulator is reliable and, with the permission of the Responsible Person, they operate and loosen the hydraulic leading screw so that the replaced insulator is gradually stressed. (2) After load transfer, the equipotential electricians shall perform the impulse test, check and confirm that the new insulator is in good condition. With the permission of the Responsible Person, they can remove the insulated transmission rope tied to the insulator, fasten it to the proper position of the load-bearing tools, remove the hydraulic leading screw, closed clamp, conductor end clamp and other load-bearing tools, and transfer them to the ground with the cooperation of the ground electrician. The transfer process shall be smooth, free from collision and winding, and the knot shall be used correctly.	15	(1) Deduct 3 points for failure to check the connection of new insulator; deduct 2 points for failure to report after checking; deduct 1 point for starting loosening the hydraulic leading screw without the consent of Responsible Person after reporting. (2) Deduct 3 points for failure to check the stress of new insulator; deduct 2 points for failure to report after checking; deduct 1 point for starting removing the hydraulic leading screw without the consent of Responsible Person after reporting. (3) Deduct 1 point for incorrect use of knot when binding the tools. (4) Deduct 2 points/time for falling objects from heights. (5) Deduct 1 point/time for the tools colliding with each other; deduct 1 point/time for the tool colliding with the electrified body or tower body; deduct 2 points for winding insulating ropes			

Part I Standard for Training and Assessment on Live Working for Operation Maintenance of ±800kV UHV Power Transmission Line

Table (Cont'd)

S/N	Project name	Quality requirements	Score	Deduction standard	Reasons for deduction	Deduction	Scoring
11	Exit the intense electric field	(1) The equipotential electrician shall remove and put the insulated transmission rope in order after checking that there is no object left in the operation area. (2) The main safety belt shall be transferred and fastened at an appropriate position of the insulator string, returned to the insulator string through the grading ring, hooked at the appropriate position of the grading ring by the potential transfer rod, and stopped when the main belt is moved along the insulator string toward the lateral side of the cross arm to three pieces of insulators outside the grading ring; the equipotential electrician shall grasps the insulator with one hand and the potential transfer rod with the other, so as to quickly disengage from the equipotential by using the potential transfer rod. (3) They can reach the cross arm along the insulator string in the "two-span and three-short-circuit" operation mode.	5	(1) Deduct 2 points for the potential transfer without applying to the Responsible Person;deduct 1 point for starting the work without the consent after applying. (2) Deduct 1 point for the inappropriate position applied for the potential transfer. (3) Deduct 2 points for the incorrect action of equipotential electrician exiting the intense electric field and repeat discharging. (4) Deduct 1 point for failure to effectively control backup protection rope			

· 331 ·

Table (Cont'd)

S/N	Project name	Quality requirements	Score	Deduction standard	Reasons for deduction	Deduction	Scoring
12	Return to the ground	After checking that there is no object left on the tower, the equipotential electricians shall report it to the Responsible Person and then climb down the tower with insulated transmission rope after obtaining the consent of the Responsible Person.	5	(1) Deduct 2 points for failure to use the falling protector when climbing down the tower. (2) Deduct 2 points for loosing the protection of the safety belt when moving on the tower. (3) Deduct 1 point for grasping the tower nail when climbing down the tower. (4) Deduct 2 points for any objects left on the tower.			
13	End of the work	(1) The responsible person shall organize all working team members to put working apparatus and materials in order and put them in a special kit (bag) after cleaning; clean the site to ensure that "the materials are removed and the site is cleaned after construction". (2) After a post-shift meeting is held, the Responsible Person shall make work summaries and comments. Comments include the construction quality of this work and the implementation of safety measures from all working personnel. (3) Responsible Person shall report to the on-duty control personnel that the work is over, apply for the restoration of circuit re-closing and terminate the work order.	10	(1) Deduct 2 points for failure to clean the tools. (2) Deduct 2 points for missing tools. (3) Deduct 2 points for failure to hold the post-shift meeting. (4) Deduct 2 points for failure to remove the fence. (5) Deduct 2 points for failure to report to dispatcher			
	Total		100				

Module 4 Standards for Training and Assessment on Live Replacement of Arbitrary Single Insulator of Tensile Glass Insulator String of ±800kV UHV Transmission Line

I. Training Standards

(I) Training Requirements

Designation of module	Live replacement of arbitrary single insulator of tensile glass insulator string of ±800kV UHV transmission line	Type of training	Operation class
Training method	Practical operation training	Training hours	21 hours
Training objectives	1. Master the electrical significance of "two-span and three-short-circuit" operation mode in entering and exiting ±800 kV intense electric field along strain insulator string. 2. Can complete the entry of ±800kV intermediate potential operation point along the strain insulator string. 3. Can independently complete the live replacement operation of any single insulator of tensile glass insulator string of ±800kV UHV transmission line (intermediate potential operation method)		
Training venue	UHV training line		
Training content	The "two-span and three-short-circuit" operation mode is adopted to enter the intense electric field along the tensile glass insulator string, and the intermediate potential operation method is adopted for replacement operation of any single insulator of tensile glass insulator string of ±800kV UHV transmission line		
Scope of application	Maintenance personnel of UHV transmission line		

(II) Referenced Procedures and Specifications

(1) Technical Specification for Live Working of ±800kV DC Line (DL/T1242-2013)

(2) Operating Code for ±800kv DC Overhead Transmission Line (GB/T28813-2012)

(3) Maintenance Specification for ± 800kV DC Overhead Transmission Line (DL/T251-2012)

(4) Technical Guide for Live Working of ± 800kV DC Transmission Line (Q/GDW302-2009)

(5) Calculation Method of Live Working Minimum Approach Distance on AC Transmission Line (GB/T 19185-2008)

(6) Guidelines of Insulation Coordination for Live Working (DL/T867-2004)

(7) Test Guide of the Insulating Tool for Live Working (DL/T878—2004)

(8) State Grid Corporation of China on the Management Regulations of Live Working (Trial Implementation) (SGCC [2007] No.751)

(9) State Grid Corporation of China Working Regulations of Power Safety(Transmission Line Section) (Q/GDW1799.2-2013)

(10) Electrotechnical Terminology - Overhead Line (GB/T 2900.51-1998)

(11) Electrotechnical Terminology - Live Working (GB/T2900.55-2016)

(12) Live Working - Terminology for Tools, Equipment and Devices (GB/T 14286-2008)

(13) Minimum Requirements for Utilization of Tools, Devices and Equipment for Live Working (DL/T 877-2004)

(14) Preventive Test Code of Tools, Devices and Equipment for Live Working (DL/T 976-2005)

(15) Insulated Tackles for Live Working (GB/T13034-2008)

(16) Live Working-Insulating Ropes (GB 13035-2008)

(17) Shielding Clothes for Live Working (GB/T6568—2008)

(18) Screen Clothes for Live Working on 1000kV AC (GB/T25726-2010)

(III) Teaching Design for Training

To complete the work task of "live replacement operation of any single insulator of tensile glass insulator string of ±800kV UHV transmission line", each training stage shall be designed according to the standard operation procedure for work task completion. Each stage includes specific training objectives, training content, hours of training, training methods (training resources), training environment, assessment and evaluation, etc, as shown in the Table 1-4-1.

Part I Standard for Training and Assessment on Live Working for Operation Maintenance of ±800kV UHV Power Transmission Line

Table 1-4-1 Live Replacement of Arbitrary Single Insulator of Tensile Glass Insulator String of ±800kV UHV Transmission Line

Training schedule	Training objectives	Training contents	Training hours	Training methods and resources	Preparation of training conditions	Assessment and evaluation
1. Theoretical teaching	1. To master the basic method for entering and leaving ±800kV intense electric field along the insulator string. 2. Be familiar with the replacement method of strain single glass insulator of transmission line	1. The electrical significance of the "two-span and three-short-circuit" operation mode when entering and leaving the intense electric field along the insulator. 2. Method and quality standard for replacement of strain single glass insulator of transmission line	2	Training methods: Lecture. Training resources: PPT, relevant regulations and specifications	Multimedia classroom	Attendance, classroom questions and assignments
2. Preparations	Be able to complete the preparation before operation	1. Work site survey. 2. Preparation of the standardized operation card. 3. Filling of the work order. 4. Preparation of tools and materials for this operation	1	Training methods: 1. Site survey and cleaning of tools and materials shall be practiced at site. 2. Preparation of operation card and the filling of work order shall adopt lecture method. Training resources: 1. ±800kV training line. 2. UHV tools warehouse. 3. Blank work order	1. UHV training transmission line 2. Multimedia classroom	
3. Work site preparation	Can complete the preparations of work site	1. Work site re-survey. 2. Job application. 3. Work site layout. 4. Pre-shift meeting. 5. Tools and materials inspection.	1	Training methods: demonstration and role play. Resources: ±800kV practical training line	±800kV Practical training line	

· 335 ·

Standard for Professional Training and Assessment for Operation Maintenance of UHV DC Power Transmission Line

Table (Cont'd)

Training schedule	Training objectives	Training contents	Training hours	Training methods and resources	Preparation of training conditions	Assessment and evaluation
4. Trainer's demonstration	The trainees can preliminarily understand the operation process of the task through inspecting and learning from each other's work.	1. The intermediate potential electrician enters and exits the intense electric field along the strain insulator string. 2. The intermediate potential electrician assembles tools. 3. The intermediate potential electrician completes the replacement of single glass insulator	2	Training methods: Demonstration. Resources; ±800kV practical training line	±800kV Practical taining line	
5. Group training of trainees	1. Be able to complete the operation of entering and exiting ±800kV intense electric field. 2. Be able to complete the replacement of single glass insulator	1. The trainees are grouped (6 in a group) to train the skill operation of entering and exiting ±800kV intense electric field and replacing insulator. 2. Trainers guide the operation of trainees and conduct safety supervision	14	Training methods: Role play. Resources: ±800kV training line	±800kV Training line	Score the operation of trainees according to the detailed rules for skill assessment and scoring
6. End of the work	1. Enable the trainees to further distinguish the shortcomings of the operation process and facilitate the promotion in the later stage. 2. Train the trainees in the working style of safe and civilized production	1. Clean the work site. 2. Report to dispatcher the end of work. 3. Comment and summarize the work task this time at post-shift meeting	1	Training methods: Lecture and inductive method	±800kV Training line	

· 336 ·

(IV) Work Flow

1. Work task

The live replacement of arbitrary single insulator of tensile glass insulator string of ±800kV UHV transmission line.

2. Requirements for Weather and Work Site

(1) The replacement operation of any single insulator of tensile glass insulator string of ±800kV UHV transmission line shall be carried out in good weather. In case of lightning (hearing thunder or seeing lightning), snow, hail, rain, fog and so on, live working is prohibited. When the wind force is greater than level 5, it is unsuitable for live working; when the relative humidity of the air is greater than 80%, insulating tools with moisture-proof properties shall be used if live working needs to be carried out. When emergency live repair is required in bad weather, relevant personnel shall be organized to fully discuss and prepare necessary safety measures, which can be implemented after being approved by the unit. In case of sudden change of weather during live working, which may endanger the safety of person or equipment, the work shall be stopped immediately; in the case of ensuring personal safety, the normal state of equipment shall be restored as soon as possible, or other measures shall be taken.

(2) The operating personnel should be in good mental states, be familiar with the organizational and technical measures to ensure safety in work and master the methods for emergency rescue and first aid for electric shock in heights; they should hold the qualification certificate for live working within the validity period.

(3) Responsible Person should organize the relevant personnel to complete field investigation in advance, determine the operating methods, required working apparatus and necessary measures according to the results, and handle the tickets for live working.

(4) The work site should be reasonably set up with fence and warning signs. Non operating personnel is forbidden to enter.

(5) The line restart device shall be deactivated for the Project.

(6) Mode of operation: Intermediate potential operation.

(7) Safe working distance and effective insulation length during operation are shown in Table 1-4-2.

Table 1-4-2 Safe Distance for Replacement Operation of Any Single Insulator of Tensile Glass Insulator String of ±800kv UHV Transmission Line (m)

Altitude	Minimum safe distance between equipotential electrician and grounding frame	Minimum effective insulation length of insulating tools	Minimum combination gap
$H \leqslant 1000$	6.8	6.8	6.7
$1000 < H \leqslant 2000$	7.3	7.3	7.3
$2000 < H \leqslant 2500$	7.9	7.8	7.8

Note: The minimum safe distance in the table include 0.5m occupying gap of human body.

(8) When the intermediate potential operator enters ±800kV intense electric field along the strain insulator string, the number of insulators shorted by the human body shall not be more than 4. The minimum number of good insulators shall meet the requirements of Table 1-4-3 after deducting the number of insulators shorted by human body and defective insulators from the strain insulator string.

Table 1-4-3 Minimum Combined Gap and Minimum Number of Good Insulators

Altitude (m)	Structural height of single glass insulator (mm)	Minimum total length of good insulator string (m)	Minimum number of good insulators
$H \leqslant 1000$	170	6.2	37
	195		32
	205		31
	240		26
$1000 < H \leqslant 2000$	170	7.1	42
	195		37
	205		35
	240		30
$2000 < H \leqslant 2500$	170	7.55	45
	195		39
	205		37
	240		32

Note: The values in the table do not include the human body occupying gap which shall not be less than 0.5m during operation.

3. Preparations

3.1 Hazards and precontrol measures

(1) Hazard —— electric shock

Precontrol measures:

① Prior to work, the Responsible Person shall contact the on-duty control personnel to disable the DC restart protection device, and perform the licensing procedures. It is strictly prohibited to disable or resume the DC restart protection device at about the time. It shall be reported to the dispatcher in time after the end of the work.

② Before climbing the tower, operators on the tower must carefully check the dual names of the line and the number of the pole and tower, and then the tower can be climbed after all have been confirmed correct.

③ If lines lose power suddenly during work, operators shall consider it as still charged. The Responsible Person shall contact the control personnel as soon as possible, and no forced energization is allowed before the on-duty control personnel getting in touch with the Responsible Person.

④ Insulated tool and insulated ropes shall be free of damage, moisture, deformation, and failure, with the effective length no less than that specified in Table1-4-2 4. It is not allowed to use non-insulated ropes (such as cotton rope, manila rope, and steel wire rope).

⑤ The ground electrician shall wear clean and dry gloves when operating the insulating tools. When entering the work site, the live working tools shall be placed on moisture-proof canvas or insulating mat to prevent dirt and dampness of the insulating tools in use.

⑥ The intermediate potential electrician shall wear flame-retardant underwear and full set of qualified shielding clothes outside (including hat, dresses & trousers, gloves, socks and shoes). All parts shall be in excellent connection conditions.

⑦ When the intermediate potential operator moves along the insulator string, the positions of the hands and feet shall be correspondingly consistent, and the number of insulators shorted by the human body and the tools shall conform to the provisions of Table 1-4-3.

⑧ The operator enters the intense electric field by means of "two-span and three-short-circuit" operation mode (also called free operation method).

⑨ When transmitting large metal objects by using insulating ropes, ground electrician shall not touch them before ground connection.

⑩ During the live working, the Responsible Person (Supervisor) shall continuously monitor the operators and correct their nonstandard operation or actions in violation at any time. Special attention shall be paid to high-place operators to ensure that there is enough safety distance and combined gap (meeting the requirements in Table 1-4-2). It is forbidden to touch two non-connected electrified bodies or electrified body and grounding body.

(2) Hazard —— falling from high places

Precontrol measures:

① Before climbing, operators work at heights must satisfy the requirements of this operation, such as physical condition, mental state, and skill and quality.

② Double safety belts shall be used by the personnel for Work at Heights. When climbing up and down the tower, the main material shall be grasped with hands, with shackles trodden and moving at a uniform speed.

③ Before the operation, the intermediate potential operator shall carefully inspect hydraulic lead screw, closed clamp and so on to ensure that the load-bearing tools are qualified; when moving along the insulator string, the positions of hands and feet must be correspondingly consistent, and the safety belt shall be fastened to the insulator string supported by hands and moved synchronously; when replacing insulators, the load-bearing tools shall be installed reliably. Before and after load transfer, impulse test shall be conducted to determine their reliability, and it shall be reported to the Responsible Person in a timely manner. The implementation can only be carried out with the permission of the Responsible Person.

④ Supervisor shall correct irregularities and violations at any time. Special attention shall be paid to operators for work at heights to prevent them from losing the protection of the safety belt or insulated backup protection rope during transposition. The safety belt or insulated backup protection rope shall not be fastened lower than operating personnel.

(3) Hazard —— injury caused by objects falling from high place

Precontrol measures:

① Operator work at heights should put personal tools and fragmentary materials into the tools bag. It is strictly forbidden to hang objects in high place or keep in the mouth.

② When the heavy equipment is transferred by the way of up and down circulation exchange, the control rope should be fastened to prevent the transferred articles from colliding with each other and accidentally touching the bearing tools in the working state.

③ Ground operator should correctly wear a helmet and use the knots. The vertical distance from the work site should not be less than the falling radius.

④ The operation site shall be set up with fence and warning signs. It shall be noted at any time that Supervisor shall prohibit irrelevant personnel and vehicles from entering operation area.

3.2 Selection of tools, instruments and materials

Tools and materials required for replacement operation of any single insulator of tensile glass insulator string of ±800kV UHV transmission line can be seen in Table 1-4-4. Before delivering tools and instruments out of warehouse, application voltage class and test period shall be carefully checked and they shall be inspected to ensure that appearance is intact, connection is firm, rotation is flexible and meet the working task requirements. After delivering tools and instruments out of warehouse, they shall be stored in tools bag or tool kit for transportation to avoid contamination and damp. Metal tools and insulated tools shall be separately loaded and transported to avoid deformation, damage or other defects caused by mixed loading and transportation.

Table 1-4-4 Tools and Materials Required for Live Replacement of Arbitrary Single Insulator of Tensile Glass Insulator String of ±800kV UHV Transmission Line

S/N	Name	Specification and Model	Unit	Qty.	Remarks
1	Shielding clothes	Shielding efficiency ≥ 40dB (shield efficiency of shielding mask ≥ 20dB)	Set	2	PPE
2	Conductive shoes	The size depends on the wearer	Pair	2	PPE
3	Flame retardant underwear	Pure mulberry silk	Set	2	PPE
4	Safety belt of double insurance	Suspender	Nos.	2	PPE
5	Safety helmet		pcs	6	PPE
6	Goggles		Nos.	2	PPE
7	Insulated transmission rope	Φ14mm, with the length matching the lifting height	Nos.	1	Insulating tool
8	Insulated backup protection rope	Φ16mm	Nos.	2	Insulating tool
9	Insulated noose	Φ14mm	Nos.	2	Insulating tool
10	Insulated tackle	1T	Nos.	1	Insulating tool
11	Hydraulic leading screw	8T	Nos.	2	Metal tool
12	Closed clamp	Tc4	Set	1	Metal tool
13	Pin puller		pcs	1	Other tools
14	Insulation resistance tester	5000V, with electrode width 2cm and interelectrode width 2cm	Set	1	Other tools

Table (Cont'd)

S/N	Name	Specification and Model	Unit	Qty.	Remarks
15	Multimeter		Set	1	Other tools
16	Wind speed, temperature and humidity tester		pcs	1	Other tools
17	Security fence		Set	Several	Other tools
18	Warning sign	"Work Here", "Access from Here", "Slow Down", "Blocking"	Set	1	Other tools
19	Red waistcoat	"Responsible Person" "Special Supervisor"	pcs	1	Other tools
20	Moisture-proof tarpaulin	3m×3m	pcs	2	Other tools
21	Personal tools	Wrench, vice	Set	1	Other tools
22	Falling protector	Corresponding to the type of pole and tower anti-fall device	pcs	2	Other tools
23	Towel	Cotton	pcs	1	Other tools
24	Insulator	The same model with replaced insulators	pcs	1	Material

3.3 Division of labor for operators

Division of labor for operators of the task is shown in Table 1-4-5.

Table 1-4-5 Division of Labor for Live Replacement of Arbitrary Single Insulator of Tensile Glass Insulator String of ±800kV UHV Transmission Line

S/N	Post	Qty. (person)	Responsibilities
1	Responsible Person	1	Be responsible for division of operating personnel of the task, work order reading, handling disconnection and restart of the line and the formalities of work permit, getting work permits, holding pre-shift meeting, dealing with emergency situations in work, quality surveillance, and the summary after work
2	Special Supervisor	1	Be responsible for the safety control of the work site
3	Intermediate potential operator	2	Be responsible for tool installation and insulator replacement
4	Ground electrician	2	Be responsible for ground auxiliary works during the operation.

4. Working Procedures

The workflow of this task is shown in Table 1-4-6.

Table 1-4-6 Workflow of Live Replacement of Arbitrary Single Insulator of Tensile Glass Insulator String of ±800kV UHV Transmission Line

S/N	Work Content	Operation Steps and Standards	Safety Measures and Precautions	Responsible Person
1	Site re-survey	The Responsible Person shall complete the following work: (1) Check the line name, tower number and dualism number to ensure they are correct; guarantee that the base and power are intact and in normal operation, and that the crossed crossing distance meets the safety requirements; confirm the defects conditions and specifications and models of wires. (2) Check that the site meteorological conditions such as wind speed and humidity should meet the operation requirements. (3) Check that the terrain and environment should meet the operation requirements. (4) Check that the safety measures listed in the work order are in line with the actual situations on site. The measures will be added if necessary.	(1) Correctly wear helmet, working clothes, work shoes and protective gloves. (2) Operation under meteorological conditions that may endanger the safety of operators is forbidden. (3) Non-operation personnel and vehicles are strictly prohibited from entering the working site	
2	Work Permit	(1) The Responsible Person is responsible for contacting the on-duty control personnel and applying for the DC restart protection device as per the contents of the work order. (2) Live working could be started only after being approved by control personnel on duty.	Live working shall not be started without the permission of the on-duty control personnel	
3	Site layout	Install the security fence and hang the signboards correctly: (1) The security fence should take full account of falling objects from the high place and the influence on road traffic. (2) The entrance and exit of the security fence shall be set reasonably. (3) Signs such as "Access from Here", "Work Here", "Slow Down" or "Blocking" shall be properly arranged	When the influence on road traffic safety is uncontrollable, the traffic management department should be contacted in time to strengthen the on-site control of traffic safety	

S/N	Work Content	Operation Steps and Standards	Safety Measures and Precautions	Responsible Person
4	Holding a pre-shift meeting	(1) All working personnel shall line up. (2) Responsible Person will read out the work order and be clear with work task and division of personnel; explain safety measures and technical measures in work; check (inquire after) mental state of all working personnel; inform of hazards in work and precontrol measures. (3) All working personnel shall sign on the work order for confirmation.	(1) Work order shall be filled and issued with standardized licensing procedure and complete signature. (2) All working personnel shall be in good mental states. (3) All working personnel shall be clear with task division of works, safety measures and technical measures	
5	Inspection of tools and instruments	(1) All necessary tools and instruments shall be prepared as per the operation requirements and placed on the waterproof tarpaulin regularly according to the category and location. The appearance and test certificate of tools and instruments shall be checked to ensure there is no omission. (2) The surface insulation resistance of insulating tools and insulating ropes shall be tested with insulation resistance tester in correct methods, and the value shall be not less than 700MΩ. (3) The new insulator is wiped up, and it is intact in appearance inspection, without rust, cracks and breakage. The surface insulation resistance shall be tested with insulation resistance tester in correct methods, and the value shall be not less than 700MΩ. (4) The internal resistance of full shielding clothes shall be tested with a multimeter in correct methods, and the value shall be not more than 20Ω. (5) The inspector shall report to the Responsible Person that all inspection results are in conformity with the operation requirements	(1) The waterproof tarpaulin shall be enough in quantity and reasonable in position, and be clean and dry. (2) Before using metal and insulating tools, they shall be carefully checked for damage, dampness, deformation and failure, and the certificate of conformity is within the validity period. (3) Insulated tools and insulated ropes shall be tested as acceptable	

Part I Standard for Training and Assessment on Live Working for Operation Maintenance of ±800kV UHV Power Transmission Line

Table (Cont'd)

S/N	Work Content	Operation Steps and Standards	Safety Measures and Precautions	Responsible Person
6	Climbing the tower	(1) The intermediate potential electrician shall check the double name and phase of the line again, inspect and confirm that the shackles are complete and firm; fasten safety belts and attach falling protectors; perform impact inspection on safety belt, backup protection rope and falling protector in correct method; the Responsible Person shall check and confirm that the connection conditions of each connection point of the double safety belt and shielding clothes worn by the equipotential electricians are in good condition, including shoulder harness, pectoral harness, belt, back rope sling, buckle and ring. (2) The electrician shall carry toolkit and insulated transmission rope (including insulated tackle) in correct methods. (3) Main band of safety belt and backup protection rope shall obliquely across the shoulder. (4) The soles shall be cleaned, and climbing the tower shall be done with the permission of the Responsible Person. (5) The electrician shall tread on shackles, grasp the main materials, and climb the tower uniformly to the proper position of the cross arm with safety belt well fastened and then break away from the falling protector	(1) Safety belt and falling protector shall pass the impulse test. (2) The safety belt and insulated transmission rope shall be prevented from hooking up tower materials. (3) The minimum safe distance between the human body and the conductor shall conform to the regulations shown in Table 1-4-2. (4) It is forbidden to grasp shackles with hands. (5) Falling protector shall be properly used. (6) Protection of safety belt shall not be lost during transposition.	
7	Entering the intense electric field	(1) The intermediate potential electrician carries the insulated transmission rope, transposes it to the hanging point of the operation phase strain insulator string, hangs the main safety belt on the insulator string connection fittings, and leaves the safety belt backup protection rope at the proper position of the cross arm.	(1) The safety belt and insulated transmission rope shall be prevented from hooking up tower materials. (2) Protection of safety belt shall not be lost during transposition.	

Table (Cont'd)

S/N	Work Content	Operation Steps and Standards	Safety Measures and Precautions	Responsible Person
7	Entering the intense electric field	(2) The intermediate potential electrician shall recheck and confirm the good connection of all parts of the shielding clothes and the insulator string and the location of the faulty insulators. With the permission of the Responsible Person, the electrician shall grasp one string and tread the other one, and steadily move to the operation point along the insulator string in the operation mode of "two-span and three-short-circuit"; the positions of hands and feet must be correspondingly consistent, and the safety belt shall be fastened to the insulator string supported by hands and moved synchronously. (3) After reaching the operation point, the intermediate potential electrician shall fix the insulated tackle at a proper position of insulator string by insulating rope sleeve, and thread it into the insulated transmission rope. The installation shall be firm, stable and convenient for work.	(3) The safe distance between the human body and the electrified body, and the combined gap between the human body and the grounding body and the electrified body shall not be less than that specified in Table 1-4-2.	
8	Install tools and transfer conductor tension	(1) The ground electrician transfers the closed clamp, hydraulic leading screw and so on to the intermediate potential electrician by using the insulated transmission rope. The lifting process shall be smooth, free from collision and winding, and the knot shall be used correctly. (2) The intermediate potential electrician installs the closed clamp (front clamp) in the slot of the rear two insulators where insulators need be replaced, the rear clamp is installed on the steel cap of the front insulator where the insulator needs to be replaced, and the hydraulic leading screw is connected. Each part of the load-bearing tools is securely and reliably installed. (3) Equipotential electricians shall check and confirm that all parts of the load-bearing tools are installed in good condition. With the permission of the Responsible Person, they can operate the hydraulic leading screw so that it is gradually stressed and the insulators to be replaced are relaxed. The two hydraulic leading screws shall be uniformly stressed.	(1) The combined gap between the human body and the grounding body and the charged body shall not be less than that specified in Table 1-4-2. (2) Objects falling from high place shall be avoided. (3) After deducting inferior insulators, insulators shorted by human body and tools, the number of good insulators shall conform to that specified in Table 1-4-3.	

Table (Cont'd)

S/N	Work Content	Operation Steps and Standards	Safety Measures and Precautions	Responsible Person
9	Replacement of insulator	(1) The intermediate potential electrician shall perform impulse test, check and confirm that the load-bearing tools are stressed normally. With the permission of the Responsible Person, they shall tie the old insulator with the insulated transmission rope, take out the fitting pins at both ends of the old insulator, continue to operate and tighten the hydraulic leading screw until the old insulator is removed. The two hydraulic leading screws shall be stressed uniformly, and the operating handle shall not knock on the insulator. (2) The ground electrician uses the other end of the insulated transmission rope to fasten the new insulator, and transfers the new insulator to the intermediate potential electrician by means of the old lower and the new upper. The lifting process shall be smooth, free from collision and winding, and the knot shall be used correctly. (3) The intermediate potential electrician installs the new insulator, resets the fitting pin at its two ends, and confirms that it is installed in place	(1) The combined gap between the human body and the grounding body and the charged body shall not be less than that specified in Table 1-4-2. (2) Prevent objects falling from heights	
10	Remove tools	(1) The intermediate potential electricians check that the connection of the new insulator is reliable and, with the permission of the Responsible Person, they operate and loosen the hydraulic leading screw so that the replaced insulator is gradually stressed. (2) After load transfer, the intermediate potential electricians shall perform the impulse test, check and confirm that the new insulator is in good condition. With the permission of the Responsible Person, they can remove the insulated transmission rope tied to the insulator, fasten it to the proper position of the load-bearing tools, remove the hydraulic leading screw, closed clamp and other load-bearing tools, and transfer them to the ground with the cooperation of the ground electrician. The transfer process shall be smooth, free from collision and winding, and the knot shall be used correctly.	(1) The combined gap between the human body and the grounding body and the charged body shall not be less than that specified in Table 1-4-2. (2) Prevent objects falling from heights	

Table (Cont'd)

S/N	Work Content	Operation Steps and Standards	Safety Measures and Precautions	Responsible Person
11	Leaving the intense electric field	(1) The intermediate potential electricians shall remove the insulated transmission rope after checking that there is no object left in the operation area. (2) They shall return to the cross arm with the insulated transmission rope along the insulator string in the "two-span and three-short-circuit" operation mode	(1) The combined gap between the human body and the grounding body and the charged body shall not be less than that specified in Table 1-4-2. (2) Protection of safety belt shall not be lost during transposition.	
12	Evacuation from the pole and tower	The intermediate potential electricians shall check that there is no object left on the tower. After being approved by the Responsible Person, the falling protector shall be properly hung and the safety belt shall be unfastened and put straight, the electrician shall carry the insulated transmission rope correctly and step down to the ground from the tower in uniform steps by treading on shackles and grasping the main materials	(1) Protection of safety belt shall not be lost during transposition. (2) Hand slippage and step missing shall be avoided and it's forbidden to grasp the shackles with hands. (3) Falling protector shall be properly used. (4) The insulated transmission rope and safety belt shall be prevented from hooking up the tower materials or shackles	
13	End of the work	(1) The Responsible Person shall organize all working personnel to put working apparatus and materials in order and put them in a special kit (bag) after cleaning; clean the site to ensure that "the materials are removed and the site is cleaned after construction". (2) After a post-shift meeting is held, the Responsible Person shall make work summaries and comments. Comments include the construction quality of this work and the implementation of safety measures from all working personnel. (3) Responsible Person shall report to the on-duty control personnel the end of the work, apply for the restoration of line restarting and terminate the work order		

II. Assessment Standard

Table 1-4-7 Detailed Rules for Assessment and Scoring of Operation and Inspection Skills of UHVDC Transmission Line of State Grid Sichuan Electric Power Corporation

Part I Standard for Training and Assessment on Live Working for Operation Maintenance of ±800kV UHV Power Transmission Line

Fill-in Column of Examinee	No.:	Name:	Position:	Unit:		
Fill-in Column of Assessor	Grade:	Assessor:	Assessment Team Leader:	Starting time:	Closing time:	Operation Duration:
Assessment module	Live replacement of arbitrary single glass insulator of tensile glass insulator string of ±800kV UHV transmission line		Assessee	Maintenance Personnel of UHV DC Transmission Line	Assessment method	Operation
						Assessment Time Limit
						90min
Job Description	The live replacement of arbitrary single glass insulator of resisting-tensile tower glass insulator string of ±800kV transmission line					
Work specifications and requirements	1. Live working shall be carried out in good weather. In case of thunder, rain, snow or fog, no live working shall be carried out. When the wind force is greater than Level 5, live working should not be carried out. When the humidity is greater than 80%, insulating tools with moisture-proof properties shall be used if live working needs to be carried out. 2. Six persons are required for this operation, including 1 Responsible Person, 1 special Supervisor, 2 intermediate potential electricians and 2 ground electricians. 3. Responsibilities of Responsible Person (Supervisor): Be responsible for division of operating personnel of the task, work order reading, handling restart deactivation of the line and the formalities of work permit, getting work permits, holding pre-shift meeting, safety supervision in the operation process, dealing with emergency situations in work, quality surveillance, and the summary after work 4. During the live working, if thunder, rain, strong wind or any other circumstance threaten the safety of the staff, the Responsible Person or Supervisor may stop working temporarily according to the circumstances. Given conditions: 1. Work orders have been handled, safety measures have been completed (restart has been deactivated), and oral application (dispatcher or assessor) shall be made at the beginning and end of the work.					

· 349 ·

	Table (Cont'd)
Work specifications and requirements	2. The instrument shall be used safely and correctly to test the insulating tool. 3. The operation must be carried out according to the working procedures. The scores of the items to be carried out shall be deducted for the process error. In case of major hidden dangers of personal, equipment and operational safety, the assessor may order the termination of the operation (assessment)
Assessment scenario preparation	1. Tower shape: ±800kV resisting-tensile tower. 2. Required operation tools: 2 double safety belts, 2 sets of shielding clothes, 2 sheets of moisture-proof cloth, 1 multimeter, 1 insulation resistance detector, 1 anemometer, 1 2-in-one temperature and humidity detector, 2 hydraulic leading screws, 2 insulating ropes, 1 set of closed clamp, 1 tackle, 1 goggles, 1 pin puller and 1 set of hand-operated tools. 3. The work site shall be monitored, and the safety measures (fence, etc.) on the work site have been fully implemented; non-operation personnel are prohibited from entering the site, and the staff must wear safety helmets when entering the work site. 4. Examinees shall bring their own work clothes, helmets and gloves
Remarks	1. The deduction shall be done until the scores of each item are deducted completely. In case of major hidden dangers of personal, equipment and operational safety, the assessor may order the termination of the operation 2. When equipment, working environment, safety belt, safety helmet, tool, shielding clothes, etc., do not conform to the operation condition, the assessor may order the termination of the operation

Part I Standard for Training and Assessment on Live Working for Operation Maintenance of ±800kV UHV Power Transmission Line

Table 1-4-8 Standards for Assessment and Scoring of Operation and Inspection Skills of UHVDC Transmission Line of State Grid Sichuan Electric Power Corporation

S/N	Project name	Quality requirements	Score	Deduction standard	Reasons for deduction	Deduction	Scoring
1	Site re-survey	(1) The Responsible Person shall go to the work site to check the line name, pole and tower number, on-site working conditions, defective parts and so on. (2) Check that the site meteorological conditions such as wind speed and humidity should meet the operation requirements. (3) Check whether the work order is complete and unmodified, check whether the safety measures listed are consistent with the actual situation on site, and supplement it if necessary	5	(1) Deduct 1 point/item for failure to check the line name, pole and tower number, on-site working conditions and defective parts. (2) Deduct 1 point/item for failure to detect wind speed, humidity and other on-site meteorological conditions. (3) Deduct 0.5 points/part for any alteration in the work order filling; deduct 1 point for incorrect work order number; deduct 1.5 points for incomplete work order filling			
2	Work Permit	(1) The Responsible Person is responsible for contacting the on-duty control personnel (referee) and applying for stopping the line restarting as per the contents of the work order. (2) Reporting content is standardized and complete	2	(1) Deduct 2 points for failure to contact the dispatching department (referee) for deactivating the restart. (2) Deduct 1 point for non-standard or incomplete terminology reporting			

· 351 ·

Standard for Professional Training and Assessment for Operation Maintenance of UHV DC Power Transmission Line

Table (Cont'd)

S/N	Project name	Quality requirements	Score	Deduction standard	Reasons for deduction	Deduction	Scoring
3	Site layout	Install the security fence and hang the signboards correctly: (1) The security fence should take full account of falling objects from the high place and the influence on road traffic. (2) The entrance and exit of the security fence shall be set reasonably. (3) Signs such as "Access from Here", "Work Here", "Access from Here" shall be properly arranged	3	(1) Deduct 1 point for failure to arrange the fence at the work site. (2) Deduct 1 point for failure to arrange the warning board. (3) Deduct 1 point for failure to hang the tower climbing operation sign			
4	Holding a pre-shift meeting	(1) All staff and personnel shall wear safety helmets and work clothes correctly. (2) Responsible Person shall wear red vest and read out the work order and be clear with work task and division of personnel; explain safety measures and technical measures in work; check (inquire after) mental state of all working personnel; inform of hazards in work and precontrol measures. (3) All working personnel shall sign on the work order for confirmation.	3	(1) Deduct 0.5 points/person for the staff not dressing uniformly. (2) Deduct 3 points for no division of labor; deduct 1 point for unclear division of labor. (3) Deduct 1 point for the on-site Responsible Person not wearing a safety monitoring vest. (4) Deduct 1 point for the work shift member failing to sign or signing incompletely on the work order.			

Part I Standard for Training and Assessment on Live Working for Operation Maintenance of ±800kV UHV Power Transmission Line

Table (Cont'd)

S/N	Project name	Quality requirements	Score	Deduction standard	Reasons for deduction	Deduction	Scoring
5	Inspection of tools and instruments	(1) All necessary tools and instruments shall be prepared as per the operation requirements and placed on the waterproof tarpaulin regularly according to the category and location. The appearance and test certificate of tools and instruments shall be checked to ensure there is no omission. (2) The surface insulation resistance of insulating tools and insulating ropes shall be tested with insulation resistance tester in correct methods, and the value shall be not less than 700MΩ. (3) The new insulator is wiped up, and it is intact in appearance inspection, without rust, cracks and breakage. The surface insulation resistance shall be tested with insulation resistance tester in correct methods, and the value shall be not less than 700MΩ. (4) The internal resistance of full shielding clothes shall be tested with a multimeter in correct methods, and the value shall be not more than 20Ω. (5) The inspector shall report to the Responsible Person that all inspection results are in conformity with the operation requirements	7	(1) Deduct 1 point for failure to use moisture-proof tarpaulin and place tools to designed positions. (2) Deduct 0.5 points/item for failure to check the appearance of tools and pass the test certificate. (3) Deduct 1 point/item for failure to use testing instrument for testing the tools. (4) Deduct 1 point for non-standard reporting of test results; deduct 0.5 points/item for incomplete reporting			

Table (Cont'd)

S/N	Project name	Quality requirements	Score	Deduction standard	Reasons for deduction	Deduction	Scoring
6	Climbing the tower	(1) The intermediate potential electrician shall check the double name and phase of the line again, inspect and confirm that the shackles are complete and firm; fasten safety belts and attach falling protectors; perform impulse test on double safety belts, backup protection rope and falling protector in correct method; the Responsible Person shall check and confirm that the connection conditions of each connection point of the double safety belt and shielding clothes worn by the intermediate potential electricians are in good condition, including shoulder harness, pectoral harness, belt, back rope sling, buckle and ring. (2) The electrician shall carry toolkit and insulated transmission rope (including insulated tackle) in correct methods. (3) During the process of climbing the tower, the electrician shall fasten the anti-fall device, tread on shackles, grasp the main materials, and climb the tower uniformly to the proper position with safety belt well fastened and then break away from the falling protector	5	(1) Deduct 1 point/item for the intermediate potential electrician failing to check the double name of the line, pole number, phase and tower material; deduct 1 point for non-report after checking. (2) Deduct 2 points/item for failure to perform the impulse test on double safety belts and falling protector. (3) Deduct 1 point for the on-site Responsible Person failing to check the safety protection equipment of the intermediate potential electrician. (4) Deduct 0.5 points/time for hands grasping the shackles. (5) Deduct 1 point for the unreasonable suspension position of the tackle transmission rope. (6) Deduct 5 points for loosing the protection of the safety belt during transposition			

Table (Cont'd)

S/N	Project name	Quality requirements	Score	Deduction standard	Reasons for deduction	Deduction	Scoring
7	Entering the intense electric field	(1) The intermediate potential electrician carries the insulated transmission rope, transposes it to the hanging point of the operation phase strain insulator string, hangs the main safety belt on the insulator string connection fittings, and leaves the safety belt backup protection rope at the proper position of the cross arm. (2) The intermediate potential electrician shall recheck and confirm the good connection of all parts of the shielding clothes and the insulator string and the location of the faulty insulators. With the permission of the Responsible Person, the electrician shall grasp one string and tread the other one, and steadily move to the operation point along the insulator string in the operation mode of "two-span and three-short-circuit"; the positions of hands and feet must be correspondingly consistent, and the safety belt shall be fastened to the insulator string supported by hands and moved synchronously. (3) After reaching the operation point, the intermediate potential electrician shall fix the insulated tackle at a proper position of insulator string by insulating rope sleeve, and thread it into the insulated transmission rope. The installation shall be firm, stable and convenient for work.	5	(1) Deduct 2 points for irrational fastening position and irregular use of safety belt back protection rope. (2) Deduct 1 point/item for the intermediate potential electrician failing to check the connection of the shielding clothes and insulator string and the location of the faulty insulator. (3) Deduct 5 points for entering the intense electric field without permission of the Responsible Person. (4) Deduct 2 points/time for the incorrect action of intermediate potential electrician entering the intense electric field and repeat discharging. (5) Deduct 1 point for the unreasonable installation position of the insulated transmission rope. (6) Deduct 2 points/time for falling objects from heights. (7) Deduct 5 points for loosing the protection of the safety belt during transposition			

Table (Cont'd)

S/N	Project name	Quality requirements	Score	Deduction standard	Reasons for deduction	Deduction	Scoring
8	Installation tool	(1) The ground electrician shall use insulated transmission rope to transfer the closed clamp, hydraulic leading screw and other tools to the intermediate potential electrician respectively, and use the knots in a correct manner, which shall be smooth, free from collision and winding. (2) The intermediate potential electrician installs the closed clamp (front clamp) in the slot of the rear two insulators where insulators need be replaced, the rear clamp is installed on the steel cap of the front insulator where the insulator needs to be replaced, and the hydraulic leading screw is connected. Each part of the load-bearing tools is securely and reliably installed. (3) Equipotential electricians shall check and confirm that all parts of the load-bearing tools are installed in good condition. With the permission of the Responsible Person, they can operate the hydraulic leading screw so that it is gradually stressed and the insulators to be replaced are relaxed. The two hydraulic leading screws shall be uniformly stressed.	15	(1) Deduct 1 point/time for unstable lifting process, colliding and winding. (2) Deduct 2 points/time for falling objects from heights. (3) Deduct 2 points for improper installation and fixing of clamp. (4) Deduct 3 points for failure to check the installation of load-bearing tools; deduct 1 point for failure to report after checking; deduct 1 point for starting tightening the leading screw without the consent of Responsible Person after reporting. (5) Deduct 3 points/time for the number of shorted insulators exceeding 4 during the operation. (6) Deduct 15 points for collision and breakage of insulators in the clamp installation. (7) Deduct 2 points for failure to tighten leading screw in a balanced way			

Part I Standard for Training and Assessment on Live Working for Operation Maintenance of ±800kV UHV Power Transmission Line

Table (Cont'd)

S/N	Project name	Quality requirements	Score	Deduction standard	Reasons for deduction	Deduction	Scoring
9	Replacement of insulator	(1) The intermediate potential electrician shall perform impulse test, check and confirm that the load-bearing tools are stressed normally. With the permission of the Responsible Person, they shall tie the old insulator with the insulated transmission rope, take out the fitting pins at both ends of the old insulator, continue to operate and tighten the hydraulic leading screw until the old insulator is removed. The two hydraulic leading screws shall be stressed uniformly, and the operating handle shall not knock on the insulator. (2) The ground electrician uses the other end of the insulated transmission rope to fasten the new insulator, and transfers the new insulator to the intermediate potential electrician by means of the old lower and the new upper. The lifting process shall be smooth, free from collision and winding, and the knot shall be used correctly. (3) They shall install new insulator and reset fitting pins at both ends.	20	(1) Deduct 3 points for failure to check the stress of load-bearing tools; deduct 2 points for failure to report after checking; deduct 1 point for removing fitting pins at both ends of the old insulator without the consent of Responsible Person after reporting. (2) Deduct 2 points for failure to tighten the hydraulic leading screw in a balanced way. (3) Deduct 1 point/time for the operating handle hitting the insulator. (4) Deduct 1 point for wrong knot. (5) Deduct 2 points/time for falling objects from heights. (6) Deduct 1 point for new and old insulators colliding with each other. (7) Deduct 1 point for the transmission insulator and tower body colliding with each other. (8) Deduct 2 points for winding insulating ropes			

· 357 ·

Table (Cont'd)

S/N	Project name	Quality requirements	Score	Deduction standard	Reasons for deduction	Deduction	Scoring
10	Remove tools	(1) The intermediate potential electricians check that the connection of the new insulator is reliable and, with the permission of the Responsible Person, they operate and loosen the hydraulic leading screw so that the replaced insulator is gradually stressed. (2) After load transfer, the intermediate potential electricians shall perform the impulse test, check and confirm that the new insulator is in good condition. With the permission of the Responsible Person, they can remove the insulated transmission rope tied to the insulator, fasten it to the proper position of the load-bearing tools, remove the hydraulic leading screw, closed clamp and other load-bearing tools, and transfer them to the ground with the cooperation of the ground electrician. The transfer process shall be smooth, free from collision and winding, and the knot shall be used correctly.	15	(1) Deduct 3 points for failure to check the connection of new insulator; deduct 2 points for failure to report after checking; deduct 1 point for starting loosening the hydraulic leading screw without the consent of Responsible Person after reporting. (2) Deduct 3 points for failure to check the stress of new insulator; deduct 2 points for failure to report after checking; deduct 1 point for starting removing the hydraulic leading screw without the consent of Responsible Person after reporting. (3) Deduct 1 point for incorrect use of knot when binding the tools. (4) Deduct 2 points/time for falling objects from heights. (5) Deduct 1 point/time for the tools colliding with each other; deduct 1 point/time for the tool colliding with the electrified body or tower body; deduct 2 points for winding insulating ropes			

Part I Standard for Training and Assessment on Live Working for Operation Maintenance of ±800kV UHV Power Transmission Line

Table (Cont'd)

S/N	Project name	Quality requirements	Score	Deduction standard	Reasons for deduction	Deduction	Scoring
11	Leaving the intense electric field	(1) The intermediate potential electricians shall fasten the insulated transmission rope after checking that there is no object left in the operation area to make preparation for exiting the potential. (2) The intermediate potential electrician exits the equipotential according to the operation mode of "two-span and three-short-circuit".	5	(1) Deduct 2 points for exiting the intense electric field without applying to the Responsible Person; deduct 1 point for starting exiting the intense electric field without the consent after applying. (2) Deduct 2 points/time for the incorrect action of intermediate potential electrician exiting the intense electric field and repeat discharging. (3) Deduct 1 point for failure to effectively control backup protection rope			
12	Return to the ground	After checking that there is no object left on the tower, the electricians can climb down the tower with insulated transmission rope after obtaining the consent of the Responsible Person.	5	(1) Deduct 5 points for failure to use the falling protector when climbing down the tower. (2) Deduct 5 points for loosing the protection of the safety belt when moving on the tower. (3) Deduct 1 point/time for hands grasping the shackles when climbing down the tower. (4) Deduct 2 points for any objects left on the tower			

Table (Cont'd)

S/N	Project name	Quality requirements	Score	Deduction standard	Reasons for deduction	Deduction	Scoring
13	End of the work	(1) The Responsible Person shall organize all working personnel to put working apparatus and materials in order and put them in a special kit (bag) after cleaning; clean the site to ensure that "the materials are removed and the site is cleaned after construction". (2) After a post-shift meeting is held, the Responsible Person shall make work summaries and comments. Comments include the construction quality of this work and the implementation of safety measures from all working personnel. (3) Responsible Person shall report to the on-duty control personnel the end of the work, apply for the restoration of line re-starting and terminate the work order.	10	(1) Deduct 2 points for failure to clean the tools. (2) Deduct 2 points for missing tools. (3) Deduct 10 points for failure to hold the post-shift meeting. (4) Deduct 2 points for failure to remove the fence. (5) Deduct 2 points for failure to report to dispatcher			
	Total		100				

Module 5 Standards for Training and Assessment on Live Replacement of ± 800kV UHV Transmission Line Conductor Spacer

I. Training Standards

(I) Training Requirements

Designation of module	Live replacement of ± 800kV UHV transmission line conductor spacer	Type of training	Operation class
Training method	Practical operation training	Training hours	21 hours
Training objectives	1. Master the electrical significance of "two-span and three-short-circuit" operation mode in entering and exiting ± 800 kV intense electric field along strain insulator string. 2. Can complete the entry of ±800kV equipotential operation point along the strain insulator string. 3. Can independently complete the replacement of conductor spacer (equipotential operation method)		
Training venue	UHV DC training line		
Training content	The "two-span and three-short-circuit" operation mode is adopted to enter the intense electric field along the strain insulator string, and the equipotential operation method is adopted for live replacement the eight-bundle conductor spacer of ± 800kV UHV DC transmission line.		
Scope of application	Maintenance Personnel of UHV DC Transmission Line		

(II). Referenced procedures and specifications

(1) Code for Designing of ±800kV DC Overhead Transmission Line (GB/T50790-2013)

(2) Maintenance Specification for ± 800kV DC Overhead Transmission Line (DL/T251-2012)

(3) Operating Code for ±800kv DC Overhead Transmission Line (GB/T28813-2012)

(4) Technical Specification for Live Working of ±800kV DC Line (DL/T1242-2013)

(5) Technical Specification for Fittings of ±800kV DC Overhead Transmission Line

(GB/T 31235-2014)

(6) State Grid Corporation of China on the Management Regulations of Live Working (Trial Implementation) (SGCC [2007] No.751)

(7) State Grid Corporation of China Working Regulations of Power Safety(Transmission Line Section) (Q/GDW1799.2-2013)

(8) Electrotechnical Terminology- Overhead Line (GB/T 2900.51-1998)

(9) Electrotechnical Terminology- Live Working (GB/T2900.55-2016)

(10) Live Working - Terminology for Tools, Equipment and Devices (GB/T 14286-2002)

(11) Insulated Tackles for Live Working (GB/T13034-2008)

(12) Live Working-Insulating Ropes (GB 13035-2008)

(13) Minimum Requirements for Utilization of Tools, Devices and Equipment for Live Working (DL/T 877-2004)

(14) Preventive Test Code of Tools, Devices and Equipment for Live Working (DL/T 976-2005)

(15) Technical Guide for Live Working of ± 800kV UHV Transmission Line (Q/GDW302-2009)

(16) Shielding Clothes for Live Working (GB/T6568—2008)

(17) Technical Requirements and Design Guide for Live Working Tools (GB/T18037-2008)

(18) Application Rules of Infrared Diagnosis for Live Electrical Equipment (DL/T664-2016)

(III) Teaching Design for Training

To complete the work task of "live replacement of ± 800kV UHV transmission line conductor spacer", each training stage shall be designed according to the standard operation procedure for work task completion. Each stage includes specific training objectives, training content, hours of training, training methods (training resources), training environment, assessment and evaluation, etc, as shown in the Table 1-5-1.

Part I Standard for Training and Assessment on Live Working for Operation Maintenance of ±800kV UHV Power Transmission Line

Table 1-5-1 Training Content Design for Live Replacement of Conductor Spacer of ±800kV UHV Transmission Line

Training process	Training objectives	Training contents	Training hours	Training methods and resources	Preparation of training conditions	Assessment and evaluation
1. Theoretical teaching	1. To master the basic method for entering and leaving ±800kV intense electric field along the insulator string. 2. To be familiar with the methods of potential transfer. 3. To be familiar with the method for replacing the damaged conductor spacer of power transmission line. 4. To be familiar with the safe distance, hazard identification and precontrol of live working on UHV DC line	1. The electrical significance of the "two-span and three-short-circuit" operation mode when entering and leaving the intense electric field along the insulator. 2. The use method for the potential transfer rod during entering and leaving the UHV intense electric field. 3. Method and quality standard for replacement of power transmission line conductor spacer. 4. Safe distance, hazard analysis and precontrol measures for live working on UHV DC line	2	Training methods: Lecture. Training resources: PPT, relevant regulations, specifications and technical guidelines	Multimedia classroom	Attendance, classroom questions and assignments
2. Preparations	Be able to complete the preparation before operation	1. Work site survey. 2. Preparation of the standardized operation card. 3. Filling of the work order. 4. Preparation of tools and materials for this operation	1	Training methods: 1. Site survey and preparation of tools, instruments and materials shall be practiced at site. 2. Preparation of operation card and the filling of work order shall adopt lecture method. Training resources: 1. ±800kV practical training line. 2. UHV tools and instruments warehouse 3. Blank work order	1.±800kV Training line 2.Multi-Media classroom	

· 363 ·

Table (Cont'd)

Training process	Training objectives	Training contents	Training hours	Training methods and resources	Preparation of training conditions	Assessment and evaluation
3. Work site preparation	Be able to complete the preparations of work site	1. Work site re-survey. 2. Work application. 3. Work site arrangement. 4. Pre-shift meeting. 5. Tools and materials inspection. 6. The use method of special spanner for spacer	1	Training methods: demonstration and role play. Resources: ±800kV training line	±800kV practical training line	
4. Trainer's presentation	The trainees can preliminarily understand the operation process of the task through inspecting and learning from each other's work	1. The equipotential electrician enters and exits the intense electric field and potential transfer along the strain insulator string. 2. The equipotential electrician reaches the replacement position of the spacer by means of walking along the line. 3. Equipotential electrician uses the special tools to complete the replacement of conductor spacer	2	Training methods: demonstration. Resources: ±800kV training line	±800kV practical training line	

Part I Standard for Training and Assessment on Live Working for Operation Maintenance of ±800kV UHV Power Transmission Line

Table (Cont'd)

Training process	Training objectives	Training contents	Training hours	Training methods and resources	Preparation of training conditions	Assessment and evaluation
5. Group training of trainees	1. Be able to enter and leave the ±800kV intense electric field along the insulator string and operate the potential transfer. 2. Be able to complete the replacement of conductor spacer in ±800kV transmission line	1. The trainees are grouped (6 in a group) to train the skill operation for entering and leaving ±800kV intense electric field, potential transfer and replacing the conductor spacer 2. Trainers guide the operation of trainees and conduct safety supervision	14	Training methods: Role play. Resources: ±800kV practical training line	±800kV practical training line	Score the operation of trainees according to the detailed rules for skill assessment and scoring
6. End of the work	1. Enable the trainees to further distinguish the shortcomings of the operation process and facilitate the promotion in the later stage. 2. Train the trainees in the work style of safe and civilized production	1. Work site cleaning. 2. Report to dispatcher. 3. Comment and summarize the work task this time at post-shift meeting	1	Training methods: Lecture and inductive method	±800kV practical training line	

(IV) Work Flow

1. Work task

The "two-span and three-short-circuit" operation mode is adopted to enter the intense electric field and reach the operation point along the strain insulator string, and the equipotential operation method is adopted for live replacement of the eight-bundle conductor spacer of ±800kV UHV transmission line.

(This operation task is applicable to the operation point of the first spacer for side phase conductor of resisting-tensile tower of ±800kV DC single circuit transmission line in 1000m and below altitude)

2. Requirements for Weather and Work Site

(1) The live replacement for eight-bundle conductor spacer of ±800kV UHV transmission line shall be carried out in good weather. In case of lightning (hearing thunder or seeing lightning), snow, hail, rain, fog and so on, live working is prohibited. Wind Power

When the wind force is greater than level 5, live working shall not be carried out; When the relative humidity is greater than 80%, insulating tool with moisture-proof performance shall be adopted if live working is required; when live rush repair must be carried out in bad weather, relevant personnel shall be organized to fully discuss and prepare necessary safety measures, which shall be carried out after approval of the Unit.

(2) The operating personnel should be in good mental states and be familiar with the organizational and technical measures to ensure safety in work; they should hold the qualification certificate for live working within the validity period.

(3) Responsible Person should organize the relevant personnel to complete field investigation in advance, determine the operating methods, required working apparatus and necessary measures according to the results, and handle the tickets for live working.

(4) The work site should be reasonably set up with fence and warning signs. Non-operating personnel is forbidden to enter.

(5) DC restart device shall be deactivated for the Project.

(6) Safe working distance and effective insulation length during operation are shown in Table 1-5-2.

Part I　Standard for Training and Assessment on Live Working for
Operation Maintenance of ±800kV UHV Power Transmission Line

Table 1-5-2　Safe Distance for Live Replacement of Conductor Spacer of ±800kV UHV Transmission Line (m)

Voltage class	Safe distance between human body and electrified body	Minimum distance to adjacent phase conductor	Minimum effective insulation length		Minimum combination gap	Minimum distance between the exposed part of human body and electrified body during potential transfer
			Insulating bar	Insulated load-bearing tools and insulating ropes		
±800kV	6.8 (7.3)	6.8 (7.3)	6.8	6.8	6.6	0.5

Note: ① When the altitude is above 1000m and the ±800kV DC single circuit transmission line is under live working, the safe distance shall be 7.3m in the brackets, the minimum effective insulation length of insulating tools shall be 7.3m in the brackets and the combinational gap shall be 7.2m in the brackets.

② As the phase-to-phase spacing of ±800kV UHV line is large enough (generally speaking, the distance between the phase and the ground shall be controlled) and no important safety factors are taken into account, the "minimum distance to the adjacent phase conductor" is not given in the "Safety Regulation", and the actual work can be carried out with reference to the data of 750kV.

(7) When working on ±800kV transmission line, ensure that the insulators in good operation phase shall be not less than 37 pieces (the single insulator is 170mm high), 32 pieces (the single insulator is 195mm high), 31 pieces (the single insulator is 205mm high), and 26 pieces (the height of single insulator is 240mm).

3. Preparations

3.1　Hazards and precontrol measures

(1) Hazard —— electric shock

Precontrol measures:

① Before work, the Responsible Person shall contact the control personnel on duty, deactivate the line DC restart device, and perform the licensing procedures.

② Before climbing the tower, equipotential operators on tower must carefully check the names of the line, the number of the pole and tower, and phase, and then the tower can be climbed after all have been confirmed correct.

③ If lines lose power suddenly during work, operators shall consider it as still charged. The Responsible Person shall contact the control personnel as soon as possible, and no forced energization is allowed before the on-duty control personnel getting in touch with the Responsible Person.

④ The ground electrician shall wear clean and dry gloves when operating the insulating tools. Insulating tools and insulating ropes shall not be damaged, damped, deformed, and malfunctioned. It is not allowed to use non-insulating ropes (such as cotton rope, manila rope, and steel wire rope). The tools and instruments for live working at the work site shall be placed on the moisture-proof tarpaulin to prevent them from dirt and moisture in use.

⑤ The equipotential operator shall wear flame-retardant underwear and full set of shielding clothes for ±800kV live working shall be worn outside the clothes (including coverall, helmet, protective mask, gloves, conductive socks and shoes). All parts shall be in excellent connection conditions. The resistance between the farthest points of the whole shielding clothes shall not be greater than 20Ω.

⑥ Before potential transfer, equipotential operator shall obtain the approval of Responsible Person, and the minimum distance between exposed part of human body and electrified body shall not be less than 0.5m. During potential transfer, potential transfer rod shall be adopted. The operation shall be fast, and end shall not be used for power charging and discharging; During delivering tools and materials to ground potential operators, the effective length of insulating tools or insulating ropes shall not be lower than the requirements as specified in Table 1-5-2.

⑦ When transmitting large metal objects by using insulating rope, the ground potential operators shall ground the metal objects before contacting with the insulating rope.

⑧ The Special Supervisor shall continuously monitor the operators and correct their nonstandard operation or actions in violation at any time. Special attention shall be paid to operators work at heights to ensure that there is enough safe distance (meeting the requirements in Table 1-5-2). It is forbidden to contact two non-connected electrified bodies or make contact with electrified body and grounding body at the same time.

(2) Hazard —— falling from high places

Precontrol measures:

① Before climbing, operators working at heights must satisfy the requirements of this operation, such as physical condition, mental state, and skill and quality.

② Before climbing the tower, operators working at heights shall conduct visual inspection and impulse test on safety belts and falling protector to ensure that their mechanical strength meets the requirements.

③ Operators who climb the tower shall first check whether the shackles is firm, shoe soles is clean, or the falling device is firm or is provided with the falling protector; grasp the main materials, tread on shackles and climb up (down) the tower at a uniform pace.

④ Supervisor shall correct irregularities and violations at any time. Special attention shall be paid to operators working at heights to prevent them from losing the protection of the safety belt or insulated backup protection rope during transposition. The safety belt shall be attached to a secure part and shall not be fastened lower than operating personnel.

⑤ When the equipotential operators move on the insulator string in parallel, they usually grasp one string with both hands and step on the other string with both feet to enter the intense electric field at a constant speed. During the movement, the backup protection rope holds two strings of insulators, and excessive hand waving and big step are not allowed.

⑥ Equipotential operators must fasten the safety belt and make the backup protection rope cover all the sub-conductors when they are walking along the conductor; they shall control the center of gravity during walking along the line to prevent the conductor from turning over.

(3) Hazard —— injury caused by objects falling from high place

Precontrol measures:

① Operator work at heights should put personal tools and fragmentary materials into the tools bag. It is strictly forbidden to hang objects in high place or keep in the mouth.

② Ground operator should correctly wear a safety helmet and use the knots to deliver the tools, instruments and materials. The vertical distance from the operation point should not be less than the falling radius.

③ The work site shall be set up with fence and warning signs. The Supervisor shall maintain vigilance at all times and prohibit non-staff members and vehicles from entering the operation area.

3.2 Selection of tools, instruments and materials

See Table 1-5-3 for tools, instruments and materials required by live replacement of eight-bundle conductor spacer of ±800kV UHV transmission line. Before delivering tools and instruments out of warehouse, application voltage class and test cycle shall be carefully checked and they shall be inspected to ensure that appearance is intact, connection is firm, rotation is flexible,and they meet the work task requirements. After delivering tools and instruments out of warehouse, they shall be stored in tools bag or tool kit for transportation to avoid contamination and damp. Metal tools and insulating tools shall be separately loaded and transported to avoid deformation, damage or other defects caused by mixed loading and transportation.

Table 1-5-3　Tools, Instruments and Materials Required by Live Replacement of Conductor Spacer of ± 800kV UHV Transmission Line

S/N	Name	Specification and Model	Unit	Qty.	Remarks
1	Insulated transmission rope	TJS-12, with the length matching the lifting height	Nos.	2	Insulating tool
2	Insulated backup protection rope	TJS-16, with buffer provided	Nos.	2	Insulating tool
3	Insulated tackle	JH10-0.5	pcs	2	Insulating tool
4	Insulated noose	TJS-14	Nos.	2	Insulating tool
5	Potential transfer rod	0.4m	Nos.	1	Insulating tool
6	Insulation jacks		Nos.	4	Insulating tool
7	I-type shielding clothes (coverall, helmet, protective mask, gloves and conductive socks)	Shielding efficiency ≥60dB (shield efficiency of shielding mask ≥20dB)	Set	2	PPE
8	Conductive shoes	The size depends on the wearer	Pair	2	PPE
9	Flame retardant underwear	Pure mulberry silk	Set	2	PPE
10	Safety belt of double insurance	Full-body harness	Nos.	2	PPE
11	Falling protector	Corresponding to the type of pole and tower falling protector	pcs	2	PPE
12	Safety helmet		pcs	6	PPE
13	Special spanner for spacer	Used for eight-bundle spacer	Nos.	1	Special tools
14	Insulation resistance tester	5000V, with electrode width 2cm and interelectrode width 2cm	Set	1	Other tools
15	Wind speed, temperature and humidity tester	HT-8321	Set	1	Other tools
16	Multimeter		pcs		Other tools
17	Interphone	As required by the work	Set	2	Other tools
18	Moisture-proof tarpaulin	2m×4m	pcs	2	Other tools
19	Security fence		Set	Several	Other tools
20	Warning sign	"Work Here", "Access from Here" "Access from Here"	Set	1	Other tools

Table (Cont'd)

S/N	Name	Specification and Model	Unit	Qty.	Remarks
21	Red waistcoat	"Responsible Person"	pcs	1	Other tools
22	Clean towel	Cotton	pcs	1	Other tools
23	Shoe covers		Pair	Several	Other tools
24	Work gloves		Pair	Several	Other tools
25	Personal tools	Tool bag, flat tong, marker pen	Set	2	Other tools
26	Eight-bundle spacer	The same model with replaced spacer	pcs	1	Material

Note: The electrical strength of the insulating tool shall meet the requirements of the "State Grid Corporation of China Working Regulations of Power Safety (Transmission Line Section)", and the test is required to be qualified and within the validity period.

3.3 Division of labor for operators

Division of labor for operators of the task is shown in Table 1-5-4.

Table 1-5-4 Division of Labor for Live Replacement of Conductor Spacer of ± 800kV UHV Transmission Line

S/N	Post	Qty. (person)	Responsibilities
1	Responsible Person	1	Be responsible for division of operating personnel of the task, work order reading, handling re-closing deactivation of the line and the formalities of work permit, getting work permits, holding pre-shift meeting, dealing with emergency situations in work, quality surveillance, and the summary after work
2	Specific responsible supervisor	1	Be responsible for safety supervision and control during operation
3	Equipotential electrician	1	Be responsible for the replacement of eight-bundle conductor spacer after entering the equipotential
4	Ground potential electrician on the tower	1	Assist the equipotential electrician to enter and exit the intense electric field
5	Ground electrician	2	Be responsible for taking site safety measures, arranging operation site, inspecting tools and instruments, delivering tools and materials, and cooperating with the equipotential electrician to enter and leave equipotential

4. Working Procedures

The workflow of this task is shown in Table 1-5-5.

Table 1-5-5 Workflow Sheet for Live Replacement of Conductor Spacer of ± 800kV UHV Transmission Line

S/N	Work Content	Operation Steps and Standards	Safety Measures and Precautions	Responsible Person
1	Site re-survey	The Responsible Person shall complete the following work: (1) Check the line name, the number of the pole and tower and ensure the phases are correct; guarantee that the foundation and the pole and tower are intact and in normal condition; ensure that the cross and span distance meets the safety requirements; confirm the defect conditions and the specifications and models of earth wires. (2) Check that the site meteorological conditions such as wind speed and humidity should meet the operation requirements. (3) Check that the terrain and environment should meet the operation requirements. (4) Check that the safety measures listed in the work order are in line with the actual situations on site. The measures will be added if necessary.	(1) Correctly wear helmet, working clothes, work shoes and protective gloves. (2) Operation under meteorological conditions that may endanger the safety of operators is forbidden. (3) Non-operation personnel and vehicles are strictly prohibited from entering the working site	
2	Work Permit	(1) The Responsible Person is in charge of contacting the control personnel and apply to stop line re-closing as per the content of work order. (2) Live working could be started only after being approved by control personnel on duty.	Live working shall not be started without the permission of the on-duty control personnel	

Part I Standard for Training and Assessment on Live Working for Operation Maintenance of ±800kV UHV Power Transmission Line

Table (Cont'd)

S/N	Work Content	Operation Steps and Standards	Safety Measures and Precautions	Responsible Person
3	Site layout	Install the security fence and hang the signboards correctly: (1) The security fence should take full account of falling objects from the high place and the influence on road traffic. (2) The entrance and exit of the security fence shall be set reasonably. (3) Signs such as "Access from Here", "Work Here", "Access from Here" shall be properly and well arranged	When the influence on road traffic safety is uncontrollable, the traffic management department should be contacted in time to strengthen the on-site control of traffic safety	
4	Holding a pre-shift meeting	(1) All working personnel shall line up. (2) Responsible Person will read out the work order and be clear with work task and division of personnel; explain safety measures and technical measures in work; check (inquire after) mental state of all working personnel; inform of hazards in work and precontrol measures. (3) All working personnel shall sign on the work order for confirmation.	(1) Work order shall be filled and issued with standardized licensing procedure and complete signature. (2) All working personnel shall be in good mental states. (3) All working personnel shall be clear with task division of works, safety measures and technical measures	
5	Inspection of tools and instruments	(1) The ground potential electrician and equipotential electrician on the tower shall wear the shielding clothes in a right way and pass the inspection, which shall be supervised and inspected by the Responsible Person. (2) Wear personal safety equipment correctly (proper size and easy lock), and the Responsible Person shall supervise and inspect it. (3) Measure the wind speed, wind direction and humidity, check the insulation performance of insulating tools, and make records	(1) Check carefully for damage, deformation and failure before using metal and insulating tools. Carry out segment insulation detection with such insulating tools as insulation resistance tester of 5000V or above and with the resistance no less than $700M\Omega$, and wipe them clean with the clean dry towel. (2) Use a multimeter to measure the resistance between the farthest ends of the shielding clothes and trousers, which shall not be greater than 20Ω. The Responsible Person shall check the connection of the electrician's shielding clothing.	

Table (Cont'd)

S/N	Work Content	Operation Steps and Standards	Safety Measures and Precautions	Responsible Person
5	Inspection of tools and instruments		(3) Check the tool assembly and make sure the connection is reliable. (4) The tools and instruments for live working shall be placed on the moisture-proof tarpaulin	
6	Climbing the tower	(1) After checking the line name and pole and tower number, the ground potential electrician on the tower and equipotential electrician shall impact to check the stress of safety belt and falling protector. (2) The ground potential electrician on tower carries the insulated transmission rope to climb the tower, and the equipotential electrician then climbs the tower. When they reach the operation point of the cross arm, they choose the appropriate position to fasten the safety belt, and the ground potential electrician on the tower installs the insulated tackle and the insulated transmission rope in the appropriate position of the cross arm. Then he shall cooperate with the ground electrician to separate the insulated transmission rope for lifting preparation	(1) After checking the correct line name and pole and tower number, he can climb for operation. (2) The anti-falling device installed on the tower shall be used during climbing the tower; when moving and transposing on the pole and tower, the safety protection shall not be lost, and the operators must climb and grasp the components securely. (3) The working electrician must wear full set of qualified shielding clothes which must be connected reliably. Before entering the equipotential at cross arm, the equipotential electrician shall re-check and confirm that each part of the shielding clothes are connected reliably before the next operation	
7	Entering the intense electric field	(1) The equipotential electrician shall transfer the safety belt to the insulator set fitting, and carry the potential transfer rod, insulated tackle and insulated transmission rope.	(1) The equipotential electrician must obtain the permission of the Responsible Person before entering the potential.	

Part I Standard for Training and Assessment on Live Working for
Operation Maintenance of ±800kV UHV Power Transmission Line

Table (Cont'd)

S/N	Work Content	Operation Steps and Standards	Safety Measures and Precautions	Responsible Person
7	Entering the intense electric field	(2) After the equipotential electrician confirms that all parts of the shielding clothes are well connected, it shall be reported to the responsible person for approval. The electrician shall grasp one string with both hands and step on the other string with both feet, and enter the intense electric field along the insulator string in the operation mode of "two-span and three-short-circuit". (3) When reaching the third insulators outside the grading ring on the conductor side, the operator shall stop, and carry out the potential transfer with the potential transfer rod	(2) When the equipotential electrician enters the insulator string, he shall use the safety belt and the backup protection rope alternately (holding two insulator strings with the backup protection rope, grasping one string with hands and stepping on the other string) without losing the protection of the safety belt, and adjust the insulated transmission rope and the potential transfer rod. (3) The equipotential electrician shall coordinate his hands and feet during entering the potential with uniform speed so as to avoid excessive hand waving, big step and other actions; The side phase combined gap composed of the grounding body and the electrified body shall be greater than 6.6m. (4) The minimum distance to adjacent conductors shall be greater than 6.8m. (5) Before the potential transfer, the equipotential electrician shall check whether the electrical connection between the potential transfer rod and the shielding clothes is reliable and that the minimum distance between the exposed part of the human body and the electrified body is greater than 0.5m, and shall get the permission of the Responsible Person; The protection of safety belt shall not be lost during potential transfer, and the action shall be accurate, stable and rapid when entering the intense electric field	

Table (Cont'd)

S/N	Work Content	Operation Steps and Standards	Safety Measures and Precautions	Responsible Person
8	Replace the eight-bundle spacer	(1) After entering the equipotential, the equipotential electrician shall tie the safety belt to the upper sub-conductor, and install the insulated protection rope (all sub-conductors shall be covered). (2) The equipotential electrician shall carry the insulated transmission rope and walk to the operation point for replacing the spacer along the conductor, and shall install the insulating noose at the appropriate place of the sub-conductor Position, then connect the insulated tackle and the insulated transmission rope, and then hang the insulated tackle into the insulated noose. (3) Equipotential electrician shall mark the installation points of the old spacer on conductor at 4 points symmetrically with the marker pen. (4) The equipotential electrician shall install 4 insulation jacks in the appropriate position next to the old spacer in the way of "two-two corresponding" to fix the sub-conductor reliably. (5) The equipotential electrician shall use the insulated transmission rope to fasten the old spacer by the way of slip knot, then use the special spanner for spacer to remove the old spacer, and put it on the ground with the insulated transmission rope in cooperation with the ground electrician. (6) The ground electrician shall lift the new spacer to the position of equipotential electrician. The equipotential electrician shall mark and install the new spacer correctly. After installation, the plane of the spacer shall be vertical to the sub-conductor. (7) The equipotential electrician shall remove the insulated jack, insulated tackle, insulated transmission rope and noose successively. (8) No tools, instruments or materials shall drop during the equipotential operation	(1) The equipotential electrician shall not lose the protection of safety belt. (2) Equipotential electrician shall cooperate closely with the ground electrician and follow the command of the Responsible Person. (3) The minimum distance to adjacent conductors shall be greater than 6.8m. (4) In the process of delivering, the conductor spacer shall not be bumped, and the insulated transmission ropes shall not be intertwined. (5) When delivering tools, the binding rope shall be correct and reliable so as to prevent objects falling from heights. (6) During delivering the tools, instruments and materials, the ground electrician is not allowed to stand directly below the operation point of equipotential electrician	

Part I Standard for Training and Assessment on Live Working for
Operation Maintenance of ±800kV UHV Power Transmission Line

Table (Cont'd)

S/N	Work Content	Operation Steps and Standards	Safety Measures and Precautions	Responsible Person
9	Leaving the intense electric field	(1) After inspecting that the spacer is installed firmly and there is no object left at the operation point, the equipotential electrician, with the permission of the Responsible Person, can return to the grading ring with insulated transmission ropes along the conductor for the preparation to leave the potential. (2) The equipotential electrician shall hook the grading ring with the potential transfer rod and enter the third insulator of the grading ring. With one hand grasping the insulator tightly and the other holding the potential transfer rod, the equipotential electrician shall break himself away from the equipotential with the help of potential transfer rod. (3) The equipotential electrician exits the intense electric field by means of "two-span and three-short-circuit" operation mode	(1) The equipotential electrician must obtain the permission of the Responsible Person before leaving the potential. (2) When the equipotential electrician returns to the insulator string, the safety belt and backup protection rope shall be used alternately. The protection of safety belt shall not be lost during potential transfer, and the instant action shall be accurate, stable and rapid when leaving the intense electric field. (3) The equipotential electrician shall coordinate his hands and feet during leaving the potential with uniform speed so as to avoid excessive hand waving, big step and other actions; The side phase combined gap composed of the grounding body and the electrified body shall be greater than 6.6m; The minimum distance between the exposed part of human body and the electrified body should not be less than 0.5m. (4) When moving along the insulator string, the equipotential electrician shall use the backup protection rope to hold two insulator strings with the hands grasping firmly and the feet stepping firmly. (5) The minimum distance to adjacent conductors shall be greater than 6.8m. (6) The equipotential electrician shall not lose the protection of safety belt or backup protection rope when returning to cross arm	

Table (Cont'd)

S/N	Work Content	Operation Steps and Standards	Safety Measures and Precautions	Responsible Person
10	Return to the ground	After checking that there is no object left on the tower, the equipotential electricians on the tower shall report it to the Responsible Person and then climb down the tower with insulated transmission rope after obtaining the consent of the Responsible Person	The anti-falling device installed on the tower shall be used in the process of climbing down the tower; when moving and transposition on the pole and tower, the safety protection shall not be lost, and the operators must grasp the components securely	
11	End of the work	(1) The Responsible Person shall organize all working personnel to put working apparatus and materials in order and put them in a special kit (bag) after cleaning; clean the site to ensure that "the materials are removed and the site is cleaned after construction". (2) After a post-shift meeting is held, the Responsible Person shall make work summaries and comments. Comments include the construction quality of this work and the implementation of safety measures from all working personnel. (3) Responsible Person shall report to the on-duty control personnel that the work is over, apply for the restoration of circuit re-closing and terminate the work order.	It is forbidden to restore the line restart device at the appointed time	

II. Assessment Standard

Part I Standard for Training and Assessment on Live Working for Operation Maintenance of ±800kV UHV Power Transmission Line

Table 1-5-6 Detailed Rules for Assessment and Scoring of Operation and Inspection Skills of UHV DC Transmission Line

Fill-in Column of Examinee	No.:		Name:		Position:		Unit:		Date:	MM/DD/YYYY	
Fill-in Column of Assessor	Grade:		Assessor:		Assessment Team Leader:		Starting time:		Closing time:	Operation duration:	
Assessment Module	Live replacement of ±800kV UHV transmission line conductor spacer			Assessee	Maintenance Personnel of UHV DC Transmission Line			Assessment method		Operation	Assessment Time Limit
											90min
Task Description	Enter the intense electric field along the strain insulator strings and carry out live replacement of the eight-bundle conductor spacer of ±800kV UHV transmission line (equipotential operation method)										
Work Specifications and Requirements	1. Live working shall be carried out in good weather. In case of thunder, rain, snow or fog, no live working shall be carried out. When the wind force is greater than Level 5 and the humidity is greater than 80%, live working should not be carried out. 2. Persons required for this operation include 1 Responsible Person, 1 special Supervisor, 1 ground electrician on the tower, 1 equipotential electrician and 2 ground auxiliary electricians. Enter the intense electric field along the strain insulator strings and carry out the live replacement of the eight-bundle spacer of the damaged conductor of ±800kV UHV transmission line. 3. Responsibilities of Responsible Person: Be responsible for division of operating personnel of the task, work order reading, handling re-closing deactivation of the line and the formalities of work permit, getting work permits, holding pre-shift meeting, dealing with emergency situations in work, quality surveillance, and the summary after work. 4. Special Supervisor: Be responsible for safety supervision and control during operation. 5. Equipotential electrician: Enter the intense electric field along the strain insulator strings and carry out the live replacement of the eight-bundle conductor spacer of the transmission line. 6. Responsibilities of electrician on the tower: Be responsible for assisting the equipotential electrician to enter and leave the intense electric field. 7. Responsibilities of ground electrician: Be responsible for implementing site safety measures, arranging work site, inspecting tools and instruments, delivering tools and materials, and cooperating with the equipotential electrician to enter and leave equipotential. 8. During the live working, if thunder, rain, strong wind or any other circumstance threaten the safety of the staff, the Responsible Person or Supervisor may stop working temporarily according to the circumstances.										

Table (Cont'd)

Assessment scenario preparation	Given conditions: 1. Training base: UHV DC ±800kV training line resisting-tensile tower large side phase A conductor first eight-bundle spacer, and model of the conductor is: 8×JL/G1A-630/45. 2. Work orders have been handled, safety measures have been completed (re-closing lock has been deactivated), and oral application (dispatcher or assessor) shall be made at the beginning and end of the work. 3. Security fence shall be installed at the work site, signs such as "Work here" and "Access from Here" shall be hung and safety measures have been completed. 4. The instruments and apparatus shall be used safely and correctly to test the insulating tool. 5. Anti-fall device should be used during climbing up (down) the tower to prevent falling from the heights. 6. The operation must be carried out according to Standard Operation Procedures. The relevant item scores shall be deducted if the working procedure is wrong. In case of major hidden dangers of personal, equipment and operational safety, the assessor may order the termination of the operation (assessment) and the assessment of this module is recorded as "unqualified" 1. Line: UHV DC ±800kV training line resisting-tensile tower large side phase A conductor, work content: live replacement of ±800kV UHV transmission line conductor spacer, and model of the conductor is: 8×JL/G1A-630/45. 2. Required operation tools and instruments: 2 insulated transmission ropes (TJS-12), 2 insulated backup protection ropes (TJS-16, with buffer), 2 insulated tackles (JH10-0.5), 2 insulation nooses (TJS-14), 4 insulation jacks, 1 potential transfer rod (0.4m), 2 Type-1 shielding clothes (coverall, helmet, protective mask, helmet for vibration damper, 1 set of insulation resistance tester (type 5000V), 1 set of wind speed and humidity tester (HT-8321), 1 multimeter, 2 moisture-proof tarpaulins (2m*4m), 1 red waistcoat (for Responsible Person), 2 clean towels, 2 sets of personal tools (workbasket, flat tongs, marker pen), and 1 eight-bundle spacer of the same model. 3. The work site shall be monitored, and the safety measures (fence, etc.) on the work site have been fully implemented; non-operation personnel are prohibited from entering the site, and the staff must wear safety helmets when entering the work site. 4. Examinees shall bring their own work clothes, flame retardant cotton underwear, safety helmets, gloves, and safety belts (including double-protective ropes)
Remarks	1. The total score of this module is 100 points, and score of each item is deducted until the corresponding point becomes zero. If the task is not completed within the specified time, the test shall be terminated immediately. The score of this module shall be calculated according to the actual score of the completed item, and the unfinished item shall not be scored. 2. In the process of assessment, if the equipment, operating environment, safety measures, safety protection and safe distance do not meet the requirements of the operation, or human misoperation occurs, which may endanger the safety of the operation, the assessor shall order the termination of the operation. 3. Exam participants should be organized for site survey before the examination and work orders should be handled in advance

Part I Standard for Training and Assessment on Live Working for Operation Maintenance of ±800kV UHV Power Transmission Line

Table 1-5-7 Standards for Assessment and Scoring of Operation and Inspection Skills of UHVDC Transmission Line

S/N	Project name	Quality requirements	Score	Deduction standard	Reasons for deduction	Deduction	Scoring
1	Site re-survey	(1) The Responsible Person shall go to the work site to check the line name, pole and tower number, on-site working conditions, defective parts and so on. (2) Check that the site meteorological conditions such as wind speed and humidity should meet the operation requirements. (3) Check whether the work order is complete and unmodified, check whether the safety measures listed are consistent with the actual situation on site, and supplement it if necessary	5	(1) No point will be awarded without a work order. (2) Deduct 1 point for failure to check the double title. (3) Deduct 1 point / item for failure to check on-site working conditions (meteorology) and defective parts. (4) Deduct 0.5 points for each alteration or untidy in the work order filling; deduct 1 point/ item for incorrect work order number; deduct 1 point for incomplete work order filling			
2	Work Permit	(1) The Responsible Person is in charge of contacting the on-duty control personnel and applies to stop line re-closing lock as per the content of work order. (2) Report content is standardized and complete. The voice is loud and clear. (3) Relevant licensing procedures are completed in a timely manner	2	(1) No point will be awarded for items of work without the work permit of the Dispatching Department (Assessor). (2) Deduct 0.5 points/item for non-standard and incomplete reporting terminology or the voice is not loud or clear enough. (3) Deduct 1 point for failure to apply the re-closing lock. (4) Deduct 1 point for failure to repeat permission content.			

Standard for Professional Training and Assessment for Operation Maintenance of UHV DC Power Transmission Line

Table (Cont'd)

S/N	Project name	Quality requirements	Score	Deduction standard	Reasons for deduction	Deduction	Scoring
2	Work Permit		2	(5) Deduct 0.5 points / item for failure to repeat each permission content (name, time, task and re-closing lock status of the Licensor). (6) Deduct 1 point for failure to complete work order in time			
3	Site layout	Install the security fence and hang the signboards correctly: (1) The security fence should take full account of falling objects from the high place and the influence on road traffic. (2) The entrance and exit of the security fence shall be set reasonably. (3) Signs such as "Access from Here", "Work Here", "Access from Here" shall be properly arranged	3	(1) Deduct 2 points for failure to arrange the security fence at the work site. (2) Deduct 1 point/place for failure to set the security fence at the work site properly. (3) Deduct 1.5 points for failure to hang signboards. (4) Deduct 0.5 points per piece for incomplete signboards. (5) Deduct 0.5 points/person for non-operating personnel entering the fenced area			
4	Holding a pre-shift meeting	(1) All staff and personnel shall wear safety helmets and work clothes correctly.	3	(1) Deduct 0.5 points/person/time for the incorrect wear of safety helmet by working personnel. Deduct 0.5 points/person/time for improper dressing.			

· 382 ·

Part I Standard for Training and Assessment on Live Working for Operation Maintenance of ±800kV UHV Power Transmission Line

Table (Cont'd)

S/N	Project name	Quality requirements	Score	Deduction standard	Reasons for deduction	Deduction	Scoring
4	Holding a pre-shift meeting	(2) The Responsible Person shall wear red waistcoat and read out the work order and be clear with work task and division of personnel; explain safety measures and technical measures in work; check (inquire after) mental state of all working personnel; inform of hazards in work and precontrol measures. (3) All working personnel shall sign on the work order for confirmation.	3	(2) Deduct 0.5 points/person for the Responsible Person and Special Supervisor failing to wear safety red waistcoat. (3) No point will be awarded for unclear work task and division of work. (4) Deduct 1 point for unclear division of labor. (5) Deduct 1 point for incomplete explanation of safety measures or precontrol measures. (6) Deduct 1 point for failure to inform hazards at work. (7) Deduct 1 point for failure to confirm mental status of work shift members. (8) Deduct 1 point for the work shift member failing to sign or signing incompletely on the work order.			
5	Inspection of tools and instruments	(1) The work shift members shall properly set the moisture-proof tarpaulin in the appropriate position. The tarpaulin shall be clean and dry. It is strictly prohibited to step on the tarpaulin.	7	(1) Deduct 1 point for the inappropriate position of moisture-proof tarpaulin. (2) Deduct 0.5 points/time for stepping on moisture-proof tarpaulin. (3) Deduct 1 point for failure to classify or place tools and instruments at fixed position.			

· 383 ·

Table (Cont'd)

S/N	Project name	Quality requirements	Score	Deduction standard	Reasons for deduction	Deduction	Scoring
5	Inspection of tools and instruments	(2) The tools and instruments shall be classified and placed neatly on moisture-proof tarpaulin according to the requirements of the fixed location management; the insulating tools shall not be mixed with metal tools and materials; and visual inspection shall be carried out on tools, instruments and apparatus. (3) All kinds of tools shall be qualified and within the validity period of test. The surface of insulating tools shall not be worn, deformed and damaged, and the operation shall be flexible. Carry out segment insulation detection with such insulating tools as insulation resistance meter of 5000V or above and with the resistance no less than 700MΩ, and wipe them clean with the clean dry towel. (4) The ground potential on the tower and equipotential personnel shall correctly wear full set of qualified shielding clothes and conductive shoes as required, each of which shall be connected well. They shall not wear chemical fiber clothes next to the skin in the shielding clothes and shall fasten safety belts; the	7	(4) Deduct 1 point/item for failure to check the "qualified" label and appearance of tools, instruments and apparatus. (5) Deduct 0.5 points per piece for missing items during inspecting the tools, instruments and apparatus. (6) Deduct 0.5 points per piece for incorrect inspection method of tools, instruments and apparatus. (7) Deduct 0.5 points per piece for failure to clean or wipe the hard insulating tools. (8) Deduct 0.5 points/time for holding insulating tools without wearing clean and dry cotton gloves. (9) Deduct 1 point/item for the improper usage of testing instruments to test the tools, instruments and the full set of shielding clothes. (10) Deduct 2 points/person/time for the tower electrician failing to wear the shielding clothes correctly or to check the connecting portion. (11) Deduct 1 point/ item: no visual inspection or impulse test (or incorrect way) for safety belt, backup protection rope and falling protector			

Part I Standard for Training and Assessment on Live Working for Operation Maintenance of ±800kV UHV Power Transmission Line

Table (Cont'd)

S/N	Project name	Quality requirements	Score	Deduction standard	Reasons for deduction	Deduction	Scoring
5	Inspection of tools and instruments	Responsible Person shall check and confirm carefully whether they wear them correctly and all parts are well-connected. (5) The full set of shielding clothes shall be tested with a multimeter, and the resistance value between the farthest two points shall not be more than 20Ω, and the resistance value of a single set shall not be more than 15Ω. (6) The tower electrician shall conduct the visual inspection on the safety belt, backup protection rope and falling protector, which shall pass the impulse test.	7	(12) Deduct 1 point/item for the Responsible Person failing to check or missing to check the safety protection equipment of tower electrician			
6	Climbing the tower	(1) Tower-climbing personnel shall check the double name, pole number and phase again, report to the Responsible Person and apply for climbing the tower. (2) The ground potential electrician on tower and equipotential electrician climb the tower with the insulated transmission rope one after another.	5	(1) Deduct 1 point/item for the tower electrician failing to check the double name, tower number and phase. (2) Deduct 1 point/item for the tower electrician failing to report the result or apply for climbing the tower. (3) The assessor shall order the termination of the operation (assessment) if the tower electrician fails to use anti-fall device during climbing the tower.			

Standard for Professional Training and Assessment for Operation Maintenance of UHV DC Power Transmission Line

Table (Cont'd)

S/N	Project name	Quality requirements	Score	Deduction standard	Reasons for deduction	Deduction	Scoring
6	Climbing the tower	(3) During tower climbing, the electrician must use anti-fall device, climb the tower at a uniform speed with the main materials in hand, hang the safety belt and backup protection rope on the shoulder and keep the safe distance of more than 6.8m away from the electrified body, and the Responsible Person shall strengthen the operation monitoring. (4) When climbing to an appropriate position, ground potential electrician shall fasten the safety belt, arrange the insulated transmission rope, and then cooperate with the ground electrician to make lifting preparation of the insulated transmission rope separately. (5) The Responsible Person shall carefully monitor and remind the whole process of tower climbing	5	(4) Deduct 0.5 points/time for grasping the shackles with hands during climbing the tower. (5) Deduct 1 point/ time for slipping or missing his step. (6) Deduct 1 point/ person for main band of safety belt and backup protection rope failing to be hung on the shoulder. (7) Deduct 1 point/ time for safety belt and backup protection rope twining and getting caught. (8) Deduct 1 point/ time for fastening the safety belt and backup protection rope lower than operating personnel. (9) Deduct 1 point/ time for fastening the safety belt and backup protection rope on the same components. (10) No point will be awarded for losing protection of safety belt during work at heights. (11) Deduct 1 point for installation of tackle without using insulating noose. (12) Deduct 1 point for the suspension position of the tackle transmission rope inconvenient for getting tools. (13) Deduct 1 point for the Responsible Person failing to monitor the operation in place. (14) The Assessor shall order the termination of the operation (assessment) for insufficient safe distance			

Part I Standard for Training and Assessment on Live Working for Operation Maintenance of ±800kV UHV Power Transmission Line

Table (Cont'd)

S/N	Project name	Quality requirements	Score	Deduction standard	Reasons for deduction	Deduction	Scoring
7	Entering the intense electric field	(1) Check that all parts of the shielding clothes are well connected before entering the potential, and report to the Responsible Person for approval. (2) The equipotential electrician shall transfer the safety belt to the insulator set fitting, and carry the potential transfer rod, insulated tackle and insulated transmission rope. (3) Before entering the insulator string, the equipotential electrician shall fasten the protection rope (holding two insulator strings with the backup protection rope, grasping one string with hands and stepping on the other string). (4) When reaching the third insulators outside the grading ring on the conductor side along the insulator string by the operating method of two-spar and three-short-circuit, the operator shall stop, wait for the approval of the Responsible Person and then carry out the potential transfer with the potential transfer rod.	10	(1) Deduct 2 points for the equipotential electrician failing to check the connecting parts of the shielding clothes. (2) Deduct 2 points for losing the protection of safety belt when equipotential electrician transfers to insulator set fitting. (3) Deduct 2 points for insufficient combined clearance between equipotential electrician and grounding body & electrified body. (4) Deduct 2 points for incorrect and unskilled action of equipotential electrician entering the intense electric field. (5) Deduct 2 points for repeated discharge due to insufficient distance of the exposed parts during potential transfer of equipotential electrician. (6) Deduct 1 point for being unskilled in potential transfer.			

Table (Cont'd)

S/N	Project name	Quality requirements	Score	Deduction standard	Reasons for deduction	Deduction	Scoring
7	Entering the intense electric field	(5) The side phase combined gap composed of the gaps between the equipotential electrician and the grounding body & the electrified body in the process of entering the potential shall not be less than 6.6 m, and the potential transfer rod must be used for potential transfer when entering the intense electric field. The minimum distance between the exposed part of human body and the electrified body should not be less than 0.5m. (6) Protection of safety belt shall not be lost when entering the intense electric field	10	(7) Deduct 5 points for potential transfer without using the potential transfer rod. (8) Deduct 3 points for carrying out potential transfer without permission of the Responsible Person. (9) Deduct 2 points for losing the protection of safety belt when equipotential electrician enters the intense electric field. (10) Deduct 2 points for the Responsible Person failing to monitor the operation in place			
8	Replace the eight-bundle spacer	(1) After entering the equipotential, the equipotential electrician shall tie the pole belt to the upper sub-conductor, and install the insulated protection rope (all sub-conductors shall be covered). (2) The equipotential electrician shall carry the insulated transmission rope and walk to the operation point for replacing the spacer, and shall correctly install the insulating noose at the appropriate place of the sub-conductor and hang insulated tackle and insulated transmission rope.	40	(1) Deduct 3 points for the incomplete covering of sub-conductor by insulated protection rope. (2) Deduct 2 points for failure to tie pole belt to upper sub-conductor during walking along the line. (3) Deduct 2 points for being unskilled in walking along the line. (4) Deduct 5 points for the insulated tackle hooked directly on the conductor.			

Part I Standard for Training and Assessment on Live Working for Operation Maintenance of ±800kV UHV Power Transmission Line

Table (Cont'd)

S/N	Project name	Quality requirements	Score	Deduction standard	Reasons for deduction	Deduction	Scoring
8	Replace the eight-bundle spacer	(3) The equipotential electrician shall mark the fixed point of old spacer symmetrically with marker pen. (4) The equipotential electrician shall install 4 insulation jacks in the appropriate position next to the old spacer in the way of "two-two corresponding" to fix the sub-conductor reliably. (5) The equipotential electrician shall fasten the insulated transmission rope to the old spacer by the way of slip knot, then use the special spanner for spacer to remove the old spacer, and put it on the moisture-proof tarpaulin in cooperation with the ground electrician. (6) The ground electrician shall lift the new spacer to the position of equipotential electrician. (7) The equipotential electrician shall mark and install the new spacer with the special spanner for spacer correctly. The installation quality shall be that the plane of the spacer shall be vertical to the sub-conductor. (8) No tools, instruments or materials shall drop during the operation	40	(5) Deduct 1 point/item for incorrect installation method or inappropriate position of insulating noose. (6) Deduct 1 point for failure to lock the insulated tackle after it is hooked on insulating noose. (7) Deduct 3 points for no marking. (8) Deduct 2 points per nos. for failure to fix the sub-conductor with 4 insulated jacks. (9) Deduct 5 points for removing the spacer without tying the knot. (10) Deduct 2 points for removing tools and instruments in the hurry and confusion. (11) Deduct 3 points per piece for falling object from heights. (12) No point will be awarded at this module for throwing old spacer at heights. (13) Deduct 2 points/person/time for the ground electrician standing directly below the work point of equipotential electrician. (14) Deduct 1 point for failure to put old spacer on moisture-proof tarpaulin. (15) Deduct 1 point for incorrect way the knot is tied when passing tools up and down.			

Table (Cont'd)

S/N	Project name	Quality requirements	Score	Deduction standard	Reasons for deduction	Deduction	Scoring
8	Replace the eight-bundle spacer	(9) The equipotential electrician shall remove the insulated jack, insulated tackle, insulated transmission rope and noose successively. (10) Equipotential operation process	40	(16) Deduct 5 points for untying the knot before the new spacer is installed. (17) Deduct 3 points for new spacer deviating from original spacer position. (18) Deduct 3 points for the plane of new spacer failing to be perpendicular to the sub-conductor after installation. (19) Deduct 1 point per piece for not removing insulated jack, insulated tackle, insulated transmission rope and noose. (20) Deduct 2 points for not removing insulated jack, insulated tackle, insulated transmission rope and noose successively. (21) No point will be awarded at this module for failure to replace the old spacer by the new spacer. (22) The assessor shall order the termination of the operation for insufficient safe distance between adjacent conductors			
9	Leaving the intense electric field	(1) After inspecting that the spacer is installed firmly and there is no object left at the operation point, the equipotential electrician, with the permission of the Responsible Person, can return to the grading ring with insulated transmission ropes along the conductor for the preparation to leave the potential.	10	(1) Deduct 2 points per item for objects left in operation point. (2) Deduct 3 points for failure to apply to the Responsible Person for potential transfer.			

Part I Standard for Training and Assessment on Live Working for Operation Maintenance of ±800kV UHV Power Transmission Line

Table (Cont'd)

S/N	Project name	Quality requirements	Score	Deduction standard	Reasons for deduction	Deduction	Scoring
9	Leaving the intense electric field	(2) The equipotential electrician shall hook the grading ring with the potential transfer rod and enter the third insulator of the grading ring. With one hand grasping the insulator tightly and the other holding the potential transfer rod, the equipotential electrician shall break himself away from the equipotential with the help of potential transfer rod. (3) Protection of safety belt shall not be lost when leaving the intense electric field. (4) The equipotential electrician leaves the intense electric field along the insulator string in the "two-span and three-short-circuit" operation mode and transfers to the cross arm. (5) The side phase combined gap composed of the gaps between the equipotential electrician and the grounding body & the electrified body in the process of leaving the potential shall not be less than 6.6 m, and the potential transfer rod must be used for potential transfer when leaving the intense electric field. The minimum distance between the exposed part of human body and the electrified body should not be less than 0.5m	10	(3) Deduct 2 points for carrying out potential transfer without permission of the Responsible Person. (4) Deduct 5 points for potential transfer without using the potential transfer rod. (5) Deduct 1 point for being unskilled in potential transfer. (6) Deduct 2 points for losing the protection of safety belt when equipotential electrician leaves the intense electric field. (7) Deduct 2 points for repeated discharge due to insufficient distance of the exposed parts during potential transfer of equipotential electrician. (8) Deduct 2 points for incorrect and unskilled action of equipotential electrician leaving the intense electric field. (9) Deduct 2 points for insufficient combined clearance between equipotential electrician and grounding body & electrified body. (10) Deduct 2 points for losing the protection of safety belt when equipotential electrician transfers to the cross arm. (11) Deduct 2 points for the Responsible Person failing to monitor the operation in place			

Table (Cont'd)

S/N	Project name	Quality requirements	Score	Deduction standard	Reasons for deduction	Deduction	Scoring
10	Return to the ground	(1) After checking that there is no object left on the tower, the electrician on the tower shall report it to the Responsible Person and then climb down the tower one after another with insulated transmission rope after obtaining the consent of the Responsible Person. (2) When climbing downing the tower, the electrician must use anti-fall device, climb down the tower at a uniform speed with the main materials in hand, hang the safety belt and backup protection rope on the shoulder and keep the safe distance of more than 6.8m away from the electrified body, and the Responsible Person shall strengthen the operation monitoring	5	(1) Deduct 2 points for any objects left on the tower. (2) Deduct 1 point/item for the tower electrician failing to report the inspection results of the leftovers or apply for climbing the tower. (3) The assessor shall order the termination of the operation (assessment) if the tower electrician fails to use anti-fall device during climbing down. (4) Deduct 0.5 points/time for grasping the shackles with hands during climbing down. (5) Deduct 1 point/time for slipping or missing his step during climbing down. (6) Deduct 1 point/ person for main band of safety belt and backup protection rope failing to be hung on the shoulder			
11	End of the work	(1) The Responsible Person shall organize all working personnel to put working apparatus and materials in order and put them in a special kit (bag) after cleaning; clean the site to ensure that "the materials are removed and the site is cleaned after construction".	10	(1) Deduct 0.5 points per piece for failure to clean or wipe the insulating tools and instruments. (2) Deduct 1 point per piece for failure to classify or place tools and instruments at fixed position. (3) Deduct 0.5 points/time for throwing about tools and instruments or stepping on moisture-proof tarpaulin.			

Part I Standard for Training and Assessment on Live Working for Operation Maintenance of ±800kV UHV Power Transmission Line

Table (Cont'd)

S/N	Project name	Quality requirements	Score	Deduction standard	Reasons for deduction	Deduction	Scoring
11	End of the work	(2) After a post-shift meeting is held, the Responsible Person shall make work summaries and comments. Comments include the construction quality of this work and the implementation of safety measures from all working personnel. (3) Responsible Person shall report to the on-duty control personnel that the work is over, apply for the restoration of circuit re-closing and terminate the work order.	10	(4) Deduct 1 point per piece for not removing fence, signs or leftovers. (5) Deduct 2 points for failure to hold the post-shift meeting. (6) Deduct 1 point when the team is not orderly or focused. (7) Deduct 1 point/person for members of the working group not attending post-shift meeting. (8) Deduct 1 point for unqualified comments. (9) Deduct 1 point for failure to report to the Dispatching Department (Assessor) the end of the work and apply for restoration of re-closing lock. (10) Deduct 0.5 points/item for non-standard and incomplete reporting terminology or the voice is not loud and clear enough. (11) Deduct 1 point for failure to repeat permission content. (12) Deduct 0.5 points/item for missing items during repeating the contents (name of working unit, name of the line, responsible Person, time, name of the line, work completion, device has restored to normal, the personnel have evacuated and re-closing lock can be restored). (13) Deduct 1 point/ item for failure to complete in time or error in the fill-in of work order termination procedure			
	Total		100				

Module 6   Standards for Training and Assessment on Live Replacement of ± 800kV UHV Transmission Line Vibration Damper

I. Training Standards

(I) Training Requirements

Module name	Live replacement of conductor vibration damper of ± 800kV UHV transmission line	Type of training	Operation
Training method	Practical operation training	Training hours	21 hours
Training objectives	1. Master the electrical significance of "two-span and three-short-circuit" operation mode in entering and exiting ±800 kV intense electric field along strain insulator string. 2. Be able to complete the entry of ±800kV equipotential operation point along the strain insulator string. 3. Be able to independently complete the replacement of conductor vibration damper (equipotential operation method)		
Training venue	UHV DC training line		
Training content	The "two-span and three-short-circuit" operation mode is adopted to enter the intense electric field along the strain insulator string, and the equipotential operation method is adopted for live replacement the conductor vibration damper of ± 800kV UHV DC transmission line		
Scope of application	Maintenance Personnel of UHV DC Transmission Line		

(II) Referenced Procedures and Specifications

(1) Code for Designing of ±800kV DC Overhead Transmission Line (GB/T50790-2013)

(2) Maintenance Specification for ± 800kV DC Overhead Transmission Line (DL/T251-2012)

(3) Operating Code for ±800kv DC Overhead Transmission Line (GB/T28813-2012)

(4) Technical Specification for Live Working of ±800kV DC Line (DL/T1242-2013)

(5) Technical Specification for Fittings of ±800kV DC Overhead Transmission Line (GB/T 31235-2014)

(6) State Grid Corporation of China on the Management Regulations of Live Working (Trial Implementation) (SGCC [2007] No.751)

(7) State Grid Corporation of China Working Regulations of Power Safety (Transmission Line Section) (Q/GDW1799.2-2013)

(8) Electrotechnical Terminology- Overhead Line (GB/T 2900.51-1998)

(9) Electrotechnical Terminology- Live Working (GB/T2900.55-2016)

(10) Live Working - Terminology for Tools, Equipment and Devices (GB/T 14286-2002)

(11) Insulated Tackles for Live Working (GB/T13034-2008)

(12) Live Working-Insulating Ropes (GB 13035-2008)

(13) Minimum Requirements for Utilization of Tools, Devices and Equipment for Live Working (DL/T 877-2004)

(14) Preventive Test Code of Tools, Devices and Equipment for Live Working (DL/T 976-2005)

(15) Technical Guide for Live Working of ±800kV UHV Transmission Line (Q/GDW302-2009)

(16) Shielding Clothes for Live Working (GB/T6568—2008)

(17) Technical Requirements and Design Guide for Live Working Tools (GB/T18037-2008)

(18) Application Rules of Infrared Diagnosis for Live Electrical Equipment (DL/T664-2016)

(III) Teaching Design for Training

To complete the work task of "live replacement of ±800kV UHV transmission line conductor vibration damper", each training stage shall be designed according to the standard operation procedure for work task completion. Each stage includes specific training objectives, training content, hours of training, training methods (training resources), training environment, assessment and evaluation, etc, as shown in the Table 1-6-1.

Table 1-6-1 Training Content Design for Live Replacement of Conductor Vibration Damper of ±800kV UHV Transmission Line

Training process	Training objectives	Training contents	Training hours	Training methods and resources	Preparation of training conditions	Assessment and evaluation
1. Theoretical teaching	1. To master the basic method for entering and leaving ±800kV intense electric field along the insulator string. 2. To be familiar with the methods of potential transfer. 3. To be familiar with the method for replacing the damaged conductor vibration damper of power transmission line. 4. To be familiar with the safe distance, hazard identification and pre control of live working on UHV DC line	1. The electrical significance of the "two-span and three-short-circuit" operation mode when entering and leaving the intense electric field along the insulator. 2. The use method for the potential transfer rod during entering and leaving the UHV intense electric field. 3. Method and quality standard for replacement of power transmission line conductor vibration damper. 4. Safe distance, hazard analysis and precontrol measures for live working on UHV DC line	2	Training methods: Lecture. Training resources: PPT, relevant regulations, specifications and technical guidelines	Multimedia classroom	Attendance, classroom questions and assignments
2. Preparations	Be able to complete the preparations before operation	1. Work site survey. 2. Preparation of the standardized operation card. 3. Filling of the work order. 4. Preparation of tools and materials for this operation	1	Training methods: 1. Site survey and preparation of tools, instruments and materials shall be practiced at site. 2. Preparation of operation card and the filling of work order shall adopt lecture method. Training resources: 1. ±800kV practical training line.	1. ±800kV Practical training line. 2. Multi-Media classroom	

Part I Standard for Training and Assessment on Live Working for Operation Maintenance of ±800kV UHV Power Transmission Line

Table (Cont'd)

Training process	Training objectives	Training contents	Training hours	Training methods and resources	Preparation of training conditions	Assessment and evaluation
2. Preparations	Be able to complete the preparations before operation	2. UHV tools warehouse. 3. Blank work order				
3. Work site preparation	Be able to complete the preparations of work site	1. Site re-survey. 2. Work application. 3. Work site arrangement. 4. Pre-shift meeting. 5. Inspection of tools, instruments and materials. 6. Usage of special spanner of vibration damper	1	Training methods: demonstration and role play. Resources: ±800kV training line	±800kV Practical training line	
4. Trainer's demonstration	The trainees can preliminarily understand the operation process of the task through inspecting and learning from each other's work	1. Wearing the personal tools and instruments and climbing the tower. 2. The equipotential electrician enters and exits the intense electric field and potential transfer along the strain insulator string. 3. The equipotential electrician reaches the replacement position of the vibration damper by means of walking along the line. 4. Equipotential electrician uses the special tools to complete the replacement of conductor vibration damper	2	Training methods: Demonstration. Resources: ±800kV training line	±800kV Training line.	

· 397 ·

Table (Cont'd)

Training process	Training objectives	Training contents	Training hours	Training methods and resources	Preparation of training conditions	Assessment and evaluation
5. Group training of trainees	1. Be able to enter and leave the ±800kV intense electric field along the insulator string and operate the potential transfer. 2. Be able to complete the replacement of ±800kV transmission line conductor vibration damper	1. The trainees are grouped (6 in a group) to train the skill operation for entering and leaving ±800kV intense electric field, potential transfer and replacing the conductor vibration damper. 2. Trainers guide the operation of trainees and conduct safety supervision	14	Training methods: Role play. Resources: ±800kV training line	±800kV Practical training line	Score the operation of trainees according to the detailed rules for skill assessment and scoring
6. End of the work	1. Enable the trainees to further distinguish the shortcomings of the operation process and facilitate the promotion in the later stage. 2. To train the trainees in the work style of safe and civilized production	1. Clean the site. 2. Report to dispatcher. 3. Comment and summarize the work task this time at post-shift meeting	1	Training methods: Lecture and inductive method	±800kV Training line	

· 398 ·

(IV) Work Flow

1. Work task

The "two-span and three-short-circuit" operation mode is adopted to enter the intense electric field and reach the operation point along the strain insulator string, and the equipotential operation method is adopted for live replacement the eight-bundle conductor vibration damper of ±800kV UHV transmission line.

2. Requirements for Weather and Work Site

(1) Live replacement for vibration damper of ±800kV UHV transmission line conductor shall be carried out in good weather.

In case of lightning (hearing thunder or seeing lightning), snow, hail, rain, fog and so on, live working is prohibited. When the wind force is greater than level 5, live working shall not be carried out; When the relative humidity is greater than 80%, insulating tool with moisture-proof performance shall be adopted if live working is required; When live rush repair must be carried out in bad weather, relevant personnel shall be organized to fully discuss and prepare necessary safety measures, which shall be carried out after approval of the Unit.

(2) The operating personnel should be in good mental states and be familiar with the organizational and technical measures to ensure safety in work; they should hold the qualification certificate for live working within the validity period.

(3) Responsible Person should organize the relevant personnel to complete field investigation in advance, determine the operating methods, required working apparatus and necessary measures according to the results, and handle the tickets for live working.

(4) The work site should be reasonably set up with fence and warning signs. Non-operating personnel is forbidden to enter.

(5) DC restart device shall be deactivated for the Project.

(6) Safe working distance and effective insulation length during operation are shown in Table 1-6-2.

Table 1-6-2 Safe Distance for Live Replacement of Conductor Vibration Damper of ±800kV UHV Transmission Line (m)

Voltage class	Safe distance between human body and electrified body	Minimum distance to adjacent phase conductor	Minimum effective insulation length of insulating tools	Minimum combination gap	Minimum distance between the exposed part of human body and electrified body during potential transfer
±800kV	6.8 (7.3)	6.9	6.8 (7.3)	6.6 (7.2)	0.5

Note: ① When the altitude is above 1000m and the ± 800kV DC single circuit transmission line is under live working, the safe distance shall be 7.3m in the brackets, the minimum effective insulation length of insulating tools shall be 7.3m in the brackets and the combinational gap shall be 7.2m in the brackets.

② As the phase-to-phase spacing of ±800kV UHV line is large enough (generally speaking, the distance between the phase and the ground shall be controlled) and no important safety factors are taken into account, the "minimum distance to the adjacent phase conductor" is not given in the "Safety Regulation", and the actual work can be carried out with reference to the data of 750kV.

(7) When working on ±800kV transmission line, ensure that the insulators in good operation phase shall be not less than 37 pieces (the single insulator is 170mm high), 32 pieces (the single insulator is 195mm high), 31 pieces (the single insulator is 205mm high), and 26 pieces (the height of single insulator is 240mm).

3. Preparations

3.1 Hazards and precontrol measures

(1) Hazard —— electric shock

Precontrol measures:

① Before work, the Responsible Person shall contact the control personnel on duty, deactivate the line DC restart device, and perform the licensing procedures.

② Before climbing the tower, equipotential operators on the tower must carefully check the names of the line, the number of the pole and tower, and phase, and then the tower can be climbed after all have been confirmed correct.

③ If lines lose power suddenly during work, operators shall consider it as still charged. The Responsible Person shall contact the control personnel as soon as possible, and no forced energization is allowed before the on-duty control personnel getting in touch with the Responsible Person.

④ The ground electrician shall wear clean and dry gloves when operating the insulating tools. Insulating tools and insulating ropes shall not be damaged, damped, deformed, and malfunctioned. It is not allowed to use non-insulating ropes (such as cotton rope, manila rope, and steel wire rope). The tools and instruments for live working at the work site shall be placed on the moisture-proof tarpaulin to prevent them from dirt and moisture in use.

⑤ The equipotential operator shall wear flame-retardant underwear and full set of shielding clothes for ±800kV live working shall be worn outside the clothes (including cover-all, helmet, protective mask, gloves, conductive socks and shoes). All parts shall be in excellent connection conditions. The resistance between the farthest points of the whole shielding clothes shall not be greater than 20Ω.

⑥ Before potential transfer, equipotential operator shall obtain the approval of Responsible Person, and the minimum distance between exposed part of human body and electrified body shall not be less than 0.5m. During potential transfer, potential transfer rod shall be adopted. The operation shall be fast, and end shall not be used for power charging and discharging; during delivering tools and materials to ground potential operators, the effective length of insulating tools or insulating ropes shall not be lower than the requirements as specified in Table 1-6-2.

⑦ When transmitting large metal objects by using insulating rope, the ground potential operators shall ground the metal objects before contacting with the insulating rope.

⑧ The Special Supervisor shall continuously monitor the operators and correct their nonstandard operation or actions in violation at any time. Special attention shall be paid to operators working at heights to ensure that there is enough safe distance (meeting the requirements in Table 1-6-2). It is forbidden to contact two non-connected electrified bodies or make contact with electrified body and grounding body at the same time.

(2) Hazard —— falling from high places

Precontrol measures:

① Before climbing, operators work at heights must satisfy the requirements of this operation, such as physical condition, mental state, and skill and quality.

② Before climbing the tower, operators working at heights shall conduct visual inspection and impulse test on safety belts and falling protector to ensure that their mechanical strength meets the requirements.

③ Operators who climb the tower shall first check whether the shackles are firm, shoe soles are clean, or the falling device is firm or is provided with the falling protector; grasp the main materials, tread on shackles and climb up (down) the tower at a uniform pace.

④ Supervisor shall correct irregularities and violations at any time. Special attention shall be paid to operators working at heights to prevent them from losing the protection of the safety belt or insulated backup protection rope during transposition. The safety belt shall be attached to a secure part and shall not be fastened lower than operating personnel.

⑤ When the equipotential operators move on the insulator string in parallel, they usually grasp one string with both hands and step on the other string with both feet to enter the intense electric field at a constant speed. During the movement, the backup protection rope holds two strings of insulators, and excessive hand waving and big step are not allowed.

⑥ Equipotential operators must fasten the safety belt and make the backup protection rope cover all the sub-conductors when they are walking along the conductor; they shall control the center of gravity during walking along the line to prevent the conductor from turning over.

(3) Hazard —— injury caused by objects falling from high place

Precontrol measures:

① Operator working at heights should put personal tools and fragmentary materials into the tools bag. It is strictly forbidden to hang objects in high place or keep in the mouth.

② Ground operator should correctly wear a safety helmet and use the knots to deliver the tools, instruments and materials. The vertical distance from the operation point should not be less than the falling radius.

③ The work site shall be set up with fence and warning signs. The Supervisor shall maintain vigilance at all times and prohibit non-staff members and vehicles from entering the operation area.

3.2 Selection of tools, instruments and materials

See Table 1-6-3 for tools, instruments and materials required by live replacement of conductor vibration damper of ±800kV UHV transmission line. Before delivering tools and instruments out of warehouse, application voltage class and test cycle shall be carefully checked to ensure that appearance is intact, connection is firm, rotation is flexible, and they meet the work task requirements. After delivering tools and instruments out of warehouse, they shall be stored in tools bag or tool kit for transportation to avoid contamination and damp. Metal tools and insulating tools shall be separately loaded and transported to avoid deformation, damage or other defects caused by mixed loading and transportation.

Table 1-6-3   Tools, Instruments and Materials Required by Live Replacement of Conductor Vibration Damper of ±800kV UHV Transmission Line

S/N	Name	Specification and Model	Unit	Qty.	Remarks
1	Insulated transmission rope	TJS-12, with the length matching the lifting height	Nos.	2	Insulating tool
2	Insulated backup protection rope	TJS-16, with buffer provided	Nos.	2	Insulating tool
3	Insulated tackle	JH10-0.5	pcs	2	Insulating tool
4	Insulated noose	TJS-14	Nos.	2	Insulating tool
5	Potential transfer rod	0.4m	Nos.	1	Insulating tool
6	I-type shielding clothes (coverall, helmet, protective mask, gloves and conductive socks)	Shielding efficiency ≥60dB (shield efficiency of shielding mask ≥20dB)	Set	2	PPE
7	Conductive shoes	The size depends on the wearer	Pair	2	PPE
8	Flame retardant underwear	Pure mulberry silk	Set	2	PPE
9	Safety belt of double insurance	Full-body harness	Nos.	2	PPE
10	Falling protector	Corresponding to the type of pole and tower falling protector	pcs	2	PPE
11	Safety helmet		pcs	6	PPE
12	Special spanner for vibration damper		Nos.	1	Special tools
13	Insulation resistance tester	5000V, with electrode width of 2cm and interelectrode width of 2cm	Set	1	Other tools
14	Wind speed, temperature and humidity tester	HT-8321	Set	1	Other tools
15	Multimeter		pcs		Other tools
16	Interphone	As required by the work	Set	2	Other tools
17	Moisture-proof tarpaulin	2m×4m	pcs	2	Other tools
18	Security fence		Set	Several	Other tools
19	Warning sign	"Work Here", "Access from Here" "Access from Here"	Set	1	Other tools

Table (Cont'd)

S/N	Name	Specification and Model	Unit	Qty.	Remarks
20	Red waistcoat	"Responsible Person"	pcs	1	Other tools
21	Clean towel	Cotton	pcs	1	Other tools
22	Shoe covers		Pair	Several	Other tools
23	Work gloves		Pair	Several	Other tools
24	Personal tools	Tool bag, flat tong, and marker pen	Set	2	Other tools
25	Vibration damper	The same model with replaced vibration damper	pcs	1	Material
26	Aluminum armor tape		m	Proper	Material

Note: The electrical strength of the insulating tool shall meet the requirements of the "State Grid Corporation of China Working Regulations of Power Safety (Transmission Line Section)", and the test is required to be qualified and within the validity period.

3.3 Division of labor for operators

Division of labor for operators of the task is shown in Table 1-6-4.

Table 1-6-4 Division of Labor for Live Replacement of Conductor Vibration Damper of ±800kV UHV Transmission Line

S/N	Post	Qty. (person)	Responsibilities
1	Responsible Person	1	Be responsible for division of operating personnel of the task, work order reading, handling formalities for stopping the DC restart device, getting work permits, holding pre-shift meeting, dealing with emergency situations in work, quality surveillance, and the summary after work
2	Specific responsible supervisor	1	Be responsible for safety supervision and control during operation
3	Equipotential electrician	1	Be responsible for entering the equipotential to replace the conductor vibration damper
4	Ground potential electrician on the tower	1	Be responsible for assisting the equipotential electrician to enter and exit the intense electric field
5	Ground electrician	2	Be responsible for taking site safety measures, arranging operation site, inspecting tools and instruments, delivering tools and materials, and cooperating with the equipotential electrician to enter and leave equipotential

4. Working Procedures

The workflow of this task is shown in Table 1-6-5.

Table 1-6-5 Workflow Sheet for Live Replacement of Conductor Vibration Damper of ±800kV UHV Transmission Line

S/N	Work Content	Operation Steps and Standards	Safety Measures and Precautions	Responsible Person
1	Site re-survey	The Responsible Person shall complete the following work: (1) Check the line name, the number of the pole and tower and ensure the phases are correct; guarantee that the foundation and the pole and tower are intact and in normal condition; ensure that the cross and span distance meets the safety requirements; confirm the defect conditions and the specifications and models of earth wires. (2) Check that the site meteorological conditions such as wind speed and humidity should meet the operation requirements. (3) Check that the terrain and environment should meet the operation requirements. (4) Check that the safety measures listed in the work order are in line with the actual situations on site. The measures will be added if necessary.	(1) Correctly wear helmet, working clothes, work shoes and protective gloves. (2) Operation under meteorological conditions that may endanger the safety of operators is forbidden. (3) Non-operation personnel and vehicles are strictly prohibited from entering the working site	
2	Work Permit	(1) The Responsible Person shall contact the on-duty control personnel and apply for stopping the DC restart device as per the contents of the work order. (2) Live working could be started only after being approved by control personnel on duty.	Live working shall not be started without the permission of the on-duty control personnel	

Table (Cont'd)

S/N	Work Content	Operation Steps and Standards	Safety Measures and Precautions	Responsible Person
3	Site layout	Install the security fence and hang the signboards correctly: (1) The security fence should take full account of falling objects from the high place and the influence on road traffic. (2) The entrance and exit of the security fence shall be set reasonably. (3) Signs such as "Access from Here", "Work Here", "Access from Here" shall be properly and well arranged	When the influence on road traffic safety is uncontrollable, the traffic management department should be contacted in time to strengthen the on-site control of traffic safety	
4	Holding a pre-shift meeting	(1) All working personnel shall line up. (2) Responsible Person will read out the work order and be clear with work task and division of personnel; explain safety measures and technical measures in work; check (inquire after) mental state of all working personnel; inform of hazards in work and precontrol measures. (3) All working personnel shall sign on the work order for confirmation	(1) The work order shall be filled in, issued and approved in a standardized manner, and the signature shall be complete. (2) All working personnel shall be in good mental states. (3) All working personnel shall be clear with task division of works, safety measures and technical measures	
5	Inspection of tools and instruments	(1) The ground potential electrician and equipotential electrician on the tower shall wear the shielding clothes in a right way and pass the inspection, which shall be supervised and inspected by the Responsible Person. (2) Wear personal safety appliance correctly (proper size and easy lock), and the Responsible Person shall supervise and inspect it. (3) Measure the wind speed, wind direction and humidity, check the insulation performance of insulating tools, and make records	(1) Check carefully for damage, deformation and failure before using metal and insulating tools. Carry out segment insulation detection with such insulating tools as insulation resistance tester of 2,500V or above and with the resistance no less than 700MΩ, and wipe them clean with the clean dry towel. (2) Use a multimeter to measure the resistance between the farthest ends of the shielding clothes and trousers, which shall not be greater than 20Ω. The Responsible Person shall check the connection of the electrician's shielding clothes. (3) Check the tool assembly and make sure the connection is reliable. (4) The tools and instruments for live working at site shall be placed on the moisture-proof tarpaulin	

Part I Standard for Training and Assessment on Live Working for Operation Maintenance of ±800kV UHV Power Transmission Line

Table (Cont'd)

S/N	Work Content	Operation Steps and Standards	Safety Measures and Precautions	Responsible Person
6	Climbing the tower	(1) After checking the line name and pole and tower number, the ground potential electrician on the tower and equipotential electrician shall impact to check the stress of safety belt and falling protector. (2) The ground potential electrician on tower carries the insulated transmission rope to climb the tower, and the equipotential electrician then climbs the tower. When they reach the operation point of the cross arm, they choose the appropriate position to fasten the safety belt, and the ground potential electrician on the tower installs the insulated tackle and the insulated transmission rope in the appropriate position of the cross arm. Then he shall cooperate with the ground electrician to separate the insulated transmission rope for lifting preparation	(1) After checking the correct line name and pole and tower number, he can climb the tower for operation. (2) The anti-falling device installed on the tower shall be used during climbing the tower; when moving and transposing on the pole and tower, the safety protection shall not be lost, and the operators must climb and grasp the components securely. (3) The working electrician must wear full set of qualified shielding clothes which must be connected reliably. Before entering the equipotential at cross arm, the equipotential electrician shall re-check and confirm that each part of the shielding clothes are connected reliably before the next operation	
7	Entering the intense electric field	(1) The equipotential electrician shall transfer the safety belt to the insulator set fitting, and carry the potential transfer rod, insulated tackle and insulated transmission rope. (2) After the equipotential electrician confirms that all parts of the shielding clothes are well connected, it shall be reported to the Responsible Person for approval. The electrician shall grasp one string with both hands and step on the other string with both feet, and enter the equipotential along the insulator string in the operation mode of "two-span and three-short-circuit".	(1) The equipotential electrician must obtain the permission of the Responsible Person before entering the potential. (2) When the equipotential electrician enters the insulator string, he shall use the safety belt and the backup protection rope alternately (holding two insulator strings with the backup protection rope, grasping one string with hands and stepping on the other string) without losing the protection of the safety belt, and adjust the insulated transmission rope and the potential transfer rod.	

Table (Cont'd)

S/N	Work Content	Operation Steps and Standards	Safety Measures and Precautions	Responsible Person
7	Entering the intense electric field	(3) When reaching the third insulators outside the grading ring on the conductor side, the operator shall stop, and carry out the potential transfer with the potential transfer rod	(3) The equipotential electrician shall coordinate his hands and feet during entering the intense electric field with uniform speed so as to avoid excessive hand waving, big step and other dangerous actions; the combined gap composed of the grounding body and the electrified body shall be greater than 6.6m. (4) The minimum distance to adjacent conductors shall be greater than 6.9m. (5) Before the potential transfer, the equipotential electrician shall check whether the electrical connection between the potential transfer rod and the shielding clothes is reliable and that the minimum distance between the exposed part of the human body and the electrified body shall be greater than 0.5m, and shall get the permission of the Responsible Person; the protection of safety belt shall not be lost during potential transfer, and the action shall be accurate, stable and rapid when entering the intense electric field	
8	Replacement of the sub-conduct or vibration damper	(1) After entering the equipotential, the equipotential electrician shall tie the safety belt to the upper sub-conductor, and install the insulated protection rope (all sub-conductors shall be covered). (2) The equipotential electrician shall carry the insulated transmission rope and walk to the operation point along the conductor for replacing the vibration damper, install the insulating noose at the appropriate place of the sub-conductor, and then connect the insulated tackle and insulated transmission rope and hook the insulated tackle into the insulating noose.	(1) The equipotential electrician shall not lose the protection of safety belt. (2) Equipotential electrician shall cooperate closely with the ground electrician and follow the command of the Responsible Person. (3) The minimum distance to adjacent conductors shall be greater than 6.9m.	

Part I Standard for Training and Assessment on Live Working for Operation Maintenance of ±800kV UHV Power Transmission Line

Table (Cont'd)

S/N	Work Content	Operation Steps and Standards	Safety Measures and Precautions	Responsible Person
8	Replacement of the sub-conduct or vibration damper	(3) Equipotential electrician shall mark the installation points of the old vibration damper on conductor at both ends with the marker pen. (4) The equipotential electrician shall use the insulated transmission rope to fasten the old vibration damper by the way of slip knot, then use the special spanner for vibration damper to remove the old vibration damper, and put it on ground with the insulated transmission rope in cooperation with the ground electrician. (5) The equipotential electrician shall remove the aluminum armor tape and place it in the tool kit; the new aluminum armor tape shall be wound according to the drawing mark. The winding shall be smooth and tight, and the twisting direction shall be the same as that of the outer conductor. (6) The ground electrician shall lift the new vibration damper to the position of equipotential electrician, and the equipotential electrician shall install it correctly. (7) Check the installation quality and make sure that the vibration damper is vertically downward with the hammer ball parallel to the main wire. The installation displacement should not exceed ±30mm; the both ends of the aluminum armor tape should be pressed back into the cord bracket of vibration damper, and both ends should be exposed 10mm. The spring washer at the bolt should be tight and flat, and the bolt direction should be consistent with other sub-conductor vibration damper. (8) The equipotential electrician shall remove insulated tackle, insulated transmission rope and noose successively. (9) No tools, instruments or materials shall drop during the equipotential operation	(4) During delivering, the conductor vibration damper shall not be bumped, and the insulated transmission ropes on both sides shall not be intertwined. (5) During delivering the tools, the binding rope shall be correct and reliable so as to prevent objects falling from heights. (6) During delivering the tools, instruments and materials, the ground electrician is not allowed to stand directly below the operation point of equipotential electrician	

Table (Cont'd)

S/N	Work Content	Operation Steps and Standards	Safety Measures and Precautions	Responsible Person
9	Leaving the intense electric field	(1) After inspecting that the vibration damper is installed firmly and there is no object left at the operation point, the equipotential electrician, with the permission of the Responsible Person, can return to the grading ring with insulated transmission along the conductor for the preparation to leave the potential. (2) The equipotential electrician shall hook the grading ring with the potential transfer rod and enter the third insulator of the grading ring. With one hand grasping the insulator tightly and the other holding the potential transfer rod, the equipotential electrician shall break himself away from the equipotential with the help of potential transfer rod. (3) The equipotential electrician exits the intense electric field by means of "two-span and three-short-circuit" operation mode	(1) The equipotential electrician must obtain the permission of the Responsible Person before leaving the potential. (2) When the equipotential electrician returns to the insulator string, the safety belt and backup protection rope shall be used alternately. The protection of safety belt shall not be lost during potential transfer, and the instant action shall be accurate, stable and rapid when leaving the intense electric field. (3) The equipotential electrician shall coordinate his hands and feet during leaving the potential with uniform speed so as to avoid excessive hand waving, big step and other actions; The combined gap composed of the grounding body and the electrified body shall be greater than 6.6m; The minimum distance between the exposed part of human body and the electrified body should not be less than 0.5m. (4) When moving along the insulator string, the equipotential electrician shall use the backup protection rope to hold two insulator strings with the hands grasping firmly and the feet stepping firmly. (5) The minimum distance to adjacent conductors shall be greater than 6.9m. (6) The equipotential electrician shall not lose the protection of safety belt or backup protection rope when returning to cross arm	

Table (Cont'd)

S/N	Work Content	Operation Steps and Standards	Safety Measures and Precautions	Responsible Person
10	Return to the ground	After checking that there is no object left on the tower, the equipotential electricians on the tower shall report it to the Responsible Person and then climb down the tower with insulated transmission rope after obtaining the consent of the Responsible Person.	The anti-falling device installed on the tower shall be used in the process of climbing down the tower; when moving and transposition on the pole and tower, the safety protection shall not be lost, and the operators must grasp the components securely	
11	End of the work	(1) The Responsible Person shall organize all working personnel to put working apparatus and materials in order and put them in a special kit (bag) after cleaning; clean the site to ensure that "the materials are removed and the site is cleaned after construction". (2) After a post-shift meeting is held, the Responsible Person shall make work summaries and comments. Comments include the construction quality of this work and the implementation of safety measures from all working personnel. (3) Responsible Person shall report to the on-duty control personnel the end of the work, apply for the restoration of DC restarting device and terminate the work order	It is forbidden to restore the DC restart device at the appointed time	

Standard for Professional Training and Assessment for Operation Maintenance of UHV DC Power Transmission Line

II. Assessment Standard

Table 1-6-6 Detailed Rules for Assessment and Scoring of Operation and Inspection Skills of UHVDC Transmission Line

Fill-in Column of Examinee	No.:	Name:		Position:		Unit:		Date: YYYY MM/DD	
Fill-in Column of Assessor	Grade:	Assessor:		Assessment Leader:		Team Starting time:		Closing time:	Operation Duration:
Assessment Module	Live replacement of conductor vibration damper of ±800kV UHV transmission line		Assessee		Maintenance Personnel of UHV DC Transmission Line		Assessment method	Operation	Assessment Time Limit	90min
Task description	Enter the intense electric field along the strain insulator strings and carry out live replacement of the conductor vibration damper of ±800kV UHV transmission line (equipotential operation method)								
Work specifications and requirements	1. Live working shall be carried out in good weather. In case of thunder, rain, snow or fog, no live working shall be carried out. When the wind force is greater than Level 5 and the humidity is greater than 80%, live working should not be carried out. 2. Persons required for this operation include 1 Responsible Person, 1 special Supervisor, 1 ground electrician on the tower, 1 equipotential electricians and 2 ground auxiliary electricians. Enter the intense electric field along the strain insulator strings and carry out the live replacement of the eight-bundle spacer of the damaged conductor vibration damper of ±800kV UHV transmission line. 3. Responsibilities of Responsible Person: Be responsible for division of operating personnel of the task, work order reading, handling formalities for stopping the DC restart device, getting work permits, holding pre-shift meeting, dealing with emergency situations in work, quality surveillance, and the summary after work. 4. Special Supervisor: Be responsible for safety supervision and control during operation. 5. Equipotential electrician: Enter the intense electric field along the strain insulator strings and carry out the live replacement of the vibration damper. 6. Responsibilities of electrician on tower: Be responsible for assisting equipotential to enter and exit intense electric field. 7. Responsibilities of ground electrician: Be responsible for implementing site safety measures, arranging work site, inspecting tools and instruments, delivering tools and materials, and cooperating with the equipotential electrician to enter and exit equipotential.								

Part I Standard for Training and Assessment on Live Working for Operation Maintenance of ±800kV UHV Power Transmission Line

Table (Cont'd)

Work specifications and requirements	8. During the live working, if thunder, rain, strong wind or any other circumstance threaten the safety of the staff, the Responsible Person or Supervisor may stop working temporarily according to the circumstances. Given conditions: 1. Training base: UHVDC±800kV Training line xx line 001#resisting-tensile tower large side positive pole sub-conductor vibration damper, model of vibration damper: FRT-7. 2. Work orders have been handled, safety measures have been completed (DC restart device has been deactivated), and oral application (dispatcher or assessor) shall be made at the beginning and end of the work. 3. Safety fence shall be installed and signs such as "Work here" and "Access from Here" shall be hung at the work site. Safety measures have been completed. 4. The instruments and apparatus shall be used safely and correctly to test the insulating tool. 5. Fall-arrest device should be used in the process of up and down the tower to prevent falling from heights. 6. The operation must be carried out according to Standard Operation Procedures. The relevant item scores shall be deducted for the process error. In case of major hidden dangers of personal, equipment and operational safety, the assessor may order the termination of the operation (assessment) and the assessment of this module is recorded as "unqualified"
Assessment scenario preparation	1. Line: UHVDV ±800kV Training line xx line 001# resisting-tensile tower large side positive pole conductor, work content: live replacement ±800kV conductor damaged vibration damper, model of vibration damper: FRT-7. 2. Required operation tools: 2 insulated transmission ropes (TJS-12), 2 insulated backup protection ropes (TJS-16, with buffer provided), 2 insulated tackles (JH10-0.5), 2 insulated nooses (TJS-14), 1 potential transfer rod (0.4m), 2 Type I shielding clothes (coverall, helmet, protective mask, gloves and conductive socks), 2 pairs of conductive shoes, 2 falling protectors, 1 special spanner for vibration damper, 1 set of insulation resistance tester (type 5000V), 1 set of wind speed and humidity tester (HT-8321), 1 multimeter, 2 Moisture-proof tarpaulins (2m*4m), 1 red waistcoat (for Responsible Person), 2 clean towels, 2 sets of personal tools (workbasket, flat tongs, marker pen), 1 vibration damper of same model and several aluminum armor tape. 3. The work site shall be monitored, and the safety measures (fence, etc.) on the work site have been fully implemented; non-operation personnel are prohibited from entering the site, and the staff must wear safety helmets when entering the work site. 4. Examinees shall bring their own work clothes, flame retardant cotton underwear, safety helmets, gloves, safety belts (including backup protection rope)
Remarks	1. The total score of this module is 100 points, and score of each item is deducted until the corresponding points are finished. If the task is not completed within the specified time, the test shall be terminated immediately. The score of this module shall be calculated according to the actual score of the completed item, and the unfinished item shall not be scored. 2. In the process of assessment, if the equipment, operating environment, safety measures, safety protection and safe distance do not meet the requirements of the operation, or human error occurs, which may endanger the safety of the operation, the assessor shall order the termination of the operation. 3. Exam participants should be organized for site survey before the examination and work orders should be handled in advance

Table 1-6-7 Standards for Assessment and Scoring of Operation and Inspection Skills of UHVDC Transmission Line

S/N	Project name	Quality requirements	Score	Deduction standard	Reasons for deduction	Deduction	Scoring
1	Site re-survey	(1) The Responsible Person shall go to the work site to check the line name, pole and tower number, on-site working conditions, defective parts and so on. (2) Check that the site meteorological conditions such as wind speed and humidity should meet the operation requirements. (3) Check whether the work order is complete and unmodified, check whether the safety measures listed are consistent with the actual situation on site, and supplement it if necessary	5	(1) No point will be awarded without a work order. (2) Deduct 1 point for unchecked double title. (3) Deduct 1 point/item for failure to check on-site working conditions (meteorology) and defective parts. (4) Deduct 0.5 points/place for any alteration or frowziness on the work order, and deduct 1 point for incorrect work order number. Deduct 1 point/item for missing items in work order			
2	Work Permit	(1) The Responsible Person contacts the on-duty control personnel and apply for stopping the DC restart device as per the contents of the work order. (2) Report content is standardized and complete. The voice is loud and clear. (3) Relevant licensing procedures are completed in a timely manner	3	(1) No point will be awarded for items of work without the work permit of the Dispatching Department (Assessor). (2) Deduct 0.5 points/item for non-standard and incomplete reporting terminology or the voice is not loud and clear enough. (3) Deduct 1 point for not applying for deactivating DC restart device. (4) Deduct 1 point for failure to repeat permission content. (5) Deduct 0.5 points/item for missing repeated content (Approver name, license time, work task, DC restart device status). (6) Deduct 1 point for failure to complete work order in time			

Part I Standard for Training and Assessment on Live Working for Operation Maintenance of ±800kV UHV Power Transmission Line

Table (Cont'd)

S/N	Project name	Quality requirements	Score	Deduction standard	Reasons for deduction	Deduction	Scoring
3	Site layout	Install the security fence and hang the signboards correctly: (1) The security fence should take full account of falling objects from the high place and the influence on road traffic. (2) The entrance and exit of the security fence shall be set reasonably. (3) Signs such as "Access from Here", "Work Here", "Access from Here" shall be properly arranged	4	(1) Deduct 2 points for failure to arrange the safety fence at the work site. (2) Deduct 1 point/place for not setting the safety fence at the work site properly. (3) Deduct 1.5 points for failure to hang signboards. (4) Deduct 0.5 points per piece for incomplete signboards. (5) Deduct 0.5 points/person for non-operating personnel entering the fenced area			
4	Holding a pre-shift meeting	(1) All staff and personnel shall wear safety helmets and work clothes correctly. (2) The Responsible Person shall wear red waistcoat and read out the work order and be clear with work task and division of personnel; explain safety measures and technical measures in work; check (inquire after) mental state of all working personnel; inform of hazards in work and precontrol measures. (3) All working personnel shall sign on the work order for confirmation	4	(1) Deduct 0.5 points/person for the incorrect wear of safety helmet by working personnel. Deduct 0.5 points/person/time for improper dressing. (2) Deduct 0.5 points/person for the failure to wear safety red waistcoat by Responsible Person and Special Supervisor. (3) No point will be awarded for unclear work task and division of work. (4) Deduct 1 point for unclear division of labor. (5) Deduct 1 point for incomplete explanation of safety measures or precontrol measures. (6) Deduct 1 point for failure to inform hazards at work. (7) Deduct 1 point for failure to confirm mental status of work shift members. (8) Deduct 1 point for the work shift member failing to sign or signing incompletely on the work order			

· 415 ·

Table (Cont'd)

S/N	Project name	Quality requirements	Score	Deduction standard	Reasons for deduction	Deduction	Scoring
5	Inspection of tools	(1) The work shift members shall properly set the moisture-proof tarpaulin in the appropriate position. The tarpaulin shall be clean and dry. It is strictly prohibited to tread the tarpaulin. (2) The tools and instruments shall be classified and placed neatly on moisture-proof tarpaulin according to the requirements of the fixed location management; the insulated tools shall not be mixed with metal tools and materials; and the appearance inspection shall be done on tools, instruments and apparatus. (3) All kinds of tools shall be qualified and within the effective time by test. The surface of insulated tools shall not be worn, deformed and damaged, and the operation shall be flexible. Carry out segment insulation detection with such insulated tools with insulation resistance meter of 2500V or above and with the resistance no less than 700M Ω, and wipe it off with a clean dry towel. (4) The ground potential and equipotential personnel on tower shall correctly wear a whole suit of qualified shield clothes and conductive shoes as required, with each part connected well, shall not wear chemical fiber clothes next to the skin in the shielding clothes and shall fasten safety belts; the Responsible Person shall carefully check and confirm whether they wear them correctly and all parts are well connected.	8	(1) Deduct 1 point for the inappropriate position of moisture-proof tarpaulin. (2) Deduct 0.5 points/time for stepping on moisture-proof tarpaulin. (3) Deduct 1 point for failure to classify or place tools and instruments at fixed position. (4) Deduct 1 point/item for failure to check qualified label and appearance of tools, instruments and apparatus. (5) Deduct 0.5 points per piece for missing items in inspection of tools, instruments, and apparatus. (6) Deduct 0.5 points per piece for incorrect inspection method of tools, instruments and apparatus. (7) Deduct 0.5 points per piece for not cleaning or wiping hard insulating tools. (8) Deduct 0.5 points/time for holding insulating tools without wearing clean and dry cotton gloves. (9) Deduct 1 point/item for the improper usage of testing instruments to test the tools, instruments and the full set of shielding clothes. (10) Deduct 2 points/person for electrician climbing the tower failing to wear the shielding clothes correctly or to check the connecting portion.			

S/N	Project name	Quality requirements	Score	Deduction standard	Reasons for deduction	Deduction	Scoring
5	Inspection of tools	(5) The full set of shielding clothes shall be tested with a multimeter, and the resistance value between the farthest two points shall not be more than 20Ω, and the resistance value of a single set shall not be more than 15Ω. (6) The electrician climbing the tower shall conduct the visual inspection on the safety belt, backup protection rope and falling protector, which shall pass the impulse test.	8	(11) Deduct 1 point/item for failing to conduct visual inspection or impulse test (or incorrect way) for safety belt, backup protection rope and falling protector (12) Deduct 1 point/item for the Responsible Person failing to check or missing to check the safety protection equipment of electrician climbing the tower			
6	Climbing the tower	(1) Tower-climbing personnel shall check the double name, pole number and phase again, report to the Responsible Person and apply for climbing the tower. (2) The ground potential electrician on tower and equipotential electrician carry the insulated transmission rope to climb the tower one after another. (3) During tower climbing, the electrician must use anti-fall device, climb the tower at a uniform speed with the main materials in hand, hang the safety belt and backup protection rope on the shoulder and keep the safety distance of more than 6.8m away from the electrified body, and the Responsible Person shall strengthen the operation monitoring.	5	(1) Deduct 1 point/item for the tower electrician failing to check the double name, tower number and phase. (2) Deduct 1 point/item for the tower electrician failing to report the result or apply for climbing the tower. (3) The assessor shall order the termination of the operation (assessment) by electrician climbing the tower failing to use anti-fall device. (4) Deduct 0.5 points/time for grasping the shackles with hands during climbing the tower. (5) Deduct 1 point/ time for slipping or missing the step when climbing the tower. (6) Deduct 1 point/ person for failing to hang main band of safety belt and backup protection rope on shoulder.			

Table (Cont'd)

S/N	Project name	Quality requirements	Score	Deduction standard	Reasons for deduction	Deduction	Scoring
6	Climbing the tower	(4) When climbing to an appropriate position, ground potential electrician shall fasten the safety belt, arrange the insulated transmission rope, and then cooperate with the ground electrician on tower to make lifting preparation of the insulated transmission rope separately. (5) The Responsible Person shall carefully monitor and remind the whole process of tower climbing	5	(7) Deduct 1 point/ time for safety belt and backup protection rope twining and getting caught. (8) Deduct 1 point/ time for fastening the safety belt and backup protection rope lower than operating personnel. (9) Deduct 1 point/ time for fastening the safety belt and backup protection rope on the same components. (10) No point will be awarded for losing protection of safety belt during work at heights. (12) Deduct 1 point for installation of tackle without insulating noose. (13) Deduct 1 point for the suspension position of the tackle transmission rope inconvenient for getting tools. (14) Deduct 1 point for the Responsible Person failing to monitor the operation in place. (15) The Assessor shall order the termination of the operation for insufficient safe distance (Assessment)			
7	Entering the intense electric field	(1) Check that all parts of the shielding clothes are well connected before entering the potential, and report to the Responsible Person for approval.	13	(1) Deduct 2 points for the equipotential electrician failing to check the connecting parts of the shielding clothes. (2) Deduct 2 points for losing the protection of safety belt when equipotential electrician transfers to insulator set fitting.			

Part I Standard for Training and Assessment on Live Working for Operation Maintenance of ±800kV UHV Power Transmission Line

Table (Cont'd)

S/N	Project name	Quality requirements	Score	Deduction standard	Reasons for deduction	Deduction	Scoring
7	Entering the intense electric field	(2) The equipotential electrician shall transfer the safety belt to the insulator set fitting, and carry the potential transfer rod, insulated tackle and insulated transmission rope. (3) Before entering the insulator string, the equipotential electrician shall fasten the protection rope (holding two insulator strings with the backup protection rope, grasping one string with hands and stepping on the other string). (4) When reaching the three pieces of insulators outside the grading ring on the conductor side along the insulator string by the operating method of two-span and three-short-circuit, the operator shall stop, wait for the approval of the Responsible Person and then carry out the potential transfer with the potential transfer rod. (5) The combined gap composed of the gaps between the equipotential electrician and the grounding body & the electrified body in the process of entering the potential shall not be less than 6.6m, and the potential transfer rod must be used for potential transfer when entering the intense electric field. The minimum distance between the exposed part of human body and the electrified body should not be less than 0.5m. (6) Protection of safety belt shall not be lost when entering the intense electric field.	13	(3) Deduct 2 points for insufficient combined clearance between equipotential electrician and grounding body & electrified body. (4) Deduct 2 points for incorrect and unskilled action of equipotential electrician entering the intense electric field. (5) Deduct 2 points for repeated discharge due to insufficient distance of the exposed parts during potential transfer of equipotential electrician. (6) Deduct 1 point for unskilled potential transfer. (7) Deduct 5 points for potential transfer without using the potential transfer rod. (8) Deduct 3 points for carrying out potential transfer without permission of the Responsible Person. (9) Deduct 2 points for losing the protection of safety belt when equipotential electrician enters the intense electric field. (10) Deduct 2 points for the Responsible Person failing to monitor the operation in place			

Table (Cont'd)

S/N	Project name	Quality requirements	Score	Deduction standard	Reasons for deduction	Deduction	Scoring
8	Replace the sub-conduct or vibration damper	(1) After entering the equipotential, the equipotential electrician shall tie the pole belt to the upper sub-conductor, and install the insulated protection rope (all sub-conductors shall be covered). (2) The equipotential electrician shall carry the insulated transmission rope and walk to the operation point for replacing the vibration damper, and shall correctly install the insulating noose at the appropriate place of the sub-conductor and hang insulated tackle and insulated transmission rope. (3) The equipotential electrician shall mark on both ends of installation site of vibration damper with marker pen. (4) The equipotential electrician shall use the insulated transmission rope to fasten the old vibration damper by the way of slip knot, then use the special spanner for vibration damper to remove the old vibration damper, and puts it on the moisture-proof tarpaulin on ground with the insulated transmission rope in cooperation with the ground electrician. (5) The equipotential electrician shall remove the aluminum armor tape and place it in the tool kit; the new aluminum armor tape shall be wound according to the drawing mark. The winding shall be smooth and tight, and the twisting direction shall be the same as that of the outer conductor.	30	(1) Deduct 3 points for the incomplete covering of sub-conductor by insulated protection rope. (2) Deduct 1 point for failure to tie pole belt to upper sub-conductor during walking along the line. (3) Deduct 2 points for unskilled action of walking along the line. (4) Deduct 5 points for the insulated tackle hooked directly on the conductor. (5) Deduct 1 point/item for incorrect installation method or inappropriate position of insulating noose. (6) Deduct 1 point for not locking after the insulated tackle was hooked on insulating noose. (7) Deduct 3 points for no marking. (8) Deduct 5 points for removing the vibration damper without tying the knot (9) Deduct 2 points for panic action of removing tools and instruments. (10) Deduct 3 points for each falling object from heights. (11) No point will be awarded at this module for throwing old vibration damper at heights. (12) Deduct 2 points/person/time for the ground electrician standing directly below the work point of equipotential electrician.			

Part I Standard for Training and Assessment on Live Working for Operation Maintenance of ±800kV UHV Power Transmission Line

Table (Cont'd)

S/N	Project name	Quality requirements	Score	Deduction standard	Reasons for deduction	Deduction	Scoring
8	Replace the sub-conduct or vibration damper	(6) The ground electrician shall lift the new vibration damper to the position of equipotential electrician. (7) The equipotential electrician should correctly install the new vibration damper and ensure that the vibration damper is vertically downward with the hammer ball parallel to the main wire. The installation displacement should not exceed ±30mm; The both ends of the aluminum armor tape should be pressed back into the cord bracket of vibration damper, and both ends should be exposed 10mm. The spring washer at the bolt should be tight and flat, and the bolt direction should be consistent with other sub-conductor vibration damper. (8) The equipotential electrician shall remove insulated tackle, insulated transmission rope and noose successively. (9) Do not drop tools, instruments and materials during equipotential operation	30	(13) Deduct 1 point for failure to put old vibration damper on moisture-proof tarpaulin. (14) Deduct 1 point for incorrect way the knot is tied when passing tools up and down. (15) Deduct 5 points for untying the knot before installing the new vibrator. (16) Deduct 3 points when the new vibration damper deviates from the original position and exceeds the allowable value. (17) Deduct 3 points respectively when the vibration damper is not vertically downward or the hammer ball is not parallel to the main wire after installation. (18) Deduct 2 points when the both ends of the aluminum armor tape are not pressed back into the cord bracket of vibration damper after installation. (19) Deduct 2 points when the both ends of the aluminum armor tape expose less than 10mm after installation. (20) Deduct 1 point/place for wrong threading direction of vibration damper bolt.			

Table (Cont'd)

S/N	Project name	Quality requirements	Score	Deduction standard	Reasons for deduction	Deduction	Scoring
8	Replace the sub-conduct or vibration damper		30	(21) Deduct 1 point/place for not removing insulated tackle, insulated transmission rope and noose. (22) Deduct 2 points for not removing insulated tackle, insulated transmission rope and noose in order. (23) No point will be awarded at this module for not finishing the replacement of old vibration damper by new vibration damper. (24) The assessor shall order the termination of the operation for insufficient safe distance between adjacent conductors			
9	Leaving the intense electric field	(1) After inspection, the vibration damper is installed firmly and there is no residue at the operation point, the Responsible Person shall authorize the equipotential electrician to carry the insulated transmission rope back along the conductor to the grading ring and prepare for leaving potential. (2) The equipotential electrician shall use the potential transfer rod to hook the grading ring and enters the third insulator from the grading ring. Grasping the insulator tightly with one hand and holding the potential transfer rod with the other hand, the equipotential electrician should use the potential transfer rod to quickly break away from the equipotential.		(1) Deduct 2 points/item for objects left in operation point. (2) Deduct 3 points for failure to apply to the Responsible Person for potential transfer. (3) Deduct 2 points for carrying out potential transfer without permission of the Responsible Person. (4) Deduct 5 points for potential transfer without using the potential transfer rod. (5) Deduct 1 point for unskilled potential transfer.			

Part I Standard for Training and Assessment on Live Working for Operation Maintenance of ±800kV UHV Power Transmission Line

Table (Cont'd)

S/N	Project name	Quality requirements	Score	Deduction standard	Reasons for deduction	Deduction	Scoring
9	Leaving the intense electric field	(3) Protection of safety belt shall not be lost when leaving the intense electric field. (4) The equipotential electrician leaves the intense electric field along the insulator string in the "two-span and three-short-circuit" operation mode and transfers to the cross arm. (5) The combined gap composed of the gaps between the equipotential electrician and the grounding body & the electrified body in the process of leaving the potential shall not be less than 6.6m, and the potential transfer rod must be used for potential transfer when leaving the intense electric field. The minimum distance between the exposed part of human body and the electrified body should not be less than 0.5m	13	(6) Deduct 2 points for losing the protection of safety belt when equipotential electrician leaves the intense electric field. (7) Deduct 2 points for repeated discharge due to insufficient distance of the exposed parts during potential transfer of equipotential electrician. (8) Deduct 2 points for incorrect and unskilled action of equipotential electrician leaving the intense electric field. (9) Deduct 2 points for insufficient combined clearance between equipotential electrician and grounding body & electrified body. (10) Deduct 2 points for losing the protection of safety belt when equipotential electrician transfers to the cross arm. (11) Deduct 2 points for the Responsible Person failing to monitor the operation in place			

Standard for Professional Training and Assessment for Operation Maintenance of UHV DC Power Transmission Line

Table (Cont'd)

S/N	Project name	Quality requirements	Score	Deduction standard	Reasons for deduction	Deduction	Scoring
10	Return to the ground	(1) After checking that there is no object left on tower, the electrician on tower shall report it to the Responsible Person and then climb down the tower successively with insulated transmission rope with the permission of the Responsible Person. (2) When climbing down the tower, the electrician must use anti-fall device, climb down the tower at a uniform speed with the main materials in hand, hang the safety belt and backup protection rope on the shoulder and keep the safety distance of more than 6.8m away from the electrified body, and the Responsible Person shall strengthen the operation monitoring.	5	(1) Deduct 2 points for any objects left on tower. (2) Deduct 1 point/item for the electrician climbing the tower failing to report the inspection results of the leftovers or apply for climbing down the tower. (3) The assessor shall order the termination of the operation (assessment) by electrician climbing down failing to use anti-fall device. (4) Deduct 0.5 points/time for grasping the shackles with hands during climbing down the tower. (5) Deduct 1 point/time for slipping or missing his step during climbing down. (6) Deduct 1 point/person for failing to hang main band of safety belt and back-up protection rope on shoulder			
11	End of the work	(1) The Responsible Person shall organize all working personnel to put working apparatus and materials in order and put them in a special kit (bag) after cleaning; clean the site to ensure that "the materials are removed and the site is cleaned after construction".	10	(1) Deduct 0.5 points per piece for failure to clean or wipe the insulating tools and instruments. (2) Deduct 1 point per piece for failure to classify or place tools and instruments at fixed position. (3) Deduct 0.5 points/time for throwing about tools and instruments or stepping on moisture-proof tarpaulin.			

Part I Standard for Training and Assessment on Live Working for Operation Maintenance of ±800kV UHV Power Transmission Line

Table (Cont'd)

S/N	Project name	Quality requirements	Score	Deduction standard	Reasons for deduction	Deduction	Scoring
11	End of the work	(2) In the post-shift meeting, and the Responsible Person shall make work summaries and comments. Comments include the construction quality of this work and the implementation of safety measures from all working personnel. (3) Responsible Person shall report to the on-duty control personnel the end of the work, apply for the restoration of DC restarting device and terminate the work order	10	(4) Deduct 1 point per piece for not removing fence and signs or leftovers. (5) Deduct 2 points for failure to hold the post-shift meeting. (6) Deduct 1 point when the team is not orderly or focused. (7) Deduct 1 point/person for members of the working group not attending post-shift meeting. (8) Deduct 1 point for unqualified comments. (9) Deduct 1 point for failure to report to the Dispatching Department (Assessor) the end of the work and apply for the restoration of DC restarting device. (10) Deduct 0.5 points/item for non-standard and incomplete reporting terminology or the voice is not loud and clear enough. (11) Deduct 1 point for failure to repeat permission content. (12) Deduct 0.5 points/item for missing repeated content (name of working unit, name of the Responsible Person, time, name of the line, performance report, the equipment has returned to normal status, the personnel have evacuated and DC restart device can be restored). (13) Deduct 1 point/ item for failure to complete work order termination procedure in time or to fill it incorrectly			
	Total		100				

Module 7 Standards for Training and Assessment on Live Repair of ± 800kV UHV Transmission Line Bundle Conductor

I. Training Standard

(I) Training Requirements

Designation of module	Live Repair of ± 800kV UHV Transmission Line Bundle Conductor	Type of training	Operation
Training method	Practical operation training	Training hours	14 hours
Training objectives	1. Master the electrical significance of "two-span and three-short-circuit" operation mode in entering and leaving ±800kV electric field along strain insulator string; 2. Be able to complete the entry of ±800kV equipotential operation point along the strain insulator string; 3. Be able to repair the conductor with pre-twisted armor rod independently (equipotential operation method)		
Training venue	UHV DC practical training line		
Training content	The "two-span and three-short-circuit" operation mode is adopted to enter the electric field along the strain insulator string, and the equipotential operation method is adopted for live repair of the ±800kV transmission line bundle conductor		
Scope of application	Maintenance personnel for ±800kV UHV transmission line		

(II) Referenced Procedures and Specifications

(1) Code for Designing of ±800kV DC Overhead Transmission Line (GB/T50790-2013)

(2) Maintenance Specification for ± 800kV DC Overhead Transmission Line (DL/T251-2012)

(3) Operating Code for ±800kv DC Overhead Transmission Line (GB/T28813-2012)

(4) Technical Specification for Live Working of ±800kV DC Line (DL/T1242-2013)

(5) Technical Specification for Fittings of ±800kV DC Overhead Transmission Line (GB/T 31235-2014)

(6) State Grid Corporation of China on the Management Regulations of Live Working (Trial Implementation) (SGCC [2007] No.751)

(7) State Grid Corporation of China Working Regulations of Power Safety (Transmission Line Section) (Q/GDW1799.2-2013)

(8) Electrotechnical Terminology- Overhead Line (GB/T 2900.51-1998)

(9) Electrotechnical Terminology- Live Working (GB/T2900.55-2016)

(10) Live Working -Terminology for Tools, Equipment and Devices (GB/T 14286-2002)

(11) Minimum Requirements for Utilization of Tools, Devices and Equipment for Live Working (DL/T 877-2004)

(12) Preventive Test Code of Tools, Devices and Equipment for Live Working (DL/T 976-2005)

(13) Technical Guide for Live Working of ±800kV UHV Transmission Line (Q/GDW302-2009)

(14) Shielding Clothes for Live Working (GB/T6568-2008)

(15) Technical Requirements and Design Guide for Live Working Tools (GB/T18037-2008)

(III) Teaching Design for Training

To complete the work task of "live repair of ±800kV UHV transmission line bundle conductor ", each training stage shall be designed according to the standard operation procedure for work task completion. Each stage includes specific training objectives, training content, hours of training, training methods (training resources), training environment, assessment and evaluation, etc, as shown in the Table 1-7-1.

Table 1-7-1 Training Content Design for Live Repair of Bundle Conductor of ±800kV UHV Transmission Line

Training schedule	Training objectives	Training content	Training hours	Training methods and resources	Preparation of training conditions	Assessment and evaluation
1. Theoretical teaching	1. Preliminarily master the basic method for entering and leaving ±800kV electric field along the insulator string; 2. Be familiar with the methods for potential transfer. 3. Be familiar with the method for repairing the damaged conductor of transmission line.	1. The electrical significance for entering and leaving the electric field along the insulator by means of the "two-span and three-short-circuit" operation mode; 2. The method to use the potential transfer rod for the UHV transmission line entering and exiting the electric field; 3. Repair method and quality standard for transmission line conductor	2	Training methods: Lecture Training Resources: PPT, relevant regulations and specifications	Multimedia classroom	Attendance, classroom questions and assignments
2. Preparations	Be able to complete the preparation before operation	1. Work site survey; 2. Preparation of the standardized operation card; 3. Filling of the work order; 4. Preparation of tools and materials for this operation	1	Training methods: 1. Work site survey and cleaning of tools and materials shall be practiced at site; 2. Preparation of operation card and the filling of work order shall adopt lecture method Training resources: 1. UHV practical training line (±800kV practical training line); 2. UHV tools warehouse; 3. Blank work order	1. UHV transmission line for practical training; 2. Multimedia classroom	

Part I Standard for Training and Assessment on Live Working for Operation Maintenance of ±800kV UHV Power Transmission Line

Table (Cont'd)

Training schedule	Training objectives	Training content	Training hours	Training methods and resources	Preparation of training conditions	Assessment and evaluation
3. Work site preparation	Be able to complete the preparations of work site	1. Work site re-survey; 2. Application for work; 3. Work site arrangement 4. Pre-shift Meeting 5. Inspection of tools	1	Training method: demonstration and role play Resource: UHV practical training line (±800kV practical training line)	UHV practical training line (±800kV practical training line)	
4. Trainer's demonstration	The trainees can preliminarily understand the operation process of the task through inspecting and learning from each other's work	1. The equipotential electrician enters and exits the electric field along the strain insulator string; 2. The equipotential electrician reaches the repair position of the conductor by means of walking along the line; 3. Equipotential electrician uses the preformed armor rods to complete the repair of conductor	1	Training method: Demonstration Resource: UHV practical training line	UHV practical training line (±800kV practical training line)	

Table (Cont'd)

Training schedule	Training objectives	Training content	Training hours	Training methods and resources	Preparation of training conditions	Assessment and evaluation
5. Group training of trainees	Through training: 1. Students are able to complete the operation of entering and exiting ±800kV electric field; 2. Students are able to complete the repair of ±800kV transmission line conductor	1. The trainees are grouped (6 in a group) to train the skill operation of entering and exiting ±800kV transmission training electric field and repair conductor; 2. Trainers guide the operation of trainees and conduct safety supervision	8	Training method: Role play Resource: UHV practical training line	UHV practical training line (±800kV practical training line)	Score the operation of trainees according to the detailed rules for skill assessment and scoring
6. End of the work	Through training: 1. Trainees can further understand the shortcomings during the operation process for later improvement; 2. Train the trainees in the working style of safe and civilized production	1. Cleaning at work site; 2. Report to dispatcher; 3. Comment and summarize the current work task at post-shift meeting	1	Training method: Lecture and inductive method	UHV practical training line (±800kV practical training line)	

(IV) Work Flow

1. Work task

The "two-span and three-short-circuit" operation mode is adopted to enter the electric field and reach the operation point along the strain insulator string, and the equipotential operation method is adopted for live repair of the bundle conductor of ±800kV UHV transmission line.

2. Requirements for Weather and Work Site

(1) Live repair of bundle conductor of ±800kV UHV transmission line shall be carried out in good weather.

In case of lightning (hearing thunder or seeing lightning), snow, hail, rain, fog and so on, live working is prohibited. When the wind force is greater than level 5, or the relative humidity of the air is greater than 80%, it is unsuitable for live working; when emergency live repair is required in bad weather, relevant personnel shall be organized to fully discuss and prepare necessary safety measures, which can be implemented after being approved by the unit.

(2) The operating personnel should be in good mental states and be familiar with the organizational and technical measures to ensure safety in work; they should hold the qualification certificate for live working within the validity period.

(3) Responsible Person should organize the relevant personnel to complete field investigation in advance, determine the operating methods, required working apparatus and necessary measures according to the results, and handle the tickets for live working.

(4) The work site should be reasonably set up with fence and warning signs. Non operating personnel is forbidden to enter.

(5) DC restart device shall be deactivated for the Project.

(6) Safe working distance and effective insulation length during operation are shown in Table 1-7-2.

Standard for Professional Training and Assessment for Operation Maintenance of UHV DC Power Transmission Line

Table 1-7-2 Safe Distance for Live Repair of Bundle Conductor of ±800kV UHV Transmission Line (m)

Voltage class	Safe distance between human body and electrified body	Minimum effective insulation length		Minimum combination gap	Minimum distance between the exposed part of human body and electrified body during potential transfer
		Insulating bar	Insulated load-bearing tools and insulating ropes		
±800kV	6.8	6.8	6.8	6.6	0.5

(7) For operation on ±800kV transmission line, the number of insulators in good operation phase shall be no less than 32 pieces.

3. Preparations

3.1 Hazards and precontrol measures

(1) Hazard —— electric shock

Precontrol measures:

① Before work, the Responsible Person shall contact the on-duty control personnel, deactivate the line DC restart device, and perform the licensing procedures.

② Before climbing the tower, ground potential operators on tower must carefully check the name of the line, the number of the pole and tower, and phase, and then the tower can be climbed after all have been confirmed correct.

③ If lines lose power suddenly during work, operators shall consider it as still charged. The Responsible Person shall contact the control personnel as soon as possible, and no forced energization is allowed before the on-duty control personnel getting in touch with the Responsible Person.

④ Insulating tool and insulating ropes shall be free of damage, moisture, deformation, and failure. It is not allowed to use non-insulating ropes (such as cotton rope, manila rope, and steel wire rope).

⑤ The equipotential operator shall wear flame-retardant underwear and full set of shielding clothes outside (including hat, dresses & trousers, gloves, socks and shoes). All parts shall be in excellent connection conditions.

⑥ Before potential transfer, equipotential operator shall obtain the approval of Responsible Person, and the minimum distance between exposed part of human body and electrified body shall not be lower than 0.5m. During potential transfer, the operation shall be fast, and end shall not be used for power charging and discharging; during delivering tools and materials to ground potential operators, the effective length of insulating tools or insulating ropes shall not be lower than the requirements as specified in Table 1-7-2.

⑦ When transmitting large metal objects by using insulating ropes, ground potential operators shall not touch them before ground connection.

⑧ The Special Supervisor shall continuously monitor the operators and correct their nonstandard operation or actions in violation at any time. Special attention shall be paid to operators working at heights to ensure that there is enough safe distance (meeting the requirements in Table 1-7-2). It is forbidden to contact two non-connected electrified bodies or make contact with electrified body and grounding body at the same time.

(2) Hazard —— falling from high places

Precontrol measures:

① Before climbing, operators working at heights must satisfy the requirements of this operation, such as physical condition, mental state, and skill and quality.

② Supervisors shall correct the nonstandard or illegal actions at any time. Special attention shall be paid to operators to prevent them from losing the protection of safety belt or insulated backup protection rope during transposition, and it is forbidden to fasten the safety belt or insulated backup protection rope in a position lower than the operating personnel.

(3) Hazard —— injury caused by objects falling from high place

Precontrol measures:

① Operator working at heights should put personal tools and fragmentary materials into the tools bag. It is strictly forbidden to hang objects in high place or keep in the mouth.

② Ground operator should correctly wear a helmet and use the knots. The vertical distance from the operation point should not be less than the falling radius.

③ The work site shall be set up with fence and warning signs. It shall be noted at any time that Supervisor shall prohibit irrelevant personnel and vehicles from entering operation area.

3.2 Selection of tools, instruments and materials

See Table 1-7-3 for tools, instruments and materials required by live repair of bundle conductor of ±800kV transmission line. Before delivering tools and instruments out of warehouse, application voltage class and test cycle shall be carefully checked to ensure that appearance is intact, connection is firm, rotation is flexible,and they meet the work task requirements. After delivering tools and instruments out of warehouse, they shall be stored in tools bag or tool kit for transportation to avoid contamination and damp. Metal tools and insulated

tools shall be separately loaded and transported to avoid deformation, damage or other defects caused by mixed loading and transportation.

Table 1-7-3 Tools, Instruments and Materials Required by Live Repair of Bundle Conductor of ± 800kV UHV Transmission Line

S/N	Name	Specification and Model	Unit	Qty.	Remarks
1	Insulated transmission rope	TJS-12	Nos.	2	
2	Insulated protection rope	TJS-16	Nos.	2	
3	Insulated tackle	JH10-1	Nos.	2	
4	Safety helmet		pcs	6	
5	Potential transfer rod		Nos.	1	
6	Insulation resistance meter	5000V	pcs	1	
7	Anemometer		pcs	1	
8	Temperature and humidity meter		pcs	1	
9	Multimeter		pcs	1	
10	Moisture-proof canvas	2m×4m	pcs	2	
11	Insulated noose		Nos.	4	
12	Shielding clothes	Shielding efficiency ≥60dB (shield efficiency of shielding mask ≥20dB)	Set	2	
13	Falling protector	Corresponding to the type of pole and tower falling protector	pcs	2	
14	Safety belt		Nos.	2	
15	Security fence		Set	Several	
16	Warning sign	"Work Here", "Access from Here" "Access from Here"	Set	1	
17	Red waistcoat	"Responsible Person"	pcs	1	
18	Preformed armor rod		Set	1	
19	Conductive paste		Box	1	
20	Sandpaper		Sheet	1	
21	Clean towel		pcs	1	
22	Interphone		Set	4	

3.3 Division of labor for operators

Division of labor for operators of the task is shown in Table 1-7-4.

Table 1-7-4 Division of Labor for Live Repair of Bundle Conductor of ±800kV UHV Transmission Line

S/N	Post	Qty. (person)	Responsibilities
1	Responsible Person	1	Be responsible for division of operating personnel of the task, work order reading, stopping the DC restart device, getting work permits, holding pre-shift meeting, dealing with emergency situations in work, quality surveillance, and the summary after work
2	Specific responsible supervisor	1	Be responsible for the safety control of the work site
3	Equipotential electrician	1	Be responsible for entering the equipotential to repair the conductor
4	Ground potential electrician on the tower	1	Be responsible for assisting the equipotential electrician to enter and exit the electric field
5	Ground electrician	2	Be responsible for transferring tools and materials and co-operating with equipotential electrician in entering and exiting the equipotential

4. Working Procedures

The workflow of this task is shown in Table 1-7-5.

Table 1-7-5 Workflow for Live Repair of Bundle Conductor of ±800kV UHV Transmission Line

S/N	Work Content	Operation Steps and Standards	Safety Measures and Precautions	Responsible Person
1	Site re-survey	The Responsible Person shall complete the following work: (1) Check the line name, the number of the pole and tower and ensure the phases are correct; guarantee that the foundation and the pole and tower are intact and in normal condition; ensure that the cross and span distance meets the safety requirements; confirm the defect conditions and the specifications and models of earth wires.	(1) Correctly wear helmet, working clothes, work shoes and protective gloves. (2) Operation under meteorological conditions that may endanger the safety of operators is forbidden. (3) Non-operation personnel and vehicles are strictly prohibited from entering the working site	

Table (Cont'd)

S/N	Work Content	Operation Steps and Standards	Safety Measures and Precautions	Responsible Person
1	Site re-survey	(2) Check that the site meteorological conditions such as wind speed and humidity should meet the operation requirements. (3) Check that the terrain and environment should meet the operation requirements. (4) Check that the safety measures listed in the work order are in line with the actual situations on site. The measures will be added if necessary.		
2	Work Permit	(1) The Responsible Person shall contact the on-duty control personnel and apply for stopping the DC transmission line restart device as per the contents of the work order. (2) Live working could be started only after being approved by control personnel on duty.	Live working shall not be started without the permission of the on-duty control personnel	
3	Site layout	Install the security fence and hang the signboards correctly: (1) The security fence should take full account of falling objects from the high place and the influence on road traffic. (2) The entrance and exit of the security fence shall be set reasonably. (3) Signs such as "Access from Here", "Work Here", "Access from Here" shall be properly arranged	When the influence on road traffic safety is uncontrollable, the traffic management department should be contacted in time to strengthen the on-site control of traffic safety	
4	Holding a pre-shift meeting	(1) All working personnel shall line up. (2) Responsible Person will read out the work order and be clear with work task and division of personnel; explain safety measures and technical measures in work; check (inquire after) mental state of all working personnel; inform of hazards in work and precontrol measures. (3) All working personnel shall sign on the work order for confirmation.	(1) Work order shall be filled and issued with standardized licensing procedure and complete signature. (2) All working personnel shall be in good mental states. (3) All working personnel shall be clear with task division of works, safety measures and technical measures	

Part I　Standard for Training and Assessment on Live Working for Operation Maintenance of ±800kV UHV Power Transmission Line

Table (Cont'd)

S/N	Work Content	Operation Steps and Standards	Safety Measures and Precautions	Responsible Person
5	Inspection tool	(1) The ground electrician and equipotential electrician on tower shall wear the shielding clothes in a right way and pass the inspection, which shall be supervised and inspected by the Responsible Person. (2) Wear personal safety equipment correctly (proper size and easy lock), and the Responsible Person shall supervise and inspect it. (3) Measure the wind speed, wind direction and humidity, check the insulation performance of insulating tools, and make records	(1) Check carefully for damage, deformation and failure before using metal and insulating tools. Carry out segment insulation detection with such insulated tools with insulation resistance meter of 2500V or above and with the resistance no less than 700MΩ, and wipe it off with a clean dry towel. (2) Use a multimeter to measure the resistance between the farthest ends of the shielding clothes and trousers, which shall not be greater than 20Ω. The Responsible Person shall check the connection of the electrician's shielding clothes. (3) Check the tool assembly and make sure the connection is reliable. (4) The tools and instruments for live working shall be placed on the moisture-proof canvas	
6	Climbing the tower	(1) After checking the line name and pole and tower number, the ground potential electrician on tower and equipotential electrician shall impact to check the stress of safety belt and falling protector. (2) The ground potential electrician on tower carries the insulated transmission rope to climb the tower, and the equipotential electrician then climbs the tower. When they reach the operation point of the cross arm, they choose the appropriate position to fasten the safety belt, and the ground potential electrician on tower installs the insulated tackle and the insulated transmission rope in the appropriate position of the cross arm. Then he shall cooperate with the ground electrician to separate the insulated transmission rope for lifting preparation	(1) After checking the correct line name and pole and tower number, he can climb for operation. (2) The anti-falling device installed on tower shall be used during climbing the tower; when moving and transposing on the pole and tower, the safety protection shall not be lost, and the operators must climb and grasp the components securely. (3) The working electrician must wear full set of qualified shielding clothes which must be connected reliably. Before the cross arm enters the equipotential, the equipotential electrician shall check and confirm that each part of the shielding clothes are connected reliably before the next operation	

Table (Cont'd)

S/N	Work Content	Operation Steps and Standards	Safety Measures and Precautions	Responsible Person
7	Entering the intense electric field	(1) The equipotential electrician shall transfer the safety belt to the insulator set fitting, and carry the potential transfer rod, insulated tackle and insulated transmission rope. (2) After the equipotential electrician confirms that all parts of the shielding clothes are well connected, it shall be reported to the Responsible Person for approval. The electrician shall grasp one string with both hands and step on the other string with both feet, and enter the equipotential along the insulator string in the operation mode of "two-span and three-short-circuit". (3) When reaching the third insulator outside the grading ring on the conductor side, the operator shall stop, and carry out the potential transfer with the potential transfer rod	(1) The equipotential electrician must obtain the permission of the Responsible Person before entering the potential. (2) Before entering the insulator string, the equipotential electrician shall fasten the protection rope (holding insulator strings with the backup protection rope and stepping on a insulator string) and adjust the insulated transmission rope and the potential transfer rod. (3) The combined gap composed of the gaps between the equipotential electrician and the grounding body and the electrified body in the process of entering the potential shall not be less than 6.6m	
8	Surface treatment of damaged conductor	(1) After entering the equipotential, the equipotential electrician shall tie the safety belt to the upper sub-conductor, and install the insulated protection rope (all sub-conductors shall be covered). (2) The equipotential electrician shall carry the insulated transmission rope and walk to the operation point along the conductor, and shall install the insulated tackle and the insulated transmission rope at the sub-conductor. (3) The electrician shall check the damage of the conductor and deal with the damage point with abrasive paper (0#) to grind the burr of the damaged part. (4) Equipotential electrician shall clean the surface of the grinded conductor with a cleaning cloth, and apply the live paste evenly on the damaged part of the conductor	(1) Equipotential electrician shall not use much strength to grind the damaged point of the conductor in case of larger damage. (2) The surface should be thoroughly cleaned after the conductor is grinded. (3) Apply the live paste evenly on the surface of the conductor	

Table (Cont'd)

S/N	Work Content	Operation Steps and Standards	Safety Measures and Precautions	Responsible Person
9	Conductor repair	(1) The ground electrician shall pass the preformed armor rods to the equipotential electrician by using an insulated transmission rope. (2) Equipotential electrician uses the preformed armor rods to repair and strengthen the damaged part of conductor	(1) The specification and model of preformed armor rod should match the conductor. (2) The center of the preformed armor rod should be located at the most serious damage. (3) The length of the preformed armor rod shall be enough to fully cover the damaged part. The unilateral length of the end of the preformed armor rod from the edge of damaged part shall not be less than 100mm (4) Bind preformed armor rod firmly to ensure no dumping, missing or loose strands	
10	Leaving the electric field	(1) After inspection, the damaged strip line has been well strengthened and there is no residue at the operation point, the Responsible Person shall authorize the equipotential electrician to carry the insulated transmission rope back to the grading ring and prepare for leaving potential by means of walking along the line. (2) The equipotential electrician shall hook the grading ring with the potential transfer rod and enter the third insulator of the grading ring. With one hand grasping the insulator tightly and the other holding the potential transfer rod, the equipotential electrician shall break himself away from the equipotential with the help of potential transfer rod. (3) The equipotential electrician exits the equipotential by means of "two-span and three-short-circuit" operation mode	(1) The equipotential electrician must obtain the permission of the Responsible Person before leaving the potential. (2) The combined gap composed of the gaps between the equipotential electrician and the grounding body and the electrified body in the process of exiting the potential shall not be less than 6.6m (3) The equipotential electrician shall move along the insulator string with the hands grasping firmly and the feet stepping firmly	

Table (Cont'd)

S/N	Work Content	Operation Steps and Standards	Safety Measures and Precautions	Responsible Person
11	Return to the ground	After checking that there is no object left on the tower, the equipotential electrician on the tower shall report it to the Responsible Person and then climb down the tower with insulated transmission rope after obtaining the consent of the Responsible Person	The anti-falling device installed on the tower shall be used in the process of climbing down the tower; when moving and transposition on the pole and tower, the safety protection shall not be lost, and the operators must grasp the components securely	
12	End of the work	(1) The Responsible Person shall organize all working personnel to put working apparatus and materials in order and put them in a special kit (bag) after cleaning; clean the site to ensure that "the materials are removed and the site is cleaned after construction". (2) After a post-shift meeting is held, the Responsible Person shall make work summaries and comments. Comments include the construction quality of this work and the implementation of safety measures from all working personnel. (3) Responsible Person shall report to the on-duty control personnel the end of the work, apply for the restoration of DC transmission line restarting device and terminate the work order	It is forbidden to restore the DC transmission line restarting device at the appointed time	

II. Assessment Standard

Table 1-7-6 Detailed Rules for Assessment and Scoring of UHVDC Skills Training of State Grid Sichuan Electric Power Corporation

Fill-in Column of Examinee	No.:		Name:		Position:		Unit:		Date:		MM/DD/YYYY
Fill-in Column of Assessor	Achievement:		Assessor:		Assessment Team Leader:		Starting time:		Closing time:		Operation Duration:
Assessment Module	Live Repair of ±800kV UHV Transmission Line Bundle Conductor		Assessee		Maintenance personnel for ±800kV UHV transmission line		Assessment method		Operation		Assessment Time Limit 60min
Job Description	Enter the electric field along the strain insulator strings and carry out live repair of the damaged conductor of ±800kV UHV transmission line.											
Work specification and requirements	1. Live working shall be carried out in good weather. In case of thunder, rain, snow or fog, no live working shall be carried out. When the wind force is greater than Level 5 and the humidity is greater than 80%, live working should not be carried out. 2. Workers required for this operation include 1 Responsible Person, 1 Special Supervisor, 1 ground electrician on the tower, 1 equipotential electrician and 2 ground auxiliary electricians. They should enter the electric field along the strain insulator strings and carry out the live repair of the damaged conductor of ±800kV UHV transmission line. 3. Responsibilities of Responsible Person: Be responsible for division of operating personnel of the task, work order reading, stopping the DC restart device for the transmission line, getting work permits, holding pre-shift meeting, dealing with emergency situations in work, quality surveillance, and the summary after work. 4. Responsibilities of Special Supervisor: Be responsible for safety control of the work site. 5. Responsibilities of equipotential electrician: Enter the electric field along the strain insulator strings and carry out the repair of the damaged conductor. 5. Responsibilities of electrician on tower: Be responsible for assisting equipotential electrician to enter and exit electric field. 6. Responsibilities of ground electrician: Be responsible for transferring tools and materials and cooperating with equipotential electrician in entering and exiting the equipotential.											

Part I Standard for Training and Assessment on Live Working for Operation Maintenance of ±800kV UHV Power Transmission Line

	Table (Cont'd)
Work specification and requirements	7. During the live working, if thunder, rain, strong wind or any other circumstance threaten the safety of the staff, the Responsible Person or Supervisor may stop working temporarily according to the circumstances. Given conditions: 1. Training base: UHV DC ±800kV line pole and tower phase A six-bundle conductor #sub-conductor, the model of the conductor: 6×JL/G3A-900/40. 2. Work orders have been handled, safety measures have been completed (DC transmission line restart device has been deactivated), and oral application (dispatcher or assessor) shall be made at the beginning and end of the work. 3. The instrument shall be used safely and correctly to test the insulating tool. 4. The operation must be carried out according to the working procedures. The scores of the items to be carried out shall be deducted for the process error. In case of major hidden dangers of personal, equipment and operational safety, the assessor may order the termination of the operation (assessment)
Assessment scenario preparation	1. Line: UHV DC ±800kV line 001#~002#tower phase A six-bundle conductor #sub-conductor, work content: live repair of bundle conductor of ±800kV transmission line, model of the conductor: 6×JL/G3A-900/40 2. Required operation tools and instruments: 1 insulated transmission rope (TJS-12), insulated protection rope (TJS-16), 1 insulated tackle (JH10-1), insulation detector, 1 potential transfer rod, insulation resistance meter (5000V), 2 suits of shielding clothes (shielding efficiency ≥60dB), 1 multimeter, 1 tarpaulin, 1 temperature and humidity meter, 1 anemometer, 2 cotton towels, 1 set of preformed armor rod, 1 0# abrasive paper, 1 wood hammer. 3. The work site shall be monitored, and the safety measures (fence, etc.) on the work site have been fully implemented; non-operation personnel are prohibited from entering the site, and the staff must wear safety helmet when entering the work site. 4. Examinees shall bring their own work clothes, flame retardant cotton underwear, safety helmets, gloves, and safety belts (including double-protective ropes)
Remarks	1. The deduction shall be done until the scores of each item are deducted completely. In case of major hidden dangers of personal, equipment and operational safety, the assessor may order the termination of the operation. 2. When equipment, working environment, safety belt, safety helmet, tool, shielding clothes, etc., do not conform to the operation condition, the assessor may order the termination of the operation

Part I Standard for Training and Assessment on Live Working for Operation Maintenance of ±800kV UHV Power Transmission Line

Table 1-7-7 Standards for Assessment and Scoring of UHVDC Skills Training of State Grid Sichuan Electric Power Corporation

S/N	Project name	Quality requirements	Score	Deduction standard	Reasons for deduction	Deduction	Scoring
1	Site re-survey	(1) The Responsible Person shall go to the work site to check the line name, pole and tower number, on-site working conditions, defective parts and so on. (2) Check that the site meteorological conditions such as wind speed and humidity should meet the operation requirements. (3) Check whether the work order is complete and unmodified, check whether the safety measures listed are consistent with the actual situation on site, and supplement it if necessary	5	(1) Deduct 1 point for failure to check the double title. (2) Deduct 1 point for failure to check on-site working conditions (meteorology) and defective parts. (3) Deduct 0.5 points/item for any alteration in the work order filling; deduct 1 point for incorrect work order number; deduct 1.5 points for incomplete work order filling			
2	Work Permit	(1) The Responsible Person is responsible for contacting the on-duty control personnel and apply for stopping the DC transmission line restart device as per the contents of the work order. (2) Reporting content should be standardized and complete	2	(1) Deduct 2 points for failure to contact the Dispatching Department (referee) for deactivating the DC transmission line restart device. (2) Deduct 0.5 points for non-standard or incomplete terminology reporting respectively			

· 443 ·

Table (Cont'd)

S/N	Project name	Quality requirements	Score	Deduction standard	Reasons for deduction	Deduction	Scoring
3	Site layout	Install the security fence and hang the signboards correctly: (1) The security fence should take full account of falling objects from the high place and the influence on road traffic. (2) The entrance and exit of the security fence shall be set reasonably. (3) Signs such as "Access from Here", "Work Here", "Access from Here" shall be properly arranged	3	(1) Deduct 0.5 points for failure to arrange the fence at the work site. (2) Deduct 0.5 points for failure to arrange the warning board. (3) Deduct 0.5 points for failure to hang the tower climbing operation sign			
4	Holding a pre-shift meeting	(1) All staff and personnel shall wear safety helmets and work clothes correctly. (2) Responsible Person shall wear red vest and read out the work order and be clear with work task and division of personnel; explain safety measures and technical measures in work; check (inquire after) mental state of all working personnel; inform of hazards in work and precontrol measures. (3) All working personnel shall sign on the work order for confirmation.	3	(1) Deduct 0.5 points for the staff not dressing uniformly. Deduct 0.5 points/person for the staff not dressing uniformly. (2) No points will be awarded to this item for no division of labor, and deduct 1 point for unclear division of labor. (3) Deduct 0.5 points for the on-site Responsible Person not wearing a safety monitoring vest. (4) Deduct 1 point for the work shift member failing to sign or signing incompletely on the work order.			

Part I Standard for Training and Assessment on Live Working for Operation Maintenance of ±800kV UHV Power Transmission Line

Table (Cont'd)

S/N	Project name	Quality requirements	Score	Deduction standard	Reasons for deduction	Deduction	Scoring
5	Inspection of tools	(1) The staff shall place the tools on the moisture-proof tarpaulin as required; the moisture-proof tarpaulin shall be clean and dry. (2) The tools shall be placed in category according to the requirements of the fixed management; the insulated tools shall not be mixed with metal tools and materials; and the appearance inspection shall be done on the tools. (3) The surface of insulating tools shall not be worn, deformed and damaged, and the operation shall be flexible. Carry out segment insulation detection with such insulated tools with insulation resistance meter of 2530V or above and with the resistance no less than 700 MΩ, and wipe it off with a clean dry towel. (4) The ground potential and equipotential personnel on tower shall correctly wear a whole suit of qualified shield clothes and conductive shoes as required, with each part connected well, shall not wear chemical fiber clothes next to the skin in the shielding clothes and shall fasten safety belts; the Responsible Person shall carefully check whether they wears correctly. (5) Tower climbing personnel shall check the double name, pole number and phase again and report them	7	(1) Deduct 1 point for failure to use moisture-proof tarpaulin and place tools to designed positions. (2) Deduct 0.5 points/item for failure to check qualified label of tool test and appearance inspection. (3) Deduct 1 point/item for failure to use testing instrument for testing the tools. (4) Deduct 2 points/person/time for the operator failing to wear the shielding clothes correctly and each part connected well. (5) Deduct 1 point for the on-site Responsible Person failing to check the safety protective equipment of the tower climbing operators. (6) Deduct 2 points/person for the tower climbing personnel failing to check the double name of the line, pole number and phase. (7) Deduct 2 points/person for the tower climbing personnel failing to report the check results			

Table (Cont'd)

S/N	Project name	Quality requirements	Score	Deduction standard	Reasons for deduction	Deduction	Scoring
6	Climbing the tower	(1) The ground potential electrician and the equipotential electrician on tower shall wear a whole suit of qualified shielding clothes, fasten the safety belt after performing the impulse test on the safety belt, and carry the insulated transmission rope to climb the tower one after another. (2) During the tower climbing process, they shall fasten the anti-fall protection device, climb the tower to an appropriate position, fasten the safety belt, arrange the insulated transmission rope, and then cooperate with the ground electrician to make lifting preparation of the insulated transmission rope separately. (3) During tower climbing, the electrician shall fasten the anti-fall protection device, climb the tower at a uniform speed, grasp the main material by hand, hang the safety belt on the shoulder and keep the safety distance of more than 6.8m away from the electrified body, and the Responsible Person shall strengthen the operation monitoring.	5	(1) Deduct 2 points respectively for failure to fasten the safety belt or to perform the impulse test on the safety belt and the backup protection rope. (2) Deduct 2 points for grasping the shackles with hands. (3) Deduct 1 point for inconvenient suspension position of tackle transmission rope for taking tools. (4) Deduct 2 points for metal tools that are difficult to ensure safe distance during transfer; deduct 2 points for tools that are not bound securely. (5) Deduct 2 points for falling object at heights. (6) Deduct 2 points for tools colliding with the tower body during the transfer process. (7) Deduct 1 point for knotting and disordered rope in tool transfer. (8) Deduct 2 points for the Responsible Person failing to monitor the operation in place. (9) Deduct 2 points for incorrect operation of electrician on tower			

Part I Standard for Training and Assessment on Live Working for Operation Maintenance of ±800kV UHV Power Transmission Line

Table (Cont'd)

S/N	Project name	Quality requirements	Score	Deduction standard	Reasons for deduction	Deduction	Scoring
7	Entering the intense electric field	(1) Before entering the insulator string, the equipotential electrician shall fasten the protection rope (holding insulator strings with the backup protection rope and stepping on a insulator string) and adjust the insulated transmission rope and the potential transfer rod. (2) After the equipotential electrician confirms that all parts of the shielding clothes are well connected, it shall be reported to the Responsible Person for approval. The equipotential electrician shall grasp one string with both hands and step on the other string with both feet, and enter the electric field along the insulator string in the operation mode of "two-span and three-short-circuit". (3) The combined gap composed of the gaps between the equipotential electrician and the grounding body & the electrified body in the process of entering the potential shall not be less than 6.6 m, and the potential transfer rod must be used for potential transfer when entering the electric field	8	(1) Deduct 2 points for insufficient distance of the exposed parts during potential transfer of equipotential electrician. (2) Deduct 2 points/time for the incorrect action of equipotential electrician entering the electric field and repeat discharging. (3) Deduct 1 point for unskillful potential transfer and 5 points for potential transfer without using the potential transfer rod. (4) Deduct 5 points for carrying out potential transfer without permission of the Responsible Person			

· 447 ·

Standard for Professional Training and Assessment for Operation Maintenance of UHV DC Power Transmission Line

Table (Cont'd)

S/N	Project name	Quality requirements	Score	Deduction standard	Reasons for deduction	Deduction	Scoring
8	Surface treatment of damaged conductor	(1) The equipotential electrician shall carry the insulated transmission rope and walk to the operation point along the conductor, and shall install the insulated tackle and the insulated transmission rope at the sub-conductor. (2) The electrician shall check the damage of the conductor and deal with the damage point with abrasive paper (0#) to grind the burr of the damaged part. (3) Equipotential electrician shall clean the surface of the grinded conductor with a cleaning cloth, and apply the live paste evenly on the surface of the damaged part of the conductor	12	(1) Deduct 2 points for the incomplete covering of sub-conductor by insulating rope. (2) Deduct 2 points for failure to report the damage situation to the Responsible Person. (3) Deduct 2 points for failure to smoothen the damaged conductor. (4) Deduct 2 points respectively for failure to remove surface oxide or apply conductive paste			
9	Conductor Repair	(1) The ground electrician shall pass the preformed armor rods to the equipotential electrician by using an insulated transmission rope. Equipotential electrician uses the preformed armor rods to repair and strengthen the damaged part of conductor. (2) The specification and model of preformed armor rod should match the conductor and its center should be located at the most serious damage. (3) The length of the preformed armor rod shall be enough to fully cover the damaged part. The unilateral length of the end of the preformed armor rod from the edge of damaged part shall not be less than 100mm. (4) Bind preformed armor rod firmly to ensure no dumping, missing or loose strands	30	(1) Deduct 3 points for the deviation of repair center exceeding 5mm. (2) Deduct 2 points for any gap at installation of preformed armor rods. (3) Deduct 8 points for deformation of preformed armor rods due to misoperation. (4) Deduct 1 point for an uneven end. (5) Deduct 0.5 points/place when preformed armor rod is not bound firmly and there are dumping, missing or loose strands			

Part I Standard for Training and Assessment on Live Working for Operation Maintenance of ±800kV UHV Power Transmission Line

Table (Cont'd)

S/N	Project name	Quality requirements	Score	Deduction standard	Reasons for deduction	Deduction	Scoring
10	Leaving the electric field	(1) After inspection, the damaged strip line has been well strengthened and there is no residue at the operation point, the Responsible Person shall authorize the equipotential electrician to carry the insulated transmission rope back to the grading ring and prepare for leaving potential by means of walking along the line. (2) The equipotential electrician shall hook the grading ring with the potential transfer rod and enter the third insulator of the grading ring. With one hand grasping the insulator tightly and the other holding the potential transfer rod, the equipotential electrician shall break himself away from the equipotential with the help of potential transfer rod. (3) The equipotential electrician exits the equipotential by means of "two-span and three-short-circuit" operation mode	10	(1) Deduct 2 points for the potential transfer without applying to the Responsible Person; deduct 1 point for starting the work without the consent after applying. (2) Deduct 2 points for insufficient distance of the exposed parts during potential transfer of equipotential electrician. (3) Deduct 2 points for the incorrect action of equipotential electrician exiting the electric field and repeat discharging. (4) Deduct 2 points for unskillful potential transfer and 3 points for potential transfer without using the potential transfer pole			
11	Return to the ground	After checking that there is no object left on the tower, the electrician on the tower shall report it to the Responsible Person and then climb down the tower with insulated transmission rope after obtaining the consent of the Responsible Person	5	(1) Deduct 2 points for failure to use the falling protector when climbing down the tower. (2) Deduct 2 points for loosing the protection of the safety belt when moving on the tower. (3) Deduct 1 point for grasping the tower nail when climbing down the tower. (4) Deduct 2 points for any objects left on the tower			

· 449 ·

Table (Cont'd)

S/N	Project name	Quality requirements	Score	Deduction standard	Reasons for deduction	Deduction	Scoring
12	End of the work	(1) The Responsible Person shall organize all working team members to put working apparatus and materials in order and put them in a special kit (bag) after cleaning; clean the site to ensure that "the materials are removed and the site is cleaned after construction". (2) After a post-shift meeting is held, the Responsible Person shall make work summaries and comments. Comments include the construction quality of this work and the implementation of safety measures from all working personnel. (3) Responsible Person shall report to the on-duty control personnel the end of the work, apply for the restoration of DC transmission line restarting device and terminate the work order	10	(1) Deduct 2 points for failure to clean the tools. (2) Deduct 2 points for missing tools. (3) Deduct 2 points for failure to hold the post-shift meeting. (4) Deduct 2 points for failure to remove the fence. (5) Deduct 2 points for failure to report to dispatcher.			
	Total		100				

Module 8　Standards for Training and Assessment on Live Treatment of Heating Defects of Drainage Plate for Conductor of ±800kV UHV Transmission Line

I. Training Standards

(I) Training Requirements

Designation of module	Live Treatment of Heating Defects of Drainage Plate for Conductor of ±800kV UHV Transmission Line	Type of training	Operation
Training method	Practical operation training	Training hours	14 hours
Training objectives	1. The "two-span and three-short-circuit" operation mode is adopted to enter and leave the ±800kV intense electric field along the strain insulator string. 2. Be able to complete the entry of ±800kV equipotential operation point along the strain insulator string. 3. Be able to independently complete the live treatment of Heating Defects of Drainage Plate for Conductor ±800kV UHV transmission line (equipotential operation method)		
Training venue	UHV DC training line		
Training content	The "two-span and three-short-circuit" operation mode is adopted to enter the intense electric field along the strain insulator string, and the equipotential operation method is adopted for live treatment of Heating Defects of Drainage Plate for Conductor ±800kV UHV transmission line		
Scope of application	Live Maintenance Personnel of UHV DC Transmission Line		

(II) Referenced Procedures and Specifications

(1) Code for Designing of ±800kV DC Overhead Transmission Line (GB/T50790-2013)

(2) Maintenance Specification for ±800kV DC Overhead Transmission Line (DL/T251-2012)

(3) Operating Code for ±800kv DC Overhead Transmission Line (GB/T28813-2012)

(4) Technical Specification for Live Working of ±800kV DC Line (DL/T1242-2013)

(5) Technical Specification for Fittings of ±800kV UHV Transmission Line (GB/T 31235-2014)

(6) State Grid Corporation of China on the Management Regulations of Live Working (Trial Implementation) (SGCC [2007] No.751)

(7) State Grid Corporation of China Working Regulations of Power Safety (Transmission Line Section) (Q/GDW1799.2-2013)

(8) Electrotechnical Terminology- Overhead Line (GB/T 2900.51-1998)

(9) Electrotechnical Terminology- Live Working (GB/T2900.55-2016)

(10) Live Working - Terminology for Tools, Equipment and Devices (GB/T 14286-2002)

(11) Insulated Tackles for Live Working (GB/T13034-2008)

(12) Live Working-Insulating Ropes (GB 13035-2008)

(13) Minimum Requirements for Utilization of Tools, Devices and Equipment for Live Working (DL/T877-2004)

(14) Preventive Test Code of Tools, Devices and Equipment for Live Working (DL/T 976-2005)

(15) Technical Guide for Live Working of ±800kV UHV Transmission Line (Q/GDW302-2009)

(16) Shielding Clothes for Live Working (GB/T6568-2008)

(17) Technical Requirements and Design Guide for Live Working Tools (GB/T18037-2008)

(18) Application Rules of Infrared Diagnosis for Live Electrical Equipment (DL/T664-2016)

(III) Teaching Design for Training

To complete the work task of "live treatment of heating defects of drainage plate for conductor of ±800kV UHV transmission line", each training stage shall be designed according to the standard operation procedure for work task completion. Each stage includes specific training objectives, training content, hours of training, training methods (training resources), training environment, assessment and evaluation, etc, as shown in the Table 1-8-1.

Part I Standard for Training and Assessment on Live Working for Operation Maintenance of ±800kV UHV Power Transmission Line

Table 1-8-1 Training Content Design for Live Treatment of Heating Defects of Drainage Plate for Conductor of ±800kV UHV Transmission Line

Training schedule	Training objectives	Training contents	Training hours	Training methods and resources	Preparation of training conditions	Assessment Evaluation
1. Theoretical teaching	1. To master the basic method for entering and leaving ±800kV DC intense electric field along the insulator string. 2. To be familiar with the methods of potential transfer. 3. To be familiar with the handling method of heating defects of transmission line conductor drainage plate	1. Adopt the "two-span three-short-circuit" operation mode to enter and leave the intense electric field along the insulator string. 2. The use method for the potential transfer rod during entering and leaving the UHV intense electric field. 3. The handling method and quality standard of heating defects of transmission line conductor drainage plate	2	Training method: Lecture. Training resources: PPT, relevant regulations and specifications	Multimedia classroom	Attendance, classroom questions and assignments
2. Preparations	Be able to complete the preparations before operation	1. Work site survey. 2. Preparation of the standardized operation card. 3. Filling of the work order. 4. Preparation of tools and materials for this operation. 5. The on-duty control personnel shall contact and apply to stop work line re-closing lock device	1	Training methods: 1. Site survey and cleaning of tools and materials shall be practiced at site. 2. Preparation of operation card and the filling of work order shall adopt lecture method. Training resources: 1. ±800kV practical training line. 2. UHV tools warehouse. 3. Blank work order	1. UHV transmission line for practical training; 2. Multimedia classroom	

Table (Cont'd)

Training schedule	Training objectives	Training contents	Training hours	Training methods and resources	Preparation of training conditions	Assessment Evaluation
3. Work site preparation	Be able to complete the preparations of work site	1. Re-closing lock has been deactivated and the dispatching permission has been obtained. 2. Work site re-survey. 3. Work site layout. 4. Pre-shift meeting. 5. Inspection of tool, instruments and materials	1	Training method: Demonstration and role play. Resources; ±800kV training line	±800kV Training line	
4. Trainer's demonstration	The trainees can preliminarily understand the operation process of the task through inspecting and learning from each other's work	1. The equipotential electrician enters and exits the intense electric field along the strain insulator string and reach the connection of jumper wire. 2. Equipotential electrician fastens the bolt of conductor jumper wire with socket spanner	2	Training methods: Demonstration. Resources; ±800kV practical training line	±800kV Practical training line	

Part I Standard for Training and Assessment on Live Working for Operation Maintenance of ±800kV UHV Power Transmission Line

Table (Cont'd)

Training schedule	Training objectives	Training contents	Training hours	Training methods and resources	Preparation of training conditions	Assessment Evaluation
5. Group training of trainees	1. Be able to complete the operation of entering and exiting ±800kV intense electric field. 2. Be able to complete the live treatment of heating of conductor jumper wire of ±800kV transmission line	1. The trainees are grouped (6 in a group) to train the skill operation of entering and exiting ±800kV intense electric field and treatment of heating of conductor jumper wire. 2. Trainers guide the operation of trainees and conduct safety supervision	7	Training methods: Role play. Resources: ±800kV practical training line	±800kV Practical training line	Score the operation of trainees according to the detailed rules for skill assessment and scoring
6. End of the work	1. Enable the trainees to further distinguish the shortcomings of the operation process and facilitate the promotion in the later stage. 2. Train the trainees in the work style of safe and civilized production	1. Cleaning up the work site. 2. Report the completion of work to the dispatch and apply for restoration of re-closing lock device. 3. Comment and summarize the work task this time at post-shift meeting	1	Training methods: Lecture and inductive method	±800kV Practical training line	

· 455 ·

(IV) Work Flow

1. Work Task

The "two-span and three-short-circuit" operation mode is adopted to enter the intense electric field along the insulator string and reach the operation point, and the equipotential operation method is adopted for live treatment of heating defects of drainage plate for conductor ±800kV UHV transmission line.

2. Requirements for Weather and Work Site

(1) Live treatment of heating defects of drainage plate for conductor ±800kV UHV transmission line shall be carried out in good weather.

In case of lightning (hearing thunder or seeing lightning), snow, hail, rain, fog and so on, live working is prohibited. When the wind force is greater than level 5, live working shall not be carried out; when the relative humidity is greater than 80%, insulating tool with moisture-proof performance shall be adopted if live working is required; when live rush repair must be carried out in bad weather, relevant personnel shall be organized to fully discuss and prepare necessary safety measures, which shall be carried out after approval of the Unit.

(2) The operating personnel should be in good mental states and be familiar with the organizational and technical measures to ensure safety in work; they should hold the qualification certificate for live working within the validity period.

(3) Responsible Person should organize the relevant personnel to complete field investigation in advance, determine the operating methods, required working apparatus and necessary measures according to the results, and handle the tickets for live working.

(4) The work site should be reasonably set up with fence and warning signs. Non-operating personnel is forbidden to enter.

(5) DC restart device shall be deactivated for the Project.

(6) Safe working distance and effective insulation length during operation are shown in Table 1-8-2.

(7) For operation on ±800kV transmission line, the number of insulators in good operation phase shall be not less than 32 pieces.

Table 1-8-2 Safe Distance for Live Repair of Conductor of ± 800kV UHV Transmission Line (m)

Altitude	Minimum safe distance between equipotential electrician and grounding frame	Minimum effective insulation length of insulating tools	Minimum combination gap
$H \leqslant 1000$	6.8	6.8	6.7
$1000 < H \leqslant 2000$	7.3	7.3	7.3
$2000 < H \leqslant 2500$	7.9	7.8	7.8

Note: The minimum safe distance in the table include 0.5m occupying gap of human body.

(8) When the intermediate potential operator enters ±800kV intense electric field along the strain insulator string, the number of insulators shorted by the human body shall not be more than 4. The minimum number of good insulators shall meet the requirements of Table 1-8-3 after deducting the number of insulators shorted by human body and defective insulators from the strain insulator string.

Table 1-8-3 Minimum Combined Gap and Minimum Number of Good Insulators

Attitude (m)	Structural height of single glass insulator (mm)	Minimum total length of good insulator string (m)	Minimum number of good insulators
$H \leqslant 1000$	170	6.2	37
	195		32
	205		31
	240		26
$1000 < H \leqslant 2000$	170	7.1	42
	195		37
	205		35
	240		30
$2000 < H \leqslant 2500$	170	7.55	45
	195		39
	205		37
	240		32

Note: The values in the table do not include the human body occupying gap which shall not be less than 0.5m during operation.

3. Preparations

3.1 Hazards and precontrol measures

(1) Hazard —— electric shock

Precontrol measures:

① Before work, the Responsible Person shall contact the on-duty control personnel, deactivate the line DC restart device, and perform the licensing procedures.

② Before climbing the tower, ground potential operators must carefully check the name of the line and then can climb the tower after confirmation.

③ If lines lose power suddenly during work, operators shall consider it as still charged. The Responsible Person shall contact the control personnel as soon as possible, and no forced energization is allowed before the on-duty control personnel getting in touch with the Responsible Person.

④ Insulating tool and insulating ropes shall be free of damage, moisture, deformation, and failure. It is not allowed to use non-insulating ropes (such as cotton rope, manila rope, and steel wire rope).

⑤ The equipotential operator shall wear flame-retardant underwear and qualified full set of shielding clothes over the underwear (including hat, mask, dresses & trousers, gloves, socks and shoes). All parts shall be in excellent connection conditions.

⑥ Before potential transfer, equipotential operator shall obtain the approval of Responsible Person, and the minimum distance between exposed part of human body and electrified body shall not be lower than 0.5m. During potential transfer, the operation shall be fast, and end shall not be used for power charging and discharging; During delivering tools and materials to ground potential operators, the effective length of insulating tools or insulating ropes shall not be lower than the requirements as specified in Table 1-8-2.

⑦The Supervisor and Responsible Person shall continuously monitor the operators and correct their nonstandard operation or actions in violation of rules and regulations at any time. Special monitoring shall be conducted to operators work at heights to ensure that there is enough safe distance (meeting the requirements in Table 1-8-2). It is forbidden to contact two non-connected electrified bodies or make contact with electrified body and grounding body at the same time.

(2) Hazard —— falling from high places

Precontrol measures:

① Before climbing, operators working at heights must satisfy the requirements of this operation, such as physical condition, mental state, and skill and quality.

The equipotential electrician shall steadily coordinate his hands and feet during entering the intense electric field with uniform speed; backup protection rope of human body shall be used in the whole process of movement.

③ Supervisors shall correct the nonstandard or illegal actions and behaviors at any time. Special monitoring shall be conducted to operators to prevent them from losing the protection of safety belt or insulated backup protection rope during transposition, and it is forbidden to fasten the safety belt or insulated backup protection rope in a position lower than the operating personnel.

(3) Hazard —— injury caused by objects falling from high place

Precontrol measures:

① Operator working at heights should put personal tools and fragmentary materials into the tools bag. It is strictly forbidden to hang objects in high place or keep in the mouth.

② Ground operator should correctly wear a helmet and use the knots. The vertical distance from the work site should not be less than the falling radius.

③ The work site shall be set up with fence and warning signs. It shall be noted at any time that Supervisor shall prohibit irrelevant personnel and vehicles from entering operation area.

3.2 Selection of tools, instruments and materials

See Table 1-8-4 for tools, instruments and materials required by live treatment of heating defects of the drainage plate for the conductor of ±800kV UHV transmission line. Before delivering tools and instruments out of warehouse, application voltage class and test cycle shall be carefully checked and they shall be inspected to ensure that appearance is intact, connection is firm, rotation is flexible, and they meet the work task requirements. After delivering tools and instruments out of warehouse, they shall be stored in tools bag or tool kit for transportation to avoid contamination and damp. Metal tools and insulating tools shall be separately loaded and transported to avoid deformation, damage or other defects caused by mixed loading and transportation.

Table 1-8-4 Tools, Instruments and Materials Required by Live Treatment of Heating Defects of Drainage Plate for Conductor of ± 800kV UHV Transmission Line

S/N	Name	Specification and Model	Unit	Qty.	Remarks
1	Insulated transmission rope	Φ12	Nos.	2	
2	Insulated backup protection rope	Φ16	Nos.	2	
3	Insulated tackle	1T	Nos.	2	
4	Socket wrench	Consistent with the model of bolt of drainage plate	Set	1	
5	Potential transfer rod		Nos.	1	
6	Insulation resistance tester	5000V	Set	1	
7	Anemometer		Set	1	
8	Temperature and humidity meter		Set	1	
9	Multimeter		Set	1	
10	Moisture-proof canvas	2m×4m	Sheet	1	
11	Shielding clothes	Shielding efficiency ≥60dB (shield efficiency of shielding mask ≥20dB)	Set	2	
12	Falling protector	Corresponding to the type of falling protector of iron tower	pcs	2	
13	Security fence		Set	Several	
14	Warning sign	"Work Here", "Access from Here" "Access from Here"	Set	1	
15	Red waistcoat	"Responsible Person" "Special Supervisor"	pcs	2	
16	Bolt	Consistent with the model of bolt of drainage plate	Nos.	Several	Standby
17	Backpack safety belt	With backup protection rope	Set	2	
18	Clean towel		pcs	1	
19	Insulated noose	Φ12	Nos.	2	
20	Safety helmet		pcs	6	
21	Tool kit		Nos.	2	
22	Interphone		Nos.	4	

3.3 Division of labor for operators

Division of labor for operators of the task is shown in Table 1-8-5.

Table 1-8-5   Division of Labor for Live Treatment of Heating Defects of Drainage Plate for Conductor of ±800kV UHV Transmission Line

S/N	Post	Qty. (person)	Responsibilities
1	Responsible Person	1	Be responsible for organizing various work on the work site
2	Specific responsible supervisor	1	Be responsible for the safety monitoring of the work site
3	Equipotential electrician	1	Be responsible for entering equipotential and treatment of heating defects of drainage plate of conductor
4	Ground potential electrician	1	Be responsible for transferring tools and materials and cooperating with equipotential electrician in entering and exiting the equipotential
5	Ground electrician	2	Be responsible for transferring tools and materials and cooperating with equipotential electrician in entering and exiting the equipotential

4. Working Procedures

The workflow of this task is shown in Table 1-8-6.

Table 1-8-6   Workflow for Live Treatment of Heating Defects of Drainage Plate for Conductor of ±800kV UHV Transmission Line

S/N	Work Content	Operation Steps and Standards	Safety Measures and Precautions	Responsible Person
1	Site re-survey	The Responsible Person shall complete the following work: (1) Check the line name on spot to ensure it is correct; guarantee that the foundation and the iron tower are intact and in normal condition, and ensure that the cross and span distance meets the safety requirements; confirm the defect conditions and the specifications and models of conductor or ground wires. (2) Check that the terrain and environment should meet the operation requirements. (3) Check that the safety measures listed in the work order are in line with the actual situations on site. The measures will be added if necessary.	(1) Correctly wear helmet, working clothes, work shoes and protective gloves. (2) Operation under meteorological conditions that may endanger the safety of operators is forbidden. (3) Non-operation personnel and vehicles are strictly prohibited from entering the working site	

S/N	Work Content	Operation Steps and Standards	Safety Measures and Precautions	Responsible Person
2	Work Permit	(1) The Responsible Person is in charge of contacting the control personnel and apply to stop line re-closing as per the content of work order. (2) Live working could be started only after being approved by control personnel on duty.	Live working shall not be started without the permission of the on-duty control personnel	
3	Site layout	Install the security fence and hang the signboards correctly: (1) The security fence should take full account of falling objects from the high place and the influence on road traffic. (2) The entrance and exit of the security fence shall be set reasonably. (3) Signs such as "Access from Here", "Work Here", "Access from Here" shall be properly arranged	When the influence on road traffic safety is uncontrollable, the traffic management department should be contacted in time to strengthen the on-site control of traffic safety	
4	Holding a pre-shift meeting	(1) All working personnel shall line up. (2) The Responsible Person will read out the work order and be clear with the work task and division of personnel; explain safety measures and technical measures in work; check (inquire after) the mental state of all working personnel; inform of hazards in work and precontrol measures. (3) All working personnel shall sign on the work order for confirmation.	(1) Work order shall be filled and issued with standardized licensing procedure and complete signature. (2) All working personnel shall be in good mental states. (3) All working personnel shall be clear with task division of works, safety measures and technical measures	

Part I Standard for Training and Assessment on Live Working for Operation Maintenance of ±800kV UHV Power Transmission Line

Table (Cont'd)

S/N	Work Content	Operation Steps and Standards	Safety Measures and Precautions	Responsible Person
5	Inspection tool	(1) The equipotential electrician and ground potential electrician shall wear the shielding clothes in a right way and pass the inspection, which shall be supervised and inspected by the Responsible Person. (2) Wear personal safety appliance correctly (proper size and easy lock), and the Responsible Person shall supervise and inspect it. (3) Measure the wind speed and humidity, check the insulation performance of insulating tools, and make records	(1) Check carefully for damage, deformation and dysfunction before using metal and insulating tools. Carry out segment insulation detection with such insulating tools as insulation resistance meter of 5000V and with the resistance no less than $700 M\Omega$, and wipe them clean with the clean dry towel. (2) Use a multimeter to measure the resistance between the farthest ends of the shielding clothes and trousers, which shall be not greater than 20Ω. The Responsible Person shall check the connection of the electrician's shielding clothes. (3) Check the tool assembly and make sure the connection is reliable. (4) The tools and instruments for live working at site shall be placed on the moisture-proof canvas	
6	Climbing the tower	(1) After checking the line name, the electrician on tower shall make impulse test on the safety belt and falling protector. (2) The electrician on tower carries the insulated transmission rope to climb the tower. When reaching the operation point of the cross arm, he shall choose the appropriate position to fasten the safety belt, and install the insulated tackle and the insulated transmission rope in the appropriate position of the cross arm. Then he shall cooperate with the ground electrician to separate the insulated transmission rope for lifting preparation	(1) After checking the correct line name, he can climb the tower for operation. (2) The falling protector installed on the tower shall be used during climbing the tower; when moving and transposing on the tower, the safety protection shall not be lost, and the operators must climb and grasp the components securely. (3) The working electrician must wear full set of qualified shielding clothes which must be connected reliably. Before entering the equipotential, the equipotential electrician shall check and confirm that each part of the shielding clothes are connected reliably before the next operation	

Table (Cont'd)

S/N	Work Content	Operation Steps and Standards	Safety Measures and Precautions	Responsible Person
7	Entering the intense electric field	(1) The ground electrician transmits the potential transfer rod to the tower. (2) The equipotential electrician shall move to the insulator set fitting with potential transfer rod, and carry the insulated tackle and insulated transmission rope. (3) After the equipotential electrician confirms that all parts of the shielding clothes are well connected, it shall be reported to the Responsible Person for approval. The equipotential electrician shall grasp one string with both hands and step on the other string with both feet, and enter the intense electric field along the insulator string in the operation mode of "two-span and three-short-circuit". (4) When reaching the third insulator outside the grading ring on the conductor side, the equipotential electrician shall stop, and carry out the potential transfer with the potential transfer rod	(1) The equipotential electrician must obtain the permission of the Responsible Person before entering the potential. (2) The safety belt and insulated transmission rope shall be prevented from hooking up tower materials. (3) Before entering the insulator string, the equipotential electrician shall fasten the backup protection rope and adjust the insulated transmission rope. (4) Protection of safety belt shall not be lost during transposition. (5) The safe distance between the human body and the electrified body, and the combined gap between the human body and the grounding body and the electrified body shall not be less than that specified in Table 2-8-2	
8	Treatment of heating defects of drainage plate of conductor	(1) After entering the equipotential, the equipotential electrician shall tie the safety belt to the sub-conductor, and ground electrician pass the socket spanner to the equipotential electrician with transmission rope. (2) Equipotential electrician fastens the connecting bolt of conductor drainage plate with socket spanner. (3) The equipotential electrician passes the socket spanner down the tower with transmission rope	(1) The equipotential electrician should pay attention to avoid excessive action range, and avoid stretching the limbs to the insulator. (2) The equipotential electrician should pay attention to avoid scalding	

Part I Standard for Training and Assessment on Live Working for Operation Maintenance of ±800kV UHV Power Transmission Line

Table (Cont'd)

S/N	Work Content	Operation Steps and Standards	Safety Measures and Precautions	Responsible Person
9	Exiting the potential	(1) After inspection, there is no residue at the operation point, the Responsible Person shall authorize the equipotential electrician to carry the insulated transmission rope and prepare for leaving potential. (2) The equipotential electrician shall hook the grading ring with the potential transfer rod and enter the third insulator of the grading ring. With one hand grasping the insulator tightly and the other holding the potential transfer rod, the equipotential electrician shall break himself away from the equipotential with the help of potential transfer rod. (3) The equipotential electrician exits the equipotential by means of "two-span and three-short-circuit" operation mode	(1) The equipotential electrician must obtain the permission of the Responsible Person before leaving the potential. (2) The combined gap composed of the gaps between the equipotential electrician and the grounding body and the electrified body in the process of exiting the potential shall not be less than 6.7m. (3) When moving along the insulator string, it shall be grasped with the hands firmly, with the feet stepping firmly.	

Standard for Professional Training and Assessment for Operation Maintenance of UHV DC Power Transmission Line

II. Assessment Standard

Table 1-8-7 Detailed Rules for Assessment and Scoring of Operation and Inspection Skills of UHV DC Transmission Line

Fill-in Column of Examinee	No.:		Name:		Position:		Unit:		Date:		MM/DD/YYYY
Fill-in Column of Assessor	Grade:		Assessor:		Assessment Team Leader:		Starting time:		Closing time:		Operating Duration:
Assessment Module	Live Treatment of Heating Defects of Drainage Plate for Conductor of ±800kV UHV Transmission Line		Assessee		Maintenance Personnel of UHV DC Transmission Line		Assessment method		Operation		Assessment Time Limit 60min
Job Description	Live Treatment of Heating Defects of Drainage Plate for Conductor of ±800kV UHV Transmission Line by the Method of Entering Intense Electric Field Along Insulator String with "Two-span and Three-short-circuit".										
Work Specifications and Requirements	1. Live working shall be carried out in good weather. In case of thunder, rain, snow or fog, no live working shall be carried out. When the wind force is greater than Level 5 and the humidity is greater than 80%, live working should not be carried out. 2. This operation requires 6 people, including one Responsible Person, one Special Supervisor, one equipotential electrician, one ground potential electrician, and two ground electricians. Heating defects of the drainage plate for the conductor of ±800kV UHV transmission line are treated by the method of entering the intense electric field along the insulator string with strain insulator string. 3. Responsibilities of Responsible Person (Supervisor): Be responsible for division of operating personnel of the task, work order reading, handling re-closing deactivation of the line and the formalities of work permit, getting work permits, holding pre-shift meeting, safety supervision in the operation process, dealing with emergency situations in work, quality surveillance, and the summary after work										

Part I Standard for Training and Assessment on Live Working for Operation Maintenance of ±800kV UHV Power Transmission Line

Table (Cont'd)

Work Specifications and Requirements	4. Responsibilities of equipotential electricians: Be responsible for the main work in this operation process, installing and dismantling the work tools according to the position of the work, and entering the equipotential for the treatment of the heating defects of the drainage plate for the conductor of the transmission line. 5. Responsibilities of ground potential electrician: Be responsible for transferring tools and materials and cooperating with equipotential electrician in entering and exiting the equipotential. 6. Responsibilities of ground electrician: Be responsible for transferring tools and materials. 7. During the live working, if thunder, rain, strong wind or any other circumstance threaten the safety of the staff, the Responsible Person or Supervisor may stop working temporarily according to the circumstances. Given conditions: 1. Side phase conductor of resisting-tensile tower of ±800kV practical training line. 2. Work orders have been handled, safety measures have been completed (re-closing has been deactivated), and oral application (dispatcher or assessor) shall be made at the beginning and end of the work. 3. The instrument shall be used safely and correctly to test the insulating tool. 4. The operation must be carried out according to the working procedures. The relevant item scores shall be deducted for the process error. In case of major hidden dangers of personal, equipment and operational safety, the assessor may order the termination of the operation (assessment)
Assessment scenario preparation	1. Tower shape: Side phase conductor of resisting-tensile tower of ±800kV practical training line. Work content: live treatment of heating defects of the drainage plate for the conductor of ±800kV UHV transmission line. 2. Required working tools: 2 x Φ12 insulated transmission ropes, 2 x Φ16 insulated protection ropes, 2 x 1T insulated tackles, 1 set of socket spanners of different types, 1 potential transfer rod, 1 5000V insulation resistance tester, 1 anemorumbometer, 1 thermohygrometer, 1 multimeter, 1 2m×4m moisture-proof canvas, 2 suits of type II shielding clothes, 2 falling protectors, 1 set of safety fence, 1 set of warning signs, 2 suits of red waistcoats, several bolts of different types, several pins of different types, 2 sets of dual fail-safe safety belts, 1 cleaning towel, 2 Φ12 insulating rope sleeves, 6 safety helmets, 1 operating rod, 2 tool kits, and 4 interphones. 3. The work site shall be monitored, and the safety measures (fence, etc.) on the work site have been fully implemented; non-operation personnel are prohibited from entering the site, and the staff must wear safety helmets when entering the work site. 4. Examinees shall bring their own work clothes, flame retardant cotton underwear, safety helmets, gloves, safety belts (including double-protective ropes).
Remarks	1. The deduction shall be completed until the scores of each item are deducted completely. In case of major hidden dangers of personal, equipment and operational safety, the assessor may order the termination of the operation 2. When equipment, working environment, safety belt, safety helmet, tool, shielding clothes, etc., do not conform to the operation condition, the assessor may order the termination of the operation

· 467 ·

Module 9 Standards for Training and Assessment on Live Replacement of Whole String of Insulators in a Straight Line of Grounding Electrode of ±800kV UHV Transmission Line

I. Training Standards

(I) Training Requirements

Designation of module	Live Replacement of Whole String of Insulators in a Straight Line of Grounding Electrode of ±800kV UHV Transmission Line	Type of training	Operation type
Training method	Practical operation training	Training hours	14 hours
Training objectives	1. Master the electrical significance of "rope ladder method" operation mode when the ±800kV DC grounding electrode tangent tower enters and exits electric field. 2. Can enter ±800kV DC grounding electrode line equipotential operation point by adopting the "rope ladder method". 3. Can independently complete the replacement of whole string of insulators in a straight line of the DC grounding electrode of ±800kV UHV transmission line (equipotential operation method)		
Training venue	UHV DC practical training line		
Training content	Live replacement of whole string of insulators on the straight line of the DC grounding electrode of the ±800kV UHV transmission line by entering the electric field from ground with the rope ladder by the equipotential operation method		
Scope of application	Maintenance personnel for ±800kV U HV DC power transmission line		

(II) Referenced Procedures and Specifications

(1) Code for Designing of ±800kV DC Overhead Transmission Line (GB/T50790-2013)

(2) Maintenance Specification for ±800kV DC Overhead Transmission Line (DL/T251-2012)

(3) Operating Code for ±800kV DC Overhead Transmission Line (GB/T28813-2012)

(4) Technical Specification for Live Working of ±800kV DC Line (DL/T1242-2013)

(5) Technical Specification for Fittings of ±800kV DC Overhead Transmission Line (GB/T 31235-2014)

(6) State Grid Corporation of China on the Management Regulations of Live Working (Trial Implementation) (SGCC [2007] No.751)

(7) State Grid Corporation of China Working Regulations of Power Safety (Transmission Line Section) (Q/GDW1799.2-2013)

(8) Electrotechnical Terminology- Overhead Line (GB/T 2900.51-1998)

(9) Electrotechnical Terminology- Live Working (GB/T2900.55-2016)

(10) Live Working - Terminology for Tools, Equipment and Devices (GB/T 14286-2002)

(11) Insulated Tackles for Live Working (GB/T13034-2008)

(12) Live Working-Insulating Ropes (GB 13035-2008)

(13) Minimum Requirements for Utilization of Tools, Devices and Equipment for Live Working (DL/T 877-2004)

(14) Preventive Test Code of Tools, Devices and Equipment for Live Working (DL/T 976-2005)

(15) Technical Guide for Live Working of ±800kV UHV Transmission Line (Q/GDW302-2009)

(16) Shielding Clothes for Live Working (GB/T6568—2008)

(17) Technical Requirements and Design Guide for Live Working Tools (GB/T18037-2008)

(18) Application Rules of Infrared Diagnosis for Live Electrical Equipment (DL/T664-2016)

(III) Teaching Design for Training

To complete the work task of "live replacement of the whole string of insulators on the straight line of the grounding electrode of ±800kV UHV transmission line", each training stage shall be designed according to the standard operation procedure for work task completion. Each stage includes specific training objectives, training content, hours of training, training methods (training resources), training environment, assessment and evaluation, etc, as shown in the Table 1-9-1.

Table 1-9-1 Training Content Design for Live Replacement of Whole String of Insulators in a Straight Line of Grounding Electrode of ±800kV UHV Transmission Line

Training schedule	Training objectives	Training contents	Training hours	Training methods and resources	Training environment	Assessment and evaluation
1. Theoretical teaching	1. Preliminarily grasp the method of using insulated rope ladders to enter and exit ±800kV DC grounding electrode line electric field. 2. Be familiar with equipment configuration and safety precautions. 3. Be familiar with the method of replacement of whole string of insulators in the straight line of the DC grounding electrode of ±800kV UHV transmission line	1. Electrical significance of using insulated rope ladders to enter and exit ±800kV DC grounding electrode line electric field. 2. Equipment allocation, hazard analysis and safety measures. 3. Be familiar with the method and quality standards of replacement of whole string of insulators in the straight line of the DC grounding electrode of ±800kV UHV transmission line	2	Training methods: Lecture. Training resources: PPT, relevant regulations and specifications	Multimedia classroom	Attendance, classroom questions and assignments
2. Preparations	Be able to complete the preparation before operation	1. Work site survey. 2. Preparation of the standardized operation card. 3. Filling of the work order. 4. Preparation of tools and materials for this operation	1	Training method: 1. Site survey and cleaning of tools and materials shall be practiced at site. 2. Preparation of operation card and the filling of work order shall adopt lecture method. Training resources: 1. ±800kV DC grounding electrode practical training line. 2. UHV tools warehouse. 3. Blank work order	1．±800kV DC grounding electrode practical training line; 2. Multimedia classroom	

Part I Standard for Training and Assessment on Live Working for Operation Maintenance of ±800kV UHV Power Transmission Line

Table (Cont'd)

Training schedule	Training objectives	Training contents	Training hours	Training methods and resources	Training environment	Assessment and evaluation
3. Work site preparation	Be able to complete the preparations of work site	1. Work site re-survey. 2. Work application. 3. Work site layout. 4. Pre-shift meeting. 5. Inspection of tools	1	Training method: Demonstration and role play. Resources: ±800kV DC grounding electrode practical training line	±800kV DC grounding electrode practical training line	
4. Trainer's demonstration	The trainees can preliminarily understand the operation process of the task through inspecting and learning from each other's work	1. Working distance measurement and conductor temperature measurement. 2. The ground potential electricians climb up the iron tower to the work site for the installation of the insulated rope ladder. 3. The equipotential electricians climb up the insulated rope ladder to enter the electric field. 4. The ground potential electricians install the tools and arrange the lifting point of the insulator string and conductor protective rope. 5. The equipotential electricians install the lifting tools at the end of the conductor, remove and restore the pins and connections at the end of the conductor.	3	Training methods: Demonstration. Resources: UHV practical training line	±800kV DC grounding electrode practical training line	

· 471 ·

Table (Cont'd)

Training schedule	Training objectives	Training contents	Training hours	Training methods and resources	Training environment	Assessment and evaluation
5. Group training of trainees	Through training: 1. Trainees are able to complete the operation of entering and exiting ±800kV DC grounding electrode line electric field. 2. Trainees can complete the method of replacement of whole string of insulators in the straight line of the DC grounding electrode of ±800kV UHV transmission line	1. The trainees are grouped (6 in a group) to train the skill operation for entering and leaving ±800kV DC grounding electrode line electric field to replace the whole string of insulators on the straight line. 2. Trainers guide the operation of trainees and conduct safety supervision	8	Training methods: Role play. Resources: ±800kV DC grounding electrode practical training line	±800kV DC grounding electrode practical training line	Score the operation of trainees according to the detailed rules for skill assessment and scoring
6. End of the work	Through training: 1. Trainees can further understand the shortcomings during the operation process for later improvement. 2. Train the trainees in the working style of safe and civilized production	1. Work site clearing. 2. Report to the controller. 3. Comment and summarize the current work task at post-shift meeting	1	Training method: Lecture and inductive method	±800kV DC grounding electrode practical training line	

(IV) Work Flow

1. Work task

Live replacement of the whole string of insulators on the straight line of the grounding electrode of ±800kV UHV transmission line by entering the electric field along the rope ladder to reach the operation point by climbing up the rope ladder by the equipotential operation method.

2. Requirements for Weather and Work Site

(1) The live replacement of the whole string of insulators on the straight line of the grounding electrode of ±800kV UHV transmission line shall be carried out in good weather. In case of lightning (hearing thunder or seeing lightning), snow, hail, rain, fog and so on at the work site, this operation shall be stopped. When wind force is greater than Level 5 or the relative humidity of air is greater than 80%, this operation should not be carried out; during the operation, pay attention to the weather changes. In case of gale or misty rain and other emergencies, take correct measures according to the procedures to ensure the safety of personnel and equipment.

(2) The operating personnel shall be in good mental states and be familiar with the organizational and technical measures to ensure safety in work; they shall hold the qualification certificate for live working within the validity period.

(3) Responsible Person shall organize the relevant personnel to complete field investigation in advance, determine the operating methods, required working apparatus and necessary measures according to the results, and handle the work order for live working.

(4) The work site shall be reasonably set up with fence and warning signs. Non operating personnel is forbidden to enter.

(5) DC restart device of the line shall be deactivated for the Project.

(6) Safe working distance and effective insulation length during operation are shown in Table 1-9-2.

Table 1-9-2 Safe Distance (m) for Live Replacement of Whole String of Insulators in a Straight Line of Grounding Electrode of ±800kV UHV Transmission Line

Voltage class	Safe distance between human body and electrified body	Minimum effective insulation length		Minimum combination gap	Minimum distance between the exposed part of human body and electrified body during potential transfer
		Insulating bar	Insulated load-bearing tools and insulating ropes		
±800kV DC grounding electrode	0.75	1.3	1.0	1.2	0.3

3. Preparations

3.1 Hazards and precontrol measures

(1) Hazard —— Preventing system overvoltage

Precontrol measures: This operation is equipotential operation. Before starting the work, the Responsible Person shall contact the on-duty controller to apply for a work order for live working on power lines, and apply to deactivate the DC restart protection device. The controller on duty shall perform the licensing procedures. After the live working is completed, report to the controller on duty in a timely manner. It is forbidden to disable and restore the DC restart device at the appointed time. If equipment loses power suddenly during live working, operators shall consider the equipment as still charged. The Responsible Person shall contact the control personnel as soon as possible, and no forced energization is allowed before the controller on duty getting in touch with the Responsible Person.

(2) Hazard —— for preventing electric shocks

Precontrol measures:

① Insulating tool and insulating ropes shall be free of damage, moisture, deformation, and failure. It is not allowed to use non-insulating ropes (such as cotton rope, manila rope, and steel wire rope).

② The equipotential operator shall wear flame-retardant underwear and full set of screen clothes outside (including hat, dresses & trousers, gloves, socks and shoes). All parts shall be in excellent connection conditions, and the resistance of the whole suit of shielding clothes shall not exceed 20Ω.

③ Before potential transfer, equipotential operator shall obtain the approval of Responsible Person, and the minimum distance between exposed part of human body and electrified body shall not be lower than 0.3m. The safe distance between the ground potential operator and the electrified body (between the equipotential operator and grounding body) is less than that specified in Table 1-9-2.

④ When transmitting large metal objects by using insulated ropes, ground potential operators shall not touch them before ground connection.

⑤ The Responsible Person and the Special Supervisor shall continuously monitor the operators and correct their nonstandard operation or actions in violation at any time. Special attention shall be paid to high-place operators to ensure that there is enough safety distance (meeting the requirements in Table 1-9-2).

⑥ When the operating temperature of the heat-resisting conductor is above 40°C, high temperature resistance live working tools shall be used, including high temperature resistance soft insulating tools, high temperature resistance hard insulating tools, and high temperature resistance heat insulation shielding clothes, and the potential transfer rod shall be used to enter the equipotential.

⑦ Insulation protective equipment shall be regarded as a conductor. During the transfer and use, the safe distance between the phase and ground and poles shall be ensured to meet the safety requirements.

(3) Hazard —— falling from high places

Precontrol measures:

① Before climbing, operators working at heights must satisfy the requirements of this operation, such as physical condition, mental state, and skill and quality.

② Supervisor shall correct irregularities and violations at any time. Special attention shall be paid to operators to prevent them from losing the protection of the safety belt or insulated backup protection rope during transposition. The safety belt or insulated backup protection rope shall not be fastened lower than operating personnel.

(4) Hazard —— injury caused by objects falling from high place

Precontrol measures:

① Operator working at heights shall put personal tools and fragmentary materials into the tools bag. It is strictly forbidden to hang objects in high place or keep something in the mouth.

② Ground operator shall correctly wear a safety helmet and use the knots. The vertical distance from the operation point shall not be less than the falling radius.

③ The work site shall be set up with fence and warning signs. It shall be noted at any time that Supervisor shall prohibit irrelevant personnel and vehicles from entering operation area.

(5) For injuries caused by conductor overheating

Precontrol measures:

① Before the operation, contact with the dispatcher to confirm the operation mode to avoid the live working when the large current of the conductor flows.

② Operators shall wear special heat-resistant shielding clothes to carry out live working to prevent the conductor from overheating and injuring people.

3.2 Selection of tools, instruments and materials

See Table 1-9-3 for tools, instruments and materials required by live replacement of whole string of insulators in a straight line of grounding electrode of ±800kV UHV transmission line. Before delivering tools and instruments out of warehouse, application voltage class and test cycle shall be carefully checked and they shall be inspected to ensure that appearance is intact, connection is firm, rotation is flexible,and they meet the work task requirements. After delivering tools and instruments out of warehouse, they shall be stored in tools bag or tool kit for transportation to avoid contamination and damp. Metal tools and insulated tools shall be separately loaded and transported to avoid deformation, damage or other defects caused by mixed loading and transportation.

Table 1-9-3 Tools, Instruments and Materials Required by Live Replacement of Whole String of Insulators in a Straight Line of Grounding Electrode of ±800kV UHV Transmission Line

S/N	Name	Specification and Model	Unit	Qty.	Remarks
1	Insulated transmission rope	TJS-14×50m	Nos.	2	Insulating tool
2	Insulated transmission rope	TJS-10×50m	Nos.	1	Insulating tool
3	Insulated noose	TJS-14	Nos.	2	Insulating tool
4	Insulated tackle	0.5T	Nos.	2	Insulating tool
5	Insulated pull-rod		Set	1	Insulating tool
6	Protection rope for conductor	TJS-18	Nos.	1	Insulating tool
7	Insulating rope ladder	30m	Nos.	2	Insulating tool
8	Insulating bar		Nos.	1	Insulating tool
9	Cross arm clamps		Set	1	Metal tool
10	Two-bundle wire lifter		Nos.	1	Metal tool
11	Somersault tackle		Nos.	1	Metal tool
12	Aluminium alloy ladder head		Nos.	1	Metal tool
13	Metal buckle		Nos.	6	Metal tool
14	Shielding clothes	500kV type I or thermal insulation type	Set	3	PPE

Table (Cont'd)

S/N	Name	Specification and Model	Unit	Qty.	Remarks
15	Human body backup protection rope		Nos.	2	PPE
16	Safety belt		Nos.	3	PPE
17	Safety helmet		pcs	8	PPE
18	Personal tools		Set	1	Other tools
19	Moisture-proof tarpaulin	3m×3m	pcs	1	Other tools
20	Multimeter		Nos.	1	Other tools
21	Anemoscope and hygroscope	HT-8321	pcs	1	Other tools
22	Megameter and test electrode	5000V	Set	1	Other tools
23	Laser distance meter		Set	1	Other tools
24	Infrared thermometer		Set	1	Other tools
25	Interphone		Set	2	Other tools
26	Red waistcoat	"Responsible Person"	pcs	1	Other tools
27	Security fence		Set	Several	Other tools
28	Insulators of the same type		pcs	6	Material

3.3 Division of labor for operators

Division of labor for operators of the task is shown in Table 1-9-4.

Table 1-9-4 Personnel Allocation for Live Replacement of Whole String of Insulators in a Straight Line of Grounding Electrode of ±800kV UHV Transmission Line

S/N	Post	Qty. (person)	Responsibilities
1	Responsible Person	1	Be responsible for various work on the work site
2	Special Supervisor	1	Be responsible for supervision of work on the tower
3	Equipotential electrician	1	Be responsible for removing and restoring the pins of the insulator at the terminal of the conductor.
4	Ground potential electrician on the tower	1	Be responsible for arranging the transmission rope, rope ladder and insulator string lifting points and installing tools
5	Ground electrician	2	Be responsible for transferring tools and materials and cooperating with equipotential electrician in entering and exiting the equipotential

4. Working Procedures

The workflow of this task is shown in Table 1-9-5.

Table 1-9-5 Work Flow for Live Replacement of Whole String of Insulators in a Straight Line of Grounding Electrode of ±800kV UHV Transmission Line

S/N	Work Content	Operation Steps and Standards			Safety Measures and Precautions	Responsible Person
1	Confirm the line name and tower number and inspect the pole and tower	The Responsible Person shall complete the following work: (1) Check the line name, the number of the pole and tower and ensure the phases are correct; guarantee that the foundation and the pole and tower are intact and in normal condition; ensure that the cross and span distance meets the safety requirements; confirm the defect conditions and the specifications and models of earth wires; (2) Check that the terrain and environment should meet the operation requirements; (3) Check that the safety measures listed in the work order are in line with the actual situations on site. The measures will be added if necessary.			(1) Correctly wear helmet, working clothes, work shoes and protective gloves; (2) Operation under meteorological conditions that may endanger the safety of operators is forbidden; (3) Non-operation personnel and vehicles are strictly prohibited from entering the working site	
2	Environmental measurement of work on site	Measurement item	Standard value	Measured value	When the wind force is greater than Level 5 and the humidity is greater than 80%, live working should not be carried out.	
		Speed	≤10m/s			
		Humidity	≤80%			
		Ambient temperature	0~38°C			
		Conductor temperature	—			
		Working distance	≥0.75m			
		Altitude	Altitude limit of 500m and 1000m			
3	Work Permit	(1) The Responsible Person is responsible for contacting the on-duty control personnel and applying for stopping the restart protection device for the line as per the contents of the work order; (2) Live working could be started only after being approved by control personnel on duty.			Live working shall not be started without the permission of the on-duty control personnel	

Table (Cont'd)

S/N	Work Content	Operation Steps and Standards	Safety Measures and Precautions	Responsible Person
4	Site layout	Install the security fence and hang the signboards correctly: (1) The security fence should take full account of falling objects from the high place and the influence on road traffic; (2) The entrance and exit of the security fence shall be set reasonably; (3) Signs such as "Access from Here", "Work Here", "Access from Here" shall be properly arranged;	When the influence on road traffic safety is uncontrollable, the traffic management department should be contacted in time to strengthen the on-site control of traffic safety	
5	Holding a pre-shift meeting	(1) All working personnel shall line up; (2) (The Responsible Person will read out the work order and be clear with the work task and division of personnel; explain safety measures and technical measures in work; check (inquire after) the mental state of all working personnel; inform of hazards in work and precontrol measures; (3) (All working personnel shall sign on the work order for confirmation	(1) The work order shall be filled in, issued and approved in a standardized manner, and the signature shall be complete; (2) All working personnel shall be in good mental states; (3) All working personnel shall be clear with task division of works, safety measures and technical measures	
6	Inspection tool	(1) The ground electrician and equipotential electrician on tower shall wear the shielding clothes in a right way and pass the inspection, which shall be supervised and inspected by the Responsible Person; (2) Wear personal safety equipment correctly (proper size and easy lock), and the Responsible Person shall supervise and inspect it; (3) Check the appearance and measure insulation performance of insulating tools, and make records	(1) (Check carefully for damage, deformation and failure before using metal and insulating tools. (Wipe the insulating tools with a clean and dry towel and carry out segment insulation detection with such insulated tools with insulation resistance meter of 2,500V or above and with the resistance no less than $700\mathrm{M}\Omega$; (2) Use a multimeter to measure the resistance between the farthest ends of the shielding clothes and trousers, which shall not be greater than 20Ω. The Responsible Person shall check the connection of the electrician's shielding clothing;	

Table (Cont'd)

S/N	Work Content	Operation Steps and Standards	Safety Measures and Precautions	Responsible Person
			(3) Check the tool assembly and make sure the connection is reliable; (4) The tools and instruments for live working shall be placed on the moisture-proof tarpaulin	
7	Climbing the tower	(1) After checking the line name and pole and tower number, the ground potential electrician on tower shall impact to check the stress of safety belt and falling protector. (2) The ground potential electricians on the tower shall carry insulated transmission ropes while climbing the tower to the cross arm, fasten safety belts and human body backup protection ropes, and arrange transmission ropes at an appropriate position	(1) After checking the correct line name and pole and tower number, he can climb for operation; (2) The anti-falling device installed on tower shall be used in the process of climbing the tower; when moving and transposition on the pole and tower, the safety protection shall not be lost, and the operators must grasp the components securely; (3) The working electrician must wear full set of qualified shielding clothes which must be connected reliably	
8	Install the somersault tackle with insulating rope	The ground electricians shall transfer the insulating bar and the somersault tackle to the tower; the ground potential electricians shall hang the somersault tackle with Φ10mm insulated transmission rope on the conductor	(1) The ground potential electrician shall not lose the protection of safety belt or human body backup protection rope when returning on the pole and tower; (2) The distance between the human body and the electrified body shall not be less than 0.75m when the somersault tackle is installed; the effective insulation length of the insulating bar shall not be less than 1.3m; (3) The somersault tackle shall be installed firmly on the conductor at a reasonable position; (4) Special Supervisor shall carefully supervise it	

Part I Standard for Training and Assessment on Live Working for Operation Maintenance of ±800kV UHV Power Transmission Line

Table (Cont'd)

S/N	Work Content	Operation Steps and Standards	Safety Measures and Precautions	Responsible Person
9	Rope ladder installation	The ground potential electrician cooperates with the ground work personnel to hang the rope ladder on the conductor and pull the rope ladder to a safe position.	(1) The rope ladder and the ladder head shall be in good appearance, without damage, missing parts and deformation; (2) The rope ladder and the ladder head shall be connected reliably; (3) The binding point for transferring the rope ladder shall be appropriate, and the human body backup protection rope placed on the ladder head shall be firm and reliable, and the tackle shall be flexible to rotate, without hooking or groove tipping; (4) Responsible Person shall carefully inspect it	
10	Entering the intense electric field	(1) The ground electrician conducts an impact check on the rope ladder to confirm that it has been reliably installed on the conductor before the insulating rope ladder is pressed down for tightening; (2) The Responsible Person reconfirms the connection of the parts of the equipotential electrician's shielding clothes; (3) The equipotential electrician cooperates with the ground electrician to tie the human body backup protection rope, and climbs the ladder to enter the intense electric field with the consent of the Responsible Person.	(1) When the ground electrician presses down the rope ladder to tighten it, the minimum distance between the conductor and the ground and spanned object shall comply with the safety regulations. The ground electrician controls the anti-falling human body backup protection rope in a reasonable way; the speed is stable during the climbing process, and the movement is standardized and skilled;	

S/N	Work Content	Operation Steps and Standards	Safety Measures and Precautions	Responsible Person
10	Entering the intense electric field		(2) Before the potential of the equipotential electrician is transferred, it must be approved by the Responsible Person. It is strictly forbidden to charge and discharge with the head during potential transfer; the distance between the exposed part of the human body and the electrified body shall not be less than 0.3m; the combined gap shall not be less than 1.2m; the equipotential electrician shall first fasten the safety belt and then lock the ladder head lock after entering the electric field; (3) The Responsible Person shall carefully supervise and remind the operator.	
11	Replacement of the whole string of insulators	(1) When the ground potential electricians installs the clamps, they shall install the pull rod in place with the cooperation of the equipotential electricians, tie the protection rope for conductor, and tie the transmission rope between the first and second pieces of insulator strings on the side of the cross arm; (2) The ground potential electrician tightens the leading screw so that the pull rod is stressed and then the impact check is carried out, so as to ensure that after the pull rod is under normal stress, the equipotential electrician pulls out the socket pin at the side of the conductor;	(1) The installation position of the clamps shall be reasonable. When installing the pull rod, the equipotential electrician and the ground potential electrician shall not operate at the same time. The insulator string binding points shall be suitable and shall be firmly tied; (2) The position and length of the protection rope for conductor shall be appropriate, and the self-locking device shall be flexible and easy to operate; (3) When shaking the screw rod, apply a constant force to avoid excessive twisting of the pull rod;	

Table (Cont'd)

S/N	Work Content	Operation Steps and Standards	Safety Measures and Precautions	Responsible Person
11	Replacement of the whole string of insulators	(3) The ground potential electrician continues to tighten the leading screw to transfer the insulator string load to the pull rod. After checking the force of the load-bearing tools again, the equipotential electrician disconnects the connection between the ball head hanging ring at the side of the conductor and socket eyes; the ground potential electrician loosens the leading screw and lowers the conductor by 200mm; (4) The ground potential electrician pulls out the socket pin at the side of the cross arm and disconnects the connection between the ball head hanging ring at the side of the cross arm and the insulator with the cooperation of the ground electrician; (5) The ground electrician transfers the new insulator to the cross arm in an alternating manner. The ground potential electrician restores the connection between the ball head hanging ring at the side of the cross arm and the insulator, and installs the pin in place; (6) The equipotential electrician restores the connection between the ball head hanging ring at the side of the conductor and the socket eyes with the cooperation of the ground potential electrician, and installs the pin in place.	(4) After the pull rod is stressed, the impact test shall be carried out to ensure that the stress is normal. The equipotential electrician and the ground potential electrician are not allowed to operate at the same time when replacing the insulator string; (5) When the old and new insulators are alternately transferred up and down, the tail rope must be controlled to avoid collision or bumping of the conductor and the pull rod; (6) After the replacement is completed, the pin placement shall be checked to avoid the pin missing or the installation not in place; (7) The Responsible Person shall carefully supervise and remind the operator.	

Table (Cont'd)

S/N	Work Content	Operation Steps and Standards	Safety Measures and Precautions	Responsible Person
12	Leaving the electric field	(1) The equipotential electrician shall check the conductor restoration and pin placement again at the conductor end, report to the Responsible Person after confirming it is correct, and apply to exit from the electric field; (2) With the consent of the Responsible Person, the equipotential electrician shall check the connection of the shielding clothes and confirm that the connection is reliable, then exit the electric field following the reverse step of entering the electric field, and go down to the ground along the insulating rope ladder	(1) The equipotential electrician must obtain the permission of the Responsible Person before leaving the potential; (2) After the equipotential electrician enters the ladder head, the safety belt shall be fastened first, and the human body backup protection measures for anti-falling shall be taken before the ladder head lock is opened; (3) The combined gap composed of the gaps between the equipotential electrician and the grounding body and the electrified body in the process of exiting the electric field shall not be less than 1.2m; (4) The ground electrician presses and tightens the rope ladder and controls the anti-falling human body backup protection rope in a reasonable manner, and no unstressed condition occurs; (5) The Responsible Person shall carefully supervise and remind the operator.	

Table (Cont'd)

S/N	Work Content	Operation Steps and Standards	Safety Measures and Precautions	Responsible Person
13	The ground potential electrician returns to the ground	After the ground potential electricians transfers the tools to the ground and check that there is no object left on tower, they shall report it to the Responsible Person and then climb down the tower with insulated transmission rope after obtaining the consent of the Responsible Person.	The anti-falling device installed on the tower shall be used in the process of climbing down the tower; when moving and transposition on the pole and tower, the safety protection shall not be lost, and the operators must grasp the components securely	
14	End of the work	(1) Responsible Person shall organize all working personnel to put working apparatus and materials in order and put them in special kit (bag) after cleaning, so that "when construction is completed, all materials are used up and the construction site is clear". (2) After a post-shift meeting is held, the Responsible Person shall make work summaries and comments. Comments include the construction quality of this work and the implementation of safety measures from all working personnel. (3) Responsible Person shall report to the on-duty control personnel the end of the work, apply for the restoration of DC restarting device of the line and terminate the work order.	It is forbidden to restore the DC transmission line restarting device at the appointed time	

II. Assessment Standard

Table 1-9-6 Detailed Rules for Assessment and Scoring of Operation and Inspection Skills of UHVDC Transmission Line of State Grid Sichuan Electric Power Corporation

Fill-in Column of Examinee	No.:		Name:		Position:		Unit:		Date:	MM/DD/YYYY
Fill-in Column of Assessor	Grade:		Assessor:		Assessment Leader:		Team Starting time:		Closing time:	Operation Duration:
Assessment Module	Live Replacement of Whole String of Insulators in a Straight Line of Grounding Electrode of ±800kV UHV Transmission Line		Assessee		Maintenance personnel for ±800kV DC grounding electrode transmission line		Assessment method		Operation	Assessment Time Limit
										60min
Job Description	Entering the Electric Field Along Rope Ladder for Live Replacement of Whole String of Insulators on Straight Line of ±800kV DC Grounding Electrode Line.									
Work specification and requirements	1. Live working shall be carried out in good weather. In case of thunder, rain, snow or fog, no live working shall be carried out. When the wind force is greater than Level 5 and the humidity is greater than 80%, live working should not be carried out. 2. Persons required for this operation include 1 Responsible Person, 1 Special Supervisor, 1 ground electrician on the tower, 1 equipotential electricians and 2 ground auxiliary electricians. Enter the electric field along the rope ladder for live replacement of whole string of insulators on straight line of ±800kV DC grounding electrode line. 3. Responsibilities of Responsible Person: Be responsible for division of operating personnel of the task, work order reading, handling formalities for stopping the DC restart device for the line, getting work permits, holding pre-shift meeting, dealing with emergency situations in work, quality surveillance, and the summary after work. 4. Responsibilities of Special Supervisor: Be responsible for safety control of the work site. 5. Responsibilities of equipotential electrician: Be responsible for removal and restoration of insulator pins at the conductor end.									

Part I Standard for Training and Assessment on Live Working for Operation Maintenance of ±800kV UHV Power Transmission Line

Table (Cont'd)

Work specification and requirements	6. Responsibilities of ground potential electrician on the tower: Be responsible for arranging the transmission rope, rope ladder and insulator string lifting points and installing tools. 7. Responsibilities of ground electrician: Be responsible for transferring tools, equipment and materials and cooperating with equipotential electrician in entering and exiting the equipotential. 8. During the live working, if thunder, rain, strong wind or any other circumstance threaten the safety of the staff, the Responsible Person or Supervisor may stop working temporarily according to the circumstances. Given conditions: 1. Training base: UHV ±800kV DC grounding electrode line tower 2. Work orders have been handled, safety measures have been completed (DC restart device has been deactivated), and oral application (dispatcher or assessor) shall be made at the beginning and end of the work. 3. The instrument shall be used safely and correctly to test the insulating tool. 4. The operation must be carried out according to the working procedures. The scores of the items to be carried out shall be deducted for the process error. In case of major hidden dangers of personal, equipment and operational safety, the assessor may order the termination of the operation (assessment)
Assessment scenario preparation	1. Line: UHV ±800kV DC grounding electrode line tower; work content: live replacement of whole string of insulators in a straight line of grounding electrode of ±800kV UHV transmission line. The type of the insulator: XZP-160 2. Required working tools: 2 insulated transmission ropes (TJS-14); 1 insulated transmission rope (TJS-10), 1 protection rope for conductor (TJS-18), 2 insulating rope sleeves (TJS-14), 2 insulated tackles (JH10-0.5); 1 insulated pull-rod, 1 insulating bar, 2 insulating rope ladders, 1 cross arm clamp (with leading screws). 1 two-bundle wire lifter, 1 somersault tackle, 1 aluminum alloy ladder head, and 3 articles of shielding clothes (type 1 or insulated type); 1 insulation resistance meter (5000V type with test electrode), and 1 multimeter; 1 wind speed and humidity gauge, 1 laser rangefinder, 1 infrared thermometer and 1 tarpaulin; 3. The work site shall be monitored, and the safety measures (fence, etc.) on the work site have been fully implemented; non-operation personnel are prohibited from entering the site, and the staff must wear safety helmet when entering the work site. 4. Examinees shall bring their own work clothes, flame retardant cotton underwear, safety belt, safety helmet, tool, shielding clothes, gloves, and safety belts (including double-protective ropes)
Remarks	1. The deduction shall be done until the scores of each item are deducted completely. In case of major hidden dangers of personal, equipment and operational safety, the assessor may order the termination of the operation. 2. When equipment, working environment, safety belt, safety helmet, tool, shielding clothes, etc., do not conform to the operation condition, the assessor may order the termination of the operation

Table 1-9-7 Standards for Assessment and Scoring of Operation and Inspection Skills of UHVDC Transmission Line of State Grid Sichuan Electric Power Corporation

S/N	Project name	Quality requirements	Score	Deduction standard	Reasons for deduction	Deduction	Scoring
1	Site re-survey	(1) The Responsible Person shall go to the work site to check the line name, pole and tower number, on-site working conditions, defective parts and so on. (2) Check that the site meteorological conditions such as wind speed and humidity should meet the operation requirements. (3) Check whether the work order is complete and unmodified, check whether the safety measures listed are consistent with the actual situation on site, and supplement it if necessary	5	(1) Deduct 1 point for failure to check the double title. (2) Deduct 1 point for failure to check on-site working conditions (meteorology) and defective parts. (3) Deduct 0.5 points/item for any alteration in the work order filling, and deduct 1 point for incorrect work order number. Deduct 1.5 points for each incomplete work order			
2	Work Permit	(1) The Responsible Person contacts the on-duty control personnel and apply for stopping the DC restart device for the line as per the contents of the work order. (2) Reporting content shall be standardized and complete	2	(1) Deduct 2 points for failure to contact the control department (referee) for disabling the DC restart device. (2) Deduct 0.5 points for non-standard or incomplete terminology reporting respectively			

Part I Standard for Training and Assessment on Live Working for Operation Maintenance of ±800kV UHV Power Transmission Line

Table (Cont'd)

S/N	Project name	Quality requirements	Score	Deduction standard	Reasons for deduction	Deduction	Scoring
3	Site layout	Install the security fence and hang the signboards correctly: (1) The security fence should take full account of falling objects from the high place and the influence on road traffic. (2) The entrance and exit of the security fence shall be set reasonably. (3) Signs such as "Access from Here", "Work Here", "Access from Here" shall be properly arranged	3	(1) Deduct 0.5 points for failure to arrange the fence at the work site. (2) Deduct 0.5 points for failure to arrange the warning board. (3) Deduct 0.5 points for failure to hang the tower climbing operation sign			
4	Holding a pre-shift meeting	(1) All staff and personnel shall wear safety helmets and work clothes correctly. (2) Responsible Person shall wear red vest and read out the work order and be clear with work task and division of personnel; explain safety measures and technical measures in work; check (inquire after) mental state of all working personnel; inform of hazards in work and precontrol measures. (3) All working personnel shall sign on the work order for confirmation.	3	(1) Deduct 0.5 points/person for the staff not dressing uniformly. Deduct 0.5 point/person for the staff not dressing uniformly. (2) Give no points to this item for no division of labor, and deduct 1 point for unclear division of labor. (3) Deduct 0.5 points for the on-site Responsible Person not wearing a safety monitoring vest. (4) Deduct 1 point for the work shift member failing to sign or signing incompletely on the work order			

· 489 ·

Standard for Professional Training and Assessment for Operation Maintenance of UHV DC Power Transmission Line

Table (Cont'd)

S/N	Project name	Quality requirements	Score	Deduction standard	Reasons for deduction	Deduction	Scoring
5	Inspection of tools	(1) The staff shall place the tools on the moisture-proof tarpaulin as required; the moisture-proof tarpaulin shall be clean and dry. (2) The tools shall be placed in category according to the requirements of the fixed management; the insulated tools shall not be mixed with metal tools and materials; and the appearance inspection shall be done on the tools. (3) The surface of insulating tools shall not be worn, deformed and damaged, and the operation shall be flexible. Carry out segment insulation detection with such insulated tools with insulation resistance meter of 2,500V or above and with the resistance no less than 700MΩ, and wipe it off with a clean dry towel. (4) The ground potential and equipotential personnel on tower shall correctly wear a whole suit of qualified shield clothes and conductive shoes as required, with each part connected well, shall not wear chemical fiber clothes next to the skin in the shielding clothes and shall fasten safety belts; the Responsible Person shall carefully check whether they wears it correctly	7	(1) Deduct 1 point for failure to use moisture-proof tarpaulin and place tools to designed positions. (2) Deduct 0.5 points/item for failure to check qualified label of tool test and appearance inspection. (3) Deduct 1 point/item for failure to use testing instrument for testing the tools. (4) Deduct 2 points/person/time for the operator failing to wear the shielding clothes correctly and each part connected well. (5) Deduct 1 point for the on-site Responsible Person failing to check the safety protective equipment of the tower climbing operators.			

· 490 ·

Table (Cont'd)

S/N	Project name	Quality requirements	Score	Deduction standard	Reasons for deduction	Deduction	Scoring
6	Climbing the tower	(1) After checking the line name and pole and tower number, the ground potential electrician on tower shall impact to check the stress of safety belt and falling protector. (2) The ground potential electricians on the tower shall carry insulated transmission ropes while climbing the tower to the cross arm, fasten safety belts and human body backup protection ropes, and arrange transmission ropes at an appropriate position	5	(1) Deduct 2 points/person for the tower climbing personnel failing to check the double name of the line, pole number and phase. (2) Deduct 2 points/person for the tower climbing personnel failing to report the check results (3) Deduct 2 points respectively for failure to fasten the safety belt or to perform the impulse test on the safety belt and the backup protection rope. (4) Deduct 2 points for grasping the shackles with hands. (5) Deduct 1 point for inconvenient suspension position of tackle transmission rope for taking tools. (6) Deduct 2 points for incorrect operation of electrician on tower. (7) Deduct 2 points for the Responsible Person failing to monitor the operation in place			

Standard for Professional Training and Assessment for Operation Maintenance of UHV DC Power Transmission Line

Table (Cont'd)

S/N	Project name	Quality requirements	Score	Deduction standard	Reasons for deduction	Deduction	Scoring
7	Install the somersault tackle with insulating rope	The ground electricians shall transfer the insulating bar and the somersault tackle to the tower; the ground potential electricians shall hang the somersault tackle with Φ10mm insulated transmission rope on the conductor	4	(1) Deduct 1 point for transferring work position without taking protection measures. (2) Deduct 2 points for the failure of the Responsible Person to remind the ground potential electrician of taking safety measures when transferring the position. (3) Deduct 1 point for improper installation position. (4) Deduct 2 points for winding of transmission rope 1 point			
8	Rope ladder installation	The ground potential electrician cooperates with the ground work personnel to hang the rope ladder on the conductor and pull the rope ladder to a safe position.	4	(1) Deduct 1 point for a object falling from high altitudes (2) Deduct 2 points for not firm installation of the rope ladder; deduct 2 points for unsuitable position. (3) Deduct 1 point for winding of transmission rope			

Part I Standard for Training and Assessment on Live Working for Operation Maintenance of ±800kV UHV Power Transmission Line

Table (Cont'd)

S/N	Project name	Quality requirements	Score	Deduction standard	Reasons for deduction	Deduction	Scoring
9	Entering the intense electric field	(1) The ground electrician conducts an impact check on the rope ladder to confirm that it has been reliably installed on the conductor before the insulating rope ladder is pressed down for tightening. (2) The Responsible Person reconfirms the connection of the parts of the equipotential electrician's shielding clothes; (3) The equipotential electrician cooperates with the ground electrician to tie the human body backup protection rope, and climbs the ladder to enter the intense electric field with the consent of the Responsible Person.	15	(1) Deduct 2 points for failure to conduct impulse test for insulating rope ladder. (2) Deduct 2 points for failure to inspect if the crossing meets the safe distance requirements before pressing down the rope ladder. (3) Deduct 2 points for failure of the equipotential electrician to apply to the Responsible Person before climbing the ladder; deduct 1 point for starting to climb the ladder without consent after application. (4) Deduct 1 point for not smooth climbing or falling (5) Deduct 1 point for failure to effectively control the backup protection rope. (6) Deduct 2 points for starting the potential transfer without applying to the Responsible Person; deduct 1 point for starting the potential transfer without the consent after applying.			

Table (Cont'd)

S/N	Project name	Quality requirements	Score	Deduction standard	Reasons for deduction	Deduction	Scoring
9	Entering the intense electric field		15	(7) Deduct 1 point for the inappropriate position applied for the potential transfer. (8) Deduct 1 point for being unskilled in potential transfer. (9) Deduct 2 points for the Responsible Person failing to monitor the operation carefully.			
10	Replacement of the whole string of insulators	(1) When the ground potential electricians installs the clamps, they shall install the pull rod in place with the cooperation of the equipotential electricians, tie the protection rope for conductor, and tie the transmission rope between the first and second pieces of insulator strings on the side of the cross arm; (2) The ground potential electrician tightens the leading screw so that the pull rod is stressed and then the impact check is carried out, so as to ensure that after the pull rod is under normal stress, the equipotential electrician pulls out the socket pin at the side of the conductor;	30	(1) Deduct 2 points for improper installation of clamps. (2) Deduct 2 points for improper position of the protection rope for conductor. (3) Deduct 2 points for wrong position of the transmission rope tied on the insulator string; deduct 3 points for not fixing of the transmission rope. (4) Deduct 2 points for failure to conduct the impact check after the leading screw is stressed; deduct 1 point for not reporting to the Responsible Person after check.			

Part I Standard for Training and Assessment on Live Working for Operation Maintenance of ±800kV UHV Power Transmission Line

Table (Cont'd)

S/N	Project name	Quality requirements	Score	Deduction standard	Reasons for deduction	Deduction	Scoring
10	Replacement of the whole string of insulators	(3) The ground potential electrician continues to tighten the leading screw to transfer the insulator string load to the pull rod. After checking the force of the load-bearing tools again, the equipotential electrician disconnects the connection between the ball head hanging ring at the side of the conductor and socket eyes; the ground potential electrician loosens the leading screw and lowers the conductor by 200mm. (4) The ground potential electrician pulls out the socket pin at the side of the cross arm and disconnects the connection between the ball head hanging ring at the side of the cross arm and the insulator with the cooperation of the ground electrician. (5) The ground electrician transfers the new insulator to the cross arm in an alternating manner. The ground potential electrician restores the connection between the ball head hanging ring at the side of the cross arm and the insulator, and installs the pin in place. (6) The equipotential electrician restores the connection between the ball head hanging ring at the side of the conductor and the socket eyes with the cooperation of the ground potential electrician, and installs the pin in place.	30	(5) Deduct 0.5 points for too large twisting of the pull rod after the leading screw is shaken. (6) Deduct 2 points for not binding the insulator before taking out the socket pin at the side of the cross arm. (7) Deduct 5 points for operating at the same time by the equipotential electrician and the ground potential electrician. (8) Deduct 0.5 points/time for collision or bumping of the pole and tower, conductor and the pull rod, etc. when the old and new insulators are alternately transferred. (9) Deduct 3 points for wrong sequence of installing insulator strings. (10) Deduct 2 points for not checking the installation of the insulator strings before loosening the leading screws; deduct 1 point for not reporting after check. (11) Deduct 2 points/item for improper installation of pins. (12) Deduct 3 points for each pin missing.			

· 495 ·

Table (Cont'd)

S/N	Project name	Quality requirements	Score	Deduction standard	Reasons for deduction	Deduction	Scoring
11	Leaving the electric field	(1) The equipotential electrician shall check the conductor restoration and pin placement again at the conductor end, report to the Responsible Person after confirming it is correct, and apply to exit from the electric field. (2) With the consent of the Responsible Person, the equipotential electrician shall check the connection of the shielding clothes and confirm that the connection is reliable, then exit the electric field following the reverse step of entering the electric field, and go down to the ground along the insulating rope ladder	8	(1) Deduct 1 point for failure to report the end of the work. (2) Deduct 2 points for failure to apply to the Responsible Person for exiting from the electric field. (3) Deduct 1 point for not checking the connection of the shielding clothes. (4) Deduct 2 points for the potential transfer without applying to the Responsible Person; deduct 1 point for starting the work without the consent after applying. (5) Deduct 1 point for the inappropriate position applied for the potential transfer (6) Deduct 2 points for the incorrect action of equipotential electrician exiting the electric field and repeat discharging. (7) Deduct 1 point for not smooth climbing or falling. (8) Deduct 1 point for failure to effectively control backup protection rope			

Part I Standard for Training and Assessment on Live Working for Operation Maintenance of ±800kV UHV Power Transmission Line

Table (Cont'd)

S/N	Project name	Quality requirements	Score	Deduction standard	Reasons for deduction	Deduction	Scoring
12	The ground potential electrician returns to the ground	After the ground potential electricians transfers the tools to the ground and check that there is no object left on tower, they shall report it to the Responsible Person and then climb down the tower with insulated transmission rope after obtaining the consent of the Responsible Person.	4	(1) Deduct 2 points for failure to use the falling protector when climbing down the tower. (2) Deduct 2 points for loosing the protection of the safety belt when moving on the tower. (3) Deduct 1 point for grasping the tower nail when climbing down the tower. (4) Deduct 2 points for any objects left on the tower			
13	End of the work	(1) The Responsible Person shall organize all working personnel to put working apparatus and materials in order and put them in a special kit (bag) after cleaning; clean the site to ensure that "the materials are removed and the site is cleaned after construction". (2) After a post-shift meeting is held, the Responsible Person shall make work summaries and comments. Comments include the construction quality of this work and the implementation of safety measures from all working personnel. (3) Responsible Person shall report to the on-duty control personnel the end of the work, apply for the restoration of DC restarting device of the line and terminate the work order.	10	(1) Deduct 2 points for failure to clean the tools. (2) Deduct 2 points for missing tools. (3) Deduct 3 points for failure to hold the post-shift meeting. (4) Deduct 1 point for unqualified comments. (5) Deduct 3 points for failure to contact the control personnel for end of the work reporting. (6) Deduct 0.5 points for each missing item from the reporting content. (Name of working unit, name of the Responsible Person, name of the line, work completion; device has restored to normal, and the personnel have evacuated). (7) Deduct 1 point for error in the fill-in of work order termination procedure.			
	Total		100				

· 497 ·

Module 10 Standards for Training and Assessment on Live Repair of ± 800kV UHV Transmission Line Grounding Electrode Line Conductor

I. Training Standards

(I) Training Requirements

Designation of module	Live repair of conductor for ±800kV UHV transmission line grounding electrode	Type of training	Operation
Training method	Practical operation training	Training hours	14 hours
Training objectives	1. Master the electrical significance of "self-climbing device" operation mode when the ±800kV DC grounding electrode tangent tower enters and exits electric field. 2. Can enter ±800kV DC grounding electrode line equipotential operation point by adopting the "self-climbing device". 3. Can independently complete the repair of ±800kV DC grounding electrode line conductor (equipotential operation method)		
Training venue	UHV DC practical training line		
Training content	Entering the electric field from the self-climbing device, the equipotential operation method is adopted to repair the grounding electrode line conductor of ±800kV UHV transmission line with electrification		
Scope of application	Maintenance personnel for ±800kV UHV DC power transmission line		

(II) Referenced Procedures and Specifications

(1) Code for Designing of ±800kV DC Overhead Transmission Line (GB/T50790-2013)

(2) Maintenance Specification for ± 800kV DC Overhead Transmission Line (DL/T251-2012)

(3) Operating Code for ±800kV DC Overhead Transmission Line (GB/T28813-2012)

(4) Technical Specification for Live Working of ±800kV DC Line (DL/T1242-2013)

(5) Technical Specification for Fittings of ±800kV UHV Transmission Line (GB/T 31235-2014)

(6) State Grid Corporation of China on the Management Regulations of Live Working (Trial Implementation) (SGCC [2007] No.751)

(7) State Grid Corporation of China Working Regulations of Power Safety (Transmission Line Section) (Q/GDW1799.2-2013)

(8) Electrotechnical Terminology- Overhead Line (GB/T 2900.51-1998)

(9) Electrotechnical Terminology- Live Working (GB/T2900.55-2016)

(10) Live Working-Terminology for Tools, Equipment and Devices (GB/T 14286-2002)

(11) Minimum Requirements for Utilization of Tools, Devices and Equipment for Live Working (DL/T 877-2004)

(12) Preventive Test Code of Tools, Devices and Equipment for Live Working (DL/T 976-2005)

(13) Technical Guide for Live Working of ±800kV UHV Transmission Line (Q/GDW302-2009)

(14) Shielding Clothes for Live Working (GB/T6568-2008)

(15) Technical Requirements and Design Guide for Live Working Tools (GB/T18037-2008)

(III) Teaching Design for Training

To complete the work task of "live repair of ±800kV UHV transmission line grounding electrode line conductor", each training stage shall be designed according to the standard operation procedure for work task completion. Each stage includes specific training objectives, training content, hours of training, training methods (training resources), training environment, assessment and evaluation, etc, as shown in the Table 1-10-1.

Table 1-10-1 Training Content Design for Live Repair of Grounding Electrode Line Conductors on ±800kV UHV Transmission Line

Training Process	Training objectives	Training contents	Training Course hours	Training methods and resources	Training environment	Assessment and evaluation
1. Theoretical teaching	1. Preliminarily grasp methods of using self-climbing device to enter and exit ± 800kV DC grounding electrode line electric field. 2. Be familiar with equipment configuration and safety precautions. 3. Be familiar with the method of and quality standards for repair of conductors in the ± 800kV DC grounding electrode line	1. Electrical significance of using self-climbing device to enter and exit ± 800kV DC grounding electrode line electric field. 2. Equipment allocation, hazard analysis and safety measures. 3. The method of and quality standards for repair of conductors in the ±800kV DC grounding electrode line	2	Training method: Lecture. Training resources: PPT, relevant regulations and specifications	Multimedia classroom	Attendance, classroom questions and assignments
2. Preparations	Be able to complete the preparations before operation	1. Work site survey. 2. Preparation of the standardized operation card. 3. Filling of the work order. 4. Preparation of tools and materials for this operation	1	Training method: 1. Work site survey and cleaning of tools and materials shall be practiced at site; 2. Preparation of operation card and the filling of work order shall adopt lecture method. Training resources: 1. ±800kV grounding electrode practical training line 2. UHV tools warehouse 3. Blank work order	1. ± 800kV grounding electrode practical training line 2. Multimedia classroom	

Part I Standard for Training and Assessment on Live Working for Operation Maintenance of ±800kV UHV Power Transmission Line

Table (Cont'd)

Training Process	Training objectives	Training contents	Training Course hours	Training methods and resources	Training environment	Assessment and evaluation
3. Preparation of work site	Be able to complete the preparations of work site	1. Work site re-survey. 2. Work application. 3. Work site layout. 4. Pre-shift meeting. 5. Tool and material inspection	1	Training methods: demonstration and role play. Resources; ±800kV grounding electrode practical training line	±800kV grounding electrode practical training line	
4. Trainer's demonstration	The trainees can preliminarily understand the operation process of the task through inspecting and learning from each other's work	1. Working distance measurement and conductor temperature measurement. 2. Ground electrician uses the UAV to spread out the hauling rope. 3. The ground electrician shall pull the backup protection rope and guide rail rope to the working position. 4. Equipotential electrician rides the self-climbing device to enter the electric field. 5. Equipotential electrician repairs the conductor	2	Training methods: Demonstration. Resources; ±800kV grounding electrode practical training line	±800kV grounding electrode practical training line	

· 501 ·

Table (Cont'd)

Training Process	Training objectives	Training contents	Training Course hours	Training methods and resources	Training environment	Assessment and evaluation
5. Group training of trainees	Through training: 1. Trainees are able to complete the operation of entering and exiting ±800kV DC grounding electrode line electric field. 2. Trainees can complete the method of repair of ±800kV DC grounding electrode line conductors	1. The trainees are grouped (6 in a group) to train the skill operation of entering and exiting the ±800kV DC electric field and repair grounding electrode line conductors. 2. Trainers guide the operation of trainees and conduct safety supervision	7	Training methods: Role play. Resources: ±800kV grounding electrode practical training line	±800kV grounding electrode practical training line	Score the operation of trainees according to the detailed rules for skill assessment and scoring
6. End of work	Through training: 1. Trainees can further understand the shortcomings during the operation process for later improvement. 2. Train the trainees in the working style of safe and civilized production	1. Cleaning up the work site. 2. Report to dispatcher. 3. Comment and summarize the work task this time at post-shift meeting	1	Training methods: Lecture and inductive method	±800kV grounding electrode practical training line	

(IV) Work Flow

1. Work task

Entering the electric field from the self-climbing device to arrive at the operation point, the equipotential operation method is adopted to repair the grounding electrode line conductor of ±800kV UHV transmission line with electrification.

2. Requirements for Weather and Work Site

(1) Live repair of ±800kV UHV transmission line grounding electrode conductor shall be carried out in good weather.

In case of lightning (hearing thunder or seeing lightning), snow, hail, rain, fog and so on at the work site, live working is prohibited. When the wind force is greater than Level 5 (10m/s) or the relative humidity of air is greater than 80%, live working should not be carried out. During the operation, pay attention to the detection of weather changes. In case of gale or misty rain and other emergencies, take correct measures according to the procedures to ensure the safety of personnel and equipment. Before carrying out the live working on the grounding electrode line, understand the basic conditions such as the line load current and operating temperature. Before and during the on-site operation, the temperature of the conductor shall be monitored by using an infrared thermometer suitable for the transmission line.

(2) The operating personnel should be in good mental states and be familiar with the organizational and technical measures to ensure safety in work; they should hold the qualification certificate for live working within the validity period.

(3) Responsible Person should organize the relevant personnel to complete field investigation in advance, determine the operating methods, required working apparatus and necessary measures according to the results, and handle the tickets for live working.

(4) The work site should be reasonably set up with fence and warning signs. Non operating personnel is forbidden to enter.

(5) DC restart device shall be deactivated for the Project.

(6) Safe working distance and effective insulation length during operation are shown in Table 1-10-2.

Table 1-10-2 Safe Distance (m) for Live Repair for Grounding Electrode Line Conductor of ±800kV UHV Transmission Line

Voltage class (altitude≤1,000)	Safe distance between human body and electrified body	Minimum effective insulation length		Minimum combination gap	Minimum distance between the exposed part of human body and electrified body during potential transfer
		Insulating bar	Insulated load-bearing tools and insulating ropes		
±800kV DC grounding electrode	6.8	6.8	6.8	6.6	0.5

(7) For operation on ±800kV transmission line, the number of insulators in good operation phase shall be no less than 32 pieces.

3. Preparations

3.1 Hazards and precontrol measures

(1) Hazard——Preventing system overvoltage

Precontrol measures: This operation is equipotential operation. Before starting the work, the Responsible Person shall contact the on-duty controller to apply for a work order for live working on power lines, and apply to deactivate the DC restart protection device. The controller on duty shall perform the licensing procedures. After the live working is completed, report to the controller on duty in a timely manner. It is forbidden to disable and restore the DC restart device at the appointed time. If equipment loses power suddenly during live working, operators shall consider the equipment as still charged. The Responsible Person shall contact the control personnel as soon as possible, and no forced energization is allowed before the controller on duty getting in touch with the Responsible Person.

(2) Hazard —— electric shock

Precontrol measures:

① Insulating tool and insulating ropes shall be free of damage, moisture, deformation, and failure. It is not allowed to use non-insulating ropes (such as cotton rope, manila rope, and steel wire rope).

② The equipotential operator shall wear flame-retardant underwear and full set of screen clothes outside (including hat, dresses & trousers, gloves, socks and shoes). All parts shall be in excellent connection conditions, and the resistance of the whole suit of shielding clothes shall not exceed 20Ω.

③ Before potential transfer, equipotential operator shall obtain the approval of Responsible Person, and the minimum distance between exposed part of human body and electrified body shall not be lower than 0.5m. The safe distance between the ground potential operator and the electrified body (between the equipotential operator and grounding body) is greater than that specified in Table 1-10-2.

④ When transmitting large metal objects by using insulated ropes, ground potential operators shall not touch them before ground connection.

⑤ The Responsible Person and the Special Supervisor shall continuously monitor the operators and correct their nonstandard operation or actions in violation at any time. Special attention shall be paid to high-place operators to ensure that there is enough safety distance (meeting the requirements in Table 1-10-2).

⑥ When the operating temperature of the heat-resisting conductor is above 40°C, high temperature resistance live working tools shall be used, including high temperature resistance soft insulating tools, high temperature resistance hard insulating tools, and high temperature resistance heat insulation shielding clothes, and the potential transfer rod shall be used to enter the equipotential.

⑦ Insulation protective equipment shall be regarded as a conductor. During the transfer and use, the safe distance between the phase and ground and poles shall be ensured to meet the safety requirements.

(3) Hazard —— falling from high places

Precontrol measures:

① Before climbing, operators working at heights must satisfy the requirements of this operation, such as physical condition, mental state, and skill and quality.

② Supervisor shall correct irregularities and violations at any time. Special attention shall be paid to operators to prevent them from losing the protection of the safety belt or insulated backup protection rope during transposition. The safety belt or insulated backup protection rope shall not be fastened lower than operating personnel.

(4) Hazard —— injury caused by objects falling from high place

Precontrol measures:

① Operator working at heights should put personal tools and fragmentary materials into the tools bag. It is strictly forbidden to hang objects in high place or keep in the mouth.

② Ground operator should correctly wear a helmet and use the knots. The vertical distance from the operation point should not be less than the falling radius.

③ The work site shall be set up with fence and warning signs. It shall be noted at any time that Supervisor shall prohibit irrelevant personnel and vehicles from entering operation area.

(5) For injuries caused by conductor overheating

Precontrol measures:

① Before the operation, contact with the dispatcher to confirm the operation mode to avoid the live working when the large current of the conductor flows.

② Operators shall wear special heat-resistant shielding clothes to carry out live working to prevent the conductor from overheating and injuring people.

3.2 Selection of tools, instruments and materials

The tools and materials required for live repair of grounding electrode line conductors for ±800kV transmission line are shown as Table 1-10-3. Before being taken out of the warehouse, the tools and instruments shall be carefully inspected for their service voltage class and test cycle, and shall be inspected to confirm that appearance is intact, connection is firm, rotation is flexible and meet the requirements of the work task. After delivering the tools and instruments out of the warehouse, they shall be stored in the tool bag or tool kit for transportation to avoid contamination and damp. Metal tools and insulated tools shall be separately loaded and transported to avoid deformation and damage caused by mixed loading and transportation.

Table 1-10-3　Tools, Instruments and Materials Required for Live Repair of Grounding Electrode Line Conductors on ±800kV UHV Transmission Line

S/N	Name	Specification and Model	Unit	Qty.	Remarks
1	Insulated transmission rope	TJS-14×50m	Nos.	1	
2	Insulated transmission rope	TJS-10×50m	Nos.	1	
3	Insulated transmission rope	TJS-14×100m	Nos.	1	
4	Insulated transmission rope	TJS-4×100m	Nos.	1	
5	Insulated noose	TJS-14	Nos.	1	
6	Insulated tackle	0.5T	Nos.	1	
7	UAV	Six rotor	Nos.	1	
8	Self-climbing device	ACTSAFE-II	Set	1	
9	Personal tools		Set	1	

Table (Cont'd)

S/N	Name	Specification and Model	Unit	Qty.	Remarks
10	Shielding clothes	500kV type I or thermal insulation type	Set	1	
11	Human body backup protection rope		Nos.	1	
12	Safety belt		Nos.	1	
13	Safety helmet		pcs	6	
14	Moisture-proof tarpaulin	3m×3m	pcs	1	
15	Multimeter		Nos.	1	
16	Wind speed and humidity gauge	HT-8321	pcs	1	
17	Megameter and test electrode	5000V	Set	1	
18	Interphone		Set	2	
19	Red waistcoat	"Responsible Person"	pcs	1	
20	Security fence		Set	Several	
21	Repair of preformed armor rods	Standard configuration	Set	2	
22	Conductive paste		Box	1	
23	Sandpaper		Sheet	1	

3.3 Division of labor for operators

Division of labor for operators of the task is shown in Table 1-10-4.

Table 1-10-4 Personnel Allocation for Live Repair of Grounding Electrode Line Conductors on ±800kV UHV Transmission Line

S/N	Post	Qty. (person)	Responsibilities
1	Responsible Person	1	Be responsible for various work on the work site
2	Equipotential electrician	1	Be responsible for installing and using the self-climbing device to enter the electric field for the repair of the conductor
3	UAV operator	1	Be responsible for the operation of UAV to extend the hauling rope
4	Ground electrician	3	Be responsible for transferring tools and materials and cooperating with equipotential electrician in entering and exiting the equipotential

4. Working Procedures

The workflow of this task is shown in Table 1-10-5.

Table 1-10-5 Work Flow Chart for Live Repair of Grounding Electrode Line Conductors on ±800kV UHV Transmission Line

S/N	Work Content	Operation Steps and Standards			Safety Measures and Precautions	Responsible Person
1	Confirm the line name and tower number and inspect the pole and tower	The Responsible Person shall complete the following work: (1) Check the line name, the number of the pole and tower and ensure the phases are correct; guarantee that the foundation and the pole and tower are intact and in normal condition; ensure that the cross and span distance meets the safety requirements; confirm the defect conditions and the specifications and models of earth wires. (2) Check that the terrain and environment should meet the operation requirements. (3) Check that the safety measures listed in the work order are in line with the actual situations on site. The measures will be added if necessary.			(1) Correctly wear helmet, working clothes, work shoes and protective gloves. (2) Operation under meteorological conditions that may endanger the safety of operators is forbidden. (3) Non-operation personnel and vehicles are strictly prohibited from entering the working site	
2	Measurement of on site work environment	Measurement item	Standard value	Measured value	When the wind force is greater than Level 5 and the humidity is greater than 80%, live working should not be carried out.	
		Speed	≤10m/s			
		Humidity	≤80%			
		Ambient temperature	0～38°C			
		Conductor temperature	—			
		Working distance	≥0.5m			
		Altitude	Altitude limit of 500m and 1000m			

Standard for Professional Training and Assessment for Operation Maintenance of UHV DC Power Transmission Line

Part I Standard for Training and Assessment on Live Working for Operation Maintenance of ±800kV UHV Power Transmission Line

Table (Cont'd)

S/N	Work Content	Operation Steps and Standards	Safety Measures and Precautions	Responsible Person
3	Work Permit	(1) The Responsible Person is responsible for contacting the on-duty control personnel and applying for stopping the DC restart protection device for the line as per the contents of the work order. (2) Live working could be started only after being approved by control personnel on duty.	Live working shall not be started without the permission of the on-duty control personnel	
4	Site layout	Install the security fence and hang the signboards correctly: (1) The security fence should take full account of falling objects from the high place and the influence on road traffic. (2) The entrance and exit of the security fence shall be set reasonably. (3) Signs such as "Access from Here", "Work Here", "Access from Here" shall be properly arranged	When the influence on road traffic safety is uncontrollable, the traffic management department shall be contacted in time to strengthen the on-site control of traffic safety	
5	Holding a pre-shift meeting	(1) All working personnel shall line up. (2) Responsible Person will read out the work order and be clear with work task and division of personnel; explain safety measures and technical measures in work; check (inquire after) mental state of all working personnel; inform of hazards in work and precontrol measures. (3) All working personnel shall sign on the work order for confirmation.	(1) Work order shall be filled and issued with standardized licensing procedure and complete signature. (2) All working personnel shall be in good mental states. (3) All working personnel shall be clear with task division of works, safety measures and technical measures	

· 509 ·

Table (Cont'd)

S/N	Work Content	Operation Steps and Standards	Safety Measures and Precautions	Responsible Person
6	Inspection tool	(1) The equipotential electrician shall wear the shielding clothes in a right way and pass the inspection, which shall be supervised and inspected by the Responsible Person. (2) Wear personal safety appliance correctly (proper size and easy lock), and the Responsible Person shall supervise and inspect it. (3) Check the appearance and measure insulation performance of insulating tools, and make records. (4) Check the proper operation of UAV and self-climbing device	(1) Check carefully for damage, deformation and dysfunction before using metal and insulating tools. Wipe the insulating tools with a clean and dry towel and carry out segment insulation detection with such insulated tools with insulation resistance meter of 2,500V or above and with the resistance no less than 700MΩ. (2) Use a multimeter to measure the resistance between the farthest ends of the shielding clothes and trousers, which shall not be greater than 20Ω. The Responsible Person shall check the connection of the electrician's shielding clothes. (3) Check the tool assembly and make sure the connection is reliable. (4) The tools and instruments for live working shall be placed on the moisture-proof tarpaulin	
7	Spreading of guide rail rope for UAV	(1) The UAV controller shall control the UAV to fly about 1.6m above the ground and hover in place. (2) The ground electrician shall hang the insulated transmission rope with a diameter of 4mm on the tripping device of the UAV. (3) The UAV operator continues to operate the UAV to ascend to a suitable position, and pulls outwards to pass through the conductor, and continues to pull the rope at approximately the same distance as the length of the conductor, and controls the tripping device to make the hauling rope drop naturally. (4) The ground electrician shall install the hauling rope as an insulated endless rope.	(1) The hovering position of the UAV shall be appropriate. The ground electrician shall hang the hauling rope with attention to the safety. (2) When the UAV is towed, it shall take off from the inside of the line to the outside and the hauling rope shall be released outwardly.	

Part I Standard for Training and Assessment on Live Working for Operation Maintenance of ±800kV UHV Power Transmission Line

Table (Cont'd)

S/N	Work Content	Operation Steps and Standards	Safety Measures and Precautions	Responsible Person
8	Hauling guide rail rope and human body backup Protection rope	(1) The ground electrician transfers the hauling rope to the operation point. (2) The ground electrician uses a hauling rope to circularly pull the guide rail rope of the self-climbing device, one end of the guide rail rope is stably installed at a firm position on the ground, and the other end is perpendicular to the ground. (3) The ground electrician uses the hauling rope to pull the backup protection rope	(1) One end of the guide rail rope must be fixed firmly on the ground. (2) The guide rail rope and the human body backup protection rope shall be kept at a certain distance to prevent entanglement	
9	Installation of self-climbing device	(1) The equipotential electrician shall check the rotation of the self-climbing device and the electric quantity of the battery again, and install the self-climbing device on the guide rail rope after confirming it is correct. (2) After the equipotential electrician completes the human body backup protection measures and operates the self-climbing device to a certain height (0.8～1m), the ground electrician conducts impulse test on the self-climbing device and the equipotential electrician. (3) After the impulse test is qualified, the equipotential electrician carries the insulated transmission rope to climb vertically	(1) The self-climbing device shall rotate properly. (2) The ground electrician shall carry out the impact test on the self-climbing device. Pay attention to the force of the self-climbing device and the guide rail rope fixed to the ground during the impact. (3) The Responsible Person shall check it carefully	

Table (Cont'd)

S/N	Work Content	Operation Steps and Standards	Safety Measures and Precautions	Responsible Person
10	Entering the intense electric field and repairing the conductor	(1) Equipotential electrician rides the self-climbing device to rise vertically to a distance of 0.3m from the conductor, check that all connections of the shielding clothes are correct, and transfer the potential with the consent of the person in charge of the working group. (2) Equipotential electrician shall fasten the safety belt for protection on the conductor and arrange insulated transmission rope where the preformed armor rods are entangled, and hang the insulated transmission rope and then report to the Responsible Person. (3) The ground electrician shall hang up the preformed armor rods and report to the Responsible Person. (4) The equipotential electrician applies for a flat conductor, and the Responsible Person shall give permits. Use the preformed armor rods to repair the conductor as required and report to the Responsible Person	(1) The ground electrician shall control the human body backup protection rope to ensure that the equipotential electrician does not lose the safety belt protection during the work at heights. (2) Before the potential of the equipotential electrician is transferred, it must be approved by the Responsible Person. It is strictly forbidden to charge and discharge with the head during potential transfer; the distance between the exposed part of the human body and the electrified body shall not be less than 0.3m; the combined gap shall not be less than 1.2m. (3) Collision shall be avoided during the transfer of the preformed armor rods, and the equipotential electrician shall avoid high falling objects when working at heights.	

Part I Standard for Training and Assessment on Live Working for Operation Maintenance of ±800kV UHV Power Transmission Line

Table (Cont'd)

S/N	Work Content	Operation Steps and Standards	Safety Measures and Precautions	Responsible Person
11	Leaving the electric field	(1) The equipotential electrician shall check the conductor repair by the preformed armor rods again, report to the Responsible Person after confirming it is correct, and apply to exit from the electric field. (2) With the consent of the Responsible Person, the equipotential electrician shall check the connection of the shielding clothes and confirm that the connection is reliable, then exit the electric field following the reverse step of entering the electric field, and take the self-climbing device down to the ground	(1) The equipotential electrician must obtain the permission of the Responsible Person before leaving the potential. (2) The combined gap composed of the gaps between the equipotential electrician and the grounding body and the electrified body in the process of exiting the electric field shall not be less than 1.2m. (3) The Responsible Person shall carefully supervise and remind the operator.	
12	End of the work	(1) The Responsible Person shall organize all working personnel to put working apparatus and materials in order and put them in a special kit (bag) after cleaning; clean the site to ensure that "the materials are removed and the site is cleaned after construction". (2) After a post-shift meeting is held, the Responsible Person shall make work summaries and comments. Comments include the construction quality of this work and the implementation of safety measures from all working personnel. (3) Responsible Person shall report to the on-duty control personnel the end of the work, apply for the restoration of DC restarting device of the line and terminate the work order.	It is forbidden to restore the DC transmission line restarting device at the appointed time	

II. Assessment standard

Table 1-10-6 Detailed Rules for Assessment and Scoring of UHVDC Skills Training of State Grid Sichuan Electric Power Corporation

Fill-in Column of Examinee	No.:		Name:		Position:		Unit:		Date:		MM/DD/YYYY
Fill-in Column of Assessor	Grade:		Assessor:		Assessment Team Leader:		Starting time:		Closing time:		Operation Duration:
Assessment module	Live repair of conductor for ±800kV UHV transmission line grounding electrode		Assessee		Maintenance personnel for ±800kV DC grounding electrode transmission line		Assessment method		Operation		Assessment Time Limit
											60min
Job Description	Entering the electric field for live repair of the damaged conductor of the ±800kV DC grounding electrode line by adopting the "self-climbing device".										
Work specification and requirements	1. Live working shall be carried out in good weather. In case of thunder, rain, snow or fog, no live working shall be carried out. When the wind force is greater than Level 5 and the humidity is greater than 80%, live working should not be carried out. 2. Workers required for this operation include 1 Responsible Person, 1 equipotential electrician, 1 operator of UAV and 3 ground auxiliary electricians. They shall enter the electric field for the live repair of the grounding electrode line conductor of ±800kV DC UHV transmission line. 3. Responsibilities of Responsible Person: Be responsible for division of operating personnel of the task, work order reading, handling formalities for stopping the DC restart device for the line, getting work permits, holding pre-shift meeting, dealing with emergency situations in work, quality surveillance, and the summary after work. 4. Responsibilities of equipotential electrician: Be responsible for installing and using the self-climbing device to enter the electric field for the repair of the conductor. 5. Responsibilities of UAV operator: Be responsible for the operation of UAV to extend the hauling rope										

Part I Standard for Training and Assessment on Live Working for Operation Maintenance of ±800kV UHV Power Transmission Line

Table (Cont'd)

Work specification and requirements	6. Responsibilities of ground electrician: Be responsible for transferring tools, equipment and materials and cooperating with equipotential electrician in entering and exiting the equipotential. 7. During the live working, if thunder, rain, strong wind or any other circumstance threaten the safety of the staff, the Responsible Person or Supervisor may stop working temporarily according to the circumstances. Given conditions: 1. Training base: UHV ±800k DC grounding electrode line 2. Work orders have been handled, safety measures have been completed (DC restart device has been deactivated), and oral application (dispatcher or assessor) shall be made at the beginning and end of the work. 3. The instrument shall be used safely and correctly to test the insulating tool. 4. The operation must be carried out according to the working procedures. The scores of the items to be carried out shall be deducted for the process error. In case of major hidden dangers of personal, equipment and operational safety, the assessor may order the termination of the operation (assessment)
Assessment scenario preparation	1. Line: UHV ±800kV DC grounding electrode line; work content: live repair of ±800kV DC grounding electrode line conductor. 2. Required working tools: 2 insulated transmission ropes (TJS-14); 1 insulated transmission rope (TJS-10), 1 insulated transmission rope (TJS-4), 1 insulated noose (TJS-14), 1 insulated tackle (JH10-0.5); 1 six-rotor UAV, 1 self-climbing device, 1 shielding clothes (type I or thermal insulation type); 1 pair of full-body safety belts, insulation resistance meter (5000V type with test electrode), 1 multimeter; 1 wind speed and humidity gauge and 1 pieces of tarpaulin. 3. The work site shall be monitored, and the safety measures (fence, etc.) on the work site have been fully implemented; non-operation personnel are prohibited from entering the site, and the staff must wear safety helmet when entering the work site. 4. Examinees shall bring their own work clothes, flame retardant cotton underwear, safety helmets, gloves, and safety belts (including double-protective ropes)
Remarks	1. The deduction shall be done until the scores of each item are deducted completely. In case of major hidden dangers of personal, equipment and operational safety, the assessor may order the termination of the operation. 2. When equipment, working environment, safety belt, safety helmet, tool, shielding clothes, etc., do not conform to the operation condition, the assessor may order the termination of the operation

Table 1-10-7 Standards for Assessment and Scoring of UHVDC Skills Training of State Grid Sichuan Electric Power Corporation

S/N	Project name	Quality requirements	Score	Deduction standard	Reasons for deduction	Deduction	Scoring
1	Site re-survey	(1) The Responsible Person shall go to the work site to check the line name, pole and tower number, on-site working conditions, defective parts and so on. (2) Check that the site meteorological conditions such as wind speed and humidity shall meet the operation requirements. (3) Check whether the work order is complete and unmodified, check whether the safety measures listed are consistent with the actual situation on site, and supplement it if necessary	5	(1) Deduct 1 point for failure to check the double title. (2) Deduct 1 point/item for failure to check on-site working conditions (meteorology) and defective parts. (3) Deduct 0.5 points/item for any alteration in the work order filling, and deduct 1 point for incorrect work order number. Deduct 1.5 points for incomplete work order filling.			
2	Work Permit	(1) The Responsible Person shall contact the on-duty control personnel and apply for stopping the transmission line DC restart device as per the contents of the work order. (2) Reporting content shall be standardized and complete	2	(1) Deduct 2 points for failure to contact the control department (referee) for disabling the DC restart device. (2) Deduct 0.5 points for non-standard or incomplete terminology reporting.			

Part I Standard for Training and Assessment on Live Working for Operation Maintenance of ±800kV UHV Power Transmission Line

Table (Cont'd)

S/N	Project name	Quality requirements	Score	Deduction standard	Reasons for deduction	Deduction	Scoring
3	Site layout	Install the security fence and hang the signboards correctly: (1) The security fence should take full account of falling objects from the heights and the influence on road traffic. (2) The entrance and exit of the security fence shall be set reasonably. (3) Signs such as "Access from Here", "Work Here", "Access from Here" shall be properly arranged.	3	(1) Deduct 0.5 points for failure to arrange the fence at the work site. (2) Deduct 0.5 points for failure to arrange the warning board. (3) Deduct 0.5 points for failure to hang the tower climbing operation sign.			
4	Holding a pre-shift meeting	(1) All staff and personnel shall wear safety helmets and work clothes correctly. (2) Responsible Person shall wear red waistcoat and read out the work order and be clear with work task and division of personnel; explain safety measures and technical measures in work; check (inquire after) mental state of all working personnel; inform of hazards in work and precontrol measures. (3) All working personnel shall sign on the work order for confirmation.	3	(1) Deduct 0.5 points for the staff not dressing uniformly. Deduct 0.5 points/person for the staff not dressing uniformly. (2) Give no points to this item for no division of labor, and deduct 1 point for unclear division of labor. (3) Deduct 0.5 points for the on-site Responsible Person not wearing a safety monitoring vest. (4) Deduct 1 point for the work shift member failing to sign or signing incompletely on the work order.			

· 517 ·

Table (Cont'd)

S/N	Project name	Quality requirements	Score	Deduction standard	Reasons for deduction	Deduction	Scoring
5	Tool inspection	(1) The staff shall place the tools on the moisture-proof canvas as required; the moisture-proof canvas shall be clean and dry. (2) The tools shall be placed in category according to the requirements of the fixed management; the insulated tools shall not be mixed with metal tools and materials; and the appearance inspection shall be done on the tools. (3) The surface of insulating tools shall not be worn, deformed and damaged, and the operation shall be flexible. Insulating tool Carry out segment insulation detection with such insulated tools with insulation resistance meter of 2,500V or above and with the resistance no less than 700MΩ, and wipe it off with a clean dry towel. (4) The ground potential and equipotential personnel on tower shall correctly wear a whole suit of qualified shield clothes and conductive shoes as required, with each part connected well, shall not wear chemical fiber clothes next to the skin in the shielding clothes and shall fasten safety belts; the Responsible Person shall carefully check whether they wear it correctly	7	(1) Deduct 1 point for failure to use moisture-proof cloth and place tools to designed positions. (2) Deduct 0.5 points/item for failure to check qualified label of tool test and appearance inspection. (3) Deduct 1 point/ item for the improper usage of testing instruments to test the tools, instruments and the full set of shielding clothes. (4) Deduct 2 points/person/time for the operator failing to wear the shielding clothes correctly and each part connected well. (5) Deduct 1 point for the on-site Responsible Person failing to check the safety protective equipment of the tower climbing operators.			

Part I Standard for Training and Assessment on Live Working for
Operation Maintenance of ±800kV UHV Power Transmission Line

Table (Cont'd)

S/N	Project name	Quality requirements	Score	Deduction standard	Reasons for deduction	Deduction	Scoring
6	Spreading of guide rail rope for UAV	(1) The UAV controller shall control the UAV to fly about 1.6m above the ground and hover in place; (2) The ground electrician shall hang the insulated transmission rope with a diameter of 4mm on the tripping device of the UAV; (3) The UAV operator continues to operate the UAV to ascend to a suitable position, and pulls outwards to pass through the conductor, and continues to pull the rope at approximately the same distance as the length of the conductor, and controls the tripping device to make the hauling rope drop naturally; (4) The ground electrician shall install the hauling rope as an insulated endless rope.	23	(1) Deduct 2 points for improper hovering position of UAV, deduct 5 points for any personnel injury. (2) Deduct 1 point/times for any tangling of hauling rope during extending process. (3) Deduct 3 points for failure of the UAV to take off from inside. (4) Deduct 2 points for horizontal towing before the UAV is elevated for enough distance. (5) Deduct 5 points for insufficient horizontal towing distance of the UAV. (6) Deduct 2 points for not releasing at one time. (7) Deduct 3 points for not unfolding and releasing the hauling rope at one time. (8) Deduct 15 points for UAV falling			
7	Hauling guide rail rope and human body backup protection rope	(1) The ground electrician transfers the hauling rope to the operation point; (2) The ground electrician uses a hauling rope to circularly pull the guide rail rope of the self-climbing device, one end of the guide rail rope is stably installed at a firm position on the ground, and the other end is perpendicular to the ground; (3) The ground electrician uses the hauling rope to pull the backup protection rope	5	(1) Deduct 2 points/ time for guide rail rope and backup protection rope twining. (2) Deduct 3 points for failure to make separate hauling of guide rail rope and backup protection rope. (3) Deduct 3 points for failure to make effective fixation of guide rail rope			

· 519 ·

Table (Cont'd)

S/N	Project name	Quality requirements	Score	Deduction standard	Reasons for deduction	Deduction	Scoring
8	Installation of self-climbing device	(1) The equipotential electrician shall check the rotation of the self-climbing device and the electric quantity of the battery again, and install the self-climbing device on the guide rail rope after confirming it is correct; (2) After the equipotential electrician completes the human body backup protection measures and operates the self-climbing device to a certain height (0.8 ~ 1m), the ground electrician conducts impulse test on the self-climbing device and the equipotential electrician. (3) After the impulse test is qualified, the equipotential electrician carries the insulated transmission rope to climb vertically.	5	(1) Deduct 2 points for failure to check the self-climbing device. (2) Deduct 3 points for not conducting impulse test; deduct 2 points for improper impulse test method. (3) Deduct 5 points for any damage to the self-climbing device caused by improper impulse test method.			
9	Entering the intense electric field and repairing the conductor	(1) Equipotential electrician rides the self-climbing device to rise vertically to a distance of 0.3m from the conductor, check that all connections of the shielding clothes are correct, and transfer the potential with the consent of the person in charge of the working group. (2) Equipotential electrician shall fasten safety belt protection on conductor, arrange insulated transmission rope at winding place of the preformed armor rods, hang the insulated transmission rope and report to the Responsible Person. (3) The ground electrician shall hang up the preformed armor rods and report to the Responsible Person.	25	(1) Deduct 3 points for the equipotential electrician failing to apply to the Responsible Person before climbing; deduct 2 points for starting to climb the ladder without consent after application. (2) Deduct 2 points for failure to effectively control the backup protection rope. (3) Deduct 2 points for the potential transfer without applying to the Responsible Person; deduct 1 point for starting the potential transfer without the consent after applying.			

Part I Standard for Training and Assessment on Live Working for Operation Maintenance of ±800kV UHV Power Transmission Line

Table (Cont'd)

S/N	Project name	Quality requirements	Score	Deduction standard	Reasons for deduction	Deduction	Scoring
9	Entering the intense electric field and repairing the conductor	(4) The equipotential electrician applies to the Responsible Person for leveling the conductor. Use the preformed armor rods to repair conductor as required and report to the Responsible Person.	25	(4) Deduct 1 point for the inappropriate position applied for the potential transfer. (5) Deduct 1 point for being unskilled in potential transfer. (6) Deduct 2 points for the Responsible Person failing to monitor the operation carefully. (7) Deduct 1 point/time for any bumping occurred during the transmission of preformed armor rods repairing strips. (8) Deduct 2 points/time for unsteady rising/declining of the self-climbing device			
10	Exit from electric field	(1) The equipotential electrician shall check the conductor repair by the preformed armor rods again, report to the Responsible Person after confirming it is correct, and apply to exit from the electric field. (2) With the consent of the Responsible Person, the equipotential electrician shall check the connection of the shielding clothes and confirm that the connection is reliable, then exit the electric field following the reverse step of entering the electric field, and take the self-climbing device down to the ground	12	(1) Deduct 1 point for failure to report the end of the work. (2) Deduct 2 points for failure to apply to the Responsible Person for exiting from the electric field. (3) Deduct 1 point for not checking the connection of the shielding clothes. (4) Deduct 2 points for the potential transfer without applying to the Responsible Person;deduct 1 point for starting the work without the consent after applying.			

· 521 ·

Table (Cont'd)

S/N	Project name	Quality requirements	Score	Deduction standard	Reasons for deduction	Deduction	Scoring
10	Exit from electric field		12	(5) Deduct 1 point for the inappropriate position applied for the potential transfer (6) Deduct 2 points for the incorrect action of equipotential electrician exiting the electric field and repeat discharging. (7) Deduct 1 point for failure to effectively control backup protection rope			
11	End of the work	(1) The Responsible Person shall organize all working personnel to put working apparatus and materials in order and put them in a special kit (bag) after cleaning; clean the site to ensure that "the materials are removed and the site is cleaned after construction". (2) After a post-shift meeting is held, the Responsible Person shall make work summaries and comments. Comments include the construction quality of this work and the implementation of safety measures from all working personnel. (3) Responsible Person shall report to the on-duty control personnel the end of the work, apply for the restoration of DC restarting device of the line and terminate the work order.	10	(1) Deduct 2 points for failure to clean the tools. (2) Deduct 2 points for missing tools. (3) Deduct 3 points for failure to hold the post-shift meeting. (4) Deduct 1 point for unqualified comments. (5) Deduct 3 points for failure to contact the control personnel for end of the work reporting. (6) Deduct 0.5 points for each missing item from the reporting content. (Name of working unit, name of the Responsible Person, name of the line, work completion; device has restored to normal, and the personnel have evacuated). (7) Deduct 1 point for error in the fill-in of work order termination procedure			
	Total		100				

Module 11 Standards for Training and Assessment on Live Replacement of ±800kV UHV Transmission Line Grounding Electrode Line Spacer

I. Training Standards

(I) Training Requirements

Designation of module	Live replacement of ±800kV UHV transmission line grounding electrode line spacer	Type of training	Operation
Training method	Practical operation training	Hours of training	16 hours
Training objectives	1. Master the electrical significance of "self-climbing device" operation mode when the ±800kV DC grounding electrode tangent tower enters and exits electric field. 2. Can enter ±800kV DC grounding electrode line equipotential operation point by adopting the "self-climbing device". 3. Can independently complete the replacement of ±800kV DC grounding electrode line spacer (equipotential operation method)		
Training venue	UHV DC practical training line		
Training content	Entering the electric field from the self-climbing device, the equipotential operation method is adopted for live replacement of the grounding electrode line spacer for ±800kV UHV transmission line		
Scope of application	Maintenance personnel for ±800kV UHV DC power transmission line		

(II) Referenced Procedures and Specifications

(1) Code for Designing of ±800kV DC Overhead Transmission Line (GB/T50790-2013)

(2) Maintenance Specification for ±800kV DC Overhead Transmission Line (DL/T251-2012)

(3) Operating Code for ±800kV DC Overhead Transmission Line (GB/T28813-2012)

(4) Technical Specification for Live Working of ±800kV DC Line (DL/T1242-2013)

(5) Technical Specification for Fittings of ±800kV DC Overhead Transmission Line (GB/T31235-2014)

(6) State Grid Corporation of China on the Management Regulations of Live Working (Trial Implementation) (SGCC [2007] No.751)

(7) State Grid Corporation of China Working Regulations of Power Safety (Power Line Section) (Q/GDW1799.2-2013)

(8) Electrotechnical Terminology- Overhead Line (GB/T2900.51-1998)

(9) Electrotechnical Terminology- Live Working (GB/T2900.55-2016)

(10) Live Working - Terminology for Tools, Equipment and Devices (GB/T14286-2002)

(11) Insulated Tackles for Live Working (GB/T13034-2008)

(12) Live Working - Insulating Ropes (GB13035-2008)

(13) Minimum Requirements for Utilization of Tools, Devices and Equipment for Live Working (DL/T877-2004)

(14) Preventive Test Code of Tools, Devices and Equipment for Live Working (DL/T976-2005)

(15) Technical Guide for Live Working of ±800kV DC Transmission Line (Q/GDW302-2009)

(16) Shielding Clothes for Live Working (GB/T6568-2008)

(17) Technical Requirements and Design Guide for Live Working Tools (GB/T18037-2008)

(18) Application Rules of Infrared Diagnosis for Live Electrical Equipment (DL/T664-2016)

(III) Teaching Design for Training

To complete the work task of "live replacement of ±800kV UHV transmission line grounding electrode line spacer", each training stage shall be designed according to the standard operation procedure for work task completion. Each stage includes specific training objectives, training content, hours of training, training methods (training resources), training environment, assessment and evaluation, etc, as shown in the Table 1-11-1.

Part I Standard for Training and Assessment on Live Working for Operation Maintenance of ±800kV UHV Power Transmission Line

Table 1-11-1 Training Content Design for Live Replacement of Grounding Electrode Line Spacers on ±800kV UHV Transmission Line

Training schedule	Training objectives	Training contents	Hours of training	Training methods and resources	Preparation of training conditions	Assessment and evaluation
1. Theoretical teaching	1. Preliminarily master the basic method for entering and exiting the ±800kV DC grounding electrode line electric field using the self-climbing device. 2. Be familiar with equipment configuration and safety precautions. 3. Be familiar with the spacer replacement method for ±800kV DC grounding electrode line	1. Electrical significance of using self-climbing device to enter and exit ±800kV DC grounding electrode line electric field. 2. Equipment allocation, hazard analysis and safety measures. 3. Replacement methods and quality standards of ±800kV DC grounding electrode line spacers	2	Training methods: Lecture. Training resources: PPT, relevant regulations and specifications	Multimedia classroom,	Attendance, classroom questions and assignments
2. Preparations	Be able to complete the preparation before operation	1. Work site survey. 2. Preparation of the standardized operation card. 3. Filling of the work order. 4. Preparation of tools and materials for this operation	1	Training methods: 1. Site survey and cleaning of tools and materials shall be practiced at site 2. Preparation of operation card and the filling of work order shall adopt lecture method Training resources: 1. ±800kV grounding electrode practical training line 2. UHV tools warehouse 3. Blank work order	1.±800kV Grounding electrode practical training line; 2.Multimedia classroom	

· 525 ·

Table (Cont'd)

Training schedule	Training objectives	Training contents	Hours of training	Training methods and resources	Preparation of training conditions	Assessment and evaluation
3. Work site preparation	Be able to complete the preparations of work site	1. Site re-survey. 2. Work application. 3. Work site layout. 4. Pre-shift meeting. 5. Inspection of tools	1	Training methods: Demonstration and role play Resources: ±800kV practical training line	±800kV Grounding electrode practical training line	
4. Trainer's demonstration	The trainees can preliminarily understand the operation process of the task through inspecting and learning from each other's work	1. Working distance measurement and conductor temperature measurement. 2. Ground electrician uses the UAV to spread out the hauling rope. 3. The ground electrician shall pull the backup protection rope and guide rail rope to the working position. 4. Equipotential electrician rides the self-climbing device to enter the electric field. 5. Equipotential electrician shall conduct the replacement of spacer	3	Training methods: Demonstration. Resources: ±800kV grounding electrode practical training line	±800kV Grounding electrode practical training line	

Part I Standard for Training and Assessment on Live Working for Operation Maintenance of ±800kV UHV Power Transmission Line

Table (Cont'd)

Training schedule	Training objectives	Training contents	Hours of training	Training methods and resources	Preparation of training conditions	Assessment and evaluation
5. Group training of trainees	Through the training: 1. Trainees are able to complete the operation of entering and exiting ±800kV DC grounding electrode line electric field. 2. Trainees can complete the method of replacement of ±800kV DC grounding electrode line spacers	1. The trainees are grouped (6 in a group) to train the skill operation of entering and exiting the ±800kV DC electric field and replacing grounding electrode line spacers. 2. Trainers guide the operation of trainees and conduct safety supervision	8	Training methods: Role play Resources: ±800kV practical training line	±800kV Grounding electrode practical training line	Score the operation of trainees according to the detailed rules for skill assessment and scoring
6. End of the work	Through the training: 1. Trainees can further understand the shortcomings during the operation process for later improvement. 2. Train the trainees in the working style of safe and civilized production	1. Cleaning up the work site. 2. Report to the controller. 3. Comment and summarize the current work task at post-shift meeting	1	Training methods: Lecture and inductive method	±800kV Grounding electrode practical training line	

(IV) Work Flow

1. Work Task

Entering the electric field from the self-climbing device to arrive at the operation point, the equipotential operation method is adopted for live replacement of the grounding electrode line spacer for ±800kV UHV transmission line.

2. Requirements for Weather and Work Site

(1) The live replacement of spacers for ±800kV DC grounding electrode line shall be carried out in good weather.

In case of lightning (hearing thunder or seeing lightning), snow, hail, rain, fog, etc., on the work site, this operation shall be prohibited. When wind force is greater than Level 5 or the relative humidity of air is greater than 80%, this operation should not be carried out; during the operation, pay attention to the weather changes. In case of gale or misty rain and other emergencies, take correct measures according to the procedures to ensure the safety of personnel and equipment.

(2) The operating personnel should be in good mental states and be familiar with the organizational and technical measures to ensure safety in work; they should hold the qualification certificate for live working within the validity period.

(3) Responsible Person should organize the relevant personnel to complete field investigation in advance, determine the operating methods, required working apparatus and necessary measures according to the results, and handle the tickets for live working.

(4) The work site should be reasonably set up with fence and warning signs. Non-operating personnel is forbidden to enter.

(5) DC restart device shall be deactivated for the Project.

(6) Safe working distance and effective insulation length during operation are shown in Table 1-11-2.

Table 1-11-2 Safe distance (m) for live replacement for grounding electrode line spacer of ±800kV UHV transmission line

Voltage class	Safe distance between human body and electrified body	Minimum effective insulation length		Minimum combination gap	Minimum distance between the exposed part of human body and electrified body during potential transfer
		Insulating bar	Insulated load-bearing tools and insulating ropes		
±800kV DC grounding electrode	0.75	1.3	1.0	1.2	0.3

3. Preparations

3.1 Hazards and precontrol measures

(1) Hazard——Preventing system overvoltage

Precontrol measures: This operation is equipotential operation. Before starting the work, the Responsible Person shall contact the on-duty controller to apply for a work order for live working on power lines, and apply to deactivate the DC restart protection device. The controller on duty shall perform the licensing procedures. After the live working is completed, report to the controller on duty in a timely manner. It is forbidden to disable and restore the DC restart device at the appointed time. If equipment loses power suddenly during live working, operators shall consider the equipment as still charged. The Responsible Person shall contact the control personnel as soon as possible, and no forced energization is allowed before the controller on duty getting in touch with the Responsible Person.

(2) Hazard —— electric shock

Precontrol measures:

① Insulating tool and insulating ropes shall be free of damage, moisture, deformation, and failure. It is not allowed to use non-insulating ropes (such as cotton rope, manila rope, and steel wire rope).

② The equipotential operator shall wear flame-retardant underwear and full set of screen clothes outside (including hat, dresses & trousers, gloves, socks and shoes). All parts shall be in excellent connection conditions, and the resistance of the whole suit of shielding clothes shall not exceed 20Ω.

③ Before potential transfer, equipotential operator shall obtain the approval of Responsible Person, and the minimum distance between exposed part of human body and electrified body shall not be lower than 0.3m. The safe distance between the ground potential operator and the electrified body (between the equipotential operator and grounding body) is less than that specified in Table 1-11-2.

④ When transmitting large metal objects by using insulated ropes, ground potential operators shall not touch them before ground connection.

⑤ The Responsible Person and the Special Supervisor shall continuously monitor the operators and correct their nonstandard operation or actions in violation at any time. Special attention shall be paid to operators working at heights to ensure that there is enough safe distance (meeting the requirements in Table 1-11-2).

⑥ When the operating temperature of the heat-resisting conductor is above 40°C, high temperature resistance live working tools shall be used, including high temperature resistance soft insulating tools, high temperature resistance hard insulating tools, and high temperature resistance heat insulation shielding clothes, and the potential transfer rod shall be used to enter the equipotential.

⑦ Insulation protective equipment shall be regarded as a conductor. During the transfer and use, the safe distance between the phase and ground and poles shall be ensured to meet the safety requirements.

(3) Hazard —— falling from high places

Precontrol measures:

① Before climbing, operators working at heights must satisfy the requirements of this operation, such as physical condition, mental state, and skill and quality.

② Supervisors shall correct the nonstandard or illegal actions at any time. Special attention shall be paid to operators to prevent them from losing the protection of safety belt or insulated backup protection rope during transposition, and it is forbidden to fasten the safety belt or insulated backup protection rope in a position lower than the operating personnel.

(4) Hazard —— injury caused by objects falling from high place

Precontrol measures:

① Operator working at heights should put personal tools and fragmentary materials into the tools bag. It is strictly forbidden to hang objects in high place or keep in the mouth.

② Ground operator should correctly wear a helmet and use the knots. The vertical distance from the operation point should not be less than the falling radius.

③ The work site shall be set up with fence and warning signs. It shall be noted at any time that Supervisor shall prohibit irrelevant personnel and vehicles from entering operation area.

(5) For injuries caused by conductor overheating

Precontrol measures:

① Before the operation, contact with the dispatcher to confirm the operation mode to avoid the live working when the large current of the conductor flows.

② Operators shall wear special heat-resistant shielding clothes to carry out live working to prevent the conductor from overheating and injuring people.

3.2 Selection of tools, instruments and materials

See Table 1-11-3 for tools, instruments and materials required for live replacement of spacer for ±800kV UHV transmission line grounding electrode. Before delivering tools and instruments out of warehouse, application voltage class and test cycle shall be carefully checked and they shall be inspected to ensure that appearance is intact, connection is firm, rotation is flexible,and they meet the work task requirements. After delivering tools and instruments out of warehouse, they shall be stored in tools bag or tool kit for transportation to avoid contamination and damp. Metal tools and insulating tools shall be separately loaded and transported to avoid deformation, damage or other defects caused by mixed loading and transportation.

Table 1-11-3 Tools, Instruments and Materials Required for Live Replacement of Grounding Electrode Line Spacers on ±800kV UHV Transmission Line

S/N	Name	Specification and Model	Unit	Qty.	Remarks
1	Insulated transmission rope	TJS-14×50m	Nos.	1	
2	Insulated transmission rope	TJS-10×50m	Nos.	1	
3	Insulated transmission rope	TJS-14×100m	Nos.	1	
4	Insulated transmission rope	TJS-4×100m	Nos.	1	
5	Insulated noose	TJS-14	Nos.	1	
6	Insulated tackle	0.5T	Nos.	1	
7	UAV	Six rotor	Nos.	1	
8	Self-climbing device	ACTSAFE-II	Set	1	
9	Personal tools		Set	1	
10	Shielding clothes	500kV type I or thermal insulation type	Set	1	
11	Human body backup protection rope		Nos.	1	
12	Safety belt		Nos.	1	
13	Safety helmet		pcs	6	
14	Moisture-proof tarpaulin	3m×3m	pcs	1	
15	Multimeter		Nos.	1	
16	Anemoscope and hygroscope	HT-8321	pcs	1	
17	Megameter and test electrode	5000V	Set	1	
18	Interphone		Set	2	
19	Red waistcoat	"Responsible Person"	pcs	1	
20	Security fence		Set	Several	
21	Spacer	For grounding electrode line	Nos.	1	

3.3 Division of labor for operators

Division of labor for operators of the task is shown in Table 1-11-4.

Table 1-11-4 Personnel Allocation for Live Replacement of Grounding Electrode Line Spacers on ±800kV UHV Transmission Line

S/N	Post	Qty. (person)	Responsibilities
1	Responsible Person	1	Be responsible for various work on the work site
2	Equipotential electrician	1	Be responsible for installing and using the self-climbing device to enter the electric field for the replacement of the spacer
3	UAV operator	1	Be responsible for the operation of UAV to extend the hauling rope
4	Ground electrician	3	Be responsible for transferring tools and materials and co-operating with equipotential electrician in entering and exiting the equipotential

4. Working procedures

The workflow of this task is shown in Table 1-11-5.

Table 1-11-5 Workflow for Live Replacement of Grounding Electrode Line Spacers on ±800kV UHV Transmission Line

S/N	Work Content	Operation Steps and Standards	Safety Measures and Precautions	Responsible Person
1	Confirm the line name and tower number and inspect the pole and tower	The Responsible Person shall complete the following work: (1) Check the line name, the number of the pole and tower and ensure the phases are correct; guarantee that the foundation and the pole and tower are intact and in normal condition; ensure that the cross and span distance meets the safety requirements; confirm the defect conditions and the specifications and models of conductor or ground wires. (2) Check that the terrain and environment should meet the operation requirements. (3) Check that the safety measures listed in the work order are in line with the actual situations on site. The measures will be added if necessary.	(1) Correctly wear helmet, working clothes, work shoes and protective gloves. (2) Operation under meteorological conditions that may endanger the safety of operators is forbidden. (3) Non-operation personnel and vehicles are strictly prohibited from entering the working site	

Part I Standard for Training and Assessment on Live Working for Operation Maintenance of ±800kV UHV Power Transmission Line

Table (Cont'd)

S/N	Work Content	Operation Steps and Standards			Safety Measures and Precautions	Responsible Person
2	Measurement of on site work environment	Measuring items	Standard value	Measured value	When the wind force is greater than Level 5 and the humidity is greater than 80%, live working should not be carried out.	
		Wind speed	≤10m/s			
		Humidity	≤80%			
		Environment temperature	0～38℃			
		Conductor temperature	—			
		Working distance	≥0.75m			
		Altitude	Altitude limit of 500m and 1000m			
3	Work Permit	(1) The Responsible Person is responsible for contacting the on-duty control personnel and applying for stopping the DC restart protection device for the line as per the contents of the work order. (2) Live working could be started only after being approved by control personnel on duty.			Live working shall not be started without the permission of the on-duty control personnel	
4	Site layout	Install the security fence and hang the signboards correctly: (1) The security fence should take full account of falling objects from the high place and the influence on road traffic. (2) The entrance and exit of the security fence shall be set reasonably. (3) Signs such as "Access from Here", "Work Here", "Access from Here" shall be properly arranged			When the influence on road traffic safety is uncontrollable, the traffic management department should be contacted in time to strengthen the on-site control of traffic safety	

Table (Cont'd)

S/N	Work Content	Operation Steps and Standards	Safety Measures and Precautions	Responsible Person
5	Holding a pre-shift meeting	(1) All working personnel shall line up. (2) The Responsible Person shall wear red waistcoat and read out the work order and be clear with work task and division of personnel; explain safety measures and technical measures in work; check (inquire after) mental state of all working personnel; inform of hazards in work and precontrol measures. (3) All working personnel shall sign on the work order for confirmation.	(1) Work order shall be filled and issued with standardized licensing procedure and complete signature. (2) All working personnel shall be in good mental states. (3) All working personnel shall be clear with task division of works, safety measures and technical measures	
6	Inspection of tools	(1) The equipotential electrician shall wear the shielding clothes in a right way and pass the inspection, which shall be supervised and inspected by the Responsible Person. (2) Wear personal safety equipment correctly (proper size and easy lock), and the Responsible Person shall supervise and inspect it. (3) Check the appearance and measure insulation performance of insulating tools, and make records. (4) Check the proper operation of UAV and self-climbing device	(1) Check carefully for damage, deformation and failure before using metal and insulating tools. Wipe the insulating tools with a clean and dry towel and carry out segment insulation detection with such insulated tools with insulation resistance meter of 2,500V or above and with the resistance no less than 700MΩ. (2) Use a multimeter to measure the resistance between the farthest ends of the shielding clothes and trousers, which shall not be greater than 20Ω. The Responsible Person shall check the connection of the electrician's shielding clothes. (3) Check the tool assembly and make sure the connection is reliable. (4) The tools and instruments for live working shall be placed on the moisture-proof tarpaulin	

Table (Cont'd)

S/N	Work Content	Operation Steps and Standards	Safety Measures and Precautions	Responsible Person
7	Spreading of guide rail rope for UAV	(1) The UAV controller shall control the UAV to fly about 1.6m above the ground and hover in place. (2) The ground electrician shall hang the insulated transmission rope with a diameter of 4mm on the tripping device of the UAV. (3) The UAV operator continues to operate the UAV to ascend to a suitable position, and pulls outwards to pass through the conductor, and continues to pull the rope at approximately the same distance as the length of the conductor, and controls the tripping device to make the hauling rope drop naturally. (4) The ground electrician shall install the hauling rope as an insulated endless rope.	(1) The hovering position of the UAV shall be appropriate. The ground electrician shall hang the hauling rope with attention to the safety. (2) When the UAV is towed, it shall take off from the inside of the line to the outside and the hauling rope shall be released outwardly.	
8	Hauling guide rail rope and human body backup protection rope	(1) The ground electrician transfers the hauling rope to the operation point. (2) The ground electrician uses a hauling rope to circularly pull the guide rail rope of the self-climbing device, one end of the guide rail rope is stably installed at a firm position on the ground, and the other end is perpendicular to the ground. (3) The ground electrician uses the hauling rope to pull the backup protection rope	(1) One end of the guide rail rope must be fixed firmly on the ground. (2) The guide rail rope and the human body backup protection rope shall be kept at a certain distance to prevent entanglement	

Table (Cont'd)

S/N	Work Content	Operation Steps and Standards	Safety Measures and Precautions	Responsible Person
9	Installation of self-climbing device	(1) The equipotential electrician shall check the rotation of the self-climbing device and the electric quantity of the battery again, and install the self-climbing device on the guide rail rope after confirming it is correct. (2) After the equipotential electrician completes the human body backup protection measures and operates the self-climbing device to a certain height (0.8 ~ 1m), the ground electrician conducts impulse test on the self-climbing device and the equipotential electrician. (3) After the impulse test is qualified, the equipotential electrician carries the insulated transmission rope to climb vertically	(1) The self-climbing device shall rotate properly. (2) The ground electrician shall carry out the impulse test on the self-climbing device. Pay attention to the force of the self-climbing device and the guide rail rope fixed to the ground during the impact. (3) The responsible person shall carefully check	
10	Entering the intense electric field and replace the spacer	(1) Equipotential electrician rides the self-climbing device to rise vertically to a distance of 0.3m from the conductor, check that all connections of the shielding clothes are correct, and transfer the potential with the consent of the person in charge of the working group. (2) Equipotential electrician shall fasten safety belt protection on the conductor and arrange insulated transmission rope at appropriate position. (3) The equipotential electrician shall dismantle the old spacer, bind the old spacer with insulated transmission rope; the ground electrician shall transfer the old spacer to the ground with insulated transmission rope, and transfer the new spacer to the equipotential electrician at the same time. (4) The equipotential electrician shall install the new spacer at the original position of the spacer	(1) The ground electrician shall control the human body backup protection rope to ensure that the equipotential electrician does not lose the safety belt protection during the work at heights. (2) Before the potential of the equipotential electrician is transferred, it must be approved by the Responsible Person. It is strictly forbidden to charge and discharge with the head during potential transfer; the distance between the exposed part of the human body and the electrified body shall not be less than 0.3m; the combined gap shall not be less than 1.2m. (3) Collision shall be avoided during the transfer of the spacers, and the equipotential electrician shall avoid high falling objects when working at heights.	

Part I   Standard for Training and Assessment on Live Working for Operation Maintenance of ±800kV UHV Power Transmission Line

Table (Cont'd)

S/N	Work Content	Operation Steps and Standards	Safety Measures and Precautions	Responsible Person
11	Leaving the electric field	(1) The equipotential electrician shall check the spacer restoration and pin placement again, report to the Responsible Person after confirming it is correct, and apply to exit from the electric field. (2) With the consent of the Responsible Person, the equipotential electrician shall check the connection of the shielding clothes and confirm that the connection is reliable, then exit the electric field following the reverse step of entering the electric field, and take the self-climbing device down to the ground	(1) The equipotential electrician must obtain the permission of the Responsible Person before leaving the potential. (2) The combined gap composed of the gaps between the equipotential electrician and the grounding body and the electrified body in the process of exiting the electric field shall not be less than 1.2m. (3) The Responsible Person shall carefully supervise and remind the operation process.	
12	End of the work	(1) The Responsible Person shall organize all working personnel to put working apparatus and materials in order and put them in a special kit (bag) after cleaning; clean the site to ensure that "the materials are removed and the site is cleaned after construction". (2) After a post-shift meeting is held, the Responsible Person shall make work summaries and comments. Comments include the construction quality of this work and the implementation of safety measures from all working personnel. (3) Responsible Person shall report to the on-duty control personnel the end of the work, apply for the restoration of DC restarting device of the line and terminate the work order.	It is forbidden to restore the DC transmission line restarting device at the appointed time	

II. Assessment Standard

Table 1-11-6 Detailed Rules for Assessment and Scoring of UHV DC Skills Training of State Grid Sichuan Electric Power Corporation

Fill-in Column of Examinee	No.:	Name:	Position:	Unit:	Date:	MM/DD/YYYY	
Fill-in Column of Assessor	Grade:	Assessor:	Assessment Team Leader:	Starting time:	Closing time:	Operation Duration:	
Assessment Module	Live replacement of spacer for ±800kV UHV transmission line grounding electrode		Assessee	Maintenance personnel for ±800kV DC grounding electrode transmission line		Assessment method	Operation	Assessment Time Limit	60min
Job Description	Enter the electric field along the rope ladder for live replacement of the ±800kV DC grounding electrode line spacer.						
Work specifications and requirements	1. Live working shall be carried out in good weather. In case of thunder, rain, snow or fog, no live working shall be carried out. When the wind force is greater than Level 5 and the humidity is greater than 80%, live working should not be carried out. 2. Workers required for this operation include 1 Responsible Person, 1 equipotential electrician, 1 operator of UAV and 3 ground auxiliary electricians. They shall enter the electric field using self-climbing device for the live replacement of the grounding electrode line spacer of ±800kV DC UHV transmission line. 3. Responsibilities of Responsible Person: Be responsible for division of operating personnel of the task, work order reading, handling formalities for stopping the DC restart device for the line, getting work permits, holding pre-shift meeting, dealing with emergency situations in work, quality surveillance, and the summary after work. 4. Responsibilities of equipotential electrician: Be responsible for installing and using the self-climbing device to enter the electric field for the replacement of the spacer. 5. Responsibilities of UAV operator: Be responsible for the operation of UAV to extend the hauling rope 6. Responsibilities of ground electrician: Be responsible for transferring tools, equipment and materials and cooperating with equipotential electrician in entering and exiting the equipotential. 7. During the live working, if thunder, rain, strong wind or any other circumstance threaten the safety of the staff, the Responsible Person or Supervisor may stop working temporarily according to the circumstances.						

Part I Standard for Training and Assessment on Live Working for
Operation Maintenance of ±800kV UHV Power Transmission Line

Table (Cont'd)

Work specifications and requirements	Given conditions: 1. Training base: UHV ±800k DC grounding electrode line 2. Work orders have been handled, safety measures have been completed (DC restart device has been deactivated), and oral application (dispatcher or assessor) shall be made at the beginning and end of the work. 3. The instrument shall be used safely and correctly to test the insulating tool. 4. The operation must be carried out according to the working procedures. The scores of the items to be carried out shall be deducted for the process error. In case of major hidden dangers of personal, equipment and operational safety, the assessor may order the termination of the operation (assessment)
Assessment scenario preparation	1. Line: UHV ±800kV DC grounding electrode line; Work content: live replacement of ±800kV DC grounding electrode line spacer. 2. Required working tools: 2 insulated transmission ropes (TJS-14); 1 insulated transmission rope (TJS-10), 1 insulated transmission rope (TJS-4), 1 insulated noose (TJS-14), 1 insulated tackle (JH10-0.5); 1 six-rotor UAV, 1 self-climbing device, 1 shielding clothes (type I or thermal insulation type); 1 pair of full-body safety belts, insulation resistance meter (5000V type with test electrode), 1 multimeter; 1 wind speed and humidity gauge and 1 pieces of tarpaulin. 3. The work site shall be monitored, and the safety measures (fence, etc.) on the work site have been fully implemented; non-operation personnel are prohibited from entering the site, and the staff must wear safety helmets when entering the work site. 4. Examinees shall bring their own work clothes, flame retardant cotton underwear, safety helmets, gloves, and safety belts (including double-protective ropes)
Remarks	1. The deduction shall be done until the scores of each item are deducted completely. In case of major hidden dangers of personal, equipment and operational safety, the assessor may order the termination of the operation. 2. When equipment, working environment, safety belt, safety helmet, tool, shielding clothes, etc., do not conform to the operation condition, the assessor may order the termination of the operation

Table 1-11-7 Standards for Assessment and Scoring of UHVDC Skills Training of State Grid Sichuan Electric Power Corporation

S/N	Project name	Quality requirements	Score	Deduction standard	Reasons for deduction	Deduction	Scoring
1	Site re-survey	(1) The Responsible Person shall go to the work site to check the line name, pole and tower number, on-site working conditions, defective parts and so on. (2) Check that the site meteorological conditions such as wind speed and humidity should meet the operation requirements. (3) Check whether the work order is complete or unmodified, check whether the safety measures listed are consistent with the actual situation on site, and supplement it if necessary	5	(1) Deduct 1 point for failure to check the double title. (2) Deduct 1 point/ item for failure to check on-site working conditions (meteorology) and defective parts. (3) Deduct 0.5 points/item for any alteration in the work order filling, and deduct 1 point for incorrect work order number. Deduct 1.5 points for each incomplete work order			
2	Work Permit	(1) The Responsible Person contacts the on-duty control personnel and apply for stopping the DC restart device for the line as per the contents of the work order. (2) Reporting content is standardized and complete	2	(1) Deduct 2 points for failure to contact the control department (referee) for disabling the DC restart device. (2) Deduct 0.5 points for non-standard or incomplete terminology reporting respectively			

Part I　Standard for Training and Assessment on Live Working for Operation Maintenance of ±800kV UHV Power Transmission Line

Table (Cont'd)

S/N	Project name	Quality requirements	Score	Deduction standard	Reasons for deduction	Deduction	Scoring
3	Site layout	Install the security fence and hang the signboards correctly: (1) The security fence should take full account of falling objects from the high place and the influence on road traffic. (2) The entrance and exit of the security fence shall be set reasonably. (3) Signs such as "Access from Here", "Work Here", "Access from Here" shall be properly arranged	3	(1) Deduct 0.5 points for failure to arrange the fence at the work site. (2) Deduct 0.5 points for failure to arrange the warning board. (3) Deduct 0.5 points for failure to hang the tower climbing operation sign			
4	Holding a pre-shift meeting	(1) All staff and personnel shall wear safety helmets and work clothes correctly. (2) Responsible Person shall wear red vest and read out the work order and be clear with work task and division of personnel; explain safety measures and technical measures in work; check (inquire after) mental state of all working personnel; inform of hazards in work and precontrol measures. (3) All working personnel shall sign on the work order for confirmation.	3	(1) Deduct 0.5 points/person for the staff not dressing uniformly. Deduct 0.5 points/person for the staff not dressing uniformly. (2) Give no points to this item for no division of labor, and deduct 1 point for unclear division of labor. (3) Deduct 0.5 points for the on-site Responsible Person not wearing a safety monitoring vest. (4) Deduct 1 point for the work shift member failing to sign or signing incompletely on the work order.			

· 541 ·

Table (Cont'd)

S/N	Project name	Quality requirements	Score	Deduction standard	Reasons for deduction	Deduction	Scoring
5	Inspection of tools and instruments	(1) The staff shall place the tools on the moisture-proof tarpaulin as required; the moisture-proof tarpaulin shall be clean and dry. (2) The tools shall be placed in category according to the requirements of the fixed management; the insulated tools shall not be mixed with metal tools and materials; and the appearance inspection shall be done on the tools. (3) The surface of insulated tools shall not be worn, deformed or damaged, and the operation shall be flexible. Carry out segment insulation detection with such insulated tools with insulation resistance meter of 2500V or above and with the resistance no less than 700MΩ, and wipe it off with a clean dry towel. (4) The ground potential and equipotential personnel on tower shall correctly wear a whole suit of qualified shield clothes and conductive shoes as required, with each part connected well, shall not wear chemical fiber clothes next to the skin in the shielding clothes and shall fasten safety belts; the Responsible Person shall carefully check whether they wears it correctly	7	(1) Deduct 1 point for failure to use moisture-proof tarpaulin and place tools to designed positions. (2) Deduct 0.5 points/item for failure to check qualified label of tool test and appearance inspection. (3) Deduct 1 point/item for failure to use testing instrument for testing the tools. (4) Deduct 2 points/person time for the operator failing to wear the shielding clothes correctly and each part connected well. (5) Deduct 1 point for the on-site Responsible Person failing to check the safety protective equipment of the tower climbing operators.			

· 542 ·

Part I Standard for Training and Assessment on Live Working for Operation Maintenance of ±800kV UHV Power Transmission Line

Table (Cont'd)

S/N	Project name	Quality requirements	Score	Deduction standard	Reasons for deduction	Deduction	Scoring
6	Spreading of guide rail rope for UAV	(1) The UAV controller shall control the UAV to fly about 1.6m above the ground and hover in place. (2) The ground electrician shall hang the insulated transmission rope with a diameter of 4mm on the tripping device of the UAV. (3) The UAV operator continues to operate the UAV to ascend to a suitable position, and pulls outwards to pass through the conductor, and continues to pull the rope at approximately the same distance as the length of the conductor, and controls the tripping device to make the hauling rope drop naturally; (4) The ground electrician shall install the hauling rope as an insulated endless rope.	23	(1) Deduct 2 points for improper hovering position of UAV, deduct 5 points for any personnel injury. (2) Deduct 1 point/times for any tangling of hauling rope during extending process. (3) Deduct 3 points for failure of the UAV to take off from inside. (4) Deduct 2 points for horizontal towing before the UAV is elevated for enough distance. (5) Deduct 5 points for insufficient horizontal towing distance of the UAV. (6) Deduct 2 points for not releasing at one time. (7) Deduct 3 points for not unfolding and releasing the hauling rope at one time. (8) Deduct 15 points for UAV falling			
7	Hauling guide rail rope and human body backup protection rope	(1) The ground electrician transfers the hauling rope to the operation point; (2) The ground electrician uses a hauling rope to circularly pull the guide rail rope of the self-climbing device, one end of the guide rail rope is stably installed at a firm position on the ground, and the other end is perpendicular to the ground; (3) The ground electrician uses the hauling rope to pull the backup protection rope.	5	(1) Deduct 2 points/ time for guide rail rope and backup protection rope twining. (2) Deduct 3 points for failure to make separate hauling of guide rail rope and backup protection rope. (3) Deduct 3 points for failure to make effective fixation of guide rail rope.			

· 543 ·

Standard for Professional Training and Assessment for Operation Maintenance of UHV DC Power Transmission Line

Table (Cont'd)

S/N	Project name	Quality requirements	Score	Deduction standard	Reasons for deduction	Deduction	Scoring
8	Installation of self-climbing device	(1) The equipotential electrician shall check the rotation of the self-climbing device and the electric quantity of the battery again, and install the self-climbing device on the guide rail rope after confirming it is correct; (2) After the equipotential electrician completes the human body backup protection measures and operates the self-climbing device to a certain height (0.8 ~ 1m), the ground electrician conducts impulse test on the self-climbing device and the equipotential electrician. (3) After the impulse test is qualified, the equipotential electrician carries the insulated transmission rope to climb vertically.	5	(1) Deduct 2 points for failure to check the self-climbing device. (2) Deduct 3 points for not conducting impulse test; deduct 2 points for improper impulse test method. (3) Deduct 5 points for any damage to the self-climbing device caused by improper impulse test method.			
9	Entering the intense electric field and replace the spacer	(1) Equipotential electrician rides the self-climbing device to rise vertically to a distance of 0.3m from the conductor, check that all connections of the shielding clothes are correct, and transfer the potential with the consent of the person in charge of the working group; (2) Equipotential electrician shall fasten safety belt protection on the conductor and arrange insulated transmission rope at appropriate position;	25	(1) Deduct 3 points for failure of the equipotential electrician to apply to the Responsible Person before climbing; deduct 2 points for starting to climb the ladder without consent after application. (2) Deduct 2 points for failure to effectively control the backup protection rope.			

Part I Standard for Training and Assessment on Live Working for Operation Maintenance of ±800kV UHV Power Transmission Line

Table (Cont'd)

S/N	Project name	Quality requirements	Score	Deduction standard	Reasons for deduction	Deduction	Scoring
9	Entering the intense electric field and replace the spacer	(3) The equipotential electrician shall dismantle the old spacer, bind the old spacer with insulated transmission rope; the ground electrician shall transfer the old spacer to the ground with insulated transmission rope, and transfer the new spacer to the equipotential electrician at the same time; (4) The equipotential electrician shall install the new spacer at the original position of the spacer.		(3) Deduct 2 points for starting the potential transfer without applying to the Responsible Person; deduct 1 point for starting the potential transfer without the consent after applying. (4) Deduct 1 point for the inappropriate position applied for the potential transfer. (5) Deduct 1 point for being unskilled in potential transfer. (6) Deduct 2 points for the Responsible Person failing to monitor the operation carefully. (7) Deduct 1 point/time for any bumping occurred during the transmission of spacers. (8) Deduct 2 points for unsteady rising/declining of self-climbing device.			
10	Exit from electric field	(1) The equipotential electrician shall check the spacer restoration and pin placement again, report to the Responsible Person after confirming it is correct, and apply to exit from the electric field;		(1) Deduct 1 point for failure to report the end of the work. (2) Deduct 2 points for failure to apply to the Responsible Person for exiting from the electric field. (3) Deduct 1 point for not checking the connection of the shielding clothes.			

· 545 ·

Table (Cont'd)

S/N	Project name	Quality requirements	Score	Deduction standard	Reasons for deduction	Deduction	Scoring
10	Exit from electric field	(2) With the consent of the Responsible Person, the equipotential electrician shall check the connection of the shielding clothes and confirm that the connection is reliable, then exit the electric field following the reverse step of entering the electric field, and take the self-climbing device down to the ground.	12	(4) Deduct 2 points for the potential transfer without applying to the Responsible Person; deduct 1 point for starting the work without the consent after applying. (5) Deduct 1 point for the inappropriate position applied for the potential transfer (6) Deduct 2 points for the incorrect action of equipotential electrician exiting the electric field and repeat discharging. (7) Deduct 1 point for failure to effectively control backup protection rope			
11	End of the work	(1) The Responsible Person shall organize all working personnel to put working apparatus and materials in order and put them in a special kit (bag) after cleaning; clean the site to ensure that "the materials are removed and the site is cleaned after construction".	10	(1) Deduct 2 points for failure to clean the tools. (2) Deduct 2 points for missing tools. (3) Deduct 3 points for failure to hold the post-shift meeting. (4) Deduct 1 point for unqualified comments.			

Part I Standard for Training and Assessment on Live Working for Operation Maintenance of ±800kV UHV Power Transmission Line

Table (Cont'd)

S/N	Project name	Quality requirements	Score	Deduction standard	Reasons for deduction	Deduction	Scoring
11	End of the work	(2) Hold a post-shift meeting, and the Responsible Person shall make work summaries and comments on the construction quality of this work and the implementation of safety measures from all working personnel. (3) Responsible Person shall report to the on-duty control personnel the end of the work, apply for the restoration of DC restarting device of the line and terminate the work order.		(5) Deduct 3 points for failure to contact the control personnel for end of the work reporting. (6) Deduct 0.5 points for each missing item from the reporting content. (Name of working unit, name of the Responsible Person, name of the line, work completion; device has restored to normal, and the personnel have evacuated). (7) Deduct 1 point for error in the fill-in of work order termination procedure			
	Total		100				

Part II

Standard for Training and Assessment on Interruption Maintenance for Operation Maintenance of ±800kV UHV Power Transmission Line

Module 1   Standard for Training and Assessment on Power-cut Replacement of Double-V Porcelain Insulator for ±800kV UHV Power Transmission Line Tangent Tower

I. Training Standard

(I) Training Requirements

Designation of module	Power-cut Replacement of Double-V Porcelain Insulator for ±800kV UHV Power Transmission Line Tangent Tower	Type of training	Operation
Training method	Practical operation training	Hours of training	21 hours
Training objectives	1. Master the operational versions and stress structures of all kinds of tools and instruments, machines and tools, as well as the technical key points of replacing the whole string insulators. 2. Acquire proficiency in the operation procedures, technical method and construction hazard of power-cut replacement of double-V porcelain insulator for ±800kV UHV power transmission line tangent tower. 3. The main operators shall be able to proficiently complete the replacement of double-V porcelain insulator on ±800kV UHV transmission line tangent tower.		
Training venue	UHV ±800kV DC practical training line		
Training content	Correctly use the operation methods of all kinds of stressed tools and instruments as well as install all kinds of tools and instruments, and use the method of power-cut replacement of double-V porcelain insulator for ±800kV transmission line tangent tower		
Scope of application	Maintenance personnel for UHV DC power transmission line		

(II) Referenced Procedures and Specifications

(1) Operating Code for Overhead Transmission Line (DL/T741-2010)

(2) Technical Code for Designing Overhead 110~500kV Transmission Line (DL/T5092-1999);

(3) State Grid Corporation of China Working Regulations of Power Safety (Power Line Section) (Q/GDW1799.2-2013)

(4) Code for Designing of ±800kV DC Overhead Transmission Line(GB50790-2013)

(5) Maintenance Specification for ± 800kV DC Overhead Transmission Line (DL/T251-2012)

(6) Operating Code for ±800kV DC Overhead Transmission Line (GB/T28813-2012)

(7) Maintenance Specification for 110(66)kV~500kV Overhead Transmission Line (State Grid Corporation of China)

(8) Guideline of Maintenance of Overhead Transmission Line State (DLT1248-2013)

(9) Maintenance Management Regulations of Electric Transmission and Transformation Equipment State (State Grid Corporation of China)

(10) Maintenance and Test Specification of Electric Transmission and Transformation Equipment State (State Grid Corporation of China)

(III) Teaching Design for Training

To complete the work task of "power-cut replacement of double-V porcelain insulator for ±800kV UHV transmission line tangent tower", each training stage shall be designed according to the standard operation procedure for work task completion. Each stage includes specific training objectives, training content, hours of training, training methods (training resources), training environment, assessment and evaluation, etc, as shown in the Table 2-1-1.

Table 2-1-1 Training Content Design for Power-cut Replacement of Double-V Porcelain Insulator for ±800kV UHV Transmission Line Tangent Tower

Training schedule	Training objectives	Training contents	Hours of training	Training methods and resources	Preparation of training conditions	Assessment and evaluation
1. Theoretical teaching	1. Master the operational versions and stress structures of all kinds of tools and instruments, machines and tools, as well as the technical key points of replacing the porcelain insulators. 2. Acquire proficiency in the operation procedures, technical method and construction hazard of replacement of double-V porcelain insulator for ±800kV power transmission line tangent tower.	1. Correctly use all kinds of stressed tools and equipment, and be familiar with the operation methods of winching machines and tools. 2. Correctly install all kinds of tools and instruments. 3. Adopt the power-cut operation method for the replacement of double-V porcelain insulator for ±800kV UHV power transmission line tangent tower.	2	Training methods: Lecture. Training resources: PPT, relevant regulations and specifications.	Multimedia classroom	Attendance, classroom questions and assignments
2. Preparations	Be able to complete the preparation before operation	1. Work site survey. 2. Preparation of the standardized operation card. 3. Filling of the work order. 4. Preparation of tools and materials for this operation	1	Training methods: 1. Site survey and cleaning of tools and materials shall be practiced at site. 2. Preparation of operation card and the filling of work order shall adopt lecture method. Training resources: 1. ±800kV practical training line. 2. UHV tools warehouse. 3. Blank work order	1. UHV transmission line for practical training 2. Multimedia classroom	

· 552 ·

Part II Standard for Training and Assessment on Interruption Maintenance for Operation Maintenance of ±800kV UHV Power Transmission Line

Table (Cont'd)

Training schedule	Training objectives	Training contents	Hours of training	Training methods and resources	Preparation of training conditions	Assessment and evaluation
3. Work site preparation	Be able to complete the preparations of work site	1. Site re-survey. 2. Work application 3. Work site layout. 4. Pre-shift meeting. 5. Inspection of tools, instruments and materials.	1	Training methods: demonstration and role play. Resources; ±800kV practical training line	±800kV practical training line	
4. Trainer's demonstration	The trainees can preliminarily understand the operation process of the task through inspecting and learning from each other's work,	1. Explain the usage of all kinds of tools and instruments. 2. Demonstrate the connection mode of tools and instruments on the tower for replacing the porcelain insulator. 3. The operators working at heights shall cooperate to demonstrate the operation procedures of replacing the porcelain insulator. 4. Replace the double-V porcelain insulator for ±800kV UHV power transmission line tangent tower under the cooperation of ground operators	2	Training methods: Demonstration method. Resources; ±800kV practical training line	±800kV practical training line	

· 553 ·

Table (Cont'd)

Training schedule	Training objectives	Training contents	Hours of training	Training methods and resources	Preparation of training conditions	Assessment and evaluation
5. Group training of trainees	1. Be able to master the usage methods and precautions of all kinds of stressed tools and instruments. 2. Master the whole operation procedures of replacing the porcelain insulator. 3. Be able to complete the replacement of double-V porcelain insulator for ±800kV transmission line tangent tower	1. The trainees will be divided into groups (4 persons for working at high altitude and 7 persons for ground co-ordination) to train the operation methods of tools, instruments and machines, and the actual operation of replacing the double-V porcelain insulator on site. 2. Trainers guide the operation of trainees and conduct safety supervision	14	Training methods: Role play. Resources: ±800kV practical training line	±800kV practical training line	Score the operation of trainees according to the detailed rules for skill assessment and scoring
6. End of the work	1. Enable the trainees to further distinguish the shortcomings of the operation process and facilitate the promotion in the later stage. 2. Train the trainees in the working style of safe and civilized production	1. Clean up the work site. 2. Report to dispatcher. 3. Comment and summarize the work task this time at post-shift meeting.	1	Training methods: Lecture and inductive method	±800kV practical training line	

(IV) Work Flow

1. Work Task

Complete the power-cut replacement of double-V porcelain insulator for ±800kV UHV power transmission line tangent tower.

2. Requirements for Weather and Work Site

(1) The power-cut replacement of double-V porcelain insulator for ±800kV UHV power transmission line tangent tower shall be carried out in good weather.

High-place operation in the open air shall be stopped in severe weather, such as heavy wind of 5 degree ad above, rainstorm, thunder, hail, heavy fog and sand storm. In special cases, if emergency maintenance needs to be performed in severe weather, arrange related personnel to fully discuss necessary safety measures, which shall be taken after approved by the company.

(2) The operators are in good mental state, and the members of the working group carefully study the work ticket and safety technical measures, and all the personnel shall be conform to the "four clear" (the operation task is clear, the danger point is clear, the operation procedure is clear, and the safety measures are clear).

(3) The Responsible Person must contact with the power cutting contact person to fulfill the Work licensing procedures, and it is strictly prohibited to stop and transfer electricity at the appointed time. Responsible Person must obtain the license work order of the licensor before checking the electricity on the line to be overhauled, setting up the ground wire and carrying out the maintenance work.

(4) After power-cut, Responsible Person shall make records carefully.

(5) Check whether there is a honeycomb on the tower before climbing the pole. It is forbidden to climb the tower if a honeycomb is found.

(6) Double safety belt shall be used by the personnel who work on the tower, and safety goggles must be worn.

3. Preparations

3.1 Hazards and precontrol measures

(1) Hazard——climbing charged line

Precontrol measures:

① Before climbing the tower for operation, Responsible Person and the members of

the working team shall check the double name and identification mark carefully (color code, distinguishing mark, etc.) to match the name of the power-cut line.

② Check the tower root and the base of pole and tower before climbing, making sure it's solid and reliable.

③ Before climbing the pole and tower, climbing tools and facilities such as safety belt, shackles and tower materials shall be inspected to ensure they are firm.

④ Before climbing the pole, the intermediate operators who do not involved the installation of the grounding wire shall carefully verify that the phase sequence, color code, name, and number of the line are consistent with the power-cut line, and confirm that the line name is not error or in opposite order.

(2) Hazard——operation in violation of safety operation procedures when climbing the tower and working on the tower may cause falling from height.

Precontrol measures:

① During climbing, in order to prevent the operator from falling, the distance between the staff members shall not be less than 1.6m.

② Before climbing, the soil on the soles shall be removed, and the kit shall be complete, and objects shall not be dropped during climbing.

③ Operators shall wear safety helmet and soft sole, and be scrupulous and when climbing pole and tower.

④ During climbing, the safety belt shall be properly packed, the long tail rope shall be placed in the kit, and the main belt shall be hung on the shoulder to prevent safety during climbing to prevent the safety belt from hooking shackles and tower materials during climbing, resulting in falling from the tower.

⑤ Operators shall not lose the protection from the safety belt and step and hold it firmly when moving on the pole and tower.

⑥ When reaching the position of operation point, fasten the safety belt (rope), it shall be firm and reliable, and shall not be hung low for high use.

⑦ The safe distance between the human body, rope, etc. and the conductor must not be less than 10.1m before the power test, and special person shall be assigned to supervise the work.

(3) Hazard——injury caused by objects falling from high place

Precontrol measures:

① The ground staff shall not stand under the position vertical to the operation point. The personnel on tower shall prevent falling objects from injury, and the tools and materials used shall be transmitted by ropes.

② Tool bags should be used when working at heights. Larger tools should be fixed on solid components and are not allowed to be randomly placed.

③ In the process of lifting with a winching, special personnel shall be assigned to command and cooperate in a unified way. Impact inspection shall be carried out immediately after the porcelain insulator string are lifted off the ground.

(4) Hazard——prevent injury caused by induced electricity

① Operators on the tower should wear a full set of shielding clothes, in order to prevent injury caused by induced electricity.

② If contact with overhead ground wire is required, reliable grounding shall be conducted before contact with overhead ground wire.

(5) Hazard——site safety monitoring

① From the beginning of the operation to the end of the operation, the safety supervisor must always carry out uninterrupted safety supervision on the operators on the site.

② The responsible person and the Supervisor must wear a safety vest.

(6) Hazard——traffic safety

① It is necessary to pay attention to the driving safety of the vehicle, drive carefully when driving, and illegal driving is not allowed.

3.2 Selection of tools, instruments and materials

See Table 2-1-2 for Tools, Instruments and Materials Required for Power-cut Replacement of Double-V Porcelain Insulator for ±800kV UHV Power Transmission Line Tangent Tower. Before delivering the tools and instruments out of the warehouse, application voltage class and test period of the tools and instruments shall be carefully checked and they shall be inspected to ensure that appearance is intact, connection is firm, rotation is flexible and meeting the requirements of the work task. After delivering the tools and instruments out of the warehouse, they shall be kept from contamination and damp. Metal tools and insulated tools shall be separately loaded and transported to avoid deformation and damage caused by mixed loading and transportation.

Table 2-1-2　Tools and Materials Required for Power-cut Replacement of Double-V Porcelain Insulator for ±800kV UHV Transmission Line Tangent Tower

S/N	Name	Specification and Model	Unit	Qty.	Remarks
1	Grounding wire	±800kV	Group	2	Insulating tool
2	Insulating gloves	10kV	Set	2	Insulating tool
3	Electricity tester	For ±800kV only	pcs	2	Insulating tool
4	Full-covered safety belt	Including 20m rear safety rope with buffer bag	Set	3	PPE
5	Winching	5T	Set	2	Mechanical tools
6	Six-hook clamp	Suitable for six-bundle conductor	Set	2	Metal tool
7	Shackle	10T	Nos.	20	Metal tool
8	Hand-operated hoist	6T	Set	2	Metal tool
9	Hand-operated hoist	12T	Set	2	Metal tool
10	Individual hand-operated tool		Set	4	Metal tool
11	Personal security wire	(diameter not less than 16mm^2)	Nos.	1	Other tools
12	Steel wire sleeve	$\phi 22$	Nos.	6	Other tools
13	Grinding rope	$\phi 6$	m	2 loops 200m per loop	Other tools
14	Interphone		Set	5	Other tools
15	Lifting rope tackle	1T	Nos.	2	Other tools
16	Transmission rope	$\phi 16$	Set	3	Other tools
17	Pin puller		pcs	2	Other tools
18	Safety vest		pcs	2	Other tools
19	Security fence		Volume	4	Other tools
20	Sole timber		pcs	Several	Other tools
21	Moisture-proof tarpaulin		Sheet	1	Other tools
22	Porcelain insulator		String	1	Material

Part II Standard for Training and Assessment on Interruption Maintenance for Operation Maintenance of ±800kV UHV Power Transmission Line

3.3 Division of labor for operators

Division of labor for operators of the task is shown in Table 2-1-3.

Table 2-1-3 Division of Labor for Power-cut Replacement of Double-V Porcelain Insulator for ± 800kV UHV Transmission Line Tangent Tower

S/N	Post	Qty. (person)	Responsibilities
1	Responsible Person		Be responsible for division of operating personnel of the task, field investigation before work, implementation of the operation plan, work order filling, field re-investigation, performing work permit procedures, holding pre-shift meeting, the implementation of safety measures on site, safety supervision in the operation process, dealing with emergency situations in work, quality surveillance, and the summary after work.
2	Safety Supervisor	2	Be responsible for the safety monitoring work during this operation
3	Operator working at heights	4	Be responsible for the operation of power-cut replacement of double-V porcelain insulator for ±800kV line
4	Ground auxiliary personnel	7	Be responsible for ground auxiliary works during the operation, including 2 winching operators

4. Working Procedures

Working process of the task is shown in Table 2-1-4.

Table 2-1-4 Workflow for Power-cut Replacement of Double-V Porcelain Insulator for ±800kV UHV Transmission Line Tangent Tower

S/N	Work Content	Operation Standard	Safety Precautions	Responsible Person
1	Site re-survey	The Responsible Person shall complete the following work: (1) Check the line name on spot to ensure it is correct; guarantee that the foundation and the iron tower are intact and in normal condition, and ensure that the cross and span distance meets the safety requirements; confirm the defect conditions and the specifications and models of conductor or ground wires. (2) Check that the terrain and environment should meet the operation requirements. (3) Check that the safety measures listed in the work order are in line with the actual situations on site. The measures will be added if necessary.	(1) Correctly wear helmet, working clothes, work shoes and protective gloves. (2) Operation under meteorological conditions that may endanger the safety of operators is forbidden. (3) Non-operation personnel and vehicles are strictly prohibited from entering the working site	

Table (Cont'd)

S/N	Work Content	Operation Standard	Safety Precautions	Responsible Person
2	Work Permit	The contact person for power off/on before the operation must contact the dispatcher to complete the work permit procedure	(1) No work shall be started without the Work Permit of approver. (2) Power transmission and cutoff at pre-determined time is not permitted.	
3	Site layout	Install the security fence and hang the signboards correctly: (1) The security fence should take full account of falling objects from the heights and the influence on road traffic. (2) The entrance and exit of the security fence shall be set reasonably. (3) Signs such as "Access from Here", "Work Here", "Slow Down" or "Blocking" shall be properly arranged	When the influence on road traffic safety is uncontrollable, the traffic management department should be contacted in time to strengthen the on-site control of traffic safety	
4	Holding a pre-shift meeting	(1) All working personnel shall line up. (2) The Responsible Person will read out the work order and be clear with the work task and division of personnel; explain safety measures and technical measures in work; check (inquire after) the mental state of all working personnel; and inform of hazards in work and precontrol measures. (3) All working personnel shall sign on the work order for confirmation.	(1) Work order shall be filled and issued with standardized licensing procedure and complete signature. (2) All working personnel shall be in good mental states. (3) All working personnel shall be clear with task division of works, safety measures and technical measures	
5	Inspection of tools and instruments	(1) All necessary tools and instruments shall be prepared as per the operation requirements and placed on the waterproof tarpaulin regularly according to the category and location. The appearance and test certificate of tools and instruments shall be checked to ensure there is no omission. (2) The inspector shall report to the Responsible Person that all inspection results are in conformity with the operation requirements	(1) The waterproof tarpaulin shall be enough in quantity and reasonable in position, and be clean and dry. (2) The appearance of tools and instruments is acceptable after inspection, there is no damage, damp, deformation and malfunction, and the certificate is valid	

Part II Standard for Training and Assessment on Interruption Maintenance for Operation Maintenance of ±800kV UHV Power Transmission Line

Table (Cont'd)

S/N	Work Content	Operation Standard	Safety Precautions	Responsible Person
6	Climbing the tower	(1) Check the double name and serial number of the line before climbing the pole and tower. For multi-circuit lines on the same tower, the Responsible Person and the members of the working team shall check the double name and identification mark carefully (color code, distinguishing mark, etc.). (2) Check the tower root and the base of pole and tower before climbing, making sure it's solid and reliable. (3) During climbing, in order to prevent the operator from falling, the distance between the staff members shall not be less than 1.6m, the safety belt shall be properly packed, the long tail rope shall be placed in the kit, and the main belt shall be hung on the shoulder, to prevent safety during climbing to prevent the safety belt from hooking shackles and tower materials during climbing, resulting in falling from the tower. (4) When the pole and tower is close to the cross arm, the Supervisor and the operator shall check the identification mark and dual tags of the power-cut line again, and can climb to the operation point only after they are confirmed to be correct	(1) Operator shall wear safety helmet and soft sole, and be scrupulous and when climbing pole and tower. (2) During climbing, the safety belt shall be properly packed, the long tail rope shall be placed in the kit, and the main belt shall be hung on the shoulder to prevent safety during climbing to prevent the safety belt from hooking shackles and tower materials during climbing, resulting in falling from the tower. (3) Operators shall not lose the protection from the safety belt and step and hold it firmly when moving on the pole and tower. (4) When reaching the position of operation point, fasten the safety belt (rope), it shall be firm and reliable, and shall not be hung low for high use. (5) The safe distance between the human body, headless rope, etc. and the conductor must not be less than 10.1m before the power test, and special person shall be assigned to supervise the work.	

Table (Cont'd)

S/N	Work Content	Operation Standard	Safety Precautions	Responsible Person
7	Inspection power and install ground wire	After the test pole is in place, fasten the safety belt to a firm and reliable component or pole, and check whether the buckle is in place correctly. Appliances such as electroscope must be transferred by transmission rope. (2) Check whether the ground wire is in good condition, and install the ground wire according to the procedures (first connect the ground terminal, then the conductor terminal). (3) When setting the ground wire, the insulating rope or insulating handle must be used for operation, and it is forbidden to install the ground wire with directly operating the metal part of the ground wire by hand. Make sure the collection of the ground wire is tightly connected to the conductor	(1) Check whether the electroscope is normal during receiving and before using. (2) It is forbidden to install the ground wire by winding the conductor	
8	for replacement of double-V porcelain insulator	After the operators working at height arriving at the operation point, the 2 operators working at height access the conductor through the porcelain insulator, and the operators on the tower use the transmission rope to transfer two sets of six-hook clamp and two sets of 12T hand hoist to the tower, fix them on both sides of the porcelain insulator and armor clamps perpendicular to the line direction, erect them firmly, and reserve sufficient space for personnel operation. After confirming that all connecting parts have been fixed, install 1 set of 12T lever hoist and insulator backup protection rope at both ends of the insulator, and 2 operators working at height tighten the two sets of 12T lever hoist at the same time. After the 12T lever hoists fully bear the pulling force of the porcelain insulator string, check again for any deformation of the cross arm and any abnormality of the connecting parts. After confirming that everything is normal, use the transmission rope to transfer the winching rope to the grounding end of the porcelain in	(1) In the process of replacing insulator string with lever hoist and six-hook clamp, it is necessary to check the connection and stress of lever hoist, wire rope noose (lifting belt) and shackle after the lever hoist begins to bear the conductor load. And the impulse test shall be conducted to confirm that it is completely reliable before continuing to tighten the lever hoist.	

Table (Cont'd)

S/N	Work Content	Operation Standard	Safety Precautions	Responsible Person
8	for replacement of double-V porcelain insulator	sulator and tie it firmly. Then, two sets of 6T lever hoists are used to tighten the conductor end insulator, and the winching rope is tightly bound at the conductor end of the porcelain insulator. The operator working at height at the conductor end removes the connecting part of the porcelain insulator and the insulator set fitting at the conductor end. The ground operator winches and tightens the winching rope at the grounding end. The operator on the iron tower removes the connecting part of the porcelain insulator and the insulator set fitting at the iron tower end. After all the connecting parts of the porcelain insulator have been removed, the porcelain insulator can be slowly lowered by two winches at the same time, and the ground operator can remove the porcelain insulator, replace with the intact porcelain insulator, and bind the porcelain insulator firmly in place before transferring it to the operators on the tower. After the porcelain insulator is lifted to the operation point, the operators on tower shall first connect the connection parts of the insulator set fitting at the conductor end and ensure that the pin is in place before continuing the connection of the porcelain insulator grounding end. Check whether the insulator set fitting and the pin are installed in place after the porcelain insulator string is installed, and the impulse test shall be conducted to confirm that the installation is correct and the connection is reliable before slowly releasing the insulator backup protection rope, then releasing the 2 sets of 12T lever hoist	(2) To avoid damaging the porcelain insulator, it shall be prevented from colliding with the porcelain insulator when using the lever hoist. (3) When using the winching rope to hoist the porcelain insulator, the ground staff and the staff on the tower must cooperate closely to prevent the winching rope from winding or the insulator string damaging the conductor due to colliding. When hanging the insulator string, the ground operators shall not stand under the vertical.	

Table (Cont'd)

S/N	Work Content	Operation Standard	Safety Precautions	Responsible Person
9	Removal of tools and instruments	Remove the six-hook clamp, lever hoist, steel wire sleeves and other tools and instruments	The upper and lower transferred tools shall not collide with each other. The binding rope shall be correct and reliable, so as to prevent objects falling from heights.	
10	End of the work	(1) Responsible person shall organize all working personnel to put working apparatus and materials in order, and clean the site to ensure "materials are removed and the site is cleaned after construction". (2) After a post-shift meeting is held, the Responsible Person shall make work summaries and comments. Comments include the construction quality of this work and the implementation of safety measures from all working personnel. (3) Responsible Person shall report the completion of the work to the work approver, resume the power transmission of the power cut line and terminate the work order		

II. Assessment Standard

Part II Standard for Training and Assessment on Interruption Maintenance for Operation Maintenance of ±800kV UHV Power Transmission Line

Table 2-1-5 Detailed Rules for Assessment and Scoring of Operation and Inspection Skills of UHV DC Transmission Line of State Grid Sichuan Electric Power Corporation

Fill-in Column of Examinee	No.:	Name:	Position:	Unit:	Date:	MM/DD/YYYY
Fill-in Column of Assessor	Grade:	Assessor:	Assessment Team Leader:	Starting time:	Closing time:	Operation Duration:
Assessment Module	Power-cut replacement of double-V porcelain insulator for ±800kV UHV DC transmission line. Assessment method: Operation. Assessment time limit: 150min					
Job description	Power-cut replacement of Grade I double-V porcelain insulator for ±800kV DC transmission line					
Work specification and requirements	1. Given conditions: C-phase V-shape porcelain insulator on ±800kV practical training line is broken, which needs to be replaced. The line is power off, test electricity and install the ground wire, the insulators to be used have been tested, the work order has been handled, and the safety measures have been taken. 2. The main operation procedures of the whole process are completed by cooperation of 1 responsible person, 2 special supervisors, 4 high-altitude operators on the tower while 7 auxiliary workers on the ground assist the exam participants to finish the up and down transfer of tools, materials and other non-technical work. 3. The exam participant should make necessary safety check before operation. 4. An oral application shall be made at the beginning of work, and an oral report also shall be made at the end of the work					

Table (Cont'd)

Assessment scenario preparation	1. Tools and instruments: 2 sets of six-hook clamp, 1 personal security wire, 6 sets of ϕ22 steel wire sleeves, 20 sets of 10T shackle, 2 sets of 12T manual hoist, 1 set of 6T lever hoist, 3 sets of transmission rope, 2 sets of 1T tackle, 2 sets of noose, 5 sets of interphone, 4 sets of double safety belts and 1 piece of moisture-proof tarpaulin. 2. Material: a string of porcelain insulator, which is with the same model. 3. Operated on the training line
Remarks	1. Personal tools shall be self-provided by the exam participant. 2. The deduction shall be done until the scores of each item are deducted completely.

Table 2-1-6 Standards for Assessment and Scoring of Operation and Inspection Skills of UHVDC Transmission Line of State Grid Sichuan Electric Power Corporation

S/N	Project name	Quality requirements	Score	Deduction standard	Reasons for deduction	Deduction	Scoring
1	Preparation for tools and materials						
1.1	Inspection of personal tools	Adjustable wrench, flat pliers, pin-pulling pliers and tool kits shall meet the quality requirements	2	Deduct 1 point for each error and missing item			
1.2	Inspection of stressed tools	The six-hook clamp, lever hoist and wire rope shall be within the test qualified period	3	Deduct 1 point for each error and missing item			
1.3	Inspection of safety tools and instruments	Double safety belts, personal safety wires and insulated gloves shall meet the quality requirements and shall be within the test qualified period	3	Deduct 1 point for each error and missing item			

Part II Standard for Training and Assessment on Interruption Maintenance for Operation Maintenance of ±800kV UHV Power Transmission Line

Table (Cont'd)

S/N	Project name	Quality requirements	Score	Deduction standard	Reasons for deduction	Deduction	Scoring
1.4	Material inspection	Check the type of porcelain insulator string, and the appearance inspection shall meet the requirements	2	Deduct 1 point for each error and missing item			
2	Venue layout						
2.1	Site fence	Layout of site fence	2	Deduct 2 points for failure to ar- range			
3	Climbing the tower and operation on cross arm						
3.1	Climbing the tower	(1) Check whether there is any abnormality. (2) Correctly carry the transmission rope (The end of the lifting rope shall be double folded, fast knot, and hung over the shoulder) (3) Correctly climb down the tower along the main material of the shackles side	6	(1) Deduct 1 point for each un- checked item. (2) Deduct 2 points for not car- rying the transmission rope, and deduct 1 points for nonstan- dard carrying method of lifting rope. (3) Deduct 1 point/time for grasping the shackles with hands. (4) Deduct 2 points for not climbing up the tower along the main material of the shackles side			

Table (Cont'd)

S/N	Project name	Quality requirements	Score	Deduction standard	Reasons for deduction	Deduction	Scoring
3.2	Access the working point on the cross arm	The safety belt protection shall not be lost from the tower body to the working point on the cross arm	3	Deduct 3 points for failure to use safety belt properly			
3.3	Install tackle and personal security wire	The transfer tackle shall be installed at the correct position for easy operation and is equipped with personal security wire.	3	Deduct 1 point for non-standard pulley installation; deducted 2 points for not hanging the personal security wire			
4	Operation on insulator string						
4.1	Access the work site	(1) Fasten the safety rope of the double safety belt to the cross arm in the proper position. (2) Access the operation point along the insulator string, and tie the pole belt to the insulator string. (3) Check the insulator string fitting pin and insulator set fitting	6	(1) Deduct 4 points for failure to use double safety belt. (2) Deduct 2 points/time for failure to use double safety belt properly. (3) Deduct 2 points for failure to check			

Part II Standard for Training and Assessment on Interruption Maintenance for Operation Maintenance of ±800kV UHV Power Transmission Line

Table (Cont'd)

S/N	Project name	Quality requirements	Score	Deduction standard	Reasons for deduction	Deduction	Scoring
4.2	Installer Appliance	(1) Correctly install steel wire rope sleeve, 6T lever hoist and 12T lever hoist. (2) Correctly install the six-hook clamp. (3) The selected models of noose of wire rope and shackle are correct and the installation is reliable. (4) The installed lever hoist and personal security wire have no effect on the operation of the personnel	9	(1) Deduct 1 point for failure to protect the tower material at each place. (2) Deduct 1 point for every incorrect installation of lever hoist, six-hook clamp. (3) Deduct 1 point for every one collision or winding after the lever hoist is stressed. (4) Deduct 1 point for every incorrect model selection of noose of wire rope and shackle. (5) Deduct 4 points for every leaving out of the lever hoist			
4.3	Tighten the porcelain insulator string	(1) At the same time, tighten the 2 sets of 12T lever hoist and ensure that the force on the lever hoist is balanced. (2) After the lever hoist is stressed, check the tackle, shackle, wire rope and other connection parts to confirm the connection is reliable, and impact test shall be taken for the insulator string. (3) After confirming that the force is correct, continue to tighten the lever hoist until the porcelain insulator string is loose.	6	(1) Deduct 1 point/time for touching the insulator. (2) Deduct 3 points for failure to carry out the impulse test. (3) Deduct 2 points for incorrect use of the lever hoist			

· 569 ·

S/N	Project name	Quality requirements	Score	Deduction standard	Reasons for deduction	Deduction	Scoring
4.4	Replacement of double-V porcelain insulator for tangent tower	(1) Tie the transmission rope at the appropriate position of the porcelain insulator. The operator working at height shall first take two sets of 6T lever hoist and then remove the conductor end of the porcelain insulator, and tie the conductor end winching rope firmly at the conductor end of the porcelain insulator. The ground winching operator shall remove the grounding end of the porcelain insulator after tightening the grounding end winching rope. (2) Two ground winching operators operate simultaneously to slowly transfer the porcelain insulators to the ground. (3) The new porcelain insulator is transferred to the tower through the winching rope, and the operator on the tower shall connect the conductor end first and install the pin in place. (4) Continue to install the porcelain insulator grounding end, install the armor clamp R pin, and check whether it is installed in place. (5) After confirming that the connection is correct, slowly release the lever hoist to make the porcelain insulator bear the stress, and carry out the impulse test	16	(1) Deduct 2 points for failure to conduct impulse test. (2) Deduct 2 points for rope winding. (3) Deduct 2 points/time for bump of transmission objects. (4) Deduct 5 points for dropping object. (5) Deduct 1 point for every uninstalled fitting pin. (6) Deduct 4 points for wrong sequence of mounting and dismounting the conductor end and the iron tower end			

Part II Standard for Training and Assessment on Interruption Maintenance for Operation Maintenance of ±800kV UHV Power Transmission Line

S/N	Project name	Quality requirements	Score	Deduction standard	Reasons for deduction	Deduction	Scoring
4.5	Remove tools and instruments	(1) Check whether the fitting pin and ball head are complete and in place, and the orientation of the socket is correct; clean the dirt on insulator string surface. (2) Remove the lever hoist, six-hook clamp and noose of wire rope, and transfer them to the ground	7	(1) Deduct 2 points for rope winding. (2) Deduct 2 points/time for bump of transmission objects. (3) Deduct 1 point for incorrect position of fitting pin. (4) Deduct 1 point for incorrect socket orientation. (5) Deduct 1 point for no cleaning of insulator string			
4.6	Clean the tools and instruments on the tower	Confirm that there are no objects left behind	4	Deduct 4 points for any object left behind			
4.7	From conductor to tower body	(1) Access the cross arm of the iron tower along the porcelain insulator. (2) Safety belt protection shall not be lost during climbing the rope ladder	8	(1) Deduct 4 points for loosing the protection of the safety belt (2) Deduct 4 points for incorrect use of rope ladder			
5	Climbing down the tower	(1) Correctly climb down the tower along the main material of the shackles side (2) Correctly carry the transmission rope (The end of the lifting rope shall be double folded, fast knot, and hung over the shoulder)	8	(1) Deduct 4 points for not carrying transmission rope, and deduct 2 points for nonstandard carrying method of transmission rope; (2) Deduct 1 point/time for grasping the shackles with hands; (3) Deduct 4 points for not climbing the tower along the main material of the shackles side			

Table (Cont'd)

S/N	Project name	Quality requirements	Score	Deduction standard	Reasons for deduction	Deduction	Scoring
6	Other requirements						
6.1	Tower operation	(1) It is prohibited to drop articles down from high space. (2) Cooperate with both hands during operation. (3) No floating objects. (4) No holding object in mouth	5	(1) Deduct 5 points for each falling object from high place. (2) Deduct 2 points for non-coordination action. (3) Deduct 2 points for floating articles. (4) Deduct 2 points for holding object in mouth			
6.2	Dress	Correctly wear working clothes, rubber work shoes, safety helmet and protective gloves	2	Deduct 2 points for each missing item			
6.3	Clean site	Clean up work site after completion and meet the requirements of civilized production	2	Deduct 2 points for failure to clean up work site			
6.4	Completion time	Complete operation as required within the specified time	3	Deduct 1 point if the time exceeds 10 minutes, terminate the operation when the time reaches 480 minutes, and only record the score of the completed part			
	Total		100				

Module 2 Standards for Training and Assessment on Power-cut Replacement of Whole Strain Insulators String on ±800kV UHV Transmission Line

I. Training Standard

(I) Training Requirements

Designation of module	Power-cut replacement of the whole strain insulator string on ±800kV UHV transmission line.	Type of training	Operation
Training method	Practical operation training	Hours of training	32 hours
Training objectives	1. Master the operational versions and stress structures of all kinds of tools and instruments, machines and tools, as well as the technical key points of replacing the whole string insulators. 2. Acquire proficiency in the operation procedures, technical method and construction work hazard of power-cut replacement of whole strain insulator string on ±800kV UHV transmission line. 3. The main operator working at height shall be able to proficiently complete the replacement of whole strain insulator string on ±800kV UHV transmission line.		
Training venue	UHV DC practical training line		
Training content	Correctly use all kinds of stressed tools and instruments; be familiar with the operation methods of tools and machines, such as the winching, etc; correctly install all kinds of tools and instruments by using the "pulley block" operation method; and replace whole strain insulator string on ±800kV UHV transmission line by using the power-cut operation method.		
Scope of application	Maintenance personnel for UHV DC power transmission line		

(II) Referenced Procedures and Specifications

(1) Operating Code for Overhead Transmission Line (DL/T741-2010)

(2) Technical Code for Designing Overhead 110~500kV Transmission Line (DL/T5092-1999);

(3) State Grid Corporation of China Working Regulations of Power Safety (Power Line Section) (Q/GDW1799.2-2013)

(4) Code for Designing of ±800kV DC Overhead Transmission Line (GB50790-2013)

(5) Maintenance Specification for ± 800kV DC Overhead Transmission Line (DL/T251-2012)

(6) Operating Code for ±800kV DC Overhead Transmission Line (GB/T28813-2012)

(7) Maintenance Specification for 110(66)kV~500kV Overhead Transmission Line (State Grid Corporation of China)

(8) Guideline of Maintenance of Overhead Transmission Line State (DLT1248-2013)

(9) Maintenance Management Regulations of Electric Transmission and Transformation Equipment State (State Grid Corporation of China)

(10) Maintenance and Test Specification of Electric Transmission and Transformation Equipment State (State Grid Corporation of China)

(III) Teaching Design for Training

To complete the power-cut replacement of whole strain insulator string on ±800kV UHV transmission line is the Work task during the design. Each training stage shall be designed according to the standard operation procedure for work task completion. Each stage includes specific training objectives, training content, hours of training, training methods (training resources), training environment, assessment and evaluation, etc, as shown in Table 2-2-1.

Part II Standard for Training and Assessment on Interruption Maintenance for Operation Maintenance of ±800kV UHV Power Transmission Line

Table 2-2-1 Training Content Design for Power-cut Replacement of Whole Strain Insulator String on ±800kV UHV Transmission Line

Training schedule	Training objectives	Training contents	Hours of training	Training methods and resources	Training conditions	Assessment and evaluation
1. Theoretical teaching	1. Master the operational versions and stress structures of all kinds of tools and instruments, machines and tools, as well as the technical key points of replacing the whole string insulators. 2. Acquire proficiency in the operation procedures, technical method and construction work hazard of replacement of whole strain insulator string on ±800kV transmission line.	1. Correctly use all kinds of stressed tools and equipment, and be familiar with the operation methods of winching machines and tools. 2. Correctly install all kinds of tools and instruments by using the "pulley block" operation method. 3. Correct use of winching to avoid transmission of insulator by jumper string. 4. Replace whole strain insulator string on ±800kV transmission line by using the power-cut operation method.	2	Training methods: Lecture. Training resources: PPT, relevant regulations and specifications.	Multimedia classroom	Attendance, classroom questions and assignments
2. Preparations	Be able to complete the preparations before operation	1. Work site survey. 2. Preparation of the standardized operation card. 3. Filling of the work order. 4. Preparation of tools and materials for this operation		Training methods: 1. Site survey and cleaning of tools and materials shall be practiced at site. 2. Preparation of operation card and the filling of work order shall adopt lecture method. Training resources: 1. ±800kV practical training line 2. UHV tools warehouse. 3. Blank work order.	1. UHV transmission line for practical training 2. Multimedia classroom	

Table (Cont'd)

Training schedule	Training objectives	Training contents	Hours of training	Training methods and resources	Training conditions	Assessment and evaluation
3. Work site preparation	Be able to complete the preparations of work site	1. Work site re-survey. 2. Work application. 3. Work site layout. 4. Pre-shift meeting. 5. Inspection of tools, instruments and materials.	3	Training methods: demonstration and role play. Resources; ±800kV practical training line.	±800kV practical training line	
4. Trainer's demonstration	The trainees can preliminarily understand the operation process of the task through inspecting and learning from each other's work.	1. Explain the usage of all kinds of tools and instruments. 2. Demonstrate the connection mode of tools and instruments on the tower for replacing the whole insulator string. 3. The operators working at heights shall cooperate to demonstrate the operation procedures of replacing the whole insulator string. 4. Replace the whole strain insulator string on ±800kV line by winching	7	Training method: Demonstration method. Resources: ±800kV practical training line.	±800kV practical training line.	

Part II Standard for Training and Assessment on Interruption Maintenance for Operation Maintenance of ±800kV UHV Power Transmission Line

Table (Cont'd)

Training schedule	Training objectives	Training contents	Hours of training	Training methods and resources	Training conditions	Assessment and evaluation
5. Group training of trainees	1. Be able to master the usage methods and precautions of various stressed tools and instruments, winching machines, etc. 2. Master the whole operation procedures of replacing the whole insulator string. 3. Be able to assist to complete the replacement of the whole strain insulator string on ±800kV transmission line.	1. The trainees will be divided into groups (A group is with 6 persons for working at high altitude and 9 persons for ground coordination) to train the operation methods of tools, instruments and machines, and the actual operation of replacing the whole insulator string on site. 2. Trainers guide the operation of trainees and conduct safety supervision.	14	Training methods: Role play. Resources: ±800kV practical training line	±800kV practical training line	Score the operation of trainees according to the detailed rules for skill assessment and scoring
6. End of the work	1. Enable the trainees to further distinguish the shortcomings of the operation process and facilitate the promotion in the later stage. 2. Train the trainees in the working style of safe and civilized production.	1. Clean the work site. 2. Report to dispatcher. 3. Comment and summarize the work task this time at post-shift meeting.		Training methods: Lecture and inductive method	±800kV practical training line	

· 577 ·

(IV) Work Flow

1. Work task

Complete the power-cut replacement of the whole strain insulator string on ±800kV UHV transmission line.

2. Requirements for Weather and Work Site

(1) Power-cut replacement of ± 800kV UHV transmission line conductor tension-resistance whole strain insulator string shall be carried out in good weather.

High-place operation in the open air shall be stopped in severe weather, such as heavy wind of 5 degree ad above, rainstorm, thunder, hail, heavy fog and sand storm. In special cases, if emergency maintenance needs to be performed in severe weather, arrange related personnel to fully discuss necessary safety measures, which shall be taken after approved by the company.

(2) The operators are in good mental state, and the members of the working group carefully study the work ticket and safety technical measures, and all the personnel shall be conform to the "four clear" (the operation task is clear, the danger point is clear, the operation procedure is clear, and the safety measures are clear).

(3) The Responsible Person must contact with the power cutting contact person to fulfill the Work licensing procedures, and it is strictly prohibited to stop and transfer electricity at the appointed time. Responsible Person must obtain the license work order of the licensor before checking the electricity on the line to be overhauled, setting up the ground wire and carrying out the maintenance work.

(4) After power-cut, Responsible Person shall make records carefully.

(5) Check whether there is a honeycomb on the tower before climbing the pole. It is forbidden to climb the tower if a honeycomb is found.

(6) Double safety belt shall be used by the personnel who work on the tower, and safety goggles must be worn.

3. Preparations

3.1 Hazards and precontrol measures

(1) Hazard——climbing live line

Precontrol measures:

① Before climbing the tower for operation, Responsible Person and the members of

the working team shall check the double name and identification mark carefully (color code, distinguishing mark, etc.) to match the name of the power-cut line.

② Check the tower root and the base of pole and tower before climbing, making sure it's solid and reliable.

③ Before climbing the pole and tower, climbing tools and facilities such as safety belt, shackles and tower materials shall be inspected to ensure they are firm.

④ Before climbing the pole, the intermediate operators who do not involved the installation of the grounding wire shall carefully verify that the phase sequence, color code, name, and number of the line are consistent with the power-cut line, and confirm that the line name is not error or in opposite order.

(2) Hazard——Operation in violation of safety operation procedures when climbing the tower and working on the tower may cause falling from height.

Precontrol measures:

① During climbing, in order to prevent the operator from falling, the distance between the staff members shall not be less than 1.6m.

② Before climbing, the soil on the soles shall be removed, and the kit shall be complete, and objects shall not be dropped during climbing.

③ Operators shall wear safety helmet and soft sole, and be scrupulous and when climbing pole and tower.

④ During climbing, the safety belt shall be properly packed, the long tail rope shall be placed in the kit, and the main belt shall be hung on the shoulder to prevent safety during climbing to prevent the safety belt from hooking shackles and tower materials during climbing, resulting in falling from the tower.

⑤ Operators shall not lose the protection from the safety belt and step and hold it firmly when moving on the pole and tower.

⑥ When reaching the position of operation point, fasten the safety belt (rope), it shall be firm and reliable, and shall not be hung low for high use.

⑦ The safe distance between the human body, headless rope, etc. and the conductor must not be less than 10.1m before the power test, and special person shall be assigned to supervise the work.

(3) Hazard——injury caused by objects falling from high place

Precontrol measures:

① The ground staff shall not stand under the position vertical to the operation point. The personnel on tower shall prevent falling objects from injury, and the tools and materials used shall be transmitted by ropes.

② Tool bags should be used when working at heights. Larger tools should be fixed on solid components and are not allowed to be randomly placed.

③ In the process of lifting with a winching, special personnel shall be assigned to command and cooperate in a unified way. Impact inspection shall be carried out immediately after the insulator string are lifted off the ground.

(4) Hazard——prevent injury caused by induced electricity

① Operators on the tower should wear a full set of shielding suits, in order to prevent injury caused by induced electricity.

② If contact with overhead ground wire is required, reliable grounding shall be conducted before contact with overhead ground wire.

(5) Hazard——Site safety monitoring

① From the beginning of the operation to the end of the operation, the safety supervisor must always carry out uninterrupted safety supervision on the operators on the site.

② The person in charge of the work and the Supervisor must wear a safety vest.

(6) Hazard——Traffic safety

① It is necessary to pay attention to the driving safety of the vehicle, drive carefully when driving, and illegal driving is not allowed.

3.2 Selection of tools, instruments and materials

See Table 2-2-2 for Tools, Instruments and Materials Required for Power-cut Replacement of Tension Whole Strain Insulator String on ±800kV UHV Transmission Line. Before delivering the tools and instruments out of the warehouse, application voltage class and test period of the tools and instruments shall be carefully checked and they shall be inspected to ensure that appearance is intact, connection is firm, rotation is flexible and meeting the requirements of the work task. After delivering the tools and instruments out of the warehouse, they shall be kept from contamination and damp. Metal tools and insulated tools shall be separately loaded and transported to avoid deformation and damage caused by mixed loading and transportation.

Part II Standard for Training and Assessment on Interruption Maintenance for Operation Maintenance of ±800kV UHV Power Transmission Line

Table 2-2-2 Tools, Instruments and Materials Required for Power-cut Replacement of Whole Strain Insulator String on ±800kV UHV Transmission Line

S/N	Name	Specification and Model	Unit	Qty.	Remarks
1	Grounding wire	Only for ±800kV	Set	2	Insulating tool
2	Insulating gloves	10kV	Set	2	Insulating tool
3	electricity tester	Only for ±800kV	pcs	2	Other tools
4	Full-covered Safety Belt	Including 24m rear safety rope with buffer bag	Set	8	PPE
5	Personal security wire	(the diameter should be less than $16mm^2$)	Nos.	2	Other tools
6	Steel wire sleeve	$\phi 24$	Nos.	20	Other tools
7	Iron tackle	15T	Nos.	12	Metal tool
8	Shackle	18T	Nos.	12	Metal tool
9	Iron pulley	5T	Nos.	8	Metal tool
10	Steel wire strands	$\phi 24$	m	150	Other tools
11	Grinding rope	$\phi 17.5$	m	600	Other tools
12	Winching	5T	Set	3	Motor tool
13	Hand-operated hoist	9T	Set	2	Metal tool
14	Hand-operated hoist	6T	Set	2	Metal tool
15	Individual hand-operated tool		Set	8	Other tools
16	Interphone		Set	10	Other tools
17	Lifting rope tackle	1T	Nos.	2	Metal tool
18	Transfer rope	$\phi 16$	Set	2	Other tools
19	Pin puller		pcs	3	Metal tool
20	Safety vest		pcs	3	Other tools
21	Goggles		Set	6	Other tools
22	Security fence		Volume	5	Other tools
23	Sole timber		pcs	Several	Other tools
24	Moisture-proof tarpaulin		Sheet	1	Other tools
25	Steel chisel		Nos.	2	Other tools
26	Iron hammer		pcs	2	Other tools
27	Glass insulator	U550BP/240T	pcs	81	Material

3.3 Division of labor for operators

Division of labor for operators of the task is shown in Table 2-2-3.

Table 2-2-3 Division of Operating Personnel Involving in Power-cut Replacement of Tension Whole Strain Insulator String on ±800kV UHV Transmission Line

S/N	Post	Qty. (person)	Responsibilities
1	Responsible Person	1	Be responsible for division of operating personnel of the task, field investigation before work, implementation of the operation plan, work order filling, field re-investigation, performing work permit procedures, holding pre-shift meeting, the implementation of safety measures on site, safety supervision in the operation process, dealing with emergency situations in work, quality surveillance, and the summary after work.
2	Safety Supervisor	2	Be responsible for the safety monitoring work during this operation
3	Operator working at heights	6	Be responsible for the operation of power-cut replacement of whole strain insulator string on ±800kV
4	Ground auxiliary personnel	5	Be responsible for ground auxiliary works during the operation.
5	Winching operator	2	Be responsible for the winching work during this operation
6	Signal commander	2	Be responsible for commanding the start and stop of 2 sets of winches

4. Working procedures

Working process of the task is shown in Table 2-2-4.

Table 2-2-4 Working Procedures for Power-cut Replacement of Tension Whole Strain Insulator String on ±800kV AC Transmission Line

S/N	Work Content	Operation Standard	Safety Precautions	Responsible Person
1	Work Permit	The contact person for power off/on before the operation must contact the dispatcher to complete the work permit procedure.	(1) No work shall be started without the Work Permit of approver. (2) Power transmission and cutoff at pre-determined time is not permitted.	

Part II Standard for Training and Assessment on Interruption Maintenance for Operation Maintenance of ±800kV UHV Power Transmission Line

Table (Cont'd)

S/N	Work Content	Operation Standard	Safety Precautions	Responsible Person
2	Site layout	Install the security fence and hang the signboards correctly: (1) The security fence should take full account of falling objects from the heights and the influence on road traffic. (2) The entrance and exit of the security fence shall be set reasonably. (3) Signs such as "Access from Here", "Work Here", "Slow Down" or "Blocking" shall be properly arranged	When the influence on road traffic safety is uncontrollable, the traffic management department should be contacted in time to strengthen the on-site control of traffic safety	
3	Holding a pre-shift meeting	(1) All working personnel shall line up. (2) The Responsible Person will read out the work order and be clear with the work task and division of personnel; explain safety measures and technical measures in work; check (inquire after) the mental state of all working personnel; and inform of hazards in work and pre-control measures. (3) All working personnel shall sign on the work order for confirmation.	(1) Work order shall be filled and issued with standardized licensing procedure and complete signature. (2) All working personnel shall be in good mental states. (3) All working personnel shall be clear with task division of works, safety measures and technical measures	
4	Inspection of tools and instruments	(1) All necessary tools and instruments shall be prepared as per the operation requirements and placed on the waterproof tarpaulin regularly according to the category and location. The appearance and test certificate of tools and instruments shall be checked to ensure there is no omission. (2) The inspector shall report to the Responsible Person that all inspection results are in conformity with the operation requirements	(1) The waterproof tarpaulin shall be enough in quantity and reasonable in position, and be clean and dry. (2) The appearance of tools and instruments is acceptable after inspection, there is no damage, damp, deformation and malfunction, and the certificate is valid	

Table (Cont'd)

S/N	Work Content	Operation Standard	Safety Precautions	Responsible Person
5	Climb pole and tower	(1) Check the double name and serial number of the line before climbing the pole and tower. For multi-circuit lines on the same tower, the Responsible Person and the members of the working team shall check the double name and identification mark carefully (color code, distinguishing mark, etc.). (2) Check the tower root and the base of pole and tower before climbing, making sure it's solid and reliable. (3) During climbing, in order to prevent the operator from falling, the distance between the staff members shall not be less than 1.6m, the safety belt shall be properly packed, the long tail rope shall be placed in the kit, and the main belt shall be hung on the shoulder to prevent safety during climbing to prevent the safety belt from hooking shackles and tower materials during climbing, resulting in falling from the tower. (4) When the pole and tower is close to the cross arm, the Supervisor and the operator shall check the identification mark and dual tags of the power-cut line again, and can enter the cross arm at the power-cut line side only after they are confirmed to be correct	(1) Operator shall wear safety helmet and soft sole, and be scrupulous and when climbing pole and tower. (2) During climbing, the safety belt shall be properly packed, the long tail rope shall be placed in the kit, and the main belt shall be hung on the shoulder to prevent safety during climbing to prevent the safety belt from hooking shackles and tower materials during climbing, resulting in falling from the tower. (3) Operators shall not lose the protection from the safety belt and step and hold it firmly when moving on the pole and tower. (4) When reaching the position of operation point, fasten the safety belt (rope), it shall be firm and reliable, and shall not be hung low for high use. (5) The safe distance between the human body, headless rope, etc. and the conductor must not be less than 10.1m before the power test, and special person shall be assigned to supervise the work.	
6	Inspection power and install ground wire	After the pole is in place, fasten the safety belt to a firm and reliable component or pole, and check whether the buckle is in place correctly. Appliances such as test pen (electroscope) must be transferred by rope. Check whether the ground wire is in good condition, and install the ground wire according to the procedures (first connect the ground terminal, then the conductor terminal).	(1) Check whether the test pole is normal during receiving and before using. (2) It is forbidden to install the ground wire by winding the conductor	

Table (Cont'd)

S/N	Work Content	Operation Standard	Safety Precautions	Responsible Person
6	Inspection power and install ground wire	When setting the ground wire, the insulating rope or insulating handle must be used for operation, and it is forbidden to install the ground wire with directly operating the metal part of the ground wire by hand. Make sure the collet of the ground wire is tightly connected to the conductor		
7	Replacement of tension whole strain insulator string	After the operators working at heights arrived at the operating point, 3 operators working at heights should install three 15T iron tackles on the solid tower material between the two strings of tension insulators at the cross-arm end; the 3 operators working at heights should install three 15T iron tackles on the corresponding live side yoke plate, fix the wire sleeve and the lever block hook on one side of the three iron tackles on the cross arm side, and then sequentially introduce the steel wire rope connected by the lever block chain in the six tackles to form a set of pulleys, with enough space for the operation; and after confirming that the lever block and the pulley block is firmly connected, the operators should tighten the 9T lever block; after the tension of the insulator string is completely withstood by the 9T lever block, the cross arm should be inspected again for deformation and the joint for no abnormalities, so as to confirm that everything is normal. Three high-altitude personnel shall pass the grinding rope on No. 1 winching through the 5T pulley on the cross arm, and then fix it on the third piece of cross arm end of insulator string. After it is fixed firmly, the grinding rope on No. 2 winching shall be passed through the two 5T pulleys on the insulator set fitting between the cross arm end and the live end in turn, and then fix the grinding rope on No. 2 winching at the third	(1) In the process of replacing insulator string with lever hoist, it is necessary to check the connection and stress of lever hoist, wire rope noose (lifting belt) and shackle after the lever hoist begins to bear the conductor load. And the impulse test shall be conducted to confirm that it is completely reliable before continuing to tighten the lever hoist. (2) To avoid damaging the insulator, it shall be prevented from colliding with the insulator when using the lever hoist. (3) When using the transmission rope to hoist the whole insulator string, the ground staff and the staff on the tower must cooperate closely to prevent the lifting rope from winding or the insulator string damaging the conductor due to colliding. When hoisting the insulator string, the ground personnel shall not stand under the vertical position of it.	

Table (Cont'd)

S/N	Work Content	Operation Standard	Safety Precautions	Responsible Person
7	Replacement of tension whole strain insulator string	piece of insulator on the live end. After all connecting parts have been connected firmly, the winching can be started. The ground grinding operator shall start the No. 2 winching. Firstly, tighten the live end of the insulator string. and then slowly release the insulator string at the live end after the high-altitude personnel taking down the connecting part of the insulator string at the live end. Secondly, start the No. 1 winching after the insulator string is vertical to the ground. After the insulator string is tensioned and lifted up at one end, the high-altitude personnel take down the connecting part of the insulator string at the cross arm end, and then slowly loosen the two ends at the same time to the ground, the strands should be used to control the insulator strings in the process of transmission to the ground to prevent collision with the jumper string, and the ground personnel cooperate to take down the insulator after putting down the insulator string. After the ground personnel take off the insulator and replace the good insulator, start the No. 1 winching firstly and then start the No. 2 winching. The installation sequence is opposite to the removal sequence. Install the cross arm end and then fasten the live end. After the insulator is in place, the operators on the tower shall first connect the insulator set fitting at the tower end and ensure that the pin is in place before continuing the connection between the metal fitting and the iron tower and conductor. Check whether the insulator set fitting and the pin are installed in place after the insulator string is installed, and the impulse test shall be conducted to confirm that the installation is correct and the connection is reliable before slowly releasing the 9T lever hoist	(4) When the whole insulator string is lifted by winching, there shall be a special person for command. The two winches shall cooperate closely. Upon the insulator strings leaving the ground, check again if the insulator strings are bound firmly, and carry out the impulse test. The lifting can be continued only after confirmation	

Part II Standard for Training and Assessment on Interruption Maintenance for Operation Maintenance of ±800kV UHV Power Transmission Line

Table (Cont'd)

S/N	Work Content	Operation Standard	Safety Precautions	Responsible Person
8	Removal of tools and instruments	Remove the ropes, lever hoist, steel wire sleeves, tackles and other tools and instruments in order		
9	End of the work	(1) Responsible person shall organize all working personnel to put working apparatus and materials in order, and clean the site to ensure "materials are removed and the site is cleaned after construction". (2) After a post-shift meeting is held, the Responsible Person shall make work summaries and comments. Comments include the construction quality of this work and the implementation of safety measures from all working personnel. (3) Responsible Person shall report the completion of the work to the work approver, resume the power transmission of the power cut line and terminate the work order		

II. Assessment Standard

Table 2-2-5 Detailed Rules for Assessment and Scoring of Operation and Inspection Skills of UHVDC Transmission Line of State Grid Sichuan Electric Power Corporation

Fill-in Column of Examinee	No.:	Name:	Position:	Unit:	Date:	MM/DD/YYYY
Fill-in Column of Assessor	Grade:	Assessor:	Assessment Team Leader:	Starting time:	Closing time:	Operation Duration:
Assessment module	Power-cut replacement of tension-resistance whole strain insulator string of ±800kV UHV transmission line tangent tower. Assessee: Maintenance personnel for UHV DC transmission line. Assessment method: Operation. Assessment time limit: 360min					
Job Description	Replace the ±800kV special pressure transmission line C phase right string of glass insulators					
Work specifications and requirements	1. Given conditions: C-phase whole right glass insulator string on ±800kV DC practical training line is aging, which needs to be replaced. The line is power off, test electricity and install the ground wire, the insulators to be used have been tested, the work order has been handled, and the safety measures have been taken. 2. The main operation procedures of the whole process are completed by 1 Responsible Person, 1 special Supervisor, 6 electricians on the tower, 7 auxiliary workers on the ground, 2 winching operators to assist the exam participant to finish the up and down transfer of tools, materials and other non-technical work. 3. The exam participant should make necessary safety check before operation. 4. The tools used to replace the whole insulator string shall meet the stress requirements. 5. An oral application shall be made at the beginning of work, and an oral report also shall be made at the end of the work					
Assessment scenario preparation	1. Tools and instruments: 150m of ϕ24 wire thread insert, 12 15T iron tackle, 8 5T iron tackle, 12 18T shackle, 150m of ϕ24 wire rope, a bundle of ϕ16 grinding rope, 2 9T lever hoists, 2 6T lever hoists, 2 5T winching, 2 sets of lifting rope tackle, 2 nooses, several intercoms, and 6 double safety belts. 2. Material: a string of glass insulator, which is with the same model. 3. Operated on the training line					
Remarks	1. Personal tools shall be self-provided by the exam participant. 2. The deduction shall be done until the scores of each item are deducted completely.					

Part II Standard for Training and Assessment on Interruption Maintenance for Operation Maintenance of ±800kV UHV Power Transmission Line

Table 2-2-6 Standards for Assessment and Scoring of Operation and Inspection Skills of UHVDC Transmission Line of State Grid Sichuan Electric Power Corporation

S/N	Project name	Quality requirements	Score	Deduction standard	Reasons for deduction	Deduction	Scoring
1	Preparation for tools and materials						
1.1	Inspection of personal tools	Adjustable wrench, flat pliers, pin-pulling pliers and tool kits shall meet the quality requirements	2	Deduct 1 point for each error and missing item			
1.2	Inspection of tools and instruments	Test the machine with winching, check whether the gear is normal and check the lever hoist	2	Deduct 1 point for each error and missing item			
1.3	Inspection of wire rope and tackle	Check the wire rope and tackle, confirm the connection is reliable and the force meets the requirements.	2	Deduct 1 point for each error and missing item			
1.4	Safety belt	The double safety belt meets the quality requirements within the test period	1	Deduct 1 point for no check			
1.5	Inspection of special tools	The appearance inspection of personal security wire and insulating gloves shall meet the requirements within the test cycle	1	Deduct 0.5 points for one item mistake or omission			
1.6	Material inspection	Clean the insulators and check the number of insulator strings, the appearance are in line with the requirements	2	Deduct 1 point for each error and missing item			
2	Venue layout						

Table (Cont'd)

S/N	Project name	Quality requirements	Score	Deduction standard	Reasons for deduction	Deduction	Scoring
2.1	Arrangement of winching site	Arrangement of winching site and angle tower pulley	2	Deduct 1 point for each error and missing item			
2.2	Site fence	Layout of site fence	1	Deduct 1 point for failure to lay out			
3	Climbing the tower and operation on cross arm						
3.1	Climbing the tower	(1) Check whether there is no abnormality. (2) Correctly carry the transmission rope (The end of the lifting rope shall be double folded, fast knot, and hung over the shoulder) (3) Correctly climb down the tower along the main material of the shackles side	6	(1) Deduct 1 point for each unchecked item. (2) Deduct 2 points for not carrying the transmission rope, and deduct 1 point for nonstandard carrying method of lifting rope. (3) Deduct 1 point/time for grasping the shackles with hands. (4) Deduct 2 points for not climbing the tower along the main material of the shackles side			
3.2	Access the working point on the cross arm	The safety belt protection shall not be lost from the tower body to the working point on the cross arm.	3	Deduct 3 points for failure to use safety belt properly.			

Part II Standard for Training and Assessment on Interruption Maintenance for Operation Maintenance of ±800kV UHV Power Transmission Line

Table (Cont'd)

S/N	Project name	Quality requirements	Score	Deduction standard	Reasons for deduction	Deduction	Scoring
3.3	Installation of transfer tackle	The transfer tackle shall be installed at the correct position for easy operation.	1	Deduct 1 point for non standard installation			
4	Operation on insulator string						
4.1	Enter into the working point site	(1) Fasten the safety rope of the double safety belt to the cross arm in the proper position. (2) Access the operation point along the insulator string, and tie the pole belt to the insulator string. (3) Check the insulator string fitting pin	5	(1) Deduct 5 points for failure to use double safety belt. (2) Deduct 3 points/time for failure to use double safety belt properly. (3) Deduct 3 points for failure to check			
4.2	Installation of 3-3 tackle block	(1) Make sure that the 3-3 tackle blocks at both the conductor end and tower- end are installed correctly. (2) The wire rope shall be threaded in the correct direction without winding. (3) The selected models of noose of wire rope and shackle are correct and the installation is reliable	6	(1) Deduct 1 point for every one wrong installation of tackle block. (2) Deduct 1 point for every one incorrect direction of the wire rope (3) Deduct 1 point for every one collision or winding after the steel wire rope is stressed. (4) Deduct 1 point for every incorrect model selection of noose of wire rope and shackle.			

Table (Cont'd)

S/N	Project name	Quality requirements	Score	Deduction standard	Reasons for deduction	Deduction	Scoring
4.3	Installation of lever hoist	(1) Transfer the lever hoist to the tower and install it correctly. (2) Connect the lever hoist with the 3-3 tackle block to make it slightly stressed	5	(1) Deduct 1 point for rope winding (2) Deduct 2 points for incorrect position (3) Deduct 1 point for one touch of insulator.			
4.4	Tighten the insulator string	(1) Tighten the lever hoist. After the chain block is stressed, check the tackle, shackle, wire rope and other connection parts to confirm the connection is reliable, and impact test shall be taken for the insulator string. (2) After confirming that the force is correct, continue to tighten the lever hoist until the insulator string is loose.	5	(1) Deduct 3 points for failure to carry out the impulse test. (2) Deduct 2 points for incorrect use of the lever hoist.			
4.5	Replacement of insulator string	(1) Correctly install a 6T auxiliary lever hoist on the loose insulator string. (2) Connect the wear-proof lifting rope at a proper position on the insulator string. (3) Tighten up the 9T lever hoist, conduct impulse test after being forced, and tighten the 6T lever hoist and take out R pin when there is no abnormality. (4) Remove the insulator set fitting at both ends of the insulator string and tighten up the wear-proof lifting rope. (5) Check the connection part of the rope and conduct impulse test after the rope is stressed.	18	(1) Deduct 2 points each time for wrong installation position or operation of the lever hoist (2) Deduct 2 points for incorrect position of lifting tackle by wear-proof rope (3) Deduct 3 points for failure to carry out the impulse test. (4) Deduct 2 points for winding of transmission rope (5) Deduct 2 points/time for bump of transmission objects			

Part II Standard for Training and Assessment on Interruption Maintenance for Operation Maintenance of ±800kV UHV Power Transmission Line

Table (Cont'd)

S/N	Project name	Quality requirements	Score	Deduction standard	Reasons for deduction	Deduction	Scoring
4.5	Replacement of insulator string	(6) Remove 6T lever hoist and transfer insulator string to the ground by wear-proof lifting rope. (7) Transfer the new insulator string to the pole and tower, and install the 9T lever hoist. (8) Tighten the 9T lever hoist, and then tighten the 6T lever hoist, install the R pins at both ends, and check whether they are installed in place.		(6) Deduct 5 points for dropping object. (7) Deduct 2 points for not installing the locking pin.			
4.6	Remove tools and appliances	(1) Check whether the locking pin and ball head are in position, the bowl is oriented correctly, and clean the insulator string. (2) After the insulator string of 3-3 tackle block is released and the insulator is stressed, the impulse test shall be conducted on the insulator string. (3) Remove the lever hoist, tackle block and wire rope, and transfer them to the ground.	10	(1) Deduct 2 points for rope winding. (2) Deduct 2 points/time for bump of transmission objects. (3) Deduct 5 points for not transferring pole belt to the insulator string. (4) Deduct 2 points for incorrect position of fitting pin. (5) Deduct 2 points for incorrect orientation of insulator big mouth (6) Deduct 3 points for failure to carry out the impulse test. (7) Deduct 1 point for no cleaning of insulator string			

Standard for Professional Training and Assessment for Operation Maintenance of UHV DC Power Transmission Line

Table (Cont'd)

S/N	Project name	Quality requirements	Score	Deduction standard	Reasons for deduction	Deduction	Scoring
4.7	Clean the tools and instruments on the tower Tools and instruments	Confirm that there are no leftovers	4	Deduct 4 points for any object left behind			
4.8	From the insulator string to the tower	The safety belt protection shall not be lost from tower to the insulator string	5	Deduct 5 points for loosing the protection of the safety belt.			
5	Step down the tower	(1) Correctly climb down the tower along the main material of the shackles side (2) Correctly carry the transmission rope (The end of the lifting rope shall be double folded, fast knot, and hung over the shoulder)	8	(1) Deduct 4 points for not carrying the transmission rope, and deduct 2 points for nonstandard carrying method of lifting rope. (2) Deduct 1 point/time for grasping the shackles with hands. (3) Deduct 4 points for not climbing the tower along the main material of the shackles side			
6	Other requirements						
6.1	Tower operation	(1) It is prohibited to drop articles down from high space. (2) Cooperate with both hands during operation. (3) No floating objects. (4) No things in the mouth	5	(1) Deduct 5 points for each falling object from high place. (2) Deduct 2 points for non-coordination action. (3) Deduct 2 points for floating articles. (4) Deduct 2 points for holding object in mouth			

Part II Standard for Training and Assessment on Interruption Maintenance for Operation Maintenance of ±800kV UHV Power Transmission Line

Table (Cont'd)

S/N	Project name	Quality requirements	Score	Deduction standard	Reasons for deduction	Deduction	Scoring
6.2	Dress	Correctly wear working clothes, rubber work shoes, safety helmet and protective gloves	2	Deduct 2 points for missing one item			
6.3	Clean site	Clean up work site after completion and meet the requirements of civilized production	2	Deduct 2 points for failure to clean up work site			
6.4	Completion time	Complete operation as required within the specified time	2	Deduct 1 point if the time exceeds 10 minutes, terminate the operation when the time reaches 480 minutes, and only record the score of completed part			
	Total		100				

Module 3 Training and Assessment Standard for Power-cut Repair of ±800kV UHV Transmission Line Overhead Ground Wire

I. Training Standard

(I) Training Requirements

Designation of module	Power-cut Repair of ±800kV UHV Transmission Line Overhead Ground Wire	Type of training	Operation Type
Training method	Practical operation training	Training class hours	14 hours
Training objectives	1. Be able to use the hanging wheel to reach the specified work position along the overhead ground wire of ±800kV UHV transmission line; 2. Be able to independently complete the operation of repairing overhead ground wire with preformed armor rods repair strip		
Training venue	UHV DC practical training line		
Training content	Use the hanging wheel to reach the specified work position along the overhead ground wire of ±800kV UHV transmission line, and repair the overhead ground wire of ±800kV UHV transmission line with preformed armor rods repair strip		
Scope of application	Maintenance personnel for UHV DC Transmission Line		

(II) Referenced Procedures and Specifications

(1) Operating Code for Overhead Transmission Line (DL/T741-2010)

(2) Technical Code for Designing Overhead 110~500kV Transmission Line (DL/T5092-1999)

(3) State Grid Corporation of China Working Regulations of Power Safety (Transmission Line Section) (Q/GDW1799.2-2013)

(4) Code for Designing of ±800kV DC Overhead Transmission Line (GB50790-2013)

(5) Maintenance Specification for ± 800kV DC Overhead Transmission Line (DL/T251-2012)

(6) Operating Code for ±800kV DC Overhead Transmission Line (GB/T28813-2012)

(7) Maintenance Specification for 110(66)kV~500kV Overhead Transmission Line (State Grid Corporation of China)

(8) Guideline of Maintenance of Overhead Transmission Line State (DLT1248-2013)

(9) Maintenance Management Regulations of Electric Transmission and Transformation Equipment State (State Grid Corporation of China)

(10) Maintenance and Test Specification of Electric Transmission and Transformation Equipment State (State Grid Corporation of China)

(III) Teaching Design for Training

To complete the power-cut replacement of ±800kV UHV transmission line overhead ground wire is the work task during the design. Each training stage shall be designed according to the standard operation procedure for work task completion. Each stage includes specific training objectives, training content, hours of training, training methods (training resources), training environment, assessment and evaluation, etc, as shown in Table 2-3-1.

Table 2-3-1 Training Content Design for Power-cut Repair of ±800kV UHV Transmission Line Overhead Ground Wire

Training schedule	Training objectives	Training contents	Training class hours	Training method and resources	Preparation of training conditions	Assessment
1. Theory Teaching	1. Master the basic method of using hanging wheel to reach the specified work position along the overhead ground wire of ±800kV UHV transmission line. 2. Familiar with the repair methods for the damage of overhead ground wire	1. Classification, structure and precautions of the hanging wheel. 2. Master the basic method of using hanging wheel to reach the specified work position along the overhead ground wire of ±800kV UHV transmission line. 3. Repair method and quality standard for transmission line overhead ground wire	2	Training method: Lecture. Training resources: PPT, relevant regulations and specifications	Multimedia Classrooms	Attendance, classroom questions and assignments
2. Preparation Work	Can complete the preparation before operation Work	1. Work site survey. 2. Preparation of the standardized operation card. 3. Fill in the first work order of transmission line. 4. Preparation of tools and instruments and materials for this operation	1	Training method: 1. Site survey and cleaning of tools and instruments and materials shall be practiced at site. 2. Preparation of operation card and the filling of work order shall adopt lecture method. Training resources: 1. ±800kV practical training line. 2. UHV tools and instruments warehouse. 3. Blank work order	1. UHV practical training line. 2. Multimedia Classrooms	

· 598 ·

Part II Standard for Training and Assessment on Interruption Maintenance for Operation Maintenance of ±800kV UHV Power Transmission Line

Table (Cont'd)

Training schedule	Training objectives	Training contents	Training class hours	Training method and resources	Preparation of training conditions	Assessment
3. Work site preparation	Can complete the preparations on work site	1. Work site re-survey. 2. Work application. 3. Work site layout. 4. Pre-shift meeting. 5. Tools and instruments and material inspection	1	Training methods: Demonstration and role play. Resources: ±800kV practical training line	±800kV practical training line	
4. Trainer's demonstration	The trainees can preliminarily understand the operation process of the task through inspecting and learning from each other's work	1. Laying of ground wire 2. Installation the hanging wheel. 3. Master the basic method of using hanging wheel to reach the specified work position along the overhead ground wire of ±800kV UHV transmission line. 4. Repair the overhead ground wire with preformed armor rods repair strip	2	Training method: Demonstration. Resources: ±800kV practical training line	±800kV practical training line	

Table (Cont'd)

Training schedule	Training objectives	Training contents	Training class hours	Training method and resources	Preparation of training conditions	Assessment
5. Group training of trainees	1. Be able to complete the correct installation of the hanging wheel. 2. Be able to use the hanging wheel to reach the specified work position along the overhead ground wire of ±800kV UHV DC transmission line; 3. Be able to complete the repair of ±800kV transmission line overhead ground wire	1. The trainees are grouped (6 in a group) to train the skills of using hanging wheel and repairing overhead ground wire. 2. Trainers guide the operation of trainees and conduct safety supervision	7	Training methods: Role play. Resources: ±800kV practical training line	±800kV practical training line	Score the operation of trainees according to the detailed rules for skill assessment and scoring
6. End of the work	1. Enable the trainees to further distinguish the shortcomings of the operation process and facilitate the promotion in the later stage. 2. Train the trainees in the work awareness of safe and civilized production	1. Cleaning of work site. 2. Report to dispatcher. 3. Comment and summarize the work task this time at post-shift meeting	1	Training methods: Lecture and inductive method	±800kV practical training line	

(IV) Work Flow

1. Work task

Complete Power-cut Repair of ±800kV UHV Transmission Line Overhead Ground Wire task.

2. Requirements for Weather and Work Site

(1) It should be carried out in good weather for power-cut repair of ±8001kV UHV transmission line overhead ground wire.

High-place operation in the open air shall be stopped in severe weather, such as heavy wind of 5 degree ad above, rainstorm, thunder, hail, heavy fog and sand storm. In special cases, if emergency maintenance needs to be performed in severe weather, arrange related personnel to fully discuss necessary safety measures, which shall be taken after approved by the company.

(2) The operators are in good mental state, and the members of the working group carefully study the work ticket and safety technical measures, and all the personnel shall be conform to the "four clear" (the operation task is clear, the danger point is clear, the operation procedure is clear, and the safety measures are clear).

(3) The Responsible Person must contact with the dispatcher to fulfill the Work permit procedures, and it is strictly prohibited to stop and transfer electricity at the appointed time. Responsible Person must obtain the license work order of the licensor before checking the electricity on the line to be overhauled, setting up the ground wire and carrying out the maintenance work.

(4) After power-cut, Responsible Person shall make records carefully.

(5) Check whether there is a honeycomb on the tower before climbing the tower. It is forbidden to climb the tower if a honeycomb is found.

(6) Double safety belt shall be used by the personnel who work on the tower, and safety goggles must be worn.

3. Preparations

3.1 Hazards and precontrol measures

(1) Hazard——climbing live line

Precontrol measures:

① Before climbing the tower, the Responsible Person and the members of the work

Standard for Professional Training and Assessment for Operation Maintenance of UHV DC Power Transmission Line

team shall carefully check that the double name and identification mark (color code, identification mark, etc.) are consistent with the name of the power-cut line.

② Check the root and base, etc. of iron tower before climbing the tower to keep it solid and reliable.

③ Before climbing the pole and tower, climbing tools and facilities such as safety belt, shackles and tower materials shall be inspected to ensure they are firm.

④ Before climbing the pole, the intermediate operators who do not involved the installation of the grounding wire shall carefully verify that the phase sequence, color code, name, and number of the line are consistent with the power-cut line, and confirm that the line name is not error or in opposite order.

(2) Hazard——Operation in violation of safety operation procedures when climbing the tower and working on the tower may cause falling from height.

Precontrol measures:

① In order to prevent tower climbing personnel from colliding with each other, the distance between tower climbing operators shall not be less than 1.6m during climbing the tower.

② Before climbing the tower, the soil on the sole shall be removed, the tool kit shall be completed, and do not carry a heavy load during climbing.

③ Operators shall wear safety helmet and soft work shoes, and climb evenly when climbing the tower.

④ During climbing the tower, the safety belt shall be properly packed, the safety rope at the back shall be placed in the kit, and the main belt shall be hung on the shoulder to prevent safety during climbing to prevent the safety belt from hooking shackles and tower materials during climbing, resulting in falling from the tower.

⑤ Operators shall not lose the protection from the safety belt and step and hold it firmly when moving on the tower.

⑥ When reaching the position of operation point, fasten the safety belt. And the safety belt and backup protection rope shall be firmly fastened on the component respectively, and shall not be hung low for high use.

⑦ The safe distance between the human body, transmission rope, etc. and the conductor must not be less than 10.1m before the power test, and special person shall be assigned to supervise the work.

(3) Hazard —— injury caused by objects falling from high place

Precontrol measures:

① The ground staff shall not stand under the position vertical to the operation point. The personnel on tower shall prevent falling objects from injury, and the tools and materials used shall be transmitted by ropes.

② Tool bags should be used when working at heights. Larger tools should be fixed on solid components and are not allowed to be randomly placed.

(4) Hazard —— prevent injury caused by induced electricity

① Before connecting the overhead ground wire, the ground terminal of the ground wire shall be reliably grounded.

② When the ground wire is placed, the operator shall wear insulating gloves and hold the insulating part.

(5) Hazard —— Site safety monitoring

① The safety supervisor shall continuously monitor the operators during the operation.

② The Responsible Person and the Supervisor must have an clear identification which is suitable for their identity.

(6) Hazard —— Traffic safety

① When driving, attention should be paid to the safety of the vehicle, driving the vehicle carefully, and illegal driving is prohibited.

3.2 Selection of tools, instruments and materials

See Table 2-3-2 for Tools, Instruments and Materials Required for the Power-cut Repair of ±800kV UHV Transmission Line Overhead Ground Wire. Before delivering the tools and instruments out of the warehouse, application voltage class and test period of the tools and instruments shall be carefully checked and they shall be inspected to ensure that appearance is intact, connection is firm, rotation is flexible and meeting the requirements of the work task. After delivering the tools and instruments out of the warehouse, they shall be kept from contamination and damp. Metal tools and insulated tools shall be separately loaded and transported to avoid deformation and damage caused by mixed loading and transportation.

Table 2-3-2 Tools, Instruments and Materials Required for the Power-cut Repair of ±800kV UHV Transmission Line Overhead Ground Wire

S/N	Name	Specification and Model	Unit	Qty.	Remarks
1	Operation hanging wheel		Set	1	Metal tool
2	Safety belt	Including 20m rear safety rope with buffer bag	Nos.	2	PPE
3	Insulating gloves		Pair	1	Insulating tool
4	Transfer rope	$\phi 16$	Nos.	1	Other tools
5	Iron tackle	0.5T	Nos.	1	Metal tool
6	Steel wire sleeve	$\phi 8$	Nos.	1	Other tools
7	Personal tools		Set	2	Other tools
8	Sandpaper		Sheet	1	Other tools
9	Steel tap		pcs	1	Other tools
10	Marker pen		Nos.		Other tools
11	Wood hammer		pcs	1	Other tools
12	Interphone		Set	3	Other tools
13	Safety Vest		pcs	2	Other tools
14	Security fence		Volume	4	Other tools
15	Moisture-proof tarpaulin		Sheet	1	Other tools
16	Hanging ground wire	±800kV	Group	3	Other tools
17	Ground wire	25mm2	Group	1	Other tools
18	electricity tester	±800kV	Nos.	1	Other tools
19	Falling protector	Type T	Nos.	1	Other tools
20	Preformed armor rods	Corresponding ground wire type	Group	1	Material

3.3 Division of labor for operators

Division of labor for operators of the task is shown in Table 2-3-3.

Part II Standard for Training and Assessment on Interruption Maintenance for Operation Maintenance of ±800kV UHV Power Transmission Line

Table 2-3-3 Division of Operating Personnel Involving in Power-cut Repair of Overhead Ground Wire on ±800kV UHV Transmission Line

S/N	Post	Qty. (person)	Responsibilities
1	Responsible Person	1	Be responsible for division of operating personnel of the task, field investigation before work, implementation of the operation plan, work order filling, field re-investigation, performing work permit procedures, holding pre-shift meeting, the implementation of safety measures on site, safety supervision in the operation process, dealing with emergency situations in work, quality surveillance, and the summary after work.
2	Safety Supervisor	1	Explain the safety measures within the scope of supervision to the supervised personnel, and inform the hazards and safety precautions before work. The Supervisor shall monitor the supervised personnel to comply with this procedures and strictly implement the on-site safety measures, and timely correct the unsafe acts and behaviors of the supervised personnel.
3	Operator working at heights	2	Be responsible for the operation of power-cut repair of ±800kV DC transmission line overhead ground wire
4	Ground auxiliary personnel	2	Be responsible for ground auxiliary works during the operation.

4. Working Procedures

Working procedures of the task if shown in Table 2-3-4.

Table 2-3-4 Work Procedures for Power-cut Repair of Overhead Ground Wire on ±800kV UHV Transmission Line

S/N	Work Content	Operation Steps and Standards	Safety Measures and Precautions	Responsible Person
1	Site re-survey	The Responsible Person shall complete the following work: (1) Check the line name, the number of the iron tower number on spot to ensure they are correct; guarantee that the foundation and the tower body are intact and in normal condition, and ensure that the cross and span distance meets the safety requirements; confirm the defect conditions and the specifications and models of ground wires.	(1) Correctly wear helmet, working clothes, work shoes and protective gloves. (2) Non-operation personnel and vehicles are strictly prohibited from entering the work site	

Table (Cont'd)

S/N	Work Content	Operation Steps and Standards	Safety Measures and Precautions	Responsible Person
1	Site re-survey	(2) Check that the terrain and environment should meet the operation requirements. (3) Check that the safety measures listed in the work order are in line with the actual situations on site. The measures will be added if necessary.		
2	Work Permit	The Responsible Person must contact with the dispatcher to fulfill the Work licensing procedures	(1) No work shall be started without the Work Permit of approver. (2) Power transmission and cut-off at pre-determined time is not permitted.	
3	Site layout	Arrange the security fence and hang the signboards correctly: (1) The security fence should take full account of falling objects from the heights and the influence on road traffic. (2) The entrance and exit of the security fence shall be set reasonably. (3) Signs such as "Access from Here", "Work Here", "Slow Down" or "Blocking" shall be properly arranged.	When the influence on road traffic safety is uncontrollable, the traffic management department should be contacted in time to strengthen the on-site control of traffic safety	
4	Holding a pre-shift meeting	(1) All working personnel shall line up. (2) The Responsible Person will read out the work order and be clear with the work task and division of personnel; explain safety measures and technical measures in work; check and inquire the mental state of all working personnel; and inform of hazards in work and precontrol measures. (3) All working personnel shall sign on the work order for confirmation.	(1) Work order shall be filled and issued with standardized licensing procedure and complete signature. (2) All working personnel shall be in good mental states. (3) All working personnel shall be clear with task division of works, safety measures and technical measures	

Table (Cont'd)

S/N	Work Content	Operation Steps and Standards	Safety Measures and Precautions	Responsible Person
5	Inspection of tools and instruments	(1) All necessary tools and instruments shall be prepared as per the operation requirements and placed on the waterproof tarpaulin regularly according to the category and location. The appearance and test certificate of tools and instruments shall be checked to ensure there is no omission. (2) The inspector shall report to the Responsible Person that all inspection results are in conformity with the operation requirements	(1) The waterproof tarpaulin shall be reasonable in position, and be clean and dry. (2) The appearance of tools and instruments is acceptable after inspection, there is no damage, damp, deformation and malfunction, and the certificate is valid	
6	Climbing the tower	(1) Check the double name and serial number of the line before climbing the tower. For multi-circuit lines on the same tower, the Responsible Person and the members of the working team shall check the double name and identification mark carefully (color code, distinguishing mark, etc.). (2) Check the tower body, etc. of iron tower before climbing the tower to keep it solid and reliable. (3) During climbing the tower, in order to prevent the operators from colliding with each other, the distance between the operators shall not be less than 1.6m, the safety belt shall be properly packed, the backup protection rope shall be placed in the kit, and the main belt shall be hung on the shoulder. (4) When climbing the tower to the cross arm, see the walking passage clearly, and the safe distance between the walking passage and the conductor shall not be less than 10.1m	(1) Operators shall wear safety helmet and soft work shoes, and climb evenly when climbing the tower. (2) During climbing the tower, the safety belt shall be properly packed, the safety rope at the back shall be placed in the kit, and the main belt shall be hung on the shoulder to prevent safety during climbing to prevent the safety belt from hooking shackles and tower materials during climbing, resulting in falling from the tower. (3) Operators shall not lose the protection from the safety belt and step and hold it firmly when moving on the tower. (4) When reaching the position of operation point, the safety belt and backup protection rope shall be firmly fastened on the component respectively, and shall not be hung low for high use. (5) The safe distance between the human body, transmission rope, etc. and the conductor must not be less than 10.1m before the power test, and special person shall be assigned to supervise the work.	

Table (Cont'd)

S/N	Work Content	Operation Steps and Standards	Safety Measures and Precautions	Responsible Person
7	Electricity testing and installing ground wire	(1) After climbing the tower to the designated position, fasten the safety belt to a firm component and check whether the safety belt is fastened. The installation position of the tackle is correct and convenient for operation. Tools and instruments must be transferred by ropes. (2) Before using the electroscope to test electricity, wear insulating gloves for self inspection and the signal shall be normal. When telescopic electroscope is used, all the insulation rods of each section should be pulled out in place to ensure the effective insulation length of the insulation rods. When testing electricity, the operator should hold the insulation handle of electroscope to ensure enough safe distance between human body and conductor. Lower layer first then upper layer, near side first then far side. The electrical inspection of power line shall be performed phase by phase. (3) The conductor ground wire shall be installed immediately after there is no voltage. The installation position of the ground terminal shall be polished with sandpaper. The ground terminal shall be installed first, then the conductor end. (4) When installing the conductor ground wire, the insulating gloves with throw and hang methods must be used for operation, and it is forbidden to install the ground wire with directly operating the metal part of the conductor ground wire by hand. Make sure the hook of the ground wire is tightly connected to the conductor	(1) Check whether the electroscope is normal during receiving and before using. (2) The ground wire shall be in good contact with the conductor and tower material. It is forbidden to install grounding wires by winding wires. (3) An enough safe distance shall be kept between human body and the conductor	

Part II Standard for Training and Assessment on Interruption Maintenance for Operation Maintenance of ±800kV UHV Power Transmission Line

Table (Cont'd)

S/N	Work Content	Operation Steps and Standards	Safety Measures and Precautions	Responsible Person
8	Installing ground wire	(1) After climbing the tower to the specified position, fasten the safety belt to a firm component and check whether the buckle is fastened. Use ropes to transfer tools and instruments. (2) Check whether the ground wire is in good condition, polish the tower material ground terminal with sandpaper, and install the conductor terminal after installing the ground terminal. (3) When installing the ground wire, the insulating gloves must be used for operation, and it is forbidden to install the ground wire with directly operating the metal part of the ground wire by hand. Make sure the hook of the ground wire is tightly connected to the ground wire	(1) It is forbidden to install the ground wire by winding the conductor (2) When the ground wire is placed, the operator shall wear insulating gloves and hold the insulating rod	
9	Installation the hanging wheel	(1) The operator shall check the corrosion of insulator set fitting before entering the ground wire. (2) Check whether the ground wire insulator is in good condition and whether the pin is complete. (3) Conduct impulse test on the ground wire. (4) Auxiliary personnel on the tower shall cooperate with installing the hanging wheel on the ground wire	(1) The operator shall check the corrosion of insulator set fitting before entering the ground wire. (2) The ground wire shall be conducted the impulse test before the installation of hanging wheel	
10	Access to operation point	Correctly use the work hanging wheel with uniform moving speed and no dangerous action; when moving on the ground wire, the safety belt protection shall not be lost.	(1) Safety belt protection shall not be lost during the whole process. (2) The hanging wheel shall not move too fast	

Table (Cont'd)

S/N	Work Content	Operation Steps and Standards	Safety Measures and Precautions	Responsible Person
11	Repair the broken ground wire	(1) Fix the hanging wheel after arriving at the operation point; (2) The operator shall polish the damaged part of the ground wire. (3) Measure the length of the preformed armor rods, and use a steel tape to measure the position of 1/2 length of the preformed armor rods at one end of the conductor damage, then paint and print. (4) The center of the preformed armor rods shall be installed at the seriously damaged part of the conductor, and without any gap. (5) Tap the end of the preformed armor rods with a wooden hammer, and the end shall be flat	(1) Correctly use the work hanging wheel with uniform moving speed and no dangerous action; when moving on the ground wire, the safety belt protection shall not be lost. (2) Before repairing the broken ground wire, the damaged point shall be polished. (3) The preformed armor rods shall be in the same direction as the broken ground wire. The winding shall be smooth and tight. The damaged part shall be in the middle of the preformed armor rods. (4) When winding the preformed armor rods, it is not allowed to pry it with force. To prevent it from deforming, both ends shall be kept flat during winding	
12	Access to cross arm	(1) The operator moves along the ground wire by the hanging wheel to return to the cross arm. (2) Protection of safety belt shall not be lost during entering the cross arm. (3) Use the transmission rope to transfer the hanging wheel to the ground	(1) Safety belt protection shall not be lost during the whole process. (2) The hanging wheel shall not move too fast	
13	Clean the tools and instruments on the iron tower	(1) Remove the tools and instruments in turn, such as the hanging wheel, ground wire, conductor ground wire, steel wire sleeves and tackle. (2) Remove the ground wire on the ground and conductor. Remove the lead (ground) wire terminal firstly, and then the ground terminal. (3) Confirm that there is no object left behind	(1) Remove the ground wire on the ground and conductor. The operator shall wear insulating gloves and hold the insulating rod. (2) An enough safe distance shall be kept between human body and the conductor. (3) Objects falling from heights shall be avoided	

Part II Standard for Training and Assessment on Interruption Maintenance for Operation Maintenance of ±800kV UHV Power Transmission Line

Table (Cont'd)

S/N	Work Content	Operation Steps and Standards	Safety Measures and Precautions	Responsible Person
14	Step down the tower	(1) Correctly climb down the tower along the main material of the shackles side (2) Correctly carry the transmission rope (The end of the transmission rope shall be double folded, fast knot, and hung over the shoulder)	Prevent falling from heights.	
15	End of the work	(1) Responsible person shall organize all working personnel to put working apparatus and materials in order, and clean the site to ensure "materials are removed and the site is cleaned after construction". (2) After a post-shift meeting is held, the Responsible Person shall make work summaries and comments. Comments include the construction quality of this work and the implementation of safety measures from all working personnel. (3) Responsible Person shall report the completion of the work to the work approver, resume the power transmission of the power cut line and terminate the work order	There must be no object left on site.	

II. Assessment Standard

Table 2-3-5 Detailed Rules for Assessment and Scoring of Operation Maintenance Skills of UHV DC Transmission Line

Fill-in Column of Examinee	No.:	Name:	Position:	Unit:	Date:	Year Month Day
Fill-in Column of Assessor	Achievement:	Assessor:	Assessment Team Leader:	Starting time:	Closing time:	Operation Duration:
Assessment Module	Power-cut repair of overhead ground wire for ±800kV UHV transmission line tangent tower. Assessee: Maintenance personnel for UHV DC transmission line. Assessment method: Operation. Assessment time limit: 100min					
Job Description	Power-cut Repair of ±800kV UHV Transmission Line Overhead Ground Wire					
Work specification and requirements	1. Given conditions: ±800kV DC practical training line left overhead ground wire needs to be repaired. The line is power off, test electricity and install the ground wire, the used tools and instruments have been tested, the work order has been handled, and the safety measures have been taken. 2. The exam participant should make necessary safety check before operation. 3. The operation hanging wheel shall meet the stress requirements. 4. The main operation procedures of the whole process are completed by 1 Responsible Person, 1 special Supervisor, 1 electrician on tower, 1 auxiliary workers on the tower and 2 auxiliary workers on the ground. Responsibilities of Responsible Person: Be responsible for the labor division of operating personnel of the work task, field investigation before work, preparation of the operation plan, work order filling, field re-investigation, performing work permit procedures, holding pre-shift meeting, implementation of the site safety measures, safety supervision in the operation process, dealing with emergency situations in work, quality surveillance of the work, and the summary after the work. Responsibility of Safety Supervisor: Explain the safety measures within the scope of supervision to the supervised personnel, and inform the hazards and safety precautions before work. Supervise the supervised personnel to comply with this procedures and implement the on-site safety measures, and timely correct the unsafe acts and behaviors of the supervised personnel. Responsibilities of operators on the tower: Be responsible for power-cut repair of ±800kV DC transmission line overhead ground wire. Auxiliary worker on the tower: be responsible for transferring tools and materials to match the installation of hanging wheel. Responsibilities of ground electricians: Assist the exam participant to complete the up and down transfer of tools and materials, and other auxiliary work.					

Part II Standard for Training and Assessment on Interruption Maintenance for Operation Maintenance of ±800kV UHV Power Transmission Line

Table (Cont'd)

Work specification and requirements	Given conditions: 1. Training line: UHV ±800kV DC practical training line left overhead ground wire, model of overhead ground wire: LBGJ-150-20AC. 2. The operation must be carried out according to the working procedure. The relevant item scores shall be deducted for the process error. In case of major hidden dangers of personal, equipment and operational safety, the assessor may order the termination of the assessment
Assessment scenario preparation	1. Line: UHV ±800kV DC practical training line left overhead ground wire. Model of overhead ground wire: LBGJ-150-20AC. 2. Required work tools and instruments: 1 set of work hanging wheel, 2 sets of safety belt, 3 nos. of walkie-talkies, 1 nos. of conductor ground wire, 1 nos. of ground wire, 1 pair of insulating gloves, 1 nos. of transmission rope, 1 nos. of 0.5T tackle, 1 nos. of steel wire sleeve, 1 sheet of sandpaper, 2 sets of personal tools and 1 nos. of wooden hammer. 3. Material: a set of preformed armor rods corresponding to the model of ground wire
Remarks	1. Personal tools shall be self-provided by the examinee. 2. The deduction shall be done until the scores of each item are deducted completely.

Table 2-3-6 Standard for Evaluation of Assessment and Scoring of Operation Maintenance Skills of UHV DC Transmission Line

S/N	Project name	Quality requirements	Score	Deduction standard	Reasons for deduction	Deduction	Scoring
1	Uniform	Correctly wear working clothes, work shoes, safety helmet and protective gloves	4	Deduct 1 point for missing one item			
2	Personal tools Check	Adjustable wrench, flat pliers, and tool kits shall meet the quality requirements	3	Deduct 1 point for failure of inspection			
3	Safety belt inspection	The safety belt meets the quality requirements within the test period	2	(1) Deduct 1 point for no appearance inspection. (2) Deduct 1 point for not checking the factory certificate and test certificate.			

Table (Cont'd)

S/N	Project name	Quality requirements	Score	Deduction standard	Reasons for deduction	Deduction	Scoring
4	Operation tools Check	The operation hanging wheel shall meet the requirements within the test cycle	2	(1) Deduct 1 point for no appearance inspection. (2) Deduct 1 point for not checking the factory certificate and test certificate.			
5	Material inspection	Confirm that the preformed armor rods are corresponding to the ground wire, and there shall be no damage in the appearance inspection	2	(1) Deduct 1 point for choosing the wrong preformed armor rods. (2) Deduct 1 point for no inspection			
6	Climbing the tower	(1) Check the line name, the number of the iron tower number on spot to ensure they are correct; guarantee that the foundation and the tower body are intact and in normal condition, and ensure that the cross and span distance meets the safety requirements; confirm the defect conditions and the specifications and models of ground wires. (2) Correctly carry the transmission rope (The lifting rope head is double folded, fast knot, and hung across over the shoulder). (3) Correctly climb on the tower along the main material of the shackles side (4) After climbing the tower to the specified position, fasten the safety belt to a firm component and check whether the buckle is fastened	9	(1) Deduct 2 points if the double title of the line is not confirmed; deduct 1 point for each item if the base, tower body and crossing point are not checked; deduct 1 point if the ground wire defect is not confirmed. (2) Deduct 2 points for not carrying the transmission rope, and deduct 1 point for nonstandard carrying method of lifting rope. (3) Deduct 1 point/time for grasping the shackles with hands. (4) Deduct 2 points for not climbing up the tower along the main material of the shackles side. (5) Deduct 2 points for non standard use of safety belt. (6) Deduct 1 point each time when stepping on air or sliding			

Part II Standard for Training and Assessment on Interruption Maintenance for Operation Maintenance of ±800kV UHV Power Transmission Line

Table (Cont'd)

S/N	Project name	Quality requirements	Score	Deduction standard	Reasons for deduction	Deduction	Scoring
7	Power test and installation of conductor and grounding wire	(1) After climbing the tower to the designated position, fasten the safety belt to a firm component and check whether the safety belt is fastened. The installation position of the tackle is correct and convenient for operation. Tools and instruments must be transferred by ropes. (2) Before using the electroscope, the operator shall check the sound and light signal again. When telescopic electroscope is used, all the insulation rods of each section should be pulled out in place to ensure the effective insulation length of the insulation rods. When testing electricity, the operator should hold the insulation handle of electroscope to ensure enough safe distance between human body and conductor. Lower layer first then upper layer, near side first then far side. The electrical inspection of power line shall be performed phase by phase. (3) The conductor ground wire shall be installed immediately after there is no voltage. The tower material position of the ground terminal shall be polished with sandpaper. The ground terminal shall be installed first, then the conductor end. The installation sequence is middle phase first, then two side phase.	8	(1) Deduct 2 points for non standard installation of tackle. (2) Deduct 2 points for non standard use of safety belt. (3) Deduct 2 points if no insulating gloves are worn for power test and installation of grounding wire. (4) Deduct 2 points for nonstandard use of electroscope, and 2 points for incorrect sequence for electricity testing. (5) Deduct 2 points if the installation sequence of both ends of the ground wire is wrong; deduct 1 point if the connection is not firm; deduct 1 point if the installed end is not inspected. (6) Deduct 2 points for winding ground wire.			

Table (Cont'd)

S/N	Project name	Quality requirements	Score	Deduction standard	Reasons for deduction	Deduction	Scoring
7	Power test and installation of conductor and grounding wire	(4) When installing the conductor ground wire, the insulating gloves must be used for operation, and it is forbidden to install the ground wire with directly operating the metal part of the conductor ground wire by hand. Make sure the hook of the ground wire is tightly connected to the conductor					
8	Hanging ground wire	(1) After climbing the tower to the specified position, fasten the safety belt to a firm component and check whether the buckle is fastened. Use ropes to transfer tools and instruments. (2) Check whether the ground wire is in good condition, polish the tower material ground terminal with sandpaper, and install the conductor terminal after installing the ground terminal. (3) When installing the ground wire, the insulating gloves must be used for operation, and it is forbidden to install the ground wire with directly operating the metal part of the ground wire by hand. Make sure the hook of the ground wire is tightly connected to the ground wire	6	(1) Deduct 2 points for non standard use of safety belt. (2) Deduct 2 points if no insulating gloves are worn for installation of grounding wire. (3) Deduct 2 points if the installation sequence of both ends of the ground wire is wrong; deduct 1 point if the connection is not firm; deduct 1 point if the installed end is not inspected. (4) Deduct 2 points for winding ground wire.			

Part II Standard for Training and Assessment on Interruption Maintenance for Operation Maintenance of ±800kV UHV Power Transmission Line

Table (Cont'd)

S/N	Project name	Quality requirements	Score	Deduction standard	Reasons for deduction	Deduction	Scoring
9	Installation the hanging wheel for operation	(1) The operator shall check the corrosion of insulator set fitting before entering the ground wire. (2) Check whether the ground wire insulator is in good condition and whether the pin is complete. (3) Conduct impulse test on the ground wire. (4) Auxiliary personnel on the tower shall cooperate with the installation work of hanging wheel	9	(1) Deduct 2 points if the insulator set fitting is not inspected for corrosion; (2) Deduct 1 point for each item for not checking if the ground wire insulator is not in good condition and pin is complete; (3) Deduct 2 points for failure to conduct impulse test to the ground wire. (4) Deduct 2 points for the installation of working hanging wheel is not standard			
10	Access to operation point	Correctly use the work hanging wheel with uniform moving speed and no dangerous action; when moving on the ground wire, the safety belt protection shall not be lost.	9	(1) Deduct 3 points per time for loosing the protection of the safety belt. (2) Deduct 2 points for the hanging wheel speed moves too fast; (3) Deduct 3 points per time each time for there is dangerous action in the process of movement			
11	Repair the broken ground wire	(1) Fix the hanging wheel after arriving at the operation point. (2) Repair and polish and level the damaged ground wire; (3) Measure the length of the preformed armor rods, and use a steel tape to measure the position of 1/2 length of the preformed armor rods at one end of the conductor damage, then paint and print;	16	(1) Deduct 3 points for failure to fix the hanging wheel in proper position. (2) Deduct 3 points if the ground wire is not polished and flat. (3) Deduct 2 points for proper marking. (4) Deduct 4 points if the installation sequence of preformed armor rods is not standard.			

· 617 ·

Table (Cont'd)

S/N	Project name	Quality requirements	Score	Deduction standard	Reasons for deduction	Deduction	Scoring
11	Repair the broken ground wire	(4) The center of the preformed armor rods shall be installed at the seriously damaged part of the conductor, and without any gap; (5) Tap the end of the preformed armor rods with a wooden hammer, and the end shall be flat		(5) Deduct 2 points if the end of preformed armor rods is not smooth. (6) Deduct 1 point for each installation of preformed armor rods with gap			
12	Access to cross arm	(1) The operator moves along the ground wire by the hanging wheel to return to the cross arm; (2) Protection of safety belt shall not be lost during entering the cross arm; (3) Use the transmission rope to transfer the hanging wheel to the ground	8	(1) Deduct 2 points for improper control of the hanging wheel moving speed. (2) Deduct 2 points per time for losing the protection of safety belt when accessing to the cross arm. (3) Deduct 2 points for nonstandard transmission of the hanging wheel			
13	Clean the tools and instruments on the iron tower	(1) Remove the tools and instruments in turn, such as the hanging wheel, ground wire, conductor ground wire, steel wire sleeves and tackle; (2) Remove the ground wire on the ground and conductor. Remove the lead (ground) wire terminal firstly, and then the ground terminal; (3) Confirm that there is no object left behind	7	(1) Deduct 3 points for the removal in the wrong order. (2) Deduct 4 points for any object left behind			

Part II Standard for Training and Assessment on Interruption Maintenance for Operation Maintenance of ±800kV UHV Power Transmission Line

Table (Cont'd)

S/N	Project name	Quality requirements	Score	Deduction standard	Reasons for deduction	Deduction	Scoring
14	Step down the tower	(1) Carry the transmission rope and climb down the tower evenly along the shackles. (2) Correctly carry the transmission rope (The end of the transmission rope shall be double folded, fast knot, and hung over the shoulder)	8	(1) Deduct 4 points for not carrying the transmission rope, and deduct 2 points for nonstandard carrying method of lifting rope. (2) Deduct 1 point/time for grasping the shackles with hands. (3) Deduct 4 points for not climbing the tower along the main material of the shackles side			
15	Place tools and instruments	Clean up work site after completion and arrange the tools neatly as required	2	Deduct 2 points for not placing tools and instruments as required			
16	Housekeeping	(1) It is prohibited to drop articles down from high space. (2) No floating objects. (3) No holding object in mouth	5	(1) Deduct 5 points for each falling object at high place. (2) Deduct 2 points for floating articles. (3) Deduct 2 points for holding object in mouth			
17	Completion time	Complete operation as required within the specified time.		Complete as required within the specified time, terminate the operation after 10 minutes, and only record the score of the completed part			
18	Total		100				

Module 4 Training and Assessment Standard for Power-cut Replacement of ±800kV UHV Transmission Line Overhead Ground Wire

I. Training Standard

(I) Training Requirements

Designation of module	Power-cut replacement of ±800kV UHV transmission line overhead ground wire	Type of training	Operation type
Training method	Practical operation training	Training class hours	21 hours
Training objectives	1. Understand the type, structure and installation way of overhead ground wire. 2. It can inspect and install the overhead ground wire of the ±800kV special pressure transmission line. 3. Be able to complete power-cut replacement of ±800kV UHV transmission line overhead ground wire		
Training venue	UHV DC practical training line		
Training content	Power-cut replacement of ±800kV UHV DC transmission line overhead ground wire		
Scope of application	Maintenance personnel for UHV DC power transmission line		

(II) Referenced Procedures and Specifications

(1) Technical Specification for Live Working of ±800kV DC Line (DL/T1242-2013)

(2) Operating Code for ±800kv DC Overhead Transmission Line (GB/T28813-2012)

(3) Maintenance Specification for ±800kV DC Overhead Transmission Line (DL/T251-2012)

Part II Standard for Training and Assessment on Interruption Maintenance for Operation Maintenance of ±800kV UHV Power Transmission Line

(4) Technical Guide for Live Working of ±800kV DC Transmission Line (Q/GDW302-2009)

(5) Shielding Clothes for Live Working (GB/T6568-2008)

(6) Guidelines of Insulation Coordination for Live Working (DL/T867-2004)

(7) Test Guide of the Insulating Tool for Live Working (DL/T878—2004)

(8) State Grid Corporation of China on the Management Regulations of Live Working (Trial Implementation) (SGCC [2007] No.751)

(9) State Grid Corporation of China Working Regulations of Power Safety (Transmission Line Section) (Q/GDW1799.2-2013)

(10) Electrotechnical Terminology - Overhead Line (GB/T 2900.51-1998)

(11) Electrotechnical Terminology - Live Working (GB/T2900.55-2016)

(12) Live Working - Terminology for Tools, Equipment and Devices (GB/T 14286-2008)

(13) Minimum Requirements for Utilization of Tools, Devices and Equipment for Live Working (DL/T877-2004)

(14) Preventive Test Code of Tools, Devices and Equipment for Live Working (DL/T 976-2005)

(15) Insulated Tackles for Live Working (GB/T13034-2008)

(16) Live Working-Insulating Ropes (GB 13035-2008)

(III) Teaching Design for Training

The design is to complete the work task of "power-cut replacement of ±800kV UHV transmission line overhead ground wire". Each training stage shall be designed according to the standard operation procedure for work task completion. Each stage includes specific training objectives, training contents, hours of training, training methods (training resources), training environment, examination and evaluation, etc., as shown in Table 2-4-1.

Table 2-4-1 Training Content Design for Power-cut Replacement of ±800kV UHV DC Transmission Line Overhead Ground Wire

Training schedule	Training objectives	Training contents	Training class hours	Training method and resources	Preparation of training conditions	Assessment
1. Theory Teaching	1. Be familiar with the type, structure and installation of overhead ground wire. 2. Be familiar with the operation procedure of electricity testing and setting ground wire. 3. Be familiar with the operation procedure of power-cut replacement of ±800kV UHV DC transmission line overhead ground wire	1. Type, structure and installation way of overhead ground wire. 2. Operation procedure of electricity testing and setting ground wire. 3. Operation procedure of power-cut replacement of ±800kV UHV DC transmission line overhead ground wire	2	Training method: Lecture. Training resources: PPT, relevant regulations and specifications	Multimedia classroom	Attendance, classroom questions and assignments
2. Preparation Work	Be able to complete the preparation before operation Work	1. Work site survey. 2. Preparation of the standardized operation card. 3. Filling of the work order. 4. Preparation of tools and materials for this operation	1	Training method: 1. Site survey and cleaning of tools and materials shall be practiced at site. 2. Preparation of operation card and the filling of work order shall adopt lecture method. Training resources: 1. ±800kV practical training line. 2. UHV tools and instruments warehouse. 3. Blank work order	1. UHV transmission line for practical training; 2. Multi-media classroom.	

Part II Standard for Training and Assessment on Interruption Maintenance for Operation Maintenance of ±800kV UHV Power Transmission Line

Table (Cont'd)

Training schedule	Training objectives	Training contents	Training class hours	Training method and resources	Preparation of training conditions	Assessment
3. Work site preparation	Be able to complete the preparations of work site	1. Work site re-survey. 2. Permit to work. 3. Site layout. 4. Pre-shift meeting. 5. Inspection of tools and instruments and materials	2	Training methods: demonstration and role play. Training resources: ±800kV practical training line	±800kV practical training line	
4. Trainer's demonstration	The trainees can preliminarily understand the operation process of the task through inspecting and learning from each other's work	1. Electricity testing and setting ground wire. 2. Tangent tower threadbare. 3. Resisting-tensile tower thread slack and connecting new and old ground wire. 4. Installation of crossover and accessories	8	Training method: Demonstration. Training resources: ±800kV UHV DC practical training line	±800kV practical training line	
5. Group training of trainees	1. Operators can test electricity and install ground wire. 2. Be able to complete power-cut replacement of ±800kV UHV DC transmission line overhead ground wire	1. The trainees are grouped (22 in a group) to train the skills of electricity testing, installing ground wire and replacing overhead ground wire 2. Trainers guide the operation of trainees and conduct safety supervision	7	Training methods: Role play. Training resource: practical training line	±800kV practical training line	Score the operation of trainees according to the detailed rules for skill assessment and scoring
6. End of the work	1. Enable the trainees to further distinguish the shortcomings of the operation process and facilitate the promotion in the later stage. 2. Train the trainees in the working style of safe and civilized production	1. Clean the work site. 2. Report to dispatcher. 3. Comment and summarize the work task this time at post-shift meeting	1	Training method: Lecture and inductive method	±800kV practical training line	

(IV) Work Flow

1. Work Task

Complete Power-cut Replacement of ± 800kV UHV Transmission Line Overhead Ground Wire.

2. Requirements for Weather and Work Site

(1) It should be carried out in good weather for power-cut replacement of ±800kV UHV transmission line overhead ground wire.

In case of thunder (hear thunder, see lightning), snow, hail, rain, fog, etc., and the wind force is greater than level 5, any working shall not be conducted. When emergency power-cur repair is required in bad weather, relevant personnel shall be organized to fully discuss and prepare necessary safety measures, which can be implemented after being approved by the unit.

(2) The operators are in good mental state, and the members of the working group carefully study the work ticket and safety technical measures, and all the personnel shall be conform to the "four clear" (the operation task is clear, the danger point is clear, the operation procedure is clear, and the safety measures are clear).

(3) The Responsible Person must contact with the power cutting contact person to fulfill the Work licensing procedures, and it is strictly prohibited to stop and transfer electricity at the appointed time. Responsible Person must obtain the license work order of the licensor before checking the electricity on the line to be overhauled, setting up the ground wire and carrying out the maintenance work.

(4) After power-cut, Responsible Person shall make records carefully.

(5) Check whether there is a honeycomb on the tower before climbing the pole. It is forbidden to climb the tower if a honeycomb is found.

(6) Double safety belt shall be used by the personnel who work on the tower.

3. Preparations

3.1 Hazards and precontrol measures

(1) Hazard - climbing live line

Precontrol measures:

① Before climbing the tower for operation, Responsible Person and the members of the working team shall check the double name and identification mark carefully (color code, distinguishing mark, etc.) to match the name of the power-cut line.

② Check the tower root and the base of pole and tower before climbing, making sure it's solid and reliable.

③ Before climbing the pole and tower, climbing tools and facilities such as safety belt, shackles and tower materials shall be inspected to ensure they are firm.

④ Before climbing the pole, the intermediate operators who do not involved the installation of the grounding wire shall carefully verify that the phase sequence, color code, name, and number of the line are consistent with the power-cut line, and confirm that the line name is not error or in opposite order.

(2) Hazard —— falling from high places

Precontrol measures:

① During climbing, in order to prevent the operator from falling, the distance between the staff members shall not be less than 1.6m.

② Before climbing, the soil on the soles shall be removed, and the kit shall be complete, and objects shall not be dropped during climbing.

③ Operators shall wear safety helmet and soft sole, and be scrupulous and when climbing pole and tower.

④ During climbing, the safety belt shall be properly packed, the long tail rope shall be placed in the kit, and the main belt shall be hung on the shoulder to prevent safety during climbing to prevent the safety belt from hooking shackles and tower materials during climbing, resulting in falling from the tower.

⑤ Operators shall not lose the protection from the safety belt and step and hold it firmly when moving on the pole and tower.

⑥ When reaching the position of operation point, fasten the safety belt (rope), it shall be firm and reliable, and shall not be hung low for high use.

⑦ The safe distance between the human body, headless rope, etc. and the conductor must not be less than 10.1m before the power test, and special person shall be assigned to supervise the work.

(3) Hazard —— injury caused by objects falling from high place

Precontrol measures:

① The ground staff shall not stand under the position vertical to the operation point. The personnel on tower shall prevent falling objects from injury, and the tools and materials used shall be transmitted by ropes.

② Tool bags should be used when working at heights. Larger tools should be fixed on solid components and are not allowed to be randomly placed.

③ In the process of lifting with a winching, special personnel shall be assigned to command and cooperate in a unified way. Impact inspection shall be carried out immediately after the insulator string are lifted off the ground.

(4) Hazard —— Prevent injury caused by induced electricity

① In order to prevent injury caused by induced electricity, electricity testing shall be carried out and the ground wire shall be installed on the maintenance line.

② If contact with overhead ground wire is required, reliable grounding shall be conducted before contact with overhead ground wire.

(5) Hazard —— Site safety monitoring

① From the beginning of the operation to the end of the operation, the safety supervisor must always carry out uninterrupted safety supervision on the operators on the site.

② The Responsible Person and the Supervisor must wear a safety vest.

(6) Hazard ——Traffic safety

① When driving, attention should be paid to the safety of the vehicle, driving the vehicle carefully, and illegal driving is prohibited.

3.2 Selection of tools, instruments and materials

See Table 2-4-2 for tools and instruments and materials required by power-cut replacement for UHV transmission line overhead ground wire of ±800kV line. Before delivering tools and instruments out of warehouse, application voltage class and test cycle shall be carefully checked and they shall be inspected to ensure that appearance is intact, connection is firm, rotation is flexible, and they meet the working task requirements. After delivering tools and instruments out of warehouse, they shall be stored in tools bag or tool kit for transportation to avoid contamination and damp. Metal tools and insulated tools shall be separately loaded and transported to avoid deformation, damage or other defects caused by mixed loading and transportation.

Table 2-4-2 Tools and Instruments and Materials Required for the Power-cut Replacement of ±800kV

S/N	Name	Specification and Model	Unit	Qty.	Remarks
1	Operation hanging wheel		Set	1	Other tools
2	Grinding rope	ϕ16	Roll	1	Other tools

Part II Standard for Training and Assessment on Interruption Maintenance for Operation Maintenance of ±800kV UHV Power Transmission Line

Table (Cont'd)

S/N	Name	Specification and Model	Unit	Qty.	Remarks
3	Lever blocks	6T	pcs	3	Metal tool
4	Winching	5T	Set	3	Maneuvering tools
5	Lifting rope tackle	1T	Set	3	Metal tool
6	Noose		Nos.	3	Other tools
7	Wire clamp		Nos.	2	Metal tool
8	Wire lifter		Set	1	Metal tool
9	Single-wheel tackle		Nos.	3	Metal tool
10	Paying-off rack		Set	2	Metal tool
11	Paying-off disk		Set	2	Metal tool
12	Steel wire sleeve	$\phi 18$	Nos.	8	Other tools
13	Iron tackle	10T	Nos.	8	Metal tool
14	Shackle	10T	Nos.	8	Metal tool
15	Shackle	8T	Nos.	6	Metal tool
16	Hydraulic press (including hydraulic pliers)		Set	2	Maneuvering tool
17	Interphone		Nos.	10	Other tools
18	Full-covered Safety Belt	Including 20m rear safety rope with buffer bag	Set	7	PPE
19	Safety Vest		pcs	3	Other tools
20	Security fence		Volume	5	Other tools
21	Sole timber		pcs	Several	Other tools
22	Ground Wire	Al-clad steel strand JLB20A, 150	Roll	3	Material
23	Strain clamp	Same model	Set	2	Material

3.3 Division of labor for operators

Division of labor for operators of the task is shown in Table 2-4-3.

Table 2-4-3 Division of Operating Personnel Involving in Power-cut Replacement of Overhead Ground Wire on ±800kV UHV Transmission Line

S/N	Post	Qty. (person)	Responsibilities
1	Responsible Person	1	Be responsible for division of operating personnel of the task, field investigation before work, implementation of the operation plan, work order filling, field re-investigation, performing work permit procedures, holding pre-shift meeting, the implementation of safety measures on site, safety supervision in the operation process, dealing with emergency situations in work, quality surveillance, and the summary after work.
2	Safety Supervisor	1	Be responsible for the safety monitoring work during this operation
3	Operator working at heights	7	Be responsible for the operation of power-cut replacement of ±800kV DC transmission line overhead ground wire
4	Ground auxiliary personnel	10	Be responsible for ground auxiliary works during the operation.
5	Winching operator	2	Be responsible for the winching work during this operation
6	Winching signal command worker	2	Be responsible for commanding the start and stop of 2 sets of winches

4. Working Procedures

Working process of the task is shown in Table 2-4-4.

Table 2-4-4 Work Procedures for Power-cut Replacement of Overhead Ground Wire on ±800kV UHV Transmission Line

S/N	Work Content	Operation Steps and Standards	Safety Measures and Precautions	Responsible Person
1	Site re-survey	The Responsible Person shall complete the following work: (1) Check the line title, the number of the pole and tower and ensure the phases are correct; guarantee that the base and the pole and tower are intact and in normal condition; ensure that the cross and span distance meets the safety requirements; confirm the defect conditions and the specifications and models of earth wires.	(1) Correctly wear helmet, working clothes, work shoes and protective gloves. (2) Operation under meteorological conditions that may endanger the safety of operators is forbidden. (3) Non-operation personnel and vehicles are strictly prohibited from entering the work site	

Part II Standard for Training and Assessment on Interruption Maintenance for Operation Maintenance of ±800kV UHV Power Transmission Line

Table (Cont'd)

S/N	Work Content	Operation Steps and Standards	Safety Measures and Precautions	Responsible Person
1	Site re-survey	(2) Check that the site meteorological conditions such as wind speed should meet the operation requirements. (3) Check that the terrain and environment should meet the operation requirements. (4) Check that the safety measures listed in the work order are in line with the actual situations on site. The measures will be added if necessary.		
2	Work Permit	The contact person for power off/on before the operation must contact the dispatcher to complete the work licensing procedures	No work shall be started without the Work Permit of approver	
3	Site layout	Install the security fence and hang the signboards correctly: (1) The security fence should take full account of falling objects from the heights and the influence on road traffic. (2) The entrance and exit of the security fence shall be set reasonably. (3) Signs such as "Access from Here", "Work Here", "Slow Down" or "Blocking" shall be properly arranged	When the influence on road traffic safety is uncontrollable, the traffic management department should be contacted in time to strengthen the on-site control of traffic safety	
4	Holding a pre-shift meeting	(1) All working personnel shall line up. (2) The Responsible Person will read out the work order and be clear with the work task and division of personnel; explain safety measures and technical measures in work; check (inquire after) the mental state of all working personnel; and inform of hazards in work and pre-control measures. (3) All working personnel shall sign on the work order for confirmation.	(1) Work order shall be filled and issued with standardized licensing procedure and complete signature. (2) All working personnel shall be in good mental states. (3) All working personnel shall be clear with task division of works, safety measures and technical measures	

Table (Cont'd)

S/N	Work Content	Operation Steps and Standards	Safety Measures and Precautions	Responsible Person
5	Inspection tool	(1) All necessary tools and instruments shall be prepared as per the operation requirements and placed on the waterproof tarpaulin regularly according to the category and location. The appearance and test certificate of tools and instruments shall be checked to ensure there is no omission. (2) The inspector shall report to the Responsible Person that all inspection results are in conformity with the operation requirements	(1) The waterproof tarpaulin shall be enough in quantity and reasonable in position, and be clean and dry. (2) The appearance of tools and instruments is acceptable after inspection, there is no damage, damp, deformation and malfunction, and the certificate is valid	
6	Climbing the tower	(1) Check the name and serial number of the line before climbing the pole and tower. For multi-circuit lines on the same tower, the Responsible Person and the members of the working team shall check the double name and identification mark carefully (color code, distinguishing mark, etc.). (2) Check the tower root and the base of pole and tower before climbing, making sure it's solid and reliable. (3) During climbing, in order to prevent the operator from falling, the distance between the staff members shall not be less than 1.6m, the safety belt shall be properly packed, the long tail rope shall be placed in the kit, and the main belt shall be hung on the shoulder to prevent safety during climbing to prevent the safety belt from hooking shackles and tower materials during climbing, resulting in falling from the tower. (4) When the pole and tower is close to the cross arm, the Supervisor and the operator shall check the identification mark and dual tags of the power-cut line again, and can enter the cross arm at the power-cut line side only after they are confirmed to be correct	(1) Operator shall wear safety helmet and soft sole, and be scrupulous and when climbing pole and tower. (2) During climbing, the safety belt shall be properly packed, the long tail rope shall be placed in the kit, and the main belt shall be hung on the shoulder to prevent safety during climbing to prevent the safety belt from hooking shackles and tower materials during climbing, resulting in falling from the tower. (3) Operators shall not lose the protection from the safety belt and step and hold it firmly when moving on the pole and tower. (4) When reaching the position of operation point, fasten the safety belt (rope), it shall be firm and reliable, and shall not be hung low for high use. (5) The safe distance between the human body, headless rope, etc. and the conductor must not be less than 9.5m before the power test, and special person shall be assigned to supervise the work	

Part II Standard for Training and Assessment on Interruption Maintenance for Operation Maintenance of ±800kV UHV Power Transmission Line

Table (Cont'd)

S/N	Work Content	Operation Steps and Standards	Safety Measures and Precautions	Responsible Person
7	Inspection power and install ground wire	(1) After the pole is in place, fasten the safety belt to a firm and reliable component or pole, and check whether the buckle is in place correctly. Tools and instruments such as test pen (electroscope) must be transferred by rope. (2) Standing at the correct position during electricity inspection: a position suitable for work and free from electric shock hazard; Insulating gloves must be worn when checking electricity. The power inspection of the line shall be carried out phase by phase, according to the sequence of first near then far, prior lower layer and posterior upper layer. (3) Check whether the ground wire is in good condition, and install the ground wire according to the procedures (first connect the ground terminal, then the conductor terminal). (4) When setting the ground wire, the insulating rope or insulated handle must be used for operation, and it is forbidden to install the ground wire with directly operating the metal part of the ground wire by hand. Make sure the collet of the ground wire is tightly connected to the conductor	(1) Before electricity test, test shall be carried out on electrical equipment to verify that the electroscope is in good condition. When testing on electrical equipment, the electroscope cannot be verified to be in good condition with high pressure generator, etc. (2) Wear insulating gloves during high voltage electrical test. Completely pull the telescopic insulating rod of the electroscope, hold the handle with hands and do not stretch out from the guard ring, and human body keeps a safe distance away from the electroscope. Do not perform electrical test at outdoor in thunderstorm weather. (3) The grounding wire shall be installed with ground terminal connected first and then conductor terminal, and must be in good contact. It is in opposite order when dismantling. It is forbidden to install the ground wire by winding the conductor	
8	Tangent tower threadbare	(1) The operator shall install the single-sheave tackle in the suitable position of the tangent tower ground wire cross arm and remove the anti-vibration hammer on the ground wire. (2) The operator should lift the ground wire with wire lifter and 6T lever hoist. (3) After removing part of the armor clamp for connection, turn the ground wire into the single-sheave tackle fixed on the ground wire cross arm		

Table (Cont'd)

S/N	Work Content	Operation Steps and Standards	Safety Measures and Precautions	Responsible Person
9	Installation of wire clamp	(1) After measuring and recording the position of anti-vibration hammer on the ground wire, the operator shall remove the anti-vibration damper. (2) After the operator uses a hanging wheel to clamp the wire clamp to a suitable position on the ground wire, tighten the grinding rope. (3) After tightening the ground wire at both ends of the ground wire through a 6T lever hoist, take off the ground wire insulator set fitting	After tightening the ground wire, an impact test shall be carried out, and the next operation can be carried out only after the impact test is confirmed to be correct	
10	Resisting-tensile tower thread slack and connecting new and old ground wire	(1) The operator loosens the 6T lever hoist at both ends to remove the lever hoist after the grinding rope is stressed. (2) Start the winching and use the winching to loosen the ground wires at both ends to the ground. (3) After loosening to the ground, the ground personnel shall cut off the strain clamp at the ground wire end, polish and crimp the new and old ground wires	(1) During paying-off, the operation of the take-up and winching should be smooth, keep the traction balance of the ground wire, and prevent the ground wire from slipping. (2) When using the winching, the number of twining on the winching shall not be less than 5, and it shall be rolled from below and be neat, and the number of tail rope control personnel shall not be less than 2. (3) When laying off and stringing, personnel shall neither stand beside, cross the stressed hauling rope, the interior angle side of the ground wire, in the dispatched ground wire circles, nor below the vertical direction of the hauling rope and overhead line in case of any injury due to accidental deviation from lines.	

Table (Cont'd)

S/N	Work Content	Operation Steps and Standards	Safety Measures and Precautions	Responsible Person
10	Resisting-tensile tower thread slack and connecting new and old ground wire		(4) Personnel are forbidden to place any part of the body directly above the hydraulic press when ground wire is crimped. The crimping machine shall have fixed facilities and shall be placed stably during operation. The thread-holding personnel on both sides shall be aligned with the position and their fingers shall not extend into the stamper	
11	Crossover and accessories Installation	(1) Start 2 winching on the traction field and tension field at the same time under unified command by specially-assigned person. (2) After the new ground wire arrives at the traction field, the tension field shall be crimped with the strain clamp. After crimping, the crimped ground wire shall be connected to the insulator armor clamp by winding and winching. (3) If the traction field tightening the ground wire, and mark it when the sag of the ground wire is consistent with the original sag. (4) Loosen the ground wire to the ground, crimp the strain clamp at the other end, then tighten the grinding rope. Connect the ground wire to the insulator string armor clamp and the new ground wire to the cable clamp by the tangent tower, fasten the bolts, and install the anti-vibration hammer according to the previously marked position.	(1) During the pay-off construction, special personnel shall be assigned to watch over temporary stay wires, cross spans, tangent towers and mechanical tools, and check that the mechanical conditions are in good condition. (2) When paying-off, communication should be kept smooth with unified signals and commands. (3) During construction, it is prohibited to cross or linger under the ground wire. (4) In the process of crossover, special personnel shall check the stress of ground anchor, steering tackle and grinding rope at any time and make timely adjustment. (5) During paying-off, the operation of the take-up and winching should be smooth, keep the traction balance of the ground wire, and prevent the ground wire from slipping.	

Table (Cont'd)

S/N	Work Content	Operation Steps and Standards	Safety Measures and Precautions	Responsible Person
12	Removal of tools and instruments	(1) Remove tools and instruments in sequence, such as grinding rope, lever hoist, wire clamp and backup protection rope, etc. (2) Remove of ground wire	The removal sequence of ground wire: conductor terminal first and then the ground terminal	
13	End of the work	(1) Responsible person shall organize all working personnel to put working tools and instruments and materials in order, and clean the site to ensure "materials are removed and the site is cleaned after construction". (2) After a post-shift meeting is held, the Responsible Person shall make work summaries and comments. Comments include the construction quality of this work and the implementation of safety measures from all working personnel. (3) Responsible Person shall report the completion of the work to the work approver, resume the power transmission of the power cut line and terminate the work order		

Part II Standard for Training and Assessment on Interruption Maintenance for Operation Maintenance of ±800kV UHV Power Transmission Line

II. Assessment Standard

Table 2-4-5 Detailed Rules for Assessment and Scoring of Operation Maintenance Skills of UHV DC Transmission Line

Fill-in Column of Examinee	No.:	Name:	Position:	Unit:	Date:	MM/DD/YYYY
Fill-in Column of Assessor	Grade:	Assessor:	Assessment Team Leader:	Starting time:	Closing time:	Operation Duration:
The assessment module	Power-cut replacement of tension-resistance whole strain insulator string of ±800kV UHV transmission line tangent tower. Assessee: Maintenance personnel for UHV DC transmission line. Assessment method: Operation. Assessment time limit: 360min					
Job Description	Power-cut replacement of the left overhead ground wire in the strain section on ±800kV UHV transmission line					
Working specification and requirements	1. Given conditions: The left overhead wire of 001#-003# tower strain section of UHV ±800kV practical training line needs to be replaced. The line is power off, test electricity and install the ground wire, the tools and instruments and materials to be used have been tested, the work order has been handled, and the safety measures have been taken. 2. The work shall be carried out under good weather. In case of lightning (hearing thunder or seeing lightning), snow, hail, rain, fog and so on with wind force greater than Level 5, working shall be prohibited. 3. This assignment needs 1 Responsible Person, 1 Safety Supervisor, 1 electrician on tower, 6 auxiliary workers on tower, 10 auxiliary workers on the ground, 2 winching operators to assist the personnel to finish the up and down transfer of tools and instruments, materials and other non-technical work. 4. Responsibilities of Responsible Person: Be responsible for the labor division of operating personnel of the work task, field investigation before work, preparation of the operation plan, work order filling, field re-investigation, performing work licensing procedures, holding pre-shift meeting, implementation of the site safety measures, safety supervision in the operation process, dealing with emergency situations in work, quality surveillance of the work, and the summary after the work. 5. Responsibilities of Special Supervisor: Be responsible for safety control of the work site. 6. Duties of personnel working at heights: Be responsible for the operation of power-cut replacement of ±800kV UHV transmission line overhead ground wire.					

Table (Cont'd)

Working specification and requirements	7. Responsibilities of ground auxiliary personnel: Be responsible for transferring tools and materials and cooperating with electricians on tower. 8. Responsibilities of winching operators: Be responsible for the winching work during this operation. 9. If thunder, rain, strong wind or any other circumstance threaten the safety of the staff in the course of working, the Responsible Person or Supervisor may stop working temporarily according to the circumstances. Given conditions: 1. Training base: The left overhead wire of 001#-003# tower strain section of UHV ±800kV practical training line; Ground wire type: Al-clad steel strand JLB20A, 150. 2. Work orders have been handled, safety measures have been completed, and oral application (dispatcher or assessor) shall be made at the beginning and end of the work. 3. The instrument shall be used safely and correctly to test the insulating tool. 4. The operation must be carried out according to the working procedures. The scores of the items to be carried out shall be deducted for the process error. In case of major hidden dangers of personal, equipment and operational safety, the assessor may order the termination of the operation (assessment)
Assessment scenario preparation	1. Line: The left overhead wire of 001#-003# tower strain section of UHV ±800kV line. Work content: power-cut replacement of UHV ±800kV transmission line overhead ground wire; Model of ground wire: JLB20A-150. 2. Tools and instruments as required: 1 hanging wheel, a bundle of Φ16 grinding rope, 3 lever hoists (6T), 3 winchings (5T), 3 sets of lifting rope tackles, 3 nooses, 2 wire clamps, 1 wire lifter, 3 single-sheave tackles, 2 sets of pay-off racks, 2 sets of pay-off reels, 8 steel wire sleeves (Φ18), 8 iron tackles (10T), 8 shackles (10T), 6 shackles (8T), 2 sets of hydraulic presses (including hydraulic pliers), some walkie-talkies, 7 double-insurance safety belts, and several wood blocks. 3. Materials: 1 reel of ground wire of the same model, 1 set of splicing sleeve and 2 sets of strain clamps. 4. The work site shall be monitored, and the safety measures (fence, etc.) on the work site have been fully implemented; non-operation personnel are prohibited from entering the site, and the staff must wear safety helmet when entering the work site. 5. Examinees shall bring their own work clothes, safety helmets, cotton gloves, and safety belts (including double-protective ropes)
Remarks	1. The deduction shall be done until the scores of each item are deducted completely. In case of major hidden dangers of personal, equipment and operational safety, the assessor may order the termination of the operation. 2. When equipment, working environment, safety belt, safety helmet, tools and instruments, etc., do not conform to the operation condition, the assessor may order the termination of the operation

Part II   Standard for Training and Assessment on Interruption Maintenance for Operation Maintenance of ±800kV UHV Power Transmission Line

Table 2-4-6   Standard for Evaluation of Assessment and Scoring of Operation Maintenance Skills of UHV DC Transmission Line

S/N	Project name	Quality requirements	Score	Deduction standard	Reasons for deduction	Deduction	Scoring
1	Uniform	Correctly wear working clothes, work shoes, safety helmet and protective gloves	5	Deduct 1 point for each error and missing item			
2	Site layout	Install the security fence and hang the signboards correctly: (1) The security fence should take full account of falling objects from the high place and the influence on road traffic. (2) The entrance and exit of the security fence shall be set reasonably. (3) Signs such as "Access from Here", "Work Here", "Access from Here" shall be properly arranged;	3	(1) Deduct 0.5 points for failure to arrange the fence at the work site. (2) Deduct 0.5 points for failure to arrange the warning board. (3) Deduct 0.5 points for failure to hang the tower climbing operation sign			
3	Holding a pre-shift meeting	(1) Responsible Person shall wear red vest and read out the work order and be clear with work task and division of personnel; explain safety measures and technical measures in work; check (inquire after) mental state of all working personnel; inform of hazards in work and precontrol measures. (2) All working personnel shall sign or the work order for confirmation.	3	(1) Give no points to this item for no division of labor, and deduct 1 point for unclear division of labor. (2) Deduct 0.5 points for the on-site Responsible Person not wearing a safety monitoring vest. (3) Deduct 1 point for the work shift member failing to sign or signing incompletely on the work order			

· 637 ·

Standard for Professional Training and Assessment for Operation Maintenance of UHV DC Power Transmission Line

Table (Cont'd)

S/N	Project name	Quality requirements	Score	Deduction standard	Reasons for deduction	Deduction	Scoring
4	Tools and instruments Check	(1) The staff shall place the tools on the moisture-proof tarpaulin as required; the moisture-proof tarpaulin shall be clean and dry. (2) The tools shall be placed according to the requirements of the fixed location management; Inspect the appearance of tools and test certificates without omission. (3) Adjustable wrench, flat pliers, pin-pulling pliers and tool kits shall meet the quality requirements. (4) Test the machine with winching, check whether the gear is normal and check the lever hoist. (5) Check the wire rope and tackle, confirm the connection is reliable and the force meets the requirements. (6) Safety belt and personal safety line meet the quality requirements within the test period. (7) The wire lifter, wire clamp and wire tumble tool meet the requirements within the test period. (8) Confirm that the type, length and appearance of ground wire and connecting pipe meet the requirements. (9) Tower climbing personnel shall check the double name, pole number and phase again and report them	10	(1) Deduct 1 point for failure to use moisture-proof tarpaulin and place tools to designed positions. (2) Deduct 0.5 points/item for failure to check qualified label of tool test and appearance inspection. (3) Deduct 1 point for failure to test the tools. (4) Deduct 2 points for incorrect winching of the test machine; deduct 2 points if the lever hoist is not correctly checked. (5) Deduct 1 point for each mistake and omission. (6) Deduct 1 point for each improper inspection. (7) Deduct 0.5 points for each mistake or omission. (8) Deduct 1 point for each mistake or omission. (9) Deduct 3 points for failure to check the double-designation			

Part II Standard for Training and Assessment on Interruption Maintenance for Operation Maintenance of ±800kV UHV Power Transmission Line

Table (Cont'd)

S/N	Project name	Quality requirements	Score	Deduction standard	Reasons for deduction	Deduction	Scoring
5	Climbing the tower	(1) Check whether there is any abnormality. (2) After the electricians on tower wear the safety belts and conduct the impulse test, fasten the safety belts and correctly carry the lifting rope (slinging over the shoulder with the lifting rope end folded double, fast knot) and climb up the tower one after another. (3) In the process of climbing the tower, fasten the anti-falling protection device, climb the tower at a uniform speed, grasp the main material, and hang the safety belt on the shoulder; The distance between the operators shall not be less than 1.6m, and the responsible person shall strengthen the operation supervision. (4) The safety belt protection shall not be lost from the tower body to the working point on the cross arm. (5) The transfer tackle shall be installed at the correct position for easy operation.	10	(1) Deduct 1 point for each unchecked item. (2) Deduct 2 points for failure to fasten the safety belt or for failure to perform the impulse test on the safety belt and the backup protection rope. Deduct 1 point for not carrying lifting rope, and deduct 1 point for nonstandard carrying method of lifting rope. (3) Deduct 2 points for grasping shackles with hands and deduct 1 point for not climbing the tower along the main material of the shackles side. (4) Deduct 2 points for failure to use safety belt properly. (5) Deduct 1 point for inconvenient suspension position of tackle transmission rope for taking tools. (6) Deduct 2 points for metal tools that are difficult to ensure safe distance during transfer; deduct 2 points for tools that are not bound securely. (7) Deduct 2 points for falling object at heights when transferring.			

Table (Cont'd)

S/N	Project name	Quality requirements	Score	Deduction standard	Reasons for deduction	Deduction	Scoring
5	Climbing the tower			(8) Deduct 2 points for tools colliding with the tower body during the transfer process. (9) Deduct 1 point for knotting and disordered rope in tool transfer.			
6	Inspection power and install ground wire	(1) After the pole is in place, fasten the safety belt to a firm and reliable component or pole, and check whether the buckle is in place correctly. Appliances such as test pen (electroscope) must be transferred by rope. (2) Check whether the ground wire is in good condition, and install the ground wire according to the procedures (first connect the ground terminal, then the conductor terminal). The ground collet shall be firmly connected on the cross arm. (3) When ground wire is installed, it must be operated with insulating rope or insulated handle, and it is forbidden to install the ground wire by directly operating the metal part of the ground wire by hand. Make sure the collection of the ground wire is tightly connected to the conductor	11	(1) Deduct 5 points for failure to carry out power test and installation of ground wire. (2) Deduct 3 points for the installation of the ground wire in the wrong order. (3) Deduct 3 points for unfirm connection of the ground wire. (4) Deduct 2 points for failure to wear insulating gloves each time			

Part II Standard for Training and Assessment on Interruption Maintenance for Operation Maintenance of ±800kV UHV Power Transmission Line

Table (Cont'd)

S/N	Project name	Quality requirements	Score	Deduction standard	Reasons for deduction	Deduction	Scoring
7	Tangent tower threadbare	(1) The operator shall install the single-sheave tackle in the suitable position of the tangent tower ground wire cross arm and remove the anti-vibration hammer on the ground wire. (2) The operator should lift the ground wire with wire lifter and 6T lever hoist. (3) After removing part hardware fittings for connection, turn the ground wire into the single-sheave tackle fixed on the ground wire cross arm.	6	(1) Deduct 2 points for failure to protect ground wire and tower material when lifting ground wire. (2) Deduct 2 points for improper operation of lever hoist. (3) Deduct 2 points in case of nonstandard use of rope ladder in climbing up and down.			
8	Installation of wire clamp	(1) After measuring and recording the position of anti-vibration hammer on the ground wire, the operator shall remove the anti-vibration damper. (2) After the operator uses a hanging wheel to clamp the wire clamp to a suitable position on the ground line, tighten the grinding rope. (3) Tighten the ground wire at both ends with a 6T lever hoist, conduct impulse test and take off the ground wire insulator set fitting after confirmation Nos.	8	(1) Deduct 3 points for failure to record the position of anti-vibration damper. (2) Deduct 3 points for improper use of hanging wheel. (3) Deduct 2 points for failure to conduct impulse test before taking off the insulator set fitting			

· 641 ·

Table (Cont'd)

S/N	Project name	Quality requirements	Score	Deduction standard	Reasons for deduction	Deduction	Scoring
9	Resisting-tensile tower thread slack and connecting new and old ground wire	(1) The operator should loosen the 6T lever hoist at both ends to make the grinding rope stressed, impulse the grinding rope and then remove the lever hoist. (2) Start the winching and use the winching to loosen the ground wires at both ends to the ground. (3) Arrange the angle tackle of the new ground wire and its direction on the tension field tower. (4) The ground personnel shall cut off the strain clamp at the ground wire end, polish and crimp the new and old ground wires. The crimping process of splicing pipe shall meet the requirements.	15	(1) Deduct 2 points for failure to conduct impulse test. (2) Deduct 1 point for failure to protect the tower material during installation of steering tackle at each place. (3) Deduct 3 points for non-standard layout of the new ground wire. (4) Deduct 2 points for failure to polish the ground wire. (5) Deduct 3 points in case that the crimping process does not meet the requirements.			
10	Installation of crossover and accessories	(1) Start 2 winchings on the traction field and tension field at the same time under unified command by specially-assigned person. (2) In the process of crossover, a special persons should be assigned to guard the mid-range center to prevent ground wire falling off. (3) The operators on tower should pay attention to on the joint to prevent wire jamming. (4) The winching operator shall control the crossover speed to prevent excessive tension. (5) After the new ground wire arrives at the traction field, the tension field shall be crimped with the strain clamp. After crimping, the crimped ground wire shall be connected to the insulator armor clamp by winding and winching.	12	(1) Deduct 2 points for each time the ground wire falling to the ground during crossover. (2) Deduct 1 point for over-size strain during crossover. (3) Deduct 2 points for thread jamming due to no one has been assigned to command. (4) Deduct 2 points in case the old ground wires are not sorted out and recovered in time.			

Part II Standard for Training and Assessment on Interruption Maintenance for Operation Maintenance of ±800kV UHV Power Transmission Line

Table (Cont'd)

S/N	Project name	Quality requirements	Score	Deduction standard	Reasons for deduction	Deduction	Scoring
10	Installation of crossover and accessories	(6) If the traction field tightening the ground wire, and mark it when the sag of the ground wire is consistent with the original sag. (7) Loosen the ground wire to the ground, crimp the strain clamp at the other end, then tighten the grinding rope. Connect the ground wire to the insulator string armor clamp and the new ground wire to the cable clamp by the tangent tower, fasten the bolts, and install the anti-vibration hammer according to the previously marked position. (8) Connect the new ground wire to the cable clamp by the tangent tower, fasten bolts and install anti-vibration damper according to the original requirements.		(5) Deduct 2 points in case the sag is not measured by a specially assigned person; 2 points for inconsistent sag with other original sag. (6) Deduct 3 points in case the model of the strain clamp is not consistent with that of the ground wire. (7) Deduct 2 points for failure to conduct impulse test after connection. (8) Deduct 2 points in case that the crimping process does not meet the requirements. (9) Deduct 2 points for failure to check the bolts fastened in place. (10) Deduct 1 point for each improper installation position of spacer, and 2 points for unacceptable installation and improper position of spacer			

Table (Cont'd)

S/N	Project name	Quality requirements	Score	Deduction standard	Reasons for deduction	Deduction	Scoring
11	Return to the ground	(1) Check whether the armor clamps and accessories are complete and in place, whether the connecting part is firm. Loosen the grinding rope and check whether the tower material is damaged, check the sag of the ground wire, and check whether the connecting armor clamps meet the requirements. (2) Remove tools in sequence, such as grinding rope, lever hoist, ground wire, tackle, wire clamp and backup protection rope, etc. and transfer to the ground. When removing the grounding wire, the conductor terminal and ground terminal shall be removed successively. First remove the upper layer, then the lower layer, first the distant end, then the near end. (3) After checking that there is no object left on the tower, the equipotential electrician on the tower shall report it to the Responsible Person and then climb down the tower with insulated transmission rope after obtaining the consent of the Responsible Person (4) Correctly climb down the main material along the shackles side, and carry the lifting rope (lifting rope head shall be double folded, fast knot, and hung over the shoulder). (5) Confirm that there is no object left behind	7	(1) Deduct 3 points for failure to check the connection of armor clamp, and 2 points for failure to check the ground wire sag. (2) Deduct 2 points for each winding of transmission rope, 2 points for each damage to tower damage. (3) Deduct 2 points for incorrect removal order of ground wire. (4) Deduct 4 points for not carrying lifting rope, and deduct 2 points for nonstandard carrying method of lifting rope. (5) Deduct 1 point for grasping shackles with hands and 4 points for not climbing the tower along the main material of the shackles side. (6) Deduct 4 points for any object left behind			

Part II Standard for Training and Assessment on Interruption Maintenance for Operation Maintenance of ±800kV UHV Power Transmission Line

Table (Cont'd)

S/N	Project name	Quality requirements	Score	Deduction standard	Reasons for deduction	Deduction	Scoring
12	End of the work	(1) The Responsible Person shall organize all working personnel to put working apparatus and materials in order and put them in a special kit (bag) after cleaning; clean the site to ensure that "the materials are removed and the site is cleaned after construction". (2) After a post-shift meeting is held, the Responsible Person shall make work summaries and comments. Comments include the construction quality of this work and the implementation of safety measures from all working personnel. (3) Responsible Person shall report the completion of the work to the work approver, resume the power transmission of the power cut line and terminate the work order (4) Complete operation as required within the specified time	10	(1) Deduct 3 points for failure to clean the work site. (2) Deduct 2 points for failure to hold the post-shift meeting. (3) Deduct 2 points for failure to report to the work approver. (4) Deduct 1 point for every extension of 1 minute. After extension of 5 minutes, incomplete items will not be scored			
	Total		100				

Module 5 Standards for Training and Assessment on Power-cut Replacement of Whole Strain Insulators String on ±800kV UHV Transmission Line Earth Electrode

I. Training Standards

(I) Training Requirements

Designation of module	The power-cut replacement of the whole strain insulator string on ±800kV UHV transmission line earth electrode.	Type of training	Operation class
Training method	Practical operation training	Training hours	14 hours
Training objectives	1. Check and use tools on power-cut replacement of whole strain Insulators string on ±800kV UHV transmission line earth electrode 2. Master the operating process and process requirements on power-cut replacement of whole strain insulators string on ±800kV UHV transmission line earth electrode 3. Master the Hazards of Power-cut Replacement of the Whole Strain Insulator String on ±800kV UHV Transmission Line Earth Electrode		
Training venue	UHV DC practical training line		
Training content	1. Tool selection and inspection on power-cut replacement of whole strain insulators string on ±800kV UHV transmission line earth electrode 2. Operating process and process requirements on power-cut replacement of whole strain insulators string on ±800kV UHV transmission line earth electrode 3. Hazards of the power-cut replacement of the whole strain insulator string on ±800kV earth electrode		
Scope of application	Maintenance personnel for UHV DC Transmission Line		

(II) Referenced Procedures and Specifications

(1) Operating Code for Overhead Transmission Line (DL/T741-2010)

(2) Technical Code for Designing 110~500kV Overhead Transmission Line (DL/T5092-1999);

(3) State Grid Corporation of China Working Regulations of Power Safety (Transmission Line Section) (Q/GDW1799.2-2013)

(4) Code for Designing of ±800kV DC Overhead Transmission Line(GB50790-2013)

(5) Maintenance Specification for ±800kV DC Overhead Transmission Line (DL/T251-2012)

(6) Operating Code for ±800kV DC Overhead Transmission Line (GB/T28813-2012)

(7) Maintenance Specification for 110(66)kV~500kV Overhead Transmission Line (State Grid Corporation of China)

(8) Guideline of Maintenance of Overhead Transmission Line State (DLT1248-2013)

(9) Maintenance Management Regulations of Electric Transmission and Transformation Equipment State (State Grid Corporation of China)

(10) Maintenance and Test Specification of Electric Transmission and Transformation Equipment State (State Grid Corporation of China)

(III) Teaching Design for Training

To complete the power-cut replacement of whole strain insulator string on ±800kV UHV transmission line earth electrode is the work task during the design. Each training stage shall be designed according to the standard operation procedure for work task completion. Each stage includes specific training objectives, training content, hours of training, training methods (training resources), training environment, assessment and evaluation, etc, as shown in Table 2-5-1.

Table 2-5-1 Training Content Design for Power-cut Replacement of Whole Strain Insulator String on ±800kV UHV Transmission Line Earth Electrode

Training schedule	Training objectives	Training contents	Training hours	Training methods and resources	Preparation of training conditions	Examination and evaluation
1. Theory Teaching	1. Check and use tools on power-cut replacement of whole strain insulators string on ±800kV UHV transmission line earth electrode 2. Master the operating process and process requirements on power-cut replacement of whole strain insulators string on ±800kV UHV transmission line earth electrode 3. Master the hazards of power-cut replacement of the whole strain insulator string on ±800kV UHV transmission line earth electrode	1. Tool selection and inspection on power-cut replacement of whole strain insulators string on ±800kV UHV transmission line earth electrode 2. Operating process and process requirements on power-cut replacement of whole strain insulators string on ±800kV UHV transmission line earth electrode 3. Hazards of the power-cut replacement of the whole strain insulator string on ±800kV UHV transmission line earth electrode	2	Training methods: Lecture. Training resources: PPT, relevant regulations and specifications	Multimedia classroom	Attendance, classroom questions and assignments
2. Preparations	Be able to complete the preparations before operation.	1. Work site survey. 2. Preparation of the standardized operation card. 3. Fill in the first work order of transmission line. 4. Preparation of tools and materials for this operation	1	Training method: 1. Site survey and cleaning of tools and materials shall be practiced at site. 2 Preparation of operation card and the filling of work order shall adopt lecture method.	1. UHV transmission line for practical training; 2. Multimedia classroom	

Part II Standard for Training and Assessment on Interruption Maintenance for Operation Maintenance of ±800kV UHV Power Transmission Line

Table (Cont'd)

Training schedule	Training objectives	Training contents	Training hours	Training methods and resources	Preparation of training conditions	Examination and evaluation
2. Preparations				Training resources: 1. ±800kV practical training line; 2. UHV tools warehouse; 3. Blank work order		
3. Work site preparation	Be able to complete the preparations of work site	1. Site re-survey. 2. Work application. 3. Work site layout. 4. Pre-shift meeting. 5. Tools and instruments and material preparation	1	Training methods: demonstration and role play. Resources: ±800kV practical training line	±800kV practical training line	
4. Trainer's demonstration	The trainees can preliminarily understand the operation process of the task through inspecting and learning from each other's work	1. Inspection of tools and instruments and explanation of the usage of all kinds of tools and instruments. 2. Install ground wire. 3. Install the operation frame, and the operators reaches the working position. 4. Use double channels for the replacement of the whole strain insulator string on ±800kV earth electrode 5. Remove tools and instruments and climb down.	1	Training method: Demonstration method. Resources: ±800kV practical training line	±800kV practical training line	

Table (Cont'd)

Training schedule	Training objectives	Training contents	Training hours	Training methods and resources	Preparation of training conditions	Examination and evaluation
5. Group training of trainees	Be able to complete the power-cut replacement of whole strain insulator string on ±800kV earth electrode	1. Trainees in groups (5 people in a group, 1 responsible person, 2 people working on tower and 2 grounding auxiliary personnel) are trained for the power-cut replacement on the ±800kV earth electrode tension-proof insulator. 2. Trainers guide the operation of trainees and conduct safety supervision	8	Training methods: Role play. Resources: ±800kV practical training line	±800kV practical training line	Score the operation of trainees according to the detailed rules for skill assessment and scoring
6. Work summary	1. Enable the trainees to further distinguish the shortcomings of the operation process and facilitate the promotion in the later stage. 2. Train the trainees in the working style of safe and civilized production	1. Work site cleaning. 2. Report to dispatcher. 3. Comment and summarize the work task this time at post-shift meeting	1	Training methods: Lecture and inductive method	±800kV practical training line	

(IV) Work Flow

1. Work Task

Complete the power-cut replacement of the whole strain insulator string on ±800kV UHV transmission line earth electrode.

2. Requirements for Weather and Work Site

(1) Power-cut replacement of ± 800kV UHV transmission line earth electrode tension-resistance whole strain insulator string shall be carried out in good weather.

High-place operation in the open air shall be stopped in severe weather, such as heavy wind of 5 degree ad above, rainstorm, thunder, hail, heavy fog and sand storm. In special cases, if emergency maintenance needs to be performed in severe weather, arrange related personnel to fully discuss necessary safety measures, which shall be taken after approved by the company.

(2) The operators are in good mental state, and the members of the working group carefully study the work ticket and safety technical measures, and all the personnel shall be conform to the "four clear" (the operation task is clear, the danger point is clear, the operation procedure is clear, and the safety measures are clear).

(3) The Responsible Person must contact with the power cutting contact person to fulfill the Work licensing procedures, and it is strictly prohibited to stop and transfer electricity at the appointed time. Responsible Person must obtain the license work order of the licensor before checking the electricity on the line to be overhauled, setting up the ground wire and carrying out the maintenance work.

(4) After power-cut, Responsible Person shall make records carefully.

(5) Check whether there is a honeycomb on the tower before climbing the tower. It is forbidden to climb the tower if a honeycomb is found.

(6) Double safety belt shall be used by the personnel who work on the tower, and safety goggles must be worn.

3. Preparations

3.1 Hazards and precontrol measures

(1) Hazard —— climbing live line

Precontrol measures:

① Before climbing the tower for operation, Responsible Person and the members of

the working team shall check the double name and identification mark carefully (color code, distinguishing mark, etc.) to match the name of the power-cut line.

② Check the tower root and the base of pole and tower before climbing, making sure it's solid and reliable.

③ Before climbing the pole and tower, climbing tools and facilities such as safety belt, shackles and tower materials shall be inspected to ensure they are firm.

④ Before climbing the pole, the intermediate operators who do not involved the installation of the grounding wire shall carefully verify that the phase sequence, color code, name, and number of the line are consistent with the power-cut line, and confirm that the line name is not error or in opposite order.

(2) Hazard —— Operation in violation of safety operation procedures when climbing the tower and working on the tower may cause falling from height.

Precontrol measures:

① During climbing, in order to prevent the operator from falling, the distance between the staff members shall not be less than 1.6m.

② Before climbing, the soil on the soles shall be removed, and the kit shall be complete, and objects shall not be dropped during climbing.

③ Operators shall wear safety helmet and soft sole, and be scrupulous and when climbing pole and tower.

④ During climbing, the safety belt shall be properly packed, the long tail rope shall be placed in the kit, and the main belt shall be hung on the shoulder to prevent safety during climbing to prevent the safety belt from hooking shackles and tower materials during climbing, resulting in falling from the tower.

⑤ Operators shall not lose the protection from the safety belt and step and hold it firmly when moving on the pole and tower.

⑥ When reaching the position of operation point, fasten the safety belt (rope), it shall be firm and reliable, and shall not be hung low for high use.

⑦ The safe distance between the human body, headless rope, etc. and the conductor must not be less than 1.6m before the power test, and special person shall be assigned to supervise the work.

(3) Hazard —— injury caused by objects falling from high place

Precontrol measures:

① The ground staff shall not stand under the position vertical to the operation point. The personnel on tower shall prevent falling objects from injury, and the tools and materials used shall be transmitted by ropes.

② Tool bags should be used when working at heights. Larger tools should be fixed on solid components and are not allowed to be randomly placed.

(4) Hazard —— Prevent injury caused by induced electricity

In order to prevent injury from induced electricity, operators on tower shall use a personal safety earthing wire

(5) Hazard —— Site safety monitoring

① From the beginning of the operation to the end of the operation, the safety supervisor must always carry out uninterrupted safety supervision on the operators on the site.

② The responsible person and the Supervisor must wear a safety vest.

(6) Hazard ——Traffic safety

It is necessary to pay attention to the driving safety of the vehicle, drive carefully when driving, and illegal driving is not allowed.

3.2 Selection of tools, instruments and materials

See Table 2-5-2 for Tools, Instruments and Materials Required for Power-cut Replacement of Tension Whole Strain Insulator String on ±800kV AC Transmission Line Earth Electrode. Before delivering the tools and instruments out of the warehouse, application voltage class and test period of the tools and instruments shall be carefully checked and they shall be inspected to ensure that appearance is intact, connection is firm, rotation is flexible and meeting the requirements of the work task. After delivering the tools and instruments out of the warehouse, they shall be kept from contamination and damp. Metal tools and insulated tools shall be separately loaded and transported to avoid deformation and damage caused by mixed loading and transportation.

Table 2-5-2 Tools, Instruments and Materials Required for Power-cut Replacement of Whole Strain Insulator String on ±800kV UHV Transmission Line Earth Electrode

S/N	Name	Specification and Model	Unit	Qty.	Remarks
1	Safety helmet		pcs	5	
2	Anemometer		Nos.	1	
3	Insulation resistance tester		Nos.	1	

Table (Cont'd)

S/N	Name	Specification and Model	Unit	Qty.	Remarks
4	Falling protector	Corresponding to the type of pole and tower falling protector	Nos.	2	
5	Safety belt		Nos.	2	
6	Transfer rope		Nos.	1	
7	Noose		Nos.	1	
8	Tackle		Nos.	2	
9	Back protection rope of the conductor		Nos.	1	
10	Wire clamp	Matches with the conductor	Nos.	2	
11	Double channels	5T	pcs	1	
12	Operation frame		Nos.	1	
13	Insulator	Be consistent with the original insulator	String	1	
14	Moisture-proof canvas	2m×4m	pcs	2	
15	Security fence		Set	Several	
16	Warning sign	"Work Here", "Access from Here", "Access from Here"	Set	1	
17	Red waistcoat	"Responsible Person"	pcs	1	
18	Cleaning towel		pcs	1	

3.3 Division of labor for operators

Division of labor for operators of the task is shown in Table 2-5-3.

Table 2-5-3 Division of Operating Personnel Involving in Power-cut Replacement of Tension Whole Strain Insulator String on ±800kV AC Transmission Line Earth Electrode

S/N	Post	Qty. (person)	Responsibilities
1	Responsible Person (Safety Supervisor)	1	Be responsible for division of operating personnel of the task, field investigation before work, implementation of the operation plan, work order filling, field re-investigation, performing work permit procedures, holding pre-shift meeting, the implementation of safety measures on site, safety supervision in the operation process, dealing with emergency situations in work, quality surveillance, and the summary after work. Be responsible for the safety monitoring work during this operation
2	Operators on tower	2	Be responsible for the operation of the power-cut replacement of the whole strain insulator string on ±800kV UHV transmission line earth electrode
3	Ground auxiliary personnel	2	Be responsible for ground auxiliary works during the operation.

4. Working procedures

The work flow of this task is shown in Table 2-5-4.

Table 2-5-4 Working Procedures for Power-cut Replacement of Tension Whole Strain Insulator String on ±800kV AC Transmission Line Earth Electrode

S/N	Work Content	Operation Standard	Safety Precautions	Responsible Person
1	Site re-survey	The Responsible Person shall complete the following work: (1) Check the line name, the number of the iron tower number on spot to ensure they are correct; guarantee that the foundation and the tower body are intact and in normal condition, and ensure that the cross and span distance meets the safety requirements; confirm the defect conditions and the specifications and models of ground wires. (2) Check that the terrain and environment should meet the operation requirements. (3) Check that the safety measures listed in the work order are in line with the actual situations on site. The measures will be added if necessary.	(1) Correctly wear helmet, working clothes, work shoes and protective gloves. (2) Non-operation personnel and vehicles are strictly prohibited from entering the work site	
2	Work Permit	The contact person for power off/on before the operation must contact the dispatcher to complete the work permit procedure.	(1) No work shall be started without the Work Permit of approver (2) Power transmission and cutoff at pre-determined time is not permitted.	
3	Site layout	Install the security fence and hang the signboards correctly: (1) The security fence should take full account of falling objects from the heights and the influence on road traffic. (2) The entrance and exit of the security fence shall be set reasonably. (3) Signs such as "Access from Here", "Work Here", "Slow Down" or "Blocking" shall be properly arranged	When the influence on road traffic safety is uncontrollable, the traffic management department should be contacted in time to strengthen the on-site control of traffic safety	

Table (Cont'd)

S/N	Work Content	Operation Standard	Safety Precautions	Responsible Person
4	Holding a pre-shift meeting	(1) All working personnel shall line up. (2) The Responsible Person will read out the work order and be clear with the work task and division of personnel; explain safety measures and technical measures in work; check (inquire after) the mental state of all working personnel; and inform of hazards in work and pre-control measures. (3) All working personnel shall sign on the work order for confirmation.	(1) The work order shall be filled in, issued and approved in a standardized manner, and the signature shall be complete (2) All working personnel shall be in good mental states. (3) All working personnel shall be clear with task division of works, safety measures and technical measures	
5	Inspection of tools and instruments	(1) All necessary tools and instruments shall be prepared as per the operation requirements and placed on the moisture-proof tarpaulin regularly according to the category and location. The appearance and test certificate of tools and instruments shall be checked to ensure there is no omission. (2) The inspector shall report to the Responsible Person that all inspection results are in conformity with the operation requirements	(1) The waterproof tarpaulin shall be enough in quantity and reasonable in position, and be clean and dry. (2) The appearance of tools and instruments is acceptable after inspection, there is no damage, damp, deformation and malfunction, and the certificate is valid.	
6	Climb the pole and tower	(1) Check the name and serial number of the line before climbing the pole and tower. Responsible Person and the members of the working team shall check the double name and identification mark carefully (color code, distinguishing mark, etc.). (2) Check the tower root and the base of pole and tower before climbing, making sure it's solid and reliable. (3) During climbing, in order to prevent the operator from falling, the distance between the staff members shall not be less than 1.6m, the safety belt shall be properly packed, the long tail rope shall be placed in the kit, and the main belt shall be hung on the shoulder to prevent safety during climbing to prevent the safety belt from hooking shackles and tower materials during climbing, resulting in falling from the tower.	(1) Operator shall wear safety helmet and soft sole, and be scrupulous and when climbing pole and tower. (2) During climbing, the safety belt shall be properly packed, the long tail rope shall be placed in the kit, and the main belt shall be hung on the shoulder to prevent safety during climbing to prevent the safety belt from hooking shackles and tower materials during climbing, resulting in falling from the tower.	

Part II Standard for Training and Assessment on Interruption Maintenance for Operation Maintenance of ±800kV UHV Power Transmission Line

Table (Cont'd)

S/N	Work Content	Operation Standard	Safety Precautions	Responsible Person
6	Climb the pole and tower	(4) When the pole and tower is close to the cross arm, the Supervisor and the operator shall check the identification mark and dual tags of the power-cut line again, and can enter the cross arm at the power-cut line side only after they are confirmed to be correct	(3) Operator shall not lose the protection from the safety belt and step and hold it firmly when moving on the pole and tower. (4) When reaching the position of operation point, fasten the safety belt (rope), it shall be firm and reliable, and shall not be hung low for high use. (5) The safe distance between the human body, headless rope, etc. and the conductor must not be less than 1.6m before grounding, and special person shall be assigned to supervise the work.	
7	Inspection power and install earth wire.	After the pole is in place, fasten the safety belt to a firm and reliable component, and check whether the buckle is in place correctly. Appliances such as electroscope must be transferred by rope. Check whether the ground wire is in good condition, and install the ground wire according to the procedures (first connect the ground terminal, then the conductor terminal). When setting the ground wire, the insulating rope or insulating handle must be used for operation, and it is forbidden to install the ground wire with directly operating the metal part of the ground wire by hand. Make sure the collet of the ground wire is tightly connected to the conductor	(1) Check whether the test pole is normal during receiving and before using. (2) It is forbidden to install the ground wire by winding the conductor	

Table (Cont'd)

S/N	Work Content	Operation Standard	Safety Precautions	Responsible Person
8	Replacement of Whole Strain Insulator String	(1) Fasten the safety rope of the double safety belt to the cross arm in the proper position. (2) Set up the transmission rope lifting point. (3) Transfer the operation frame to the cross arm. (4) Operators on the tower hang the front section of the operation frame on the conductor, and put the back end of the working frame on the cross arm and fasten it firmly. (5) One operator on the tower enters the operation point smoothly along the operation frame and ties the main belt of safety belt on the insulator string. (6) Transfer the wire clamp, double hook and steel wire sleeve to the tower. (7) Install the steel wire sleeve, double hook wire tightener, and install the wire clamp at the proper position of the conductor. (8) Tighten the double-hook wire tightener to make it slightly stressed. (9) Install one end of the conductor back-up protection steel wire sleeve on the cross arm, and fix it firmly. The other end shall be secured to the conductor by a wire clamp. (10) The backup protection wire clamp of conductor shall be installed in proper position in front of the wire clamp connecting the double-hook wire tightener. (11) Tighten the double-hook tightener, conduct impulse test and transfer the safety belt to the double hook. (12) Before and after the insulator is replaced, fix the insulator string on the double hook with a short rope. (13) Remove the M-pin, remove the insulator to be replaced and transfer it to the ground.		

Part II Standard for Training and Assessment on Interruption Maintenance for Operation Maintenance of ±800kV UHV Power Transmission Line

Table (Cont'd)

S/N	Work Content	Operation Standard	Safety Precautions	Responsible Person
8	Replacement of Whole Strain Insulator String	(14) Transfer the new insulator to the pole and tower, and install the insulator and the M-pin. (15) Check whether the M-pin and ball head are in position, the bowl is oriented correctly, and clean the insulator string. (16) After the double hook is released and the insulator is stressed, the impulse test shall be conducted on the insulator string		
9	Removal of tools and instruments	(1) Remove the double hook tightener and backup protection steel wire sleeves, and transfer them to the ground. (2) Remove the operation frame and transfer it to the ground. (3) Confirm that there are no leftovers at the working point.		
10	End of the work	(1) Responsible person shall organize all working personnel to put working apparatus and materials in order, and clean the site to ensure "materials are removed and the site is cleaned after construction". (2) After a post-shift meeting is held, the Responsible Person shall make work summaries and comments. Comments include the construction quality of this work and the implementation of safety measures from all working personnel. (3) Responsible Person shall report the completion of the work to the work approver, resume the power transmission of the power cut line and terminate the work order		

Standard for Professional Training and Assessment for Operation Maintenance of UHV DC Power Transmission Line

II. Assessment Standard

Table 2-5-5 Detailed Rules for Assessment and Scoring of UHVDC Skills Training of State Grid Sichuan Electric Power Corporation

Fill-in Column of Examinee	No.:	Name:	Position:	Unit:	Date:	MM/DD/YYYY
Fill-in Column of Assessor	Grade:	Assessor:	Assessment Team Leader:	Starting time:	Closing time:	Operation duration:
Evaluation module	Complete the power-cut replacement of the whole strain insulator string on ±800kV UHV transmission line grounding electrode.		Assessee	Maintenance personnel for ±800kV UHV DC power transmission line	Assessment method	Operation	Assessment Time Limit	60min
Job Description	Complete the power-cut replacement of the whole strain insulator string on ±800kV UHV transmission line grounding electrode.
Work specifications and requirements	Given conditions: 1. Training base: Resisting-tensile tower of ±800kV UHV DC power transmission line grounding electrode 2. The insulators used have been tested, work orders have been handled, safety measures have been completed, and oral application (dispatcher or assessor) shall be made at the beginning and end of the work. 3. The operation must be carried out according to the working procedures. The scores of the items to be carried out shall be deducted for the process error. In case of major hidden dangers of personnel, equipment and operational safety, the assessor may order the termination of the operation (assessment) 4. The whole process shall be independently completed by the reference personnel; 1 auxiliary worker on the pole and tower, 2 auxiliary workers on the ground assist the exam participant to finish the up and down transfer of tools, materials and other non-technical work. 5. An oral application shall be made at the beginning of work, and an oral report also shall be made at the end of the work
Assessment scenario preparation	1. Tools and instruments: 1 pair of operation frame, 1 5T double-hook tightener, 2 wire clamps, 1 set of endless rope tackle (1T), 1 steel wire sleeve, 1 backup protective steel wire sleeve for conductor, 3 5T shackles, 2 lifting ropes, 1 short rope (3m), shoulder&back double control safety belt and climbing tools. 2. Materials: Suspension insulators of the same type. 3. Operated on the training line.
Remarks	1. The deduction shall be done until the scores of each item are deducted completely. In case of major hidden dangers of personal, equipment and operational safety, the assessor may order the termination of the operation. 2. When equipment, working environment, safety belt, safety helmet, tool, etc., do not conform to the operation condition, the assessor may order the termination of the operation

Part II Standard for Training and Assessment on Interruption Maintenance for Operation Maintenance of ±800kV UHV Power Transmission Line

Table 2-5-6 Detailed Rules for Assessment and Scoring of UHVDC Skills Training of State Grid Sichuan Electric Power Corporation

S/N	Project name	Quality requirements	Score	Deduction standard	Reasons for deduction	Deduction	Scoring
1	Uniform	Correctly wear working clothes, work shoes, safety helmet and protective gloves	4	Deduct 1 point for missing one item			
2	Preparation for tools and materials	(1) Inspection of personal tools: adjustable wrench, flat pliers, pin-pulling pliers and tool kits shall meet the quality requirements. (2) Climbing tools, safety belts: Lifting plates/foot buckles and double safety belts shall be inspected for appearance and test cycle, and impulse test shall be conducted. (3) Inspection of special tools: the appearance inspection shall meet the requirements, the double hook tightener shall be in the test cycle, and the double hook tightener shall be adjusted properly	8	(1) Deduct 1 point for no appearance inspection. (2) Deduct 1 point for not checking the factory certificate and test certificate. (3) Deduct 1 point for missing one item			
3	Material inspection	Clean the insulator, and the appearance inspection shall meet the requirements	2	(1) Deduct 2 points for choosing the wrong insulator. (2) Deduct 1 point for no inspection			

Table (Cont'd)

S/N	Project name	Quality requirements	Score	Deduction standard	Reasons for deduction	Deduction	Scoring
4	Climbing the tower	(1) Check the line name, the number of the iron tower number on spot to ensure they are correct; guarantee that the foundation and the tower body are intact and in normal condition, and ensure that the cross and span distance meets the safety requirements; confirm the defect conditions and the specifications and models of ground wires. (2) The climbing action of the pole (tower) shall be standard and without dangerous action. (3) Correctly carry the lifting rope (The end of the lifting rope shall be double folded, fast knot, and hung over the shoulder) (4) The safety belt protection shall not be lost from the pole and tower body to the working point on the cross arm. (5) Install the endless rope in the appropriate position above the cross arm	8	(1) Deduct 4 points if the double title of the line is not confirmed; deduct 1 point for each item if the base, tower body and crossing point are not checked; deduct 1 point if the ground wire defect is not confirmed. (2) Deduct 4 points for not carrying lifting rope, and deduct 2 points for nonstandard carrying method of lifting rope. (3) Deduct 1 point for each nonstandard climbing. (4) Deduct 5 points for failure to use safety belt properly. (5) Deduct 2 points for improper fixing position of endless rope			
5	Power test and installation of conductor and ground wire	(1) After climbing the tower to the designated position, fasten the safety belt to a firm component and check whether the safety belt is fastened. The installation position of the tackle is correct and convenient for operation. Tools and instruments must be transferred by ropes.	10	(1) Deduct 2 points for non standard installation of tackle (2) Deduct 2 points for non standard use of safety belt. (3) Deduct 2 points if no insulating gloves are worn for power test and installation of grounding wire.			

Part II Standard for Training and Assessment on Interruption Maintenance for Operation Maintenance of ±800kV UHV Power Transmission Line

Table (Cont'd)

S/N	Project name	Quality requirements	Score	Deduction standard	Reasons for deduction	Deduction	Scoring
5	Power test and installation of conductor and ground wire	(2) Before using the electroscope, the operator shall check the sound and light signal again. When telescopic electroscope is used, all the insulation rods of each section should be pulled out in place to ensure the effective insulation length of the insulation rods. When testing electricity, the operator should hold the insulation handle of electroscope to ensure enough safe distance between human body and conductor. Lower layer first then upper layer, near side first then far side. The electrical inspection of power line shall be performed phase by phase. (3) The conductor ground wire shall be installed immediately after there is no voltage. The tower material position of the ground terminal shall be polished with sandpaper. The ground terminal shall be installed first, then the conductor end. The installation sequence is middle phase first, then two side phase. (4) When installing the conductor ground wire, the insulating gloves must be used for operation, and it is forbidden to install the ground wire with directly operating the metal part of the conductor ground wire by hand. Make sure the hook of the ground wire is tightly connected to the conductor		(4) Deduct 2 points for nonstandard use of electroscope, and 2 points for incorrect sequence for electricity testing. (5) Deduct 2 points if the installation sequence of both ends of the ground wire is wrong; deduct 1 point if the connection is not firm; deduct 1 point if the installed end is not inspected. (6) Deduct 2 points for winding ground wire.			

· 663 ·

Table (Cont'd)

S/N	Project name	Quality requirements	Score	Deduction standard	Reasons for deduction	Deduction	Scoring
6	Install the operation frame and enter the working position	(1) Check the connection of insulator string is reliable. (2) Transfer the operation frame to the cross arm, and fix it firmly at the correct installation position Fasten the safety rope of the double safety belt to the cross arm in the proper position. (3) Enter the operation point smoothly along the operation frame and tie the the main belt of safety belt on cross arm and tie poles belt on the insulator string.	8	(1) Deduct 2 points for not checking the insulator string. (2) Deduct 3 points if the installation position of the operation frame is not correct, and deduct 5 points if it is not firm. (3) Deduct 5 points if the safety rope of double safety belt is not used and 3 points if the position is wrong			
7	Install stringing tool	(1) Check insulator string M pin. (2) Transfer the wire clamp, double hook and steel wire sleeve to the tower. (3) Install the steel wire sleeve, double hook wire tightener, and install the wire clamp at the proper position of the conductor. (4) Tighten the double-hook wire tightener to make it slightly stressed. (5) Install one end of the conductor backup protection steel wire sleeve on the cross arm, and fix it firmly. The other end shall be secured to the conductor by a wire clamp. (6) The backup protection wire clamp of conductor shall be installed in proper position in front of the wire clamp connecting the double-hook wire tightener	16	(1) Deduct 3 points for not checking the insulator string M pin. (2) Deduct 1 point each time the rope is wound during transmission. (3) Deduct 2 points for incorrect installation position of wire clamp. (4) Deduct 2 points for improper fixing position of wire clamp. (5) Deduct 2 points for improper tightness of steel wire sleeves for backup protection of conductor. (6) Deduct 2 points each time for impact phenomenon of conveying objects. (7) Deduct 5 points for each falling object.			

Part II Standard for Training and Assessment on Interruption Maintenance for Operation Maintenance of ±800kV UHV Power Transmission Line

Table (Cont'd)

S/N	Project name	Quality requirements	Score	Deduction standard	Reasons for deduction	Deduction	Scoring
8	Replacement of insulator	(1) Tighten the double-hook tightener, conduct impulse test and transfer the safety belt to the double hook. (2) Before and after the insulator is replaced, fix the insulator string on the double hook with a short rope. (3) Remove the M-pin, remove the insulator to be replaced and transfer it to the ground. (4) Transfer the new insulator to the pole and tower, and install the insulator and the M-pin. (5) Check whether the M-pin and ball head are in position, the bowl is oriented correctly, and clean the insulator string. (6) After the double hook is released and the insulator is stressed, the impulse test shall be conducted on the insulator string	18	(1) Deduct 5 points each time for not transferring pole belt to the load-bearing object. (2) Deduct 5 points for loosing the protection of the safety belt. (3) Deduct 2 points for not cleaning the insulator string. (4) Deduct 1 point for one touch of insulator. (5) Deduct 5 points for each M-pin not installed, 2 points for the incorrect position of M-pin, and 2 points for incorrect direction of the larger opening of insulator. (6) Deduct 5 points for no impulse test			
9	Removal of tools and instruments	(1) Remove the double hook tightener and backup protection steel wire sleeves, and transfer them to the ground. (2) Remove the operation frame and transfer it to the ground. (3) Confirm that there are no leftovers at the working point. (4) The safety belt protection shall not be lost from tower to the cross arm	10	(1) Deduct 5 points for loosing the protection of the safety belt. (2) Deduct 2 points each time for impact phenomenon of conveying objects and 1 point each time the rope is wound during transmission. (3) Deduct 5 points for any objects left			

Table (Cont'd)

S/N	Project name	Quality requirements	Score	Deduction standard	Reasons for deduction	Deduction	Scoring
10	Climb down the tower	(1) The climbing down action from the pole (tower) shall be standard and without dangerous action. (2) Correctly carry the lifting rope (The end of the lifting rope shall be double folded, fast knot, and hung over the shoulder) (3) Correctly climb down the tower along the main material of the shackles side	8	(1) Deduct 4 points for not carrying lifting rope, and deduct 2 points for nonstandard carrying method of lifting rope. (2) Deduct 1 point for each nonstandard climbing down. (3) Deduct 4 points if the safety belt is not fastened under the climbers. (4) Deduct 4 points for not climbing down the tower along the main material of the shackles side			
11	Other requirements	(1) It is prohibited to drop articles down from high space. (2) Cooperate with both hands during operation. (3) No floating objects; (4) No things in the mouth (5) No dangerous action shall occur in the process of the upper and lower towers. (6) Correctly wear working clothes, rubber work shoes, safety helmet and protective gloves. (7) Clean up work site after completion and meet the requirements of civilized production	10	(1) Deduct 5 points for each falling object at high place. (2) Deduct 2 points for non-coordination action. (3) Deduct 2 points for floating articles. (4) Deduct 2 points for things in the mouth. (5) Deduct 2 points for slipping. (6) In case of unstable suspension and disqualification, zero point shall be recorded. (7) Deduct 1 point for wrong wearing or missing wearing of labor protection articles. (8) Deduct 2 points for failure to clean up work site			
12	Total		100				

Module 6 Standards for Training and Assessment on Power-cut Replacement of ± 800kV UHV Transmission Line Grounding Electrode Conductor

I. Training Standards

(I) Training Requirements

Designation of module	Power-cut replacement of ±800kV UHV transmission line grounding electrode conductor	Type of training	Operation Class
Training method	Practical operation training	Training hours	14 hours
Training objectives	1. Master the operational versions and stress structures of all kinds of tools and instruments, machines and tools, as well as the technical key points of replacing the grounding electrode level conductor 2. Acquire proficiency in the operation procedures, technical method and construction work hazard of power-cut replacement of ±800kV UHV transmission line grounding electrode conductor. 3. Able to complete the work of power-cut replacement of ±800kV UHV transmission line grounding electrode conductor		
Training venue	UHV DC training line		
Training content	Correctly use the operation methods of all kinds of stressed tools and instruments as well as install all kinds of tools and instruments, and use method of power-cut replacement of ±800kV UHV transmission line grounding electrode conductor.		
Scope of application	Maintenance personnel for ±800kV UHV DC power transmission line		

(II) Referenced Procedures and Specifications

(1) Operating Code for Overhead Transmission Line (DL/T741-2010)

(2) Technical Code for Designing 110~500kV Overhead Transmission Line (DL/T5092-1999);

(3) State Grid Corporation of China Working Regulations of Power Safety (Transmission Line Section) (Q/GDW1799.2-2013)

(4) Code for Designing of ±800kV DC Overhead Transmission Line(GB50790-2013)

(5) Maintenance Specification for ±800kV DC Overhead Transmission Line (DL/T251-2012)

(6) Operating Code for ±800kV DC Overhead Transmission Line (GB/T28813-2012)

(7) Maintenance Specification for 110(66)kV~500kV Overhead Transmission Line (State Grid Corporation of China)

(8) Guideline of Maintenance of Overhead Transmission Line State (DLT1248-2013)

(9) Maintenance Management Regulations of Electric Transmission and Transformation Equipment State (State Grid Corporation of China)

(10) Maintenance and Test Specification of Electric Transmission and Transformation Equipment State (State Grid Corporation of China)

(III) Training Teaching Design

To complete the work task of "power-cut replacement of ±800kV UHV transmission line grounding electrode conductor", each training stage shall be designed according to the standard operation procedure for work task completion. Each stage includes specific training objectives, training content, hours of training, training methods (training resources), training environment, assessment and evaluation, etc, as shown in the Table 2-6-1.

Part II　Standard for Training and Assessment on Interruption Maintenance for Operation Maintenance of ±800kV UHV Power Transmission Line

Table 2-6-1　Training content design for power-cut replacement for grounding electrode conductor of ±800kV UHV transmission line

Training schedule	Training objectives	Training contents	Training hours	Training methods and resources	Preparation of training conditions	Assessment and evaluation
1. Theory Teaching	Be familiar with the operating steps of power-cut replacement of ±800kV transmission line grounding electrode conductor	Be familiar with the operating steps of power-cut replacement of ±800kV transmission line grounding electrode conductor	2	Training methods: Lecture. Training resources: PPT, relevant regulations and specifications	Multimedia classroom	Attendance, classroom questions and assignments
2. Preparations	Be able to complete the preparation before operation.	1. Site survey. 2. Preparation of the standardized operation card. 3. Filling of the work order. 4. Preparation of tools and materials for this operation	1	Training methods: 1. Site survey and cleaning of tools and materials shall be practiced at site. 2. Preparation of operation card and the filling of work order shall adopt lecture method. Training resources: 1. ±800kV training line. 2. UHV tools warehouse. 3. Blank work order	1. UHV training transmission line 2. Multimedia classroom	
3. Work site preparation	Be able to complete the site preparations	1. Site re-survey. 2. Work application. 3. Work site layout. 4. Pre-shift meeting. 5. Tools and instruments and materials inspection	1	Training methods: demonstration and role play. Resources: ±800kV practical training line	±800kV practical training line.	

Table (Cont'd)

Training schedule	Training objectives	Training contents	Training hours	Training methods and resources	Preparation of training conditions	Assessment and evaluation
4. Trainer demonstration	The trainees can preliminarily understand the operation process of the task through inspecting and learning from each other's work,	1. Climbing the tower 2. Test electricity and hang ground wire. 3. Operation on conductor. 4. Installation of crossover and accessories. 5. Removal of tools and instruments	1	Training methods: Demonstration method Resources: ±800kV practical training line	±800kV practical training line	
5. Group training of trainees	Be able to finish the work of power-cut replacement of ±800kV transmission line grounding electrode conductor in groups	1. The trainees are divided into groups (22 persons per group) to train the power-cut replacement of ±800kV transmission line grounding electrode conductor. 2. Trainers guide the operation of trainees and conduct safety supervision	8	Training methods: Role play. Resources: ±800kV practical training line	±800kV practical training line.	Score the operation of trainees according to the detailed rules for skill assessment and scoring
6. End of the work	1. Enable the trainees to further distinguish the shortcomings of the operation process and facilitate the promotion in the later stage. 2. Train the trainees in the working style of safe and civilized production	1. Work site cleaning. 2. Report to dispatcher. 3. Comment and summarize the work task this time at post-shift meeting	1	Training methods: Lecture and inductive method	±800kV practical training line.	

(IV) Work Flow

1. Work Task

Complete power-cut replacement of ±800kV UHV transmission line grounding electrode conductor.

2. Requirements for Weather and Work Site

(1) Power-cut replacement of ± 800kV UHV transmission line conductor grounding electrode conductor shall be carried out in good weather.

High-place operation in the open air shall be stopped in severe weather, such as heavy wind of 5 degree ad above, rainstorm, thunder, hail, heavy fog and sand storm. In special cases, if emergency maintenance needs to be performed in severe weather, arrange related personnel to fully discuss necessary safety measures, which shall be taken after approved by the company.

(2) The operators are in good mental state, and the members of the working group carefully study the work ticket and safety technical measures, and all the personnel shall be conform to the "four clear" (the operation task is clear, the danger point is clear, the operation procedure is clear, and the safety measures are clear).

(3) The Responsible Person must contact with the power cutting contact person to fulfill the Work licensing procedures, and it is strictly prohibited to stop and transfer electricity at the appointed time. Responsible Person must obtain the license work order of the licensor before checking the electricity on the line to be overhauled, setting up the ground wire and carrying out the maintenance work.

(4) After power-cut, Responsible Person shall make records carefully.

(5) Check whether there is a honeycomb on the tower before climbing the pole. It is forbidden to climb the tower if a honeycomb is found.

(6) Double safety belt shall be used by the personnel who work on the tower, and safety goggles must be worn.

3. Preparations

3.1 Hazards and precontrol measures

(1) Hazard - climbing live line

Precontrol measures:

① Responsible Person and the members of the working team shall check the double name and identification mark carefully before climbing the tower.(color code, distinguishing mark, etc.) shall be consistent with the name of the power cut line.

② Check the root and base, etc. of iron tower before climbing the tower to keep it solid and reliable.

③ Before climbing the tower, climbing tools and facilities such as safety belt, shackles and tower materials shall be inspected to ensure they are firm.

④ Before climbing the pole, the intermediate operators who do not involved the installation of the grounding wire shall carefully verify that the phase sequence, color code, name, and number of the line are consistent with the power-cut line, and confirm that the line name is not error or in opposite order.

(2) Hazard —— Operation in violation of safety operation procedures when climbing the tower and working on the tower may cause falling from height.

Precontrol measures:

① During climbing, in order to prevent the operator from falling, the distance between the staff members shall not be less than 1.6m.

② Before climbing, the soil on the soles shall be removed, and the kit shall be complete, and objects shall not be dropped during climbing.

③ Operator shall wear safety helmet and soft sole, and be scrupulous and when climbing tower.

④ During climbing, the safety belt shall be properly packed, the long tail rope shall be placed in the kit, and the main belt shall be hung on the shoulder to prevent safety during climbing to prevent the safety belt from hooking shackles and tower materials during climbing, resulting in falling from the tower.

⑤ Operators shall not lose the protection from the safety belt and step and hold it firmly when moving on the tower.

⑥ When reaching the position of operation point, fasten the safety belt (rope), it shall be firm and reliable, and shall not be hung low for high use.

⑦ The safe distance between the human body, headless rope, etc. and the conductor must not be less than 9.5m before the power test, and special person shall be assigned to supervise the work.

(3) Hazard —— injury caused by objects falling from high place

Precontrol measures:

① The ground staff shall not stand under the position vertical to the operation point.

The personnel on tower shall prevent falling objects from injury, and the tools and materials used shall be transmitted by ropes.

② Tool bags should be used when working at heights. Larger tools should be fixed on solid components and are not allowed to be randomly placed.

③ In the process of lifting with a winching, special personnel shall be assigned to command and cooperate in a unified way. Impact inspection shall be carried out immediately after the insulator string are lifted off the ground.

(4) Hazard —— Prevent injury caused by induced electricity

① Operators on the tower should wear a full set of shielding suits, in order to prevent injury caused by induced electricity.

② If contact with overhead ground wire is required, reliable grounding shall be conducted before contact with overhead ground wire.

(5) Hazard —— Site safety monitoring

① From the beginning of the operation to the end of the operation, the safety supervisor must always carry out uninterrupted safety supervision on the operators on the site.

② The person in charge of the work and the Supervisor must wear a safety vest.

(6) Hazard ——Traffic safety

It is necessary to pay attention to the driving safety of the vehicle, drive carefully when driving, and illegal driving is not allowed.

3.2 Selection of tools, instruments and materials

See Table 2-6-2 for Tools, Instruments and Materials Required for Power-cut Replacement of ±800kV UHV transmission line grounding electrode conductor. Before delivering the tools and instruments out of the warehouse, application voltage class and test period of the tools and instruments shall be carefully checked and they shall be inspected to ensure that appearance is intact, connection is firm, rotation is flexible and meeting the requirements of the work task. After delivering the tools and instruments out of the warehouse, they shall be kept from contamination and damp. Metal tools and insulated tools shall be separately loaded and transported to avoid deformation and damage caused by mixed loading and transportation.

Table 2-6-2   Tools, Instruments and Materials Required for the Power-cut Replacement of ±800kV UHV Transmission Line Grounding Electrode Conductor

S/N	Name	Specification and Model	Unit	Qty.	Remarks
1	Transfer rope		Nos.	2	
2	Protection rope		Nos.	2	
3	Tackle		Nos.	2	
4	Safety helmet		pcs	6	
5	Anemometer		pcs	1	
6	Temperature and humidity meter		pcs	1	
7	Multimeter		pcs	1	
8	Moisture-proof canvas	2m×4m	pcs	2	
9	Noose		Nos.	4	
10	Falling protector	Corresponding to the type of falling protector of the tower	pcs	2	
11	Safety belt		Nos.	2	
12	Security fence		Set	Several	
13	Warning sign	"Work Here", "Access from Here", "Access from Here"	Set	1	
14	Red waistcoat	"Responsible Person"	pcs	1	
15	Sandpaper		Sheet	1	
16	Clean towels		pcs	1	
17	Interphone		Set	4	
18	lever		Nos.	1	

3.3 Division of labor for operators

Division of labor for operators of the task is shown in Table 2-6-3.

Table 2-6-3   Personnel Allocation Table for Power-cut Replacement of ±800kV UHV Transmission Line Grounding Electrode Conductor

S/N	Post	Qty. (person)	Responsibilities
1	Responsible Person	1	Be responsible for division of operating personnel of the task, field investigation before work, implementation of the operation plan, work order filling, field re-investigation, performing work permit procedures, holding pre-shift meeting, the implementation of safety measures on site, safety supervision in the operation process, dealing with emergency situations in work, quality surveillance, and the summary after work.
2	Safety Supervisor	1	Be responsible for the safety monitoring work during this operation
3	Operator working at heights	7	Be responsible for power-cut replacement of ±800kV UHV transmission line grounding electrode conductor
4	Ground auxiliary personnel	10	Be responsible for ground auxiliary works during the operation.
5	Winching operator	2	Be responsible for the winching work during this operation
6	Signal commander	2	Be responsible for commanding the start and stop of 2 sets of winches

4. Working procedures

S/N	Work Content	Operation Standard	Safety Precautions	Responsible Person
1	Site re-survey	The Responsible Person shall complete the following work: (1) Check the line name, the number of the iron tower number on spot to ensure they are correct; guarantee that the foundation and the tower body are intact and in normal condition, and ensure that the cross and span distance meets the safety requirements; confirm the defect conditions and the specifications and models of ground wires; (2) Check that the terrain and environment should meet the operation requirements; (3) Check that the safety measures listed in the work order are in line with the actual situations on site. The measures will be added if necessary.	(1) Correctly wear helmet, working clothes, work shoes and protective gloves. (2) Non-operation personnel and vehicles are strictly prohibited from entering the working site	

Table (Cont'd)

S/N	Work Content	Operation Standard	Safety Precautions	Responsible Person
2	Work Permit	The contact person for power off/on before the operation must contact the dispatcher to complete the work permit procedure	(1) No work shall be started without the Work Permit of approver (2) Power transmission and cutoff at pre-determined time is not permitted.	
3	Site layout	Install the security fence and hang the signboards correctly: (1) The security fence should take full account of falling objects from the heights and the influence on road traffic. (2) The entrance and exit of the security fence shall be set reasonably. (3) Signs such as "Access from Here", "Work Here", "Slow Down" or "Blocking" shall be properly arranged	When the influence on road traffic safety is uncontrollable, the traffic management department should be contacted in time to strengthen the on-site control of traffic safety	
4	Holding a pre-shift meeting	(1) All working personnel shall line up. (2) The Responsible Person will read out the work order and be clear with the work task and division of personnel; explain safety measures and technical measures in work; check (inquire after) the mental state of all working personnel; and inform of hazards in work and pre-control measures. (3) All working personnel shall sign on the work order for confirmation.	(1) The work order shall be filled in, issued and approved in a standardized manner, and the signature shall be complete. (2) All working personnel shall be in good mental states. (3) All working personnel shall be clear with task division of works, safety measures and technical measures	

Part II Standard for Training and Assessment on Interruption Maintenance for Operation Maintenance of ±800kV UHV Power Transmission Line

Table (Cont'd)

S/N	Work Content	Operation Standard	Safety Precautions	Responsible Person
5	Inspection of tools and instruments	(1) All necessary tools and instruments shall be prepared as per the operation requirements and placed on the moisture-proof tarpaulin regularly according to the category and location. The appearance and test certificate of tools and instruments shall be checked to ensure there is no omission. (2) The inspector shall report to the Responsible Person that all inspection results are in conformity with the operation requirements	(1) The waterproof tarpaulin shall be enough in quantity and reasonable in position, and be clean and dry. (2) The appearance of tools and instruments is acceptable after inspection, there is no damage, damp, deformation and malfunction, and the certificate is valid.	
6	Climbing the tower	(1) Check the name and serial number of the line before climbing the tower. For multi-circuit lines on the same tower, the Responsible Person and the members of the working team shall check the double name and identification mark carefully (color code, distinguishing mark, etc.). (2) Check the root and base, etc. of iron tower before climbing the tower to keep it solid. (3) During climbing, in order to prevent the operator from falling, the distance between the staff members shall not be less than 1.6m, the safety belt shall be properly packed, the long tail rope shall be placed in the kit, and the main belt shall be hung on the shoulder to prevent safety during climbing to prevent the safety belt from hooking shackles and tower materials during climbing, resulting in falling from the tower. (4) When the tower is close to the cross arm, the Supervisor and the operator shall check the identification mark and dual tags of the power-cut line again, and can enter the cross arm at the power-cut line side only after they are confirmed to be correct.	(1) Operator shall wear safety helmet and soft sole, and be scrupulous and when climbing tower. (2) During climbing, the safety belt shall be properly packed, the long tail rope shall be placed in the kit, and the main belt shall be hung on the shoulder to prevent safety during climbing to prevent the safety belt from hooking shackles and tower materials during climbing, resulting in falling from the tower. (3) Operators shall not lose the protection from the safety belt and step and hold it firmly when moving on the tower. (4) When reaching the position of operation point, fasten the safety belt (rope), it shall be firm and reliable, and shall not be hung low for high use.	

Table (Cont'd)

S/N	Work Content	Operation Standard	Safety Precautions	Responsible Person
6	Climbing the tower		(5) The safe distance between the human body, headless rope, etc. and the conductor must not be less than 9.5m before the power test, and special person shall be assigned to supervise the work.	
7	Inspection power and install ground wire	(1) After the pole is in place, fasten the safety belt to a firm and reliable component, and check whether the buckle is in place correctly. Appliances such as test pen (electroscope) must be transferred by rope. (2) Check whether the ground wire is in good condition, and install the ground wire according to the procedures (first connect the ground terminal, then the conductor terminal). (3) When setting the ground wire, the insulating rope or insulating handle must be used for operation, and it is forbidden to install the ground wire with directly operating the metal part of the ground wire by hand. Make sure the collection of the ground wire is tightly connected to the conductor	(1) Check whether the test pole is normal during receiving and before using. (2) It is forbidden to install the ground wire by winding the conductor	

Part II Standard for Training and Assessment on Interruption Maintenance for Operation Maintenance of ±800kV UHV Power Transmission Line

II. Assessment Standard

Table 2-6-5 Detailed Rules for Assessment and Scoring of UHV DC Skills Training of State Grid Sichuan Electric Power Corporation

Fill-in Column of Examinee	No.:	Name:	Position:	Unit:	Date:	MM/DD/YYYY
Fill-in Column of Assessor	Achievement:	Assessor:	Assessment	Team Leader:	Starting time:	Closing time: Operation Duration:
Assessment module	Assessment for power-cut replacement of ±800kV UHV transmission line, Assessment for UHV ±800kV DC transmission line, Operating time 480mm for maintenance personnel on 480. conductor grounding electrode conductor object					
Job Description	Power-cut Replacement of ±800kV Transmission Line XX#-XX#Resisting-tensile Tower Section Grounding Electrode Conductor					
Work specifications and requirements	1. The whole process shall be completed independently by one exam participant; 6 auxiliary workers on tower, 10 auxiliary workers on the ground, 2 grinding operators to assist the exam participant to finish the up and down transfer of tools, materials and other non-technical work. 2. The exam participant should make necessary safety check before operation. 3. The tools used to replace the grounding electrode conductor should meet the stress requirements. 4. An oral application shall be made at the beginning of work, and an oral report also shall be made at the end of the work. Given conditions: 1. Training base: ±800kV UHV DC power transmission line tensile section grounding electrode conductor 2. Work orders have been handled, safety measures have been completed, and oral application (dispatcher or assessor) shall be made at the beginning and end of the work. 3. The operation must be carried out according to the working procedures. The scores of the items to be carried out shall be deducted for the process error. In case of major hidden dangers of personal, equipment and operational safety, the assessor may order the termination of the operation (assessment)					

· 679 ·

Table (Cont'd)

Assessment scenario preparation	1. Tools and instruments: a bundle of Φ16 grinding rope, 2 lever hoists (9T), 1 lever hoist (6T), 3 winchings (5T), 3 sets of lifting rope tackles, 3 nooses, 2 sets of rotating joints, 2 pairs of net covers, 2 wire clamps, 1 wire lifter, 1 wire tumble tool, 2 sets of special tools for spacers, 3 single-sheave tackles, 2 sets of pay-off racks, 2 sets of pay-off reels, 8 steel wire sleeves (Φ20), 8 iron tackles (15T), 8 shackles (18T), 6 shackles (10T), 2 sets of hydraulic presses (including hydraulic pliers), some walkie-talkies, 1 rope ladder, 7 double-insurance safety belts, and several wood blocks. 2. Materials: 3 reels of conductors of the same model, and 2 sets of strain clamps. 3. Operated on the training line.
Remarks	1. The deduction shall be done until the scores of each item are deducted completely. In case of major hidden dangers of personal, equipment and operational safety, the assessor may order the termination of the operation. 2. When equipment, working environment, safety belt, safety helmet, tool, shielding clothes, etc., do not conform to the operation condition, the assessor may order the termination of the operation

Table 2-6-6 Detailed Rules for Assessment and Scoring of UHV DC Skills Training of State Grid Sichuan Electric Power Corporation

S/N	Project name	Quality requirements	Score	Deduction standard	Reasons for deduction	Deduction	Scoring
1	Uniform	Correctly wear working clothes, work shoes, safety helmet and protective gloves	4	Deduct 1 point for each error and missing item			
2	Preparation for tools and materials						
2.1	Inspection of personal tools	Inspection of personal tools: adjustable wrench, flat pliers, pin-pulling pliers and tool kits which shall meet the quality requirements.	2	Deduct 1 point for each error and missing item			

Part II Standard for Training and Assessment on Interruption Maintenance
for Operation Maintenance of ±800kV UHV Power Transmission Line

Table (Cont'd)

S/N	Project name	Quality requirements	Score	Deduction standard	Reasons for deduction	Deduction	Scoring
2.2	Inspection of tools and instruments	Test-run the winching, and check whether the gear position is normal; Whether the lever hoist meets the quality requirements.	2	Deduct 1 point for each error and missing item			
2.3	Check the wire rope and tackle, confirm the connection is reliable and the force meets the requirements.	Check the wire rope and tackle, confirm the connection is reliable and the force meets the requirements.	2	Deduct 1 point for each error and missing item			
2.4	Safety belt inspection	Check whether safety belt and personal safety line meet the quality requirements within the test period.	2	Deduct 1 point for each error and missing item			
2.5	Inspection of special tools	Check whether the wire lifter, wire clamp and wire tumble tool meet the requirements within the test period.	2	Deduct 1 point for each error and missing item			
2.6	Material inspection	Confirm the conductor type, length and appearance meet the requirements.	2	Deduct 1 point for each error and missing item			
3	Site layout	Install the security fence and hang the signboards correctly: (1) The security fence should take full account of falling objects from the heights and the influence on road traffic. (2) The entrance and exit of the security fence shall be set reasonably. (3) Signs such as "Access from Here" "Work Here", "Access from Here", "Access from here" etc.		(1) Deduct 0.5 points for failure to arrange the fence at the work site. (2) Deduct 0.5 points for failure to arrange the warning board. (3) Deduct 0.5 points for failure to hang the tower climbing operation sign. (4) Deduct 2 point for each error or missing item			

Table (Cont'd)

S/N	Project name	Quality requirements	Score	Deduction standard	Reasons for deduction	Deduction	Scoring
3	Site layout	shall be properly arranged. (4) The layout of winching on the traction field shall be completed correctly, including arrangement of winching site and angle tower pulley. (5) The layout of winching on the tension field shall be completed correctly, including arrangement of winching site and angle tower pulley	10				
4	Climbing the tower	(1) Electrician on tower shall check the dual names of the line, pole number and phase, check whether the tower meets the climbing conditions and report the results to the responsible person. (2) The electrician on the tower shall fasten the safety belt, correctly conduct impact test on the safety belt, backup protection rope and falling protector, and report to the person in charge before climbing the tower. (3) When climbing the tower, the electrician shall fasten the anti-fall device, climb at constant speed, tread on shackles, and grasp the main materials. After arriving at the working point of the cross arm, fasten the safety belt on the firm and reliable component, check whether the buckle is in place correctly, select the appropriate position to arrange the tackle transfer rope.	5	(1) Deduct 2 points respectively for failure to fasten the safety belt or to perform the impulse test on the safety belt and the backup protection rope. (2) Deduct 2 points for grasping the shackles with hands. (3) Deduct 1 point for inconvenient suspension position of tackle transmission rope for taking tools. (4) Deduct 2 points for falling object at heights. (5) Deduct 2 points for tools colliding with the tower body during the transfer process. (6) Deduct 1 point for knotting and disordered rope in tool transfer.			

Part II Standard for Training and Assessment on Interruption Maintenance for Operation Maintenance of ±800kV UHV Power Transmission Line

Table (Cont'd)

S/N	Project name	Quality requirements	Score	Deduction standard	Reasons for deduction	Deduction	Scoring
4	Climbing the tower			(7) Deduct 2 points for the Responsible Person failing to monitor the operation in place. (8) Deduct 2 points for incorrect operation of electrician on tower.			
5	Inspection power and install ground wire	(1) The ground electrician shall pass the test pen(electroscope) and other tools to the tower electrician by using the rope. (2) Electricity testing and setting ground wire. When setting the ground wire, the insulating rope or insulating handle must be used for operation, and it is forbidden to install the ground wire with directly operating the metal part of the ground wire by hand. Make sure the collection of the ground wire is tightly connected to the conductor	4	(1) Deduct 4 points for failure to wear insulating gloves in electricity testing and installation of ground wire. (2) Deduct 4 points for installation of ground wire by winding the conductor			
6	Operation on the conductor	(1) The electrician on tower should inspect the rust of the hardware fittings, conduct impulse test on the insulator string, and let the insulator string enter into the conductor with the permission of the Responsible Person. (2) The electrician on tower should enter the grounding electrode conductor, remove the spacer from the conductor and mark the removal position.		(1) Deduct 3 points for failure to use double safety belt or loss of safety protection. (2) Deduct 2 points for failure to conduct impulse test for insulator string. (3) Deduct 2 points for transferring spacer without transmission rope. (4) Deduct 2 points for transferring spacer without transmission rope.			

· 683 ·

Standard for Professional Training and Assessment for Operation Maintenance of UHV DC Power Transmission Line

Table (Cont'd)

S/N	Project name	Quality requirements	Score	Deduction standard	Reasons for deduction	Deduction	Scoring
6	Operation on the conductor	(3) The electrician on tower should enter the conductor at the tangent tower along the rope ladder and install the single-sheave tackle, then remove the vibration damper, lift the grounding electrode conductor with the wire lifter and 6T lever hoist, take off the hardware fittings, and turn the grounding electrode conductor into the single-sheave tackle. (4) After removing part hardware fittings for connection, turn the ground wire into the single-sheave tackle fixed on the ground wire cross arm. (5) The operator should connect the 5T winching to the grounding electrode conductor with wire clamp, and then tighten the grounding electrode conductor with 9T lever hoist to take off the conductor. Loose the conductors at both ends to the ground for the resisting-tensile towers on both sides. (6) The operator shall remove the strain clamp at the conductor end, install the net cover on the old and new conductors, bind them firmly, and connect the old and new conductors with the rotary joint.	15	(5) Deduct 1 point for no mark. (6) Deduct 2 points for failure to protect conductor and tower material when lifting conductor. (7) Deduct 2 points in case of non-standard use of rope ladder in climbing up and down. (8) Deduct 2 points for failure to connect the winching and conductor first. (9) Deduct 2 points for incorrect position of lever hoist installed. (10) Deduct 1 points for failure to protect the tower material during installation of steering tackle at each place. (11) Deduct 3 points for non-standard layout of the new conductor. (12) Deduct 3 points for no binding of net cover. (13) Deduct 2 points for insufficient length of net cover. (14) Deduct 1 point for incorrect use of rotating joint			

Part II Standard for Training and Assessment on Interruption Maintenance for Operation Maintenance of ±800kV UHV Power Transmission Line

Table (Cont'd)

S/N	Project name	Quality requirements	Score	Deduction standard	Reasons for deduction	Deduction	Scoring
7	Crossover and accessories Installation	(1) Start 2 winchings on the traction field and tension field at the same time under unified command by specially-assigned person. (2) After the new conductor arrives at the traction field, the tension field shall be crimped with the strain clamp. After crimping, the crimped conductor shall be connected to the insulator armor clamp by winding and winching. (3) If the traction field tightening the ground wire, and mark it when the sag of the conductor is consistent with the original sag. (4) Loosen the conductor to the ground, crimp the strain clamp at the other end, then tighten the grinding rope. Connect the ground wire to the insulator string armor clamp and the new conductor to the cable clamp by the tangent tower, fasten the bolts, and install the anti-vibration hammer according to the previously marked position. (5) Remove tools in sequence, such as grinding rope, lever hoist, tackle, wire clamp and backup protection rope, etc. (6) Removal of ground wire	40	(1) Deduct 2 points for each time the conductor falling to the ground during crossover. (2) Deduct 1 point for oversize strain during crossover. (3) Deduct 2 points for thread jamming due to no one has been assigned to command. (4) Deduct 2 points in case the old wires are not sorted out and recovered in time. (5) Deduct 2 points in case the sag is not measured by a specially assigned person. (6) Deduct 2 points for inconsistent sag with other conductor. (7) Deduct 3 points in case the model of the strain clamp is not consistent with that of the conductor. (8) Deduct 2 points for failure to conduct impulse test after connection. (9) Deduct 2 points in case that the crimping process does not meet the requirements.		

Standard for Professional Training and Assessment for Operation Maintenance of UHV DC Power Transmission Line

Table (Cont'd)

S/N	Project name	Quality requirements	Score	Deduction standard	Reasons for deduction	Deduction	Scoring
7	Crossover and accessories Installation			(10) Deduct 2 points for failure to check the bolts fastened in place. (11) Deduct 1 point for each improper installation position of spacer. (12) Deduct 2 points for unacceptable installation and improper position of spacer. (13) Deduct 3 points for failure to check the connection of hardware fittings. (14) Deduct 2 points for winding transmission rope. (15) Deduct 2 points for failure to check the conductor sag. (16) Deduct 2 points for each damage to tower material.			
8	Return to the ground	After checking that there is no object left on tower, the electrician on tower shall report to the Responsible Person and then climb down the tower with transmission rope after approval of the Responsible Person	5	(1) Deduct 2 points for failure to use the falling protector when climbing down the tower. (2) Deduct 2 points for loosing the protection of the safety belt when moving on the tower. (3) Deduct 1 point for grasping the tower nail when climbing down the tower. (4) Deduct 2 points for any objects left on the tower			

· 686 ·

Part II Standard for Training and Assessment on Interruption Maintenance for Operation Maintenance of ±800kV UHV Power Transmission Line

Table (Cont'd)

S/N	Project name	Quality requirements	Score	Deduction standard	Reasons for deduction	Deduction	Scoring
9	End of the work	(1) The Responsible Person shall organize all working personnel to put working apparatus and materials in order and put them in a special kit (bag) after clearing; clean the site to ensure that "the materials are removed and the site is cleaned after construction". (2) After a post-shift meeting is held, the Responsible Person shall make work summaries and comments. Comments include the construction quality of this work and the implementation of safety measures from all working personnel	5	(1) Deduct 2 points for failure to clean the tools. (2) Deduct 2 points for missing tools. (3) Deduct 2 points for failure to hold the post-shift meeting. (4) Deduct 2 points for failure to remove the fence. (5) Deduct 2 points for failure to report to dispatcher			
	Total		100				

· 687 ·